Glutathione

Oxidative Stress and Disease

Lester Packer, PhD
Enrique Cadenas, MD, PhD
University of Southern California School of Pharmacy
Los Angeles, California

Oxidative Stress in Cancer, AIDS, and Neurodegenerative Diseases
Edited by Luc Montagnier, René Olivier, and Catherine Pasquier

Understanding the Process of Aging: The Roles of Mitochondria, Free Radicals, and Antioxidants
Edited by Enrique Cadenas and Lester Packer

Redox Regulation of Cell Signaling and Its Clinical Application
Edited by Lester Packer and Junji Yodoi

Antioxidants in Diabetes Management
Edited by Lester Packer, Peter Rösen, Hans J. Tritschler, George L. King, and Angelo Azzi

Free Radicals in Brain Pathophysiology
Edited by Giuseppe Poli, Enrique Cadenas, and Lester Packer

Nutraceuticals in Health and Disease Prevention
Edited by Klaus Krämer, Peter-Paul Hoppe, and Lester Packer

Environmental Stressors in Health and Disease
Edited by Jürgen Fuchs and Lester Packer

Handbook of Antioxidants: Second Edition, Revised and Expanded
Edited by Enrique Cadenas and Lester Packer

Flavonoids in Health and Disease: Second Edition, Revised and Expanded
Edited by Catherine A. Rice-Evans and Lester Packer

Redox–Genome Interactions in Health and Disease
Edited by Jürgen Fuchs, Maurizio Podda, and Lester Packer

Thiamine: Catalytic Mechanisms in Normal and Disease States
Edited by Frank Jordan and Mulchand S. Patel

Phytochemicals in Health and Disease
Edited by Yongping Bao and Roger Fenwick

Carotenoids in Health and Disease
Edited by Norman I. Krinsky, Susan T. Mayne, and Helmut Sies

For more information about this series, please visit:
https://www.crcpress.com/Oxidative-Stress-and-Disease/book-series/
CRCOXISTRDIS

Glutathione

Edited by
Leopold Flohé

CRC Press is an imprint of the
Taylor & Francis Group, an **informa** business

CRC Press
Taylor & Francis Group
6000 Broken Sound Parkway NW, Suite 300
Boca Raton, FL 33487-2742

First issued in paperback 2020

ISBN 13: 978-0-367-65699-7 (pbk)
ISBN 13: 978-0-8153-6532-7 (hbk)

Library of Congress Cataloging-in-Publication Data

Names: Flohé, L. (Leopold), 1938- editor.
Title: Glutathione / editor, Leopold Flohé.
Other titles: Glutathione (Flohé) | Oxidative stress and disease.
Description: Boca Raton: Taylor & Francis, 2018. | Series: Oxidative stress and disease | Includes bibliographical references.
Identifiers: LCCN 2018017686 | ISBN 9780815365327 (hardback: alk. paper)
Subjects: | MESH: Glutathione—metabolism | Glutathione—physiology | Glutathione—biosynthesis | Oxidative Stress
Classification: LCC QP552.G58 | NLM QU 68 | DDC 572/.65—dc23
LC record available at https://lccn.loc.gov/2018017686

Visit the Taylor & Francis Web site at
http://www.taylorandfrancis.com

and the CRC Press Web site at
http://www.crcpress.com

DOI: 10.1201/9781351261760

The Open Access version of chapter 11 was funded by The Swedish Childhood Cancer Foundation (grant PR2021-0071)

Contents

PART I Biosynthesis, Compartmentation, and Transport of Glutathione

PART II Glutathione-Dependent Hydroperoxide Metabolism

PART III   Conjugations and Isomerizations

PART IV   The Glutaredoxins

PART V Glutathione Derivatives and Substitutes in Pathogenic Microorganisms

Series Preface

GLUTATHIONE

The editor, Professor Leopold Flohé – Redox Pioneer 2010 (*Antioxidants & Redox Signaling* 13, 1617–1622) – is an eminent scientist who opened the field of selenium biochemistry by proving that this trace element is an integral moiety of the enzyme glutathione peroxidase. His seminal studies on glutathione peroxidase and its catalytic mechanism preceded the discoveries of other glutathione peroxidases and of peroxiredoxins. This pioneer work on glutathione peroxidase branched out to a myriad of topics of significance for redox biology, of which two are worth mentioning. First, it challenged the dogma of hydroperoxide metabolism being only the domain of heme-containing enzymes, thus placing glutathione peroxidase(s) at the center stage of peroxide metabolism. Second, it established the basis for an ever-developing integrated redox network with implications for the regulation of signaling and transcriptional pathways. Dr. Flohé also pioneered several studies on hydroperoxide metabolism (it may be viewed also as 'antioxidant defense') in pathogens of the genera *Trypanosoma*, *Leishmania*, and *Mycobacterium*, with the goal of identifying suitable drug targets, thus lending a translational value to the knowledge and technology in the field.

To edit a book on glutathione and glutathione-dependent biological phenomena is not an easy task because of the abundance of reviews and books on these subjects. To the editor's credit, *Glutathione* compiles state-of-the-art glutathione-dependent biological processes. Following an editor's introduction and historical perspective on glutathione, the book consists of five parts: biosynthesis, compartmentalization and transport of glutathione, glutathione-dependent hydroperoxide metabolism, roles of glutathione derivatives and substitutes in conjugations and isomerization, glutaredoxins, and glutathione in pathogenic microorganisms. Each section or part contains several chapters written by highly esteemed experts in the field.

In bringing *Glutathione* to fruition in the series *Oxidative Stress and Disease*, credit must be given to the editor, Professor Leopold Flohé, and to the experts in the various aspects of glutathione and glutathione-dependent biological processes, whose thorough and innovative work is the basis for this book.

Lester Packer
Enrique Cadenas

Preface

Glutathione was detected as a thiol-containing compound in 1888 by J. de Ray-Pailhade (de Rey-Pailhade, 1888) and named "philothion," a term composed of the Greek words for love and sulfur. The elucidation of its structure did not start but in the early 1920s and kept some Nobel laureates and others busy for more than a decade. It was Sir Frederick Gowland Hopkins who started with the isolation of philothion and renamed it glutathione, yet the chemical analysis of the isolated material led him to conclude that it was a dipeptide consisting of glutamic acid and cysteine (Hopkins, 1921; Hopkins and Dixon, 1922). The addition of the missing third amino acid, glycine, goes back to Hunter and Eagles (1927a,b), Pirie and Pinhey (1929), and Kendall et al. (1929). Its structure as γ-glutamyl-cysteinyl-glycine, however, had to wait for final proof by chemical synthesis in 1935 (Harrington and Mead, 1935) and 1936 (du Vigneaud and Miller, 1936), respectively.

In the 1950s, the state of the art, up to the biosynthesis of glutathione and its enzymatic regeneration from the disulfide form, was compiled in symposia held at Ridgefield (Colowik et al., 1954) and London (Crook, 1958). In 1972, a bunch of impecunious youngsters dared to invite the most famous glutathione researchers from all over the world to a small meeting to be held in the German university village of Tübingen, and, to their great surprise, they all showed up and appeared to enjoy the modest hospitality the "organizers" could afford. Despite a strong chemical touch, the clinical implications of genetic errors relevant to the regeneration of reduced glutathione were a major focus of this conference (Flohé et al., 1974). Highlights were contributed by the Kosowers, demonstrating the influence of the redox state of the glutathione system on seemingly unrelated cellular functions such as ion transport, and by Wolfgang A. Günzler providing the first unequivocal proof of the selenoprotein nature of glutathione peroxidase, now addressed to as GPx1 (Flohé, 2009). At this meeting, Alton Meister also presented his γ-glutamyl cycle, at that time still claimed to be responsible for amino acid transport in general. The meeting at the Kroc Foundation in the Santa Inez Valley in California focused on the role of glutathione S-transferases in the metabolism of xenobiotics (Arias and Jakoby, 1976). The following meeting at the Reißensburg near Ulm in Germany dealt with the function of glutathione in the liver and kidney (Sies and Wendel, 1978). At this meeting, Alton Meister seemed still to defend his original idea of a major role of the γ-glutamyl cycle in amino acid transport. Although his statements were already less apodictic, he was heavily challenged by Norman Curthoys and his colleagues from the University of Pittsburgh and my former students Rolf Hahn, Helmut Heinle, and Albrecht Wendel, whom I had left behind in Tübingen when taking a break from academia to work for a drug company. In the meantime, it had become clear from localization and organ perfusion studies that the original concept of the γ-glutamyl cycle was no longer tenable, but it had been heuristic in fertilizing the work on salvage and inter-organ trafficking of glutathione via extracellular degradation and resynthesis, as well as clarifying the biosynthesis of mercapturic acids and leukotrienes via cooperation of different cells, tissues and organs. Transport, turnover, and storage of glutathione then became the central issue of the 1981 meeting in Osaka (Sakamoto et al., 1983), and in 1982, glutathione even made its career as subject of the Fifth Karolinska Nobel Conference, which tried to bridge the gap between toxicologists and clinicians interested in glutathione-related inborn errors of metabolism (Larsson et al., 1983). Finally, the discovery of philothion by de Rey-Pailhade was celebrated in Osaka in December 1988, and the proceedings of this meeting were published one year later as "Glutathione Centennial" (Taniguchi et al., 1989).

Also in 1989, an extensive monograph on glutathione appeared in two parts of the series Coenzymes and Cofactors (Dolphin et al., 1989). It starts with a "brief" history of glutathione comprising 34 pages plus 427 references masterly written by the late Alton Meister. These two volumes also contain chapters dealing with the physicochemical properties of glutathione and related thiols, its chemical and biochemical syntheses, and detailed updates on glutathione-dependent enzymatic systems. The book remains an invaluable access to the first 100 years of glutathione research.

Surprisingly, no serious attempts to compile the state of the art in the field have since been made. This does not mean that the interest in the compound is fading away. On the contrary, the lay press has knighted glutathione as "the most important antioxidant of the organism," a label that is to characterize it as being always beneficial and indispensable for well-being. In strict chemical terms, however, this label is wrong and, in biological terms, at least an oversimplification. The mislabeling of glutathione as an antioxidant originates from the discovery of glutathione peroxidases, which efficiently reduce all kind of hydroperoxides. Yet without the aid of these enzymes, glutathione reacts with hydroperoxides only in a biologically irrelevant speed. It rather tends to react with the most abundant radical, molecular dioxygen, thereby forming superoxide radicals, initiating free radical chain reactions and oxidative tissue destruction. Moreover, hydrogen peroxide and other hydroperoxides meanwhile adopted new roles in living systems: They abandoned their ugly image as oxidative poisons and became mediators or modulators of signaling cascades, and accordingly, many of the glutathione peroxidases acquired unexpected novel functions as sensors, regulators, or modulators in metabolic regulation and differentiation. Another honor often conferred to glutathione is "the key detoxifying agent." It is based on the assumption that nature anticipated the ingenuity of industrial chemists to create uncountable toxic compounds and prophylactically developed the realm of glutathione S-transferases just to conjugate and excrete the chemical waste via the mercapturic acid pathway. Again, we have meanwhile learned that the S-transferases may also increase the toxicity of xenobiotics, they also have distinct endogenous substrates and catalyze processes, such as cis-trans isomerization, leukotriene biosynthesis, or protein glutathionylation, which have little in common with detoxification reactions that fascinated researchers in the past century. What we are experiencing is a profound change in paradigms: Glutathione is no longer just an antioxidant and detoxifying compound but a redox-active one that dominates many and highly diversified biological processes (Flohé, 2010).

It is not the aim of this book to reiterate what has meanwhile become text-book knowledge. As mentioned, the history of glutathione research has been described in detail, and the chemical properties of glutathione, as presented in the 1989 monograph (Dolphin et al., 1989), are still valid. The editor tried to select the topics in which substantial progress has been made over the past three decades. He admits that his selection is biased by the conviction that most, if not all, glutathione-dependent reactions of physiological relevance are catalyzed by enzymes or due to other specific interactions with proteins (Flohé, 2013; Berndt et al., 2014).

Accordingly, the first part of the book contains an update of glutathione biosynthesis with special emphasis on its regulation in the context of adaptive stress response (Chapter 1). It further reviews glutathione transport systems, which together with *de novo* synthesis determine the cellular and subcellular pools (Chapter 2).

Part II covers the glutathione peroxidases (Chapters 3–10), which in principle catalyze similar reactions. It starts with the attempt to solve the enigma how these enzymes can accelerate the reaction of glutathione with hydroperoxides by orders of magnitude irrespective of the involvement of a selenocysteine or cysteine in the catalytic process (Chapter 3). The remaining articles will reveal how discrete differences in substrate specificities, localization, and substrate availability modify their biological roles in most surprising ways. In fact, one and the same glutathione peroxidase gene (*gpx4*), depending on the mode and site of its expression, may inter-fere with unrelated phenomena such as spermiogenesis, chromatin compaction, and ferroptosis (Chapters 7 and 8).

Part III centers on the diversified roles of the different glutathione *S*-transferases, which for sure can detoxify xenobiotics but are gaining increasing interest as a means for synthesis of powerful mediators or hormones or as catalysts of specific glutathionylation in redox regulation (Chapters 11–14). The part ends with a chapter on nitrosoglutathione, which apparently plays a pivotal role in metabolic regulation by the NO radical (Chapter 14).

Part IV deals with the glutaredoxins, a subfamily of the redoxins, long believed to catalyze the reduction of protein disulfides by glutathione promiscuously or con-sidered to back up the thioredoxin systems in the synthesis of deoxyribonucleotides and in protecting proteins with sensitive thiol groups (Chapters 15–16). Although their glutathione binding sites are similar or identical, the diverse surroundings of the reaction centers enable them to interact with defined proteins selectively. Inverse genetics revealed that the role of a particular glutaredoxin, e.g., in the brain, cannot be substituted by another glutaredoxin or by any of the thioredoxins.

Part V seemingly breaks with the aim to limit the scope of the book to the role of glutathione to human health and disease. The thiol metabolism in the pathogens addressed is largely analogous to that of mammalian organisms. The differences, however, are pronounced enough to qualify the analogous or homologous enzymes as potential drug targets and, thus, have attracted considerable interest for the development of drugs to treat human and veterinary diseases such as trypanosomiasis, leishmaniasis (Chapters 17 and 18), tuberculosis (Chapter 19), and other bacterial infections (Chapter 20).

With some interruptions, the editor spent 50 years of his life with research on glutathione and did not get bored. His hope is to infect the next generation with his enthusiasm for this fascinating molecule, and he closes this introduction with the often-quoted reminder of the Kosowers: "Lest I forget thee, glutathione" (Kosower and Kosower, 1969).

Leopold Flohé
Potsdam, Germany

REFERENCES

Arias, I. M., and W. B. Jakoby. 1976. *Glutathione: Metabolism and Function.* Vol. 6, *Kroc Foundation Series.* New York: Raven Press.

Berndt, C., C. H. Lillig, and L. Flohé. 2014. Redox regulation by glutathione needs enzymes. *Front Pharmacol* 5:168.

Colowik, S., D. R. Schwarz, and A. Lazarow. 1954. *Glutathione. Proceedings of the symposium held at Ridgefield, Connecticut, November 1953.* New York: Acad Press.

Crook, E. M. 1958. *Glutathione. Biochemical Society Symposium no 17.* London: Biochemical Society (Great Britain).

de Rey-Pailhade, J. 1888. Sur un corps d'origine organiques hydrogenant le soufre a froid. *Compt. Rend. de l' Academie des Sciences* 106:1683–4.

Dolphin, D., R. Poulson, and O. Avramovic. 1989. *Glutathione. Chemical, Biochemical, and Medical Aspects., Coenzymes and Cofactors, Vol III.* New York: Wiley Interscience.

du Vigneaud, V., and G. L. Miller. 1936. A synthesis of glutathione. *J Biol Chem* 116:469–76.

Flohé, L. 2009. The labour pains of biochemical selenology: The history of selenoprotein biosynthesis. *Biochim Biophys Acta* 1790 (11):1389–403.

Flohé, L. 2010. Changing paradigms in thiology: From antioxidant defence towards redox regulation. *Meth Enzymol* 473:1–39.

Flohé, L. 2013. The fairytale of the GSSG/GSH redox potential. *Biochim Biophys Acta* 1830 (5):3139–42.

Flohé, L., H. C. Benöhr, H. Sies et al. 1974. *Glutathione.* Stuttgart: Thieme.

Harrington, C. R., and T. H. Mead. 1935. Synthesis of glutathione. *Biochem J* 29:1602–11.

Hopkins, F. G. 1921. On an autoxidisible constituent of the cell. *Biochem J* 15:286–305.

Hopkins, F. G., and M. Dixon. 1922. On glutathione. II. A thermostable oxidation-reduction system. *J Biol Chem* 54:527–63.

Hunter, G., and B. A. Eagles. 1927a. Glutathione. A critical study. *J Biol Chem* 72:147–66.

Hunter, G., and B. A. Eagles. 1927b. Non-protein sulfur compounds in blood. II. Glutathione. *J Biol Chem* 72:133–46.

Kendall, E. C., B. F. McKenzie, and H. L. Mason. 1929. A study of glutathione: I, Its preparation in crystalline form and its identification. *J Biol Chem* 84:657–74.

Kosower, E. M., and N. S. Kosower. 1969. Lest I forget thee, glutathione. *Nature* 224 (5215):117–20.

Larsson, A., S. Orrenius, A. Holmgren, and B. Mannervik. 1983. *Functions of Glutathione.* New York: Raven Press.

Pirie, N. W., and K. G. Pinhey. 1929. The titration curve of glutathione. *J Biol Chem* 84:321–33.

Sakamoto, Y., T. Higashi, and N. Tateishi. 1983. *Glutathione. Storage, Transport and Turnover in Mammals.* Tokyo: Japan Scientific Societies Press.

Sies, H., and A. Wendel. 1978. *Functions of Glutathione in Liver and Kidney. Proceedings in Life Sciences.* Heidelberg: Springer.

Taniguchi, N., T. Higashi, Y. Sakamoto, and A. Meister. 1989. *Glutathione Centennial. Molecular Perspectives and Clinical Implications.* San Diego: Academic Press, Inc.

Editor

Leopold Flohé, MD, PhD Dr (h.c.), is a retired professor of biochemistry. Born in 1938 in Grevenbroich, Germany, he obtained his diploma in 1967 in biochemistry and in 1968 a medical doctorate from the University of Tübingen, Tübingen, Germany, where he pioneered in selenium-catalyzed metabolism of hydroperoxides and became a professor for biochemistry in 1974. From 1976 to 1990, he served as director for research and development at Grünenthal GmbH, a pharmaceutical company located in Aachen, Germany. Thereafter, he became the scientific director of the National Research Center of Biotechnology (GBF, now HZI) in Braunschweig, Germany. After 5 years, he decided to switch back from research administration to bench-type research, spent a three-month sabbatical at the University of Berkeley, Berkeley, California, and then took the Chair of Biochemistry at the Technical University of Braunschweig. When he reached the age of obligatory retirement, he kept working as senior scientist in a start-up company, as Chairman of the COST Action CM 0801 (EU), as guest professor at the University of Magdeburg, Magdeburg, Germany, as cooperation partner with the biochemical institute of the Universidad de la República (Montevideo, Uruguay), and as Distinguished Visiting Professor at the University of Padova, Padova, Italy. His diversified scientific interests are reflected in a heterogeneous list of awards and honors: Award of the Anna Monica Foundation for work on endogenous depression (1973), Claudius Galenus Prize for production of recombinant urokinase (1985), honorary degree of the University of Buenos Aires (Argentina, 1997) for achievements in biochemical parasitology, Klaus Schwarz Commemorative Medal for selenium research (1997), Science and Humanity Price of the Oxygen Club of California for lifetime achievements (2001), Trevor Frank Slater Award and Gold Medal for free radical research (2006), Redox Pioneer #3 for work on glutathione and related subjects (2010), and an honorary degree of the Universidad de la Republica (Uruguay, 2013) for elucidation of the hydroperoxide metabolism of trypanosomes.

List of Contributors

Haike Antelmann
Department of Microbiology
Freie Universität Berlin
Berlin, Germany

Scott D. Barnett
Department of Pharmacology
University of Nevada
Reno School of Medicine
Reno, Nevada

Carsten Berndt
Department of Neurology
Life Science Center
Heinrich-Heine-Universität
  Düsseldorf
Düsseldorf, Germany

Lars Bräutigam
Science for Life Laboratory
Karolinska Institutet
Stockholm, Sweden

Iain L. O. Buxton
Department of Pharmacology
Reno School of Medicine
University of Nevada
Reno, Nevada

Marcelo A. Comini
Laboratory for Redox Biology of
  Trypanosomatids
Institut Pasteur de Montevideo
Montevideo, Uruguay

Marcus Conrad
Institute of Developmental Genetics
Helmholtz Zentrum München
Munich, Germany

Giorgio Cozza
Dipartimento di Medicina
  Molecolare
Università degli Studi di Padova
Padova, Italy

Marcel Deponte
Faculty of Chemistry –
  Biochemistry
Technische Universität Kaiserslautern
Kaiserslautern, Germany

Anna Dorothee Engelke
Department of Neurology
Life Science Center
Heinrich-Heine-Universität
  Düsseldorf
Düsseldorf, Germany

Leopold Flohé
Departamento de Bioquímica
Universidad de la República
Montevideo, Uruguay
and
Dipartimento di Medicina
  Molecolare
Università degli Studi di Padova
Padova, Italy

Henry J. Forman
USC Davis School of Gerontology
University of California at Merced
Los Angeles, California

José Pedro Friedmann Angeli
Rudolf Virchow Center for
  Experimental Biomedicine
University of Würzburg
Würzburg, Germany

Jesper Haeggström
Department of Medical Biochemistry
 and Biophysics
Karolinska Institutet
Stockholm, Sweden

Diane E. Handy
Brigham and Women's Hospital
Harvard Medical School
Boston, Massachusetts

Martin Hugo
German Institute of Human
 Nutrition
Nuthetal, Germany

Per-Johan Jakobsson
Department of Medicine
Karolinska Institutet
Stockholm, Sweden

Valerian E. Kagan
Department of Environmental and
 Occupational Health
University of Pittsburgh
Pittsburgh, Pennsylvania

Terrance J. Kavanagh
Department of Environmental &
 Occupational Health
 Science
University of Washington
Seattle, Washington

Anna P. Kipp
Institute of Nutrition
Friedrich Schiller University
 Jena
Jena, Germany

Lars-Oliver Klotz
Institute of Nutrition, Department of
 Nutrigenomics
Friedrich Schiller University Jena
Jena, Germany

Lawrence H. Lash
Department of Pharmacology
Wayne State University School of
 Medicine
Detroit, Michigan

Klaudia Lepka
Department of Neurology
Life Science Center
Heinrich-Heine-Universität
 Düsseldorf
Düsseldorf, Germany

Linda Liedgens
Faculty of Chemistry –
 Biochemistry
Technische Universität Kaiserslautern
Kaiserslautern, Germany

Nico Linzner
Department of Microbiology
Freie Universität Berlin
Berlin, Germany

Vu Van Loi
Department of Microbiology
Freie Universität Berlin
Berlin, Germany

Joseph Loscalzo
Brigham and Women's Hospital
Harvard Medical School
Boston, Massachusetts

Matilde Maiorino
Dipartimento di Medicina
 Molecolare
Università degli Studi di Padova
Padova, Italy

Yefim Manevich
Department of Cell and Molecular
 Pharmacology,
Medical University of South Carolina
Mount Pleasant, South Carolina

Bengt Mannervik
Department of Biochemistry and
  Biophysics
Stockholm University
Stockholm, Sweden

Joris Messens
Brussels Center for Redox
  Biology
Vrije Universiteit Brussel
Brussels, Belgium

Ralf Morgenstern
Institutet för Miljömedicin
Karolinska Institutet
Stockholm, Sweden

Laura Orian
Università degli Studi di Padova
Dipartimento di Scienze Chimiche
Padova, Italy

Brandán Pedre
Brussels Center for Redox Biology
Vrije Universiteit Brussel
Brussels, Belgium

Lucía Piacenza
Departamento de Bioquímica. Facultad
  de Medicina
Universidad de la República
Montevideo, Uruguay

Rafael Radi
Departamento de Bioquímica. Facultad
  de Medicina
Universidad de la República
Montevideo, Uruguay

Leonardo Astolfi Rosado
Brussels Center for Redox
  Biology
Vrije Universiteit Brussel
Brussels, Belgium

Antonella Roveri
Dipartimento di Medicina Molecolare
Università degli Studi di Padova
Padova, Italy

Lutz Schomburg
Institute for Experimental
  Endocrinology
Charité - Universitätsmedizin Berlin
Berlin, Germany

Birgitta Sjödin
Department of Biochemistry
Stockholm University
Stockholm, Sweden

Holger Steinbrenner
Institute of Nutrition, Department of
  Nutrigenomics
Friedrich Schiller University Jena
Jena, Germany

Kenneth D. Tew
Cell and Molecular Pharmacolgy &
  Experimental Therapeutics, BSB
Medical University of South Carolina
Charleston, South Carolina

Stefano Toppo
Dipartimento di Medicina
  Molecolare
Università degli Studi di Padova
Padova, Italy

Madia Trujillo
Departamento de Bioquímica.
  Facultad de Medicina
Universidad de la República
Montevideo, Uruguay

Quach Ngog Tung
Department of Microbiology
Freie Universität Berlin
Berlin, Germany

Fulvio Ursini
Dipartimento di Medicina
    Molecolare
Università degli Studi di Padova
Padova, Italy

Hongqiao Zhang
USC Davis School of Gerontology
University of California at Merced
Los Angeles, California

Part I

Biosynthesis, Compartmentation, and Transport of Glutathione

1 Biosynthesis of Glutathione and Its Regulation

Henry J. Forman and Hongqiao Zhang
University of California, Merced

Terrance J. Kavanagh
University of Washington

CONTENTS

1.1 INTRODUCTION

We are honored to participate in this volume on glutathione (GSH), a subject to which the editor, Leopold Flohé, has made many outstanding contributions (Ursini and Maiorino, 2010).

DOI: 10.1201/9781351261760-2

This chapter will review the biosynthesis of GSH and its regulation in mammals. We refer readers to reviews on GSH synthesis in plants (May et al., 1998; Noctor et al., 1998) and yeast (Wu and Moye-Rowley, 1994) for information on organisms in those taxa. Although some bacteria synthesize GSH, many use other thiols in its place (Fahey, 2013). GSH is used in antioxidant defense, cell cycle regulation, and other important roles in cells, making it important to maintain its concentrations. Although other chapters focus on the uses of GSH, we will discuss how GSH synthesis is maintained through feedback and feedforward regulation at several levels of control.

1.2 STRUCTURE OF GSH

GSH is a tripeptide composed of an *N*-terminal glutamic acid, a central cysteine, and a C-terminal glycine (Figure 1.1). There are two interesting points about the GSH structure that contribute to its uniqueness. One is that the glutamate is attached to cysteine in an amide bond through the γ-carboxyl rather than the α-carboxyl moiety. The second unique aspect of GSH is that it contains a large percentage of the cysteine in cells, as the concentration of GSH ranges from 1 to 10 mM depending on the cell type (Meister, 1988). Indeed, when cysteine is limiting, which is the usual situation, it is preferentially used for GSH synthesis; only protein synthesis has a higher priority than GSH synthesis (Stipanuk et al., 1992). Glycine, the third amino acid in GSH, is linked to cysteine in a normal peptide amide bond. Thus, GSH or γ-L-Glu-L-Cys-Gly is an unusual peptide. But you likely already knew that if you are reading this book.

1.3 GSH SYNTHESIS

The concentration of GSH is maintained by multiple mechanisms. This chapter is focused on *de novo* synthesis of GSH; however, perhaps the most important mechanism maintaining GSH is the reduction of glutathione disulfide (GSSG) by glutathione reductase (GR) using NADPH. GSSG is produced during the reduction of hydroperoxides by glutathione peroxidases. Although the pathway for GSH synthesis and the supply of the amino acids were largely worked out decades ago, today we are still faced with the challenge of how to most effectively elevate GSH when that would appear to help resist oxidative or xenobiotic stresses. Thus, this chapter

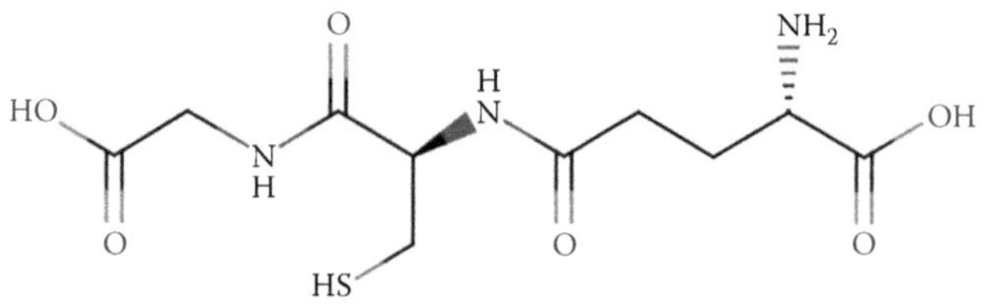

FIGURE 1.1   The structure of GSH.

combines the old with the new aspects of GSH synthesis, which are needed to understand its regulation.

1.3.1 THE γ-GLUTAMYL PATHWAY

The main synthetic pathway for GSH is called the γ-glutamyl cycle, which is illustrated in Figure 1.2. In step 1, L-glutamic acid is joined to L-cysteine in an ATP-dependent reaction catalyzed by glutamate cysteine ligase (GCL). This enzyme is composed of catalytic and modifier subunits that will be described in detail next. GCL enzymatic activity is regulated by the concentrations of substrates, the feedback inhibition by GSH, and the association of the two subunits, and at their transcription and translation levels. In step 2, glycine is rapidly added to γ-Glu-Cys by forming an amide bond with the cysteine carboxyl group in a second ATP-dependent reaction catalyzed by glutathione synthase (GS). GCL is usually considered as the rate-limiting step in *de novo* GSH synthesis, although some evidence suggests that increasing GS can also increase GSH synthesis (Lu, 2009). These two steps would be sufficient for *de novo* GSH synthesis, but as Meister and coworkers reported, GSH synthesis is part of a cycle in which the glutamic acid is used to import other amino acids from the external fluid (Orlowski and Meister, 1970). Thus, in step 3, GSH is exported from cells. As the extracellular GSH concentration is in the micromolar range, the steep gradient from inside to outside drives the export of GSH. Once outside, the cysteinyl-glycine part of GSH is exchanged in step 4 for another amino acid in a reaction catalyzed by the *exo*-enzyme γ-glutamyl transferase (GGT; also known

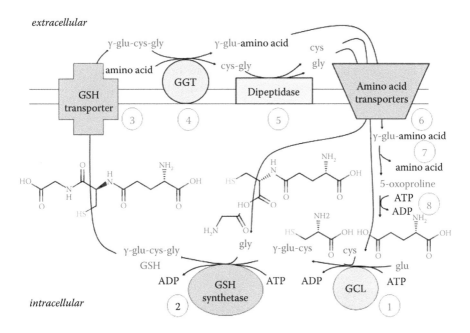

FIGURE 1.2 The γ-glutamyl pathway.

as γ-glutamyl transpeptidase). The cysteine and glycine from cysteinyl-glycine can be recovered through step 5 in which a dipeptidase cleaves the amide bond, followed by the uptake of the amino acids through specific transporters in step 6. The γ-glutamyl-amino acid formed in step 4 can also be transported by a transporter that is specific for γ-glutamyl-amino acids. Once inside, the γ-Glu-amino acid is converted by γ-glutamyl cyclotransferase in step 7 into the amino acid and oxoproline. Oxoproline is hydrolyzed to L-Glu in an ATP-dependent reaction catalyzed by oxoprolinase in step 8. Thus, all three amino acids are cycled preferentially providing for the restoration of GSH lost from the cell but also for the import of other amino acids. Although all the steps in the cycle are interesting in that deficiencies cause disease, we will largely focus on how GCL and GS are regulated.

1.3.2 Availability of Cysteine and Cystine

Normal nutrition supplies glutamic acid and glycine in far more abundance than is needed for GSH synthesis. The availability of cysteine, however, normally limits GSH synthesis (Meister, 1981). In fact, as the K_M for cysteine for GCL is near the normal physiological concentration (Richman and Meister, 1975), providing more cysteine generally increases GSH concentration. As described previously, cysteine can be recovered from the GSH that leaves cells. In the blood, cystine, the disulfide form of cysteine, is the predominant form of these two amino acids. Fortunately, the preferred amino acid acceptor of the γ-glutamyl group used by GGT is cystine (Anderson and Meister, 1983). Therefore, GGT uses GSH and cystine to produce γ-glutamyl-cystine and Cys-Gly. The γ-glutamyl-cystine is then taken up by the γ-glutamyl amino acid transporter. Once inside the cell, γ-glutamyl-cystine can be reduced to form γ-Glu-Cys by transhydrogenation with GSH to form GSSG, cysteine, and γ-glutamyl-cysteine (Anderson and Meister, 1983). GSSG is rapidly reduced in cells by GR using NADPH. The formation of γ-glutamyl-cysteine from γ-glutamyl-cystine bypasses GCL, the usually rate-limiting enzyme.

1.3.2.1 Cystine Uptake

Cystine can also be transported through the x(c)⁻ transport system in exchange for glutamate and then reduced to cysteine and used by GCL (Deneke and Fanburg, 1989). Indeed, increasing evidence suggests that the x(c)⁻ transport system is critical for maintenance of GSH in preventing ferroptosis in tumors (Jiang et al., 2015). Thus, cysteine can be recovered by transport of cysteine, cystine, γ-glutamyl-cysteine, or γ-glutamyl-cystine, which varies in importance among cells. Although cysteine can be synthesized from other sulfur-containing compounds described here, much of it is obtained from the diet. This classifies it as a semi-essential amino acid. The cysteine is oxidized to cystine during food preparation so that the majority of the cystine is taken up by the x(c)⁻ transport system.

1.3.2.2 N-Acetylcysteine

N-Acetylcysteine (NAC) is an amide formed from acetic acid and cysteine that is readily taken up by cells. Inside the cells, it is rapidly hydrolyzed to form its parent compounds, providing a source of cysteine but also potentially acidifying the cytosol through the sudden increase in acetic acid. A recent study (McCarty and DiNicolantonio, 2015) suggested that cysteine from the diet may not be sufficient in the elderly but can be remedied by supplying NAC and inducers of GCL transcription (see later). The abrogation of damage by NAC is used in many studies to imply the involvement of oxidative stress. Indeed, as it is a poor scavenger of reactive species, its effectiveness more likely involves the elevation of GSH, although, as stated previously, it acidifies the cytosol.

1.3.2.3 Methionine and the Transsulfuration Pathway

In some tissues, much of the cysteine is provided by using the essential amino acid methionine using the transsulfuration pathway (Banerjee and Zou, 2005). An excellent review by Stipanuk (2004) of the uptake of cysteine and methionine, and the transsulfuration pathway describes these important contributions in much greater detail than we can provide here. Briefly, methionine, ATP, and serine are used to produce cysteine in the scheme shown in Figure 1.3. The importance of this pathway in some tissues is apparent from the demonstration that deficiency of the final enzyme in the pathway, cystathionine γ-lyase, results in neurodegeneration in Huntington's disease (Paul et al., 2014).

1.3.3 REGULATION OF ENZYMATIC ACTIVITY AND EXPRESSION

1.3.3.1 GCL Kinetics

Alton Meister's group did most of the pioneering work on understanding the regulation of GCL activity at the enzyme kinetics level. Here, we will describe the major points from their studies (Huang et al., 1993a,b,c; Richman and Meister, 1975; Anderson and Meister, 1983). GCL activity is regulated by the availability of its substrates. But it is also regulated by the interaction between its catalytic (GCLC) and modifier (GCLM) subunits.

Let us first consider the K_M's for the substrates of the GCLC/GCLM dimer, also called the holoenzyme. The K_M for cysteine for GCL activity is 0.35 mM, which is within the intracellular concentration range for this amino acid (Richman and Meister, 1975). The K_M for glutamate for GCL activity is also within the intracellular concentration range for glutamate (Huang et al., 1993a,c; Chen et al., 2005). Similarly, the K_M for ATP (0.87 mM) (Chen et al., 2005) is within its normal physiological range. Thus, GCL activity is regulated by the intracellular concentration of all three substrates. Nonetheless, as cysteine is the one that is most limited in terms of its availability from either the diet or its synthesis, it is considered the limiting factor, particularly during oxidative stress when GSH synthesis is generally increased.

GCL activity is feedback inhibited by GSH. But this inhibition is competitive with glutamate rather than allosteric (Richman and Meister, 1975). The K_i for GSH is 2.3 mM, which is also in the range of its intracellular steady-state concentration in most cells. Thus, it is expected that GCL operates in cells at less than the maximum rate, as feedback inhibition by GSH would be significant. The combination of substrate availability and feedback inhibition suggests that GSH synthesis is well regulated at the enzyme kinetics level.

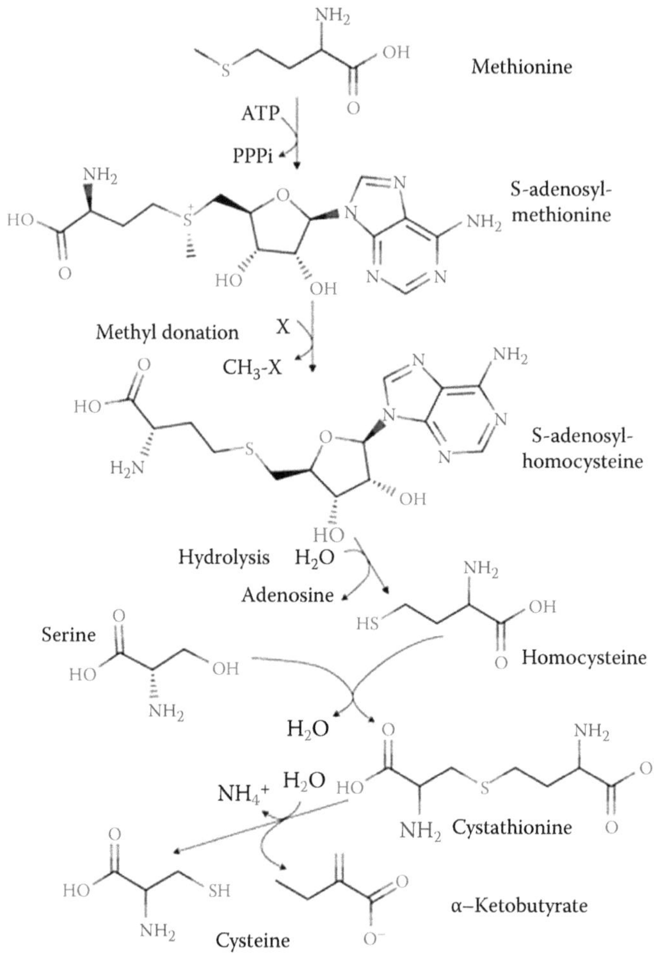

FIGURE 1.3 The transsulfuration pathway.

When we consider the role of the interaction of the two subunits, regulation at the enzymatic activity level becomes more complex. Using recombinant proteins, it was shown that GCLC has GCL catalytic activity in the absence of GCLM, which itself has no catalytic activity (Huang et al., 1993a). Nonetheless, the effect of GCLM is profound, as GCLC alone has an approximately ninefold higher K_M for glutamate (Huang et al., 1993a) allowing the holoenzyme to be sensitive to glutamate concentrations in the physiological range. The same study also demonstrated that GCLM significantly lowers the K_i for GSH. Later studies by Chen et al. (2005) showed that GCLM decreased the K_M for ATP by approximately sixfold and confirmed the reported effects on the K_M for glutamate and the K_i for GSH. Furthermore, GCLM increased the k_{cat} of GCLC by 4.4-fold (Chen et al., 2005). During apoptosis, GCLC can be cleaved to a shorter form by caspase-3, which may cause a decline in GCL activity due to altered interaction with GCLM (Franklin et al., 2003).

Although the holoenzyme is a 1:1 heterodimer, we found that the binding of GCLM and GCLC is not 100% efficient and that the ratio of the two subunits varies among tissues (Krzywanski et al., 2004). The interaction of GCLM with GCLC due to changes in their expression has physiological implications (Chen et al., 2005; Lee et al., 2006) that are described in more detail in a later section.

Phosphorylation of GCLC, which may be achieved by several kinases, has been reported to lower the catalytic V_{max} (Sun et al., 1996). But GCLC phosphorylation does not appear to alter the interaction between the two GCL subunits (Chen et al., 2005).

1.3.3.2 GCL Expression

GCL activity is mainly determined by the expression level of its catalytic subunit GCLC, while the ratio of GCLC to GCLM may also affect the maximal GCL activity (Krzywanski et al., 2004). Thus, alteration of the expression level of either GCLC or GCLM could influence GCL activity and GSH levels. The protein level of either subunit is usually well correlated with its mRNA level, which is primary regulated at the transcriptional and posttranscriptional levels (Zhang et al., 2012).

1.3.3.2.1 GCLC Expression

GCLC expression in cells is altered under many pathophysiological conditions and upon the exposure to various endogenous or exogenous stimuli. Increased expression of GCLC is observed in many human tumors such as colorectal tumor (Kuo et al., 1998), lung cancers (Oguri et al., 1998), and renal carcinoma (Li et al., 2014), and in drug-resistant tumor cells such as prostate carcinoma cells (Mulcahy et al., 1994), myeloma cells (Mulcahy et al., 1994), and ovarian cancer cells (Godwin et al., 1992). GCLC expression can be upregulated by endogenous and exogenous stimuli including methyl mercury (Woods et al., 1992), quinones (Shi et al., 1994), insulin and glucocorticoid (Lu et al., 1992), silica nanoparticles (Zhang et al., 2017), butylated hydroxyanisole (Eaton and Hamel, 1994), *tert*-butylhydroquinone (Liu et al., 1996), β-naphthoflavone (Mulcahy et al., 1997), heat shock (Kondo et al., 1993), zinc (Ha et al., 2006), melatonin (Urata et al., 1999), arsenite (Thompson et al., 2009), garlic extract (Kay et al., 2010), and lipid peroxidation products such as hydroxynonenal (Dickinson et al., 2002) and 15-deoxy-delta(12,14)-prostaglandin J2 (Levonen et al., 2001). A common feature of these stimuli is that they are electrophiles or cause the production of electrophiles in cells. Therefore, GCLC induction is generally considered as an adaptive response to oxidative stress to protect cells against further severe oxidative damage (Zipper and Mulcahy, 2000).

On the other hand, GCLC expression is depressed in some pathologic conditions and upon exposure to some stimuli. For example, GCLC expression is decreased in hepatocytes in endotoxic shock (Ko et al., 2008), lungs in chronic obstructive pulmonary disease (COPD) (Cheng et al., 2016), and the fibrotic foci of idiopathic pulmonary fibrosis (Tiitto et al., 2004). Its expression is downregulated by stimuli including transforming growth factor beta (TGFβ) (Liu et al., 2012), lipopolysaccharide (Payabvash et al., 2006; Ko et al., 2008; Tomasi et al., 2014), and silica nanoparticles (Zhang et al., 2017).

The molecular mechanism of GCLC regulation has been extensively studied. The promoter sequence and activity of *GCLC* in humans, rats, and mice have been systematically examined, and similar *cis*-elements are identified in the *GCLC* promoter region of rodents and humans, including an electrophile response element (EpRE, also known as the antioxidant response element), a 12-O-tetradecanoylphorbol 13-acetate (TPA) response element (TRE), and consensus sequences for nuclear factor-kappa B (NF-κB), SP1, and metal response element. Depending on cell types and stressors, the expression of *GCLC* is regulated through multiple upstream signaling pathways including ERK1/2 (Zipper and Mulcahy, 2000; Chen et al., 2008; Yang et al., 2011), JNK1/2 (Dickinson et al., 2002), p38MAPK (Zipper and Mulcahy, 2000), and PI3K/AKT (Kim et al., 2004; Arisawa et al., 2009). Downstream, *GCLC* expression is regulated through transcription factors nuclear factor (erythroid-derived 2)-like 2 protein (commonly called Nrf2) and activating protein 1 (AP-1) (Wild et al., 1998; Dickinson et al., 2002; Dickinson et al., 2003; Zhang et al., 2007; Zhang and Forman, 2010). Nrf2 and AP-1 members are located in the cytosol in the resting conditions, but once activated, they are translocated into the nucleus, where Nrf2 forms heterodimers with partners (such as c-Jun or small Maf proteins Maf G/F/K) and binds to EpREs, whereas AP-1 dimer binds to TREs in the *GCLC* promoter. Nrf2 and its partners, and AP-1 members (such as Jun and ATF family members) are potential regulatory targets of aforementioned kinases.

EpRE also binds transcription factors other than Nrf2 as a partner, a competitor, or a replacement. Several EpRE-binding proteins have been identified including AP-1 family proteins (c-Jun, Jun-B, Jun-D, Fra1, Fra2, c-fos, ATFs), basic leucine zipper family protein (Nrf1, Nrf2, Bach1, Bach2, Bach3), and small Maf proteins (G/F/K). A different combination of the heterodimers may have different and sometime opposite effects on *GCLC* promoter activity. Binding of Bach1, for example, is found to negatively regulate EpRE activity and *GCLC* mRNA expression (Warnatz et al., 2011). Binding of an AP-1 dimer composed of c-Jun and Fra1 to EpRE is responsible for the downregulation of *GCLC* mRNA expression by TGFβ (Jardine et al., 2002). Multiple EpREs and TREs are identified in the human *GCLC* promoter, but only one of them regulates the inducible expression, whereas the others are needed for maintaining the basal expression of *GCLC* (Zhang and Forman, 2010). A proximal TRE (−263 to −269) is critical in mediating oxidative stress induction of *GCLC*, and a distal EpRE (−3139 to −3149) is involved in its constitutive and inducible expression (Mulcahy et al., 1997).

The possible involvement of NF-κB in *GCLC* mRNA expression is controversial. NF-κB activation increases rat *Gclc* promoter activity, and mutation of the NF-κB binding site depresses rat *Gclc* promoter activity and its response to Nrf2 overexpression and oxidant exposure (Yang et al., 2005a,b), suggesting that NF-κB is required for rat *Gclc* expression. However, NF-κB activation by lipopolysaccharide (LPS) does not increase human GCLC expression in macrophages. Instead, it causes a decrease in GCLC expression (Zhang et al., 2012). No direct evidence is available on the interaction between NF-κB and *GCLC* promoters. The effect of NF-κB signaling on GCLC mRNA expression might be indirect, possibly affecting the activity of Nrf2 and AP-1.

In addition, GCLC expression can also be regulated at the posttranscription level by affecting mRNA stability (Liu et al., 1998). *GCLC* mRNA stability can be regulated through the binding of protein Hu-antigen R (HuR) to the AU-rich element

(+2785AUUUA sequence) in the 3′-untranslated region of *GCLC* mRNA. Without HuR binding, the half-life of *GCLC* mRNA is decreased from 20 h to less than 8 h (Song et al., 2005). The significance of this regulation pathway, however, needs to be further elucidated, especially under pathophysiological conditions with altered GCLC expression.

1.3.3.2.2 GCLM Expression

GCLM promoters of rodents and humans all contain EpRE and TRE *cis*-elements, and their expression, including the constitutive and inducible expression, is thus regulated through Nrf2/EpRE and AP-1/TRE signaling pathways. Upon exposure to electrophiles, GCLM is frequently induced simultaneously with GCLC via similar pathways involving Nrf2 and AP-1 activation (Wild et al., 1998; Dickinson et al., 2002; Dickinson et al., 2003; Zhang et al., 2007; Zhang and Forman, 2010). However, in some situations, the regulation of GCLM is distinct from that of GCLC. For example, ethanol increased rat GCLC expression through activating Nrf2 and AP-1 pathway, but GCLM expression did not change (Lu et al., 1999). TGFβ depressed rat GCLC expression but did not affect GCLM expression (Fu et al., 2008). In response to LPS exposure, human GCLC expression is depressed via an NF-κB-dependent pathway, but GCLM expression is increased in an Nrf2- and NF-κB-independent pathway (Zhang et al., 2012). The exact mechanisms involved in the distinct regulation of *GCLC* and *GCLM*, including the *trans*- and *cis*-elements involved, likely depend on suppressor elements, but their identities remain obscure.

1.3.3.2.3 Relative Expression of GCLC and GCLM Subunits among Tissues

Because the efficient catalytic activity of the GCL holoenzyme depends on the association of the GCLC and GCLM subunits, it is interesting to examine their relative expression in various tissues and among different cell types within those tissues. The consequence of this variable expression is that induction of GCLM in tissues where it is limiting should increase GSH synthesis, given the higher catalytic activity of the holoenzyme, relative to that of GCLC monomer alone. Krzywanski et al. (2004) showed that GCLM was in fact limiting in its expression in human bronchial epithelial cells and in the liver, kidney, and heart tissues of mice, with the ratios of GCLM to GCLC ranging from 0.06 to 1, depending on the cells/tissue type.

Chen and colleagues (2005) investigated the expression of GCL subunits in mice in which the *Gclm* gene has been knocked out (i.e., *Gclm* null mice), in order to determine the impact that the resulting low GSH in these animals had on gene and protein expression. Wild-type mice served as a control for expression in these studies, and thus, the relative level of the subunits was examined in multiple tissues at both the mRNA level (using RT-PCR) and the protein level, usually by semiquantitative Western immunoblotting. They also examined the effects of titrating recombinant GCLM protein into tissue homogenates from various organs of *Gclm* null mice, and then assessed the formation of GCL holoenzyme and its activity. This exercise yielded interesting information regarding the relative contribution of GCLM and GCLC to holoenzyme formation, and whether the levels of GCLM were limiting. One caveat to this study is that lower GSH can increase the expression of GCLC through Nrf2-mediated transcription (see above), and thus, the levels of GCLC that are present under normal (i.e., wild-type) conditions may be different. Interestingly, when the kidney was assessed, they found that GCLC did not form more holoenzyme when recombinant GCLM was added to the homogenate, suggesting a complicated interaction of the two subunits in this organ.

The Stipanuk laboratory evaluated the relative expression of GCLC and GCLM proteins in liver cells and rat liver in the context of variable cysteine supply (Lee et al., 2006). This interesting work revealed that cysteine deprivation from culture media or from the diet resulted in the formation of more holoenzyme in human HepG2 liver cells in culture and in rat liver, respectively, with the consequence that there was an increase in the catalytic activity of GCL when cysteine was missing from the diet. They also made the observation that GCLM was likely limiting since adding recombinant human GCLM to rat liver homogenates increased the proportion of GCLC in the holoenzyme form. They, like others before (Seelig et al., 1984; Seelig and Meister, 1985; Tu and Anders, 1998), also noted that a significant proportion of GCL holoenzyme is resistant to dithiothreitol-mediated dissociation and concluded that there can be covalent interactions between the two subunits.

Subsequently, the relative levels of GCLM and GCLC proteins were examined in another *Gclm* null mouse model by the Kavanagh laboratory (McConnachie et al., 2007). This study also showed variable protein expression of the two subunits in the liver and kidney, and, supporting what had been published by Chen and colleagues (2005) in their *Gclm* null model, heterozygous mice had significantly less GCL activity in the liver, kidney, brain, and lung, suggesting haplo-insufficiency for *Gclm* in these mice.

There is scant peer-reviewed published information available on the expression levels of GCL subunit mRNAs and proteins in human organs and tissues. However, recent availability of open source data sets have been published by Fagerberg et al. (2014) and the Mouse ENCODE Consortium (Yue et al., 2014). These data have been made accessible through the U.S. National Library of Medicine's National Center for Biotechnology Information (e.g., www.ncbi.nlm.nih.gov/gene/2730 and www. ncbi.nlm.nih.gov/gene/14630) and display tissue-specific mRNA expression for most human and mouse genes. The Human Protein Atlas (www.proteinatlas.org) allows one to examine the relative expression of GCLC and GCLM proteins (by semiquantitative immunohistochemistry) in various organs and in cells within those organs.

Figure 1.4 shows the relationship between mRNA and protein levels for the GCL subunits among various human organs. Figure 1.5 shows the relative expression of *GCLC* and *GCLM* mRNAs vs. semiquantitative estimates of their protein levels (i.e., none = 0, low = 1, medium = 2, high = 3) among the organs for which these data are currently available. Figure 1.6 shows the relationship between GCLC and GCLM protein levels cataloged over these organs. Three conclusions can be drawn for these analyses:

1. Despite variations between individual organs, there is a general similarity between the relative mRNA expression levels for GCLC and GCLM subunits across organs, supporting the earlier work published by Krzywanski et al. (2004). In general, the expression of *GCLC* mRNA tends to be higher than that of *GCLM* mRNA (as indicated by the slope of the regression line, which is considerably less than unity).

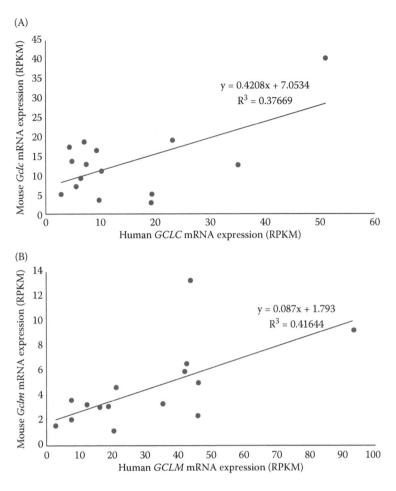

FIGURE 1.4   Comparison of mRNA expressions in human and mouse tissues. Expression values were derived from the number of reads per kilobase per million mapped reads s(RPKM) for *GCLC* (panel A) and *GCLM* (panel B). Data taken from Tables 1.1 and 1.2.

2. There is only weak evidence for a relationship between mRNA levels and protein levels for each subunit at the whole organ level. However, in future analyses, mRNA *in situ* hybridization at the single cell level may be more likely to show a stronger relationship to subunit protein levels determined by immunohistochemistry.

3. The ratio of the subunit mRNA expression is highly variable among the organs, with *GCLM* in most organs being the subunit with limiting expression, whereas protein levels tend to be more highly correlated, at least as determined by the ratio of the subunits expression to that of *GAPDH* expression (Figure 1.5). Although it would be desirable to compare the ratio of *GCLM* to *GCLC* mRNA or protein with GSH levels in those organs, such information has been difficult to obtain.

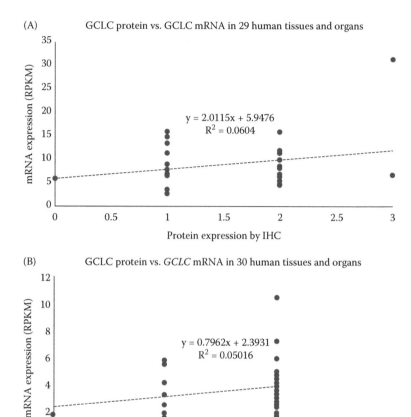

FIGURE 1.5   Relative expression of *GCLC* mRNA and protein levels (panel A) and *GCLM* mRNA and protein levels (panel B) among human organs. The mRNA expression values were derived from the number of reads/thousand/million bases (RPKM) and the semiquantitative expression levels for proteins listed in Table 1.2.

1.3.3.2.4   *Polymorphisms in GCL Subunit Genes and Their Association with Various Disease States*

In this section, we discuss genetic polymorphisms in the human *GCLC* and *GCLM* genes, several of which are fairly common.

The seminal work of Koide et al. (2003) and Nakamura et al. (2002, 2003) showing that polymorphisms in *GCLC* and *GCLM* are predictive of myocardial infarction (MI) and coronary artery vasospasm was instrumental in gaining attention for these polymorphisms not only in cardiovascular disease (CVD) but also in other diseases and conditions. Since then, many other groups have found similar risks, or none at all (see next), which seems to be dependent on location-specific risk factors (e.g., diet, culture) and perhaps race/ethnicity. In follow-on studies, Katakami et al. (2010,

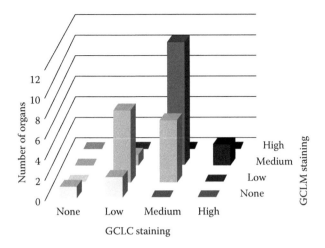

FIGURE 1.6 Comparison of human GCLC and GCLM protein expression levels across multiple organs. Data are derived from semiquantitative expression levels for proteins listed in Table 1.2.

2014) showed that the greater the number of pro-oxidant alleles for *GCLC* (-588C/T), *SOD2* (Val16Ala), *NOS3* (G894T), *CYBA* (C242T), and *MPO* (-463G/A), the higher the risk of MI and coronary heart disease in Type II diabetic male patients.

The relationship between the *GCLM* -588C/T polymorphism and dilated cardio-myopathy (DCM) was investigated by Watanabe et al. (2013). These researchers found a significant increased risk of DCM among people with the -588T allele.

Tang et al. (2015) investigated the prevalence of multiple gene polymorphisms in mothers and babies with congenital heart defects. They found that a number of poly-morphisms in *GCLC* were associated with obstructive heart defects among infants born to obese women.

In an examination of the roles that folate, homocysteine, and transsulfuration pathways play in conotruncal heart defects in children, Hobbs et al. (2014) found that multiple polymorphisms in *GCLC* were significantly associated with heart defects. In a follow-on study, this group also showed a similar enrichment in the *GCLC* rs6458939 polymorphism among children with conotruncal heart defects born to obese mothers and those using tobacco (Tang et al., 2014).

A study of Han Chinese examined the relationship between CVD and the prev-alence of *GCLC* -129C/T and *GCLM* -23G/T polymorphisms (Zuo et al., 2007). The former was a risk factor, and the latter may be protective of CVD. In contrast, Muehlhause et al. (2007) did not find an association between *GCLM* -588C/T and ischemic heart disease in a German cohort. In support of the notion that ethnicity and perhaps lifestyle factors have a major influence on the strength of the association between *GCL* polymorphisms and CVD, Skvortsova et al. (2017) found highly significant effects of the *GCLM* -588T polymorphism (odds ratio [OR] = 4.79; $p = 0.03$) and the *GCLC* -129T polymorphism (OR = 4.79, $p = 0.03$) and ischemic heart disease among Kasaks, but neither polymorphism was a risk factor for Russians.

TABLE 1.1

Mouse *Gclm* and *Gclc* mRNA Expression across Multiple Tissues

Mouse Organ or Tissue	GCLM mRNA Expression	GCLC mRNA Expression	GCLM mRNA/ GCLC mRNA
Adrenal gland	42.71	6.09	7.01
Cerebellum	14.95	10.40	1.44
Cerebral cortex	7.45	8.99	0.83
Frontal lobe	7.41	9.55	0.80
Genital fat pad	20.35	9.23	2.20
Colon	35.47	9.99	3.55
Duodenum	93.48	4.10	22.78
Heart	2.92	2.61	1.11
Kidney	46.26	34.64	1.34
Large intestine	9.49	5.75	1.65
Liver	43.87	50.50	0.87
Lung	12.47	19.12	0.65
Mammary gland	12.89	9.24	1.40
Ovary	20.86	19.01	1.10
Placenta	21.17	9.51	2.23
Small intestine	42.23	4.58	9.17
Spleen	18.97	6.79	2.79
Stomach	46.44	7.12	6.48
Subcutaneous fat	6.53	6.28	1.04
Testis	7.56	5.43	1.39
Thymus	6.29	8.00	0.79
Urinary bladder	16.27	22.90	0.71

Source: Data are taken from the mouse ENCODE transcriptome database (Yue et al. 2014), which is available at NCBI Gene (www.ncbi.nlm.nih.gov/gene/14630 for *Gclm* and www.ncbi.nlm.nih. gov/gene/14629 for *Gclc*).

Campolo and coworkers (2007) found that the *GCLC* -129C/T polymorphism T allele was associated with lower plasma α-tocopherol and cysteinyl-glycine (a product of γ-GGT) after a transsulfuration challenge test (infusing methionine), suggesting a significant homocysteine-mediated change in redox status in erythrocytes that was influenced by this polymorphism.

GCLC is known to have a guanine-adenine-guanine (GAG) trinucleotide repeat (TNR) polymorphism present in the 5′ untranslated region of the mRNA, with 5–10 tandem GAG repeats (Walsh et al., 1996; Willis et al., 2003). This TNR polymorphism has been shown to be associated with pulmonary function in adult cystic fibrosis patients who have a relatively mild mutation in the *CFTR* gene (Mckone et al., 2006). It has also been shown to influence the rate of decline in lung function among smokers, especially those who have low vitamin C intake (Siedlinski et al., 2008).

In addition to the above-mentioned effects on lung diseases, polymorphisms in a number of antioxidant genes, including those of *GCLM*, *HMOX1*, *NQO1*,

TABLE 1.2

Human GCLM and GCLC mRNA and Protein Levels across Multiple Human Tissues

Human Organ or Tissue	*GCLM* mRNA Expression	*GCLC* mRNA Expression	*GCLM* mRNA/ *GCLC* mRNA	GCLM Protein Levels	GCLC Protein Levels	GCLM Protein/ GCLC Protein
Adipose	5.50	8.00	0.69	1	2	0.50
Adrenal	10.50	8.40	1.25	2	2	1.00
Breast	4.70	7.60	0.62	2	1	2.00
Caudate	3.10	8.80	0.35	2	1	2.00
Cerebellum	2.60	4.40	0.59	2	2	1.00
Cerebral cortex	3.40	6.10	0.56	2	2	1.00
Cervix	1.80	6.40	0.28	1	1	1.00
Colon	2.40	9.80	0.24	2	2	1.00
Endometrium	1.80	5.10	0.35	2	2	1.00
Esophagus	4.20	13.20	0.32	1	1	1.00
Fallopian tube	2.90	6.50	0.45	2	3	0.67
Heart	1.90	3.10	0.61	1	1	1.00
Hippocampus	2.60	7.10	0.37	1	1	1.00
Hypothalamus	4.10	6.40	0.64	2		n/a
Kidney	5.00	6.70	0.75	2	2	1.00
Liver	6.00	15.70	0.38	2	2	1.00
Lung	4.80	5.60	0.86	2	2	1.00
Ovary	1.80	5.70	0.32	0	0	n/a
Pancreas	1.70	4.50	0.38	2	2	1.00
Prostate	3.20	13.30	0.24	2	1	2.00
Salivary gland	3.30	7.60	0.43	1	1	2.00
Skeletal muscle	5.80	2.60	2.23	1	1	1.00
Skin	1.60	6.50	0.25	1	1	1.00
Small intestine	4.40	14.60	0.30	2	1	1.00
Spleen	3.80	15.80	0.24	2	1	2.00
Stomach	3.20	11.70	0.27	2	2	2.00
Testis	1.90	4.80	0.40	2	2	1.00

Source: mRNA Data are Taken from the GTEX Database (Fagerberg et al. 2014), and Protein Data is from the Human Protein Atlas (www.proteinatlas.org).

n/a, not applicable.

GSTM1, *GSTP1*, *GC*, and *CAT*, were used to construct a score for genetic risk and based their association with oxidation of guanine (8-oxo-dG) as a biomarker of oxidative burden. There was a highly significant inverse association between black carbon exposure (a surrogate for mobile sources of air pollution) and reductions in two measures of lung function, forced vital capacity (FVC), and forced expiratory volume in 1 s (FEV1), in elderly men in the Normative Aging Study over a 5-year period (Mordukhovich et al., 2015). Tang et al. (2013) found a decline in lung function with smoking among individuals of European descent having the -129C/T polymorphism. The TT genotype was associated with a steeper rate of decline in FEV1/FVC (0.9%/year) than the CC genotypes. As mentioned previously, smoking and decreases in lung function have been associated with *GCLM* polymorphisms in the context of low ascorbate (vitamin C) intake (Siedlinski et al., 2008).

Chronic beryllium disease (CBD) can be a progressive and oftentimes a fatal lung disease (Mayer and Hamzeh, 2015; Fontenot et al., 2016). Bekris and colleagues (2006) uncovered a relationship between CBD and the 5′ GAG TNR polymorphism in *GCLC* (the 7/7 repeat genotypes were protective); in addition, the -588C/T single-nucleotide polymorphism (SNP) in *GCLM* was associated with CBD susceptibility.

The TNR polymorphism in the *GCLC* promoter has been shown to predispose to lung cancer and cancers of the digestive tract (Nichenametla et al., 2013). This same group also showed that this polymorphism influences GCLC protein levels through effects on translation (Nichenametla et al., 2011). However, de Lima Marson et al. (2013) were not able to show such a relationship between *GCLC* -129C/T polymorphism and cystic fibrosis risk (De Lima Marson et al., 2013). More recently, Marson et al. (2014) examined the relationships between *CFTR* and modifier genes in the GSH pathway, including GSTM1 deletion, GSTT1 deletion, GSTP1 +313A/G, *GCLC* -129C/T, and -3506A/G. This work showed that *GCLC* -129C/T polymorphism was associated with a higher frequency of *Pseudomonas aeruginosa* mucoid in the CC genotype, and *GCLC* -3506A/G had a higher frequency of the no-mucoid *P. aeruginosa* in the AA genotype. There was also evidence for increased colonization of *Achromobacter xylosoxidans*. This latter polymorphism was also associated with a higher Bhalla computed tomography (CT) score.

Community-acquired pneumonia with pulmonary complications was shown to be associated with *GCLC* polymorphisms (Salnikova et al., 2013). Yuniastuti et al. (2017) examined the frequency of *GCLC* -129C/T polymorphism in patients with active tuberculosis (TB). There is some evidence for -129C/T being an attributing allele in TB patients. However, it is hard to know what this means without having the frequencies of this polymorphism in noninfected controls.

In a meta-analysis of the effects of the *GCLM* -588C/T polymorphism, Hu et al. (2008) found a significant effect of this polymorphism in COPD. The authors discovered that the -588T allele was related to the risk of progressing to COPD. However, Liu et al. (2007) found no relationship between *GCLC* -129C/T and COPD. Nor did Chappell et al. (2008) find any relationship between *GCLC* or *GCLM* polymorphism and COPD risk.

Polonikov and colleagues (2007) reported a significant association between *GCL* -23G/T/-588C/T linked polymorphisms and bronchial asthma. For allergic asthma, this polymorphism represented a decreased risk, whereas for nonallergic asthma, it was actually a factor for increased risk of disease.

Associations between the *GCLC* TNR polymorphism and plasma thiols in schizophrenia patients were investigated by Gysin et al. (2011). They concluded that this could be useful in early diagnosis for patients at higher risk of oxidized redox states. In addition, Tosic and coworkers (2006) studied a number of polymorphisms in *GCLM* and in *GS*. They found that a *GCLM* intron 1 polymorphism (rs2301022) showed a strong relationship between carrying a G allele and risk for schizophrenia in two case–control studies and in a family study. Ma et al. (2010) were able to reproduce these findings in a Han Chinese population, but the significance was less than that seen by Tosic et al. (2006) in Europeans.

Xin et al. (2016) found a significant association with *GCLC* polymorphisms and low GSH in the prefrontal cortex and peripheral blood redox status, and GSH and GPx activities were negatively correlated. Low brain GSH correlated with low peripheral oxidation status and dysregulated GSH homeostasis in early psychosis patients. Also, low erythrocyte GSH is a predictor of transition to disease in at-risk patients (Lavoie et al., 2017). However, replication of these findings in a Japanese population failed (Hanzawa et al., 2011). Additionally, no association with *GCLM* polymorphisms was observed for bipolar disorder (Fullerton et al., 2010) or for the *GCLC* -129C/T polymorphism and risk of stroke (Man et al., 2010), and there was no association between *GCLM* SNPs, *GCLC* SNPs, or *GCLC* TNR polymorphisms and mental depression (Berk et al., 2011).

Because of the very important role that GSH plays in normal liver function and in the detoxification of many xenobiotics, nonalcoholic fatty liver disease (NAFLD) and the more serious nonalcoholic steatohepatitis have been examined for their relationship with *GCL* polymorphisms. Oliveira et al. (2010) investigated the effects of polymorphisms in the liver microsomal lipid transfer protein (MTP) and in GCLC. There was a highly significant effect on the risk of NAFLD of at least one T allele in the -129C/T polymorphism in GCLC (OR: 12.14; 95% CI: 2.01–73.35). On the other hand, Hashemi and colleagues (2011) did not detect any link between *GCLC* -129C/T polymorphism and NAFLD.

Regarding the effects of *GCL* polymorphisms on drug and xenobiotic metabolism, the first published evidence that polymorphisms in GCLC had a functional effect was the study by Walsh et al. (2001), who showed the GAG TNR has an influence on GSH level and resistance of a large number of cultured cancer cell lines to chemotherapeutics. This TNR polymorphism also interacts with the -129C/T SNP in a way that only the seven GAG repeat polymorphisms caused a decreased expression of GCLC (Butticaz et al., 2011). It also influences survival and differential protein expression in skin fibroblasts exposed to the model oxidant *tert*-butylhydroquinone in culture (Gysin et al., 2009). Wang et al. (2012) found that the intronic SNP rs761142 T>G was associated with lower *GCLC* mRNA expression and was predictive of an idiosyncratic drug reaction to sulfamethoxazole, which can induce hypersensitivity reactions in the skin of HIV/AIDS patients. It has been reported that the *GCLC* -129C/T polymorphism influenced the retention of methylmercury in erythrocytes (Custodio et al., 2004). Barcelos et al. (2013) found similar effects for the *GCLM* -588T/T genotype. These individuals had less Hg in their blood and hair than those with CC and CT genotypes, suggesting a protective effect of the TT genotype. Harari and coworkers (2012) found that Hg levels in the blood and urine of gold miners and merchants in Ecuador were influenced by the *GCLM* -588C/T polymorphism but not related to the degree of neurotoxicity observed in these people. There was a suggestion of an effect of *GCLM* -588T/T genotype, the levels of the omega-3 polyunsaturated acids, eicosapentaenoic acid, and docosahexaenoic acid in the blood, and their effects on methylmercury burden, but the number of subjects carrying this rare allelic combination was too small to show a significant interaction (Engstrom et al., 2011).

Vieira et al. (2011) observed that the frequency of individuals with a compromised glomerular filtration rate (i.e., <60 mL/min) was significantly higher in patients carrying the *GCLC* -129C/T polymorphism genotypes C/T+T/T (47.1%) than in those carrying the C/C genotype (31.1%).

Bekris et al. (2007) showed an association of the 5′ *GCLC* TNR polymorphism with age at disease onset, as well as GAD65 autoantibodies, among Type 1 diabetes patients. The effect of *GCLC* -129C/T polymorphism on immune response in rubber industry workers was explored by Jonsson and coworkers (2008). There was a suggestion of an increased number of circulating immunocytes (total leukocytes, neutrophils, and eosinophils) in workers carrying the *GCLC* -129T allele, but this was not significant. No effect of the *GCLM* -588C/T polymorphism was found on these parameters.

Lin et al. (2013) reported an association between *GCLC* -129C/T polymorphism (rs17883901) and risk for breast cancer among women with a body mass index (BMI) > 25 (OR: 1.91 and 95% CI: 1.09, 3.36). However, if these women had a lower BMI at the age of 20, there was no association, suggesting an interaction between BMI and *GCLC* polymorphism and age.

Khadzhieva et al. (2014) assessed the association between oxidative stress-related genetic polymorphisms and idiopathic recurrent miscarriage. In this study, the *GCLC* -129C/T polymorphism was not associated with the risk of miscarriage.

Exposure to petroleum products/distillates among pesticide workers was associated with an increase in prostate cancer in those carrying a *GCLC* polymorphism (rs1883633). Men carrying at least one minor allele showed an increase of about 3.7-fold (Koutros et al., 2011).

Taken together, there is abundant epidemiological evidence that relatively common noncoding polymorphisms in both *GCLC* and *GCLM* influence the risk of various diseases and xenobiotic exposure outcomes. The effects of relatively rare polymorphisms/exonic mutations that affect the structure and/or catalytic activity of GCLC are discussed in the following section.

1.3.3.3 GS Expression

GS is an enzyme that is constitutively expressed in all cells. In rat liver hepatoma H4IIE cells, transcription factor nuclear factor 1 (NF-1) mediates the repression of the *Gss* gene by binding to two NF-1 motifs at −1025/−1008 and −808/−791. Using electrophoresis mobility shift assay, super shift assay, and dominant-negative c-Jun transfection, 6 of the 12 potential AP-1 sites in the promoter region involved in the basal and inducible expression of rat *Gs* gene were suggested to be functional in regulating expression (Yang et al., 2002, 2005b). In human Chang liver cells, both Nrf1 and Nrf2 are involved in the regulation of *GS* gene, since overexpression of either protein increased the promoter activity (Yang et al., 2005a). Two EpREs (−168/−158 and −2130/−2120) in the promoter of human *GS* gene are responsible for the basal expression since mutation of both resulted in a significant decrease in basal promoter activity. Chromatin immunoprecipitation assays suggest that Nrf2 and c-Jun bind to the distal EpRE, whereas Nrf1 and c-Jun bind to the proximal EpRE in the promoter of human *GS* gene (Lee et al., 2005). Nrf2 involvement in the regulation of GS is further supported by evidence that Nrf2 overexpression increased the expression of GS (Shih et al., 2003). In summary, GS is regulated through a similar mechanism as GCLC and GCLM, that is, through Nrf2/EpRE and/or AP-1/TRE pathways.

1.3.4 DYSREGULATION OF GSH SYNTHESIS

1.3.4.1 Diseases Associated with Changes in GSH Synthesis

Changes in GSH synthesis and consequences for oxidative stress have been associated with a number of diseases including pulmonary (Ciofu and Lykkesfeldt, 2014; Santus et al., 2014; Matera et al., 2016), cardiovascular (Ndrepepa and Kastrati, 2016; Watanabe et al., 2016; Mistry and Brewer, 2017), renal (Stepniewska et al., 2015), hepatic (De Andrade et al., 2015; Jung, 2015; Han et al., 2016; Lu et al., 2016), gastrointestinal (Alzoghaibi, 2013), diabetes (Rabbani et al., 2016), chronic neurological disorders (including Parkinson's disease, Alzheimer's disease, Huntington's disease, amyotrophic lateral sclerosis, and Friedrich's ataxia) (Johnson et al., 2012; Gu et al., 2015; Mcbean et al., 2015), hematological disorders (Koralkova et al., 2014; Murphy et al., 2014), immune (Amir Aslani and Ghobadi, 2016; Short et al., 2016), ocular (Wells et al., 2009; Kumar et al., 2013), developmental disorders (Davis and Auten, 2010), and various forms of cancer. For a recent summary of this area, the reader is referred to the excellent review by Lu (2013). As noted previously, the levels of expression by GCLC, GCLM, and GS can have a dramatic effect on GSH synthesis, and there has been a lot of interest in the factors that control their expression.

One area that has not been explored until very recently is the influence of GSH and its redox status on epigenetic control of gene expression, including the effects on methylation, acetylation, histone–DNA interactions, and chromatin architecture. García-Giménez and colleagues (2017) have recently reviewed this interesting and evolving field, and the reader is encouraged to consult their review for more information.

1.3.5 RARE INBORN ERRORS OF METABOLISM WHERE THERE IS ABSENCE OF PROTEIN OR WHERE MUTATION CAUSES SUBSTANTIAL REDUCTION IN ENZYME ACTIVITY

There are exceedingly rare mutations in GCLC that result in either complete loss of function [based on mouse studies (Dalton et al., 2000; Shi et al., 2000); this is presumed lethal] or reduced function with only partial activity in GCL and GSH synthesis. Most of the reports on these rare nonsynonymous mutations in the coding region of *GCLC* that result in pathology of the hematological system, the central nervous system, or disturbances in neuromuscular function.

Konrad et al. (1972) reported significantly lower levels of GSH and GCL activity in two patients presenting with hemolytic anemia and spinocerebellar degeneration. These two patients had approximately 2% and 3% of normal GSH levels, but 13% and 9% of normal erythrocyte GCL activity, respectively. Beutler et al. (1990) described a 22-year-old patient with about 10% of normal erythrocyte GSH levels and 6% of normal GCL activity. The patient presented with no neurological deficits. This patient did show increased reticulocytes (11%) but relatively normal hematocrit, suggesting compensation for chronic nonspherocytic hemolytic anemia. The authors concluded that the modest amount of GSH present and GCL activity present in the blood of this patient was due to its enrichment in reticulocytes. This patient also showed only slight decreases in cultured skin fibroblast GSH levels (approximately 60% of normal controls), and a blood lymphoblast cell line derived from the patient similarly showed a decrease relative to controls (approximately half of normal controls). A follow-on study by Beutler et al. (1999) described the nature of the mutation in this case which resulted in attenuated GCL activity. Molecular analysis and DNA sequencing found a 1109A>T mutation that resulted in substitution of leucine for histidine at position 370 in GCLC. In another study of moderate deficiency in GCL activity in two individuals, Hirono and colleagues (1996) reported that hemolytic anemia was the only consequence of lower GCL activity.

Ristoff and colleagues (2000) showed in a case report that a 473C→T missense mutation resulted in an amino acid substitution of leucine for proline at position 158 in GCLC. There was hemolytic anemia and episodic jaundice in one of the probands of an extended Dutch family with consanguineous marriages [originally described by Prins et al. (1966)]. There was also a patient with similar symptoms and mutation in a more recent generation from this family. The level of GCL activity was only about 2% of normal. Although heterozygosity in this mutation results in about 50% of GCL activity, this did not result in any clinical consequences with respect to GSH levels or any adverse symptoms. Mañú Pereira et al. (2007) describe an individual with only about 1% of normal GCL activity who also presented with hemolytic anemia (with RBC GSH levels being roughly 5% of normal), neurological deficits, and muscular abnormalities. The mutation in GCLC in this case was determined to be a C-to-T transition at base 1241, which resulted in a leucine at position 414 instead of proline in the GCLC protein. More recently, Almusafri et al. (2017) described six children from two separate consanguineous families. One family had three children with a 1772G>A nucleotide substitution, resulting in a substitution of asparagine for serine at amino acid position 591 in GCLC. The second family (also three children) had a 514T>A mutation resulting in a serine-to-threonine mutation at amino acid position 172 in GCLC. In both families, the children presented with hemolytic anemia at birth, but this was not evident beyond the neonatal period (although there was some anemia into childhood). No other abnormalities were noted. Interestingly, in two of the children from the first family, erythrocyte GSH levels were 33% and 18% of normal (GSH levels were not reported for the other member of this family, nor the three children in the second family, and GCL activity was not measured in any of these children). Thus, it appears that the mutation in the first family had relatively mild effects on GSH, and this is presumably the reason that there were no noted symptoms beyond mild anemia.

1.4   DECREASED GSH CONTENT WITH AGING

Advancing age is associated with increased accumulation of oxidative damage in tissues, which is the basis of the free radical theory of aging (Harman, 1956). Age-related accumulation of oxidative damage may involve an excessive production of reactive oxidants and electrophiles, a decreased antioxidant defense, and an impaired capacity to remove or repair damage. As GSH is one of the major agents for the elimination of peroxides and other electrophiles, the decrease in GSH content observed with aging in animal models and humans may play a major role in aging, or at least, the increased susceptibility to oxidative stress associated with age-related pathologies.

Alteration of GSH content in tissues with aging is largely investigated in rodent models. In comparison with young adult rodents, total GSH content is significantly lower (in a range of 20%–50%) in many tissues of elderly animals. These tissues include the liver, kidney, lung, and cerebral cortex, and cerebellum in rats (Ravindranath et al., 1989; Favilli et al., 1994; Liu and Choi, 2000; Sasaki et al., 2001); and the liver, kidney, heart, lung epithelial lining fluid, cortex and brain stem, serum, and spleen lymphocytes in mice (Hazelton and Lang, 1980; De and Darad, 1991; Teramoto et al., 1994; Nakata et al., 1996; Wang et al., 2003; Gould et al., 2010). In some tissues including mouse heart or diaphragm, however, total GSH content may not change with age (Choi et al., 2000). Controversial data are reported on the change of total GSH content with aging in some tissues, for example, in some studies, no change of total GSH content with aging in the liver of mouse and rat is found (Ravindranath et al., 1989; Nakata et al., 1996; Toroser and Sohal, 2007). The content of reduced GSH is reportedly lower in the mouse lung (Teramoto et al., 1994) and rat liver (Sanz et al., 1997). The decrease in total GSH content with aging is also observed in humans. Compared with younger control subjects, total GSH concentration is significantly decreased in erythrocytes (Rizvi and Maurya, 2007; Sekhar et al., 2011).

In summary, although there are some controversial reports, in general there is no doubt that the total GSH content is decreased in most tissues investigated in rodent models and in erythrocytes in humans. Regarding the mechanism of age-related decrease in total GSH content, available evidence supports that it is due to an insufficient GSH synthesis. The decrease in total GSH content with aging is usually associated with decreased expression of GCL (both GCLC and GCLM) and GS in mouse and rat models (Liu and Choi, 2000; Liu and Dickinson, 2003; Wang et al., 2003; McConnachie et al., 2007), whereas the steady-state level of cysteine does not change with age (Liu and Choi, 2000; Liu and Dickinson, 2003). However, there is evidence that deficient precursors are responsible for age-related decrease in GSH content in erythrocytes. The decrease in GSH in erythrocytes with age in human subjects is associated with decreased cysteine level, and dietary supplementation with GSH precursor cysteine and glycine could restore GSH concentration and decrease oxidative stress and oxidant damage (Sekhar et al., 2011). Obviously, further evidence is needed to elucidate the underlying mechanism of age-related decreases in GSH content. The effect of aging on the ability to increase GSH in response to stimuli is a marked suppression of the inducibility of the GCL subunits in most cells and tissues. This is largely the result of decreased Nrf2 signaling in aging (Zhang et al., 2015).

1.5 SUMMARY

Synthesis of GSH, which may first appear to be simply the sequential addition of amino acids to form a tripeptide, is anything but simple. The first step, catalyzed by GCL, is unusual in the use of the γ-carboxyl group of glutamate to form a peptide bond that is resistant to hydrolysis by intracellular peptidases. The feedback inhibition is competitive, but modulated by a subunit, GCLM, that does not possess catalytic activity.

Although the genes for both subunits share some common regulatory elements, most notably Nrf2, complexity comes from mismatching expression of both the mRNAs and protein for each of the two subunits and mismatched expression of their ratio. This is likely related to the requirements for GSH use that result in variation in GSH concentration among cell types and tissues that work through both feedback on GCL and signaling for the expression and turnover of the subunit mRNAs and proteins.

Adding to the complexity is the revelation that GCL, long thought to be the rate-limiting enzyme for GSH *de novo* synthesis with cysteine availability as the limiting factor, is that GS is also regulated and that differences in its expression can alter GSH concentration.

The next level of complexity comes from the epidemiological evidence that relatively common noncoding polymorphisms in both *GCL* genes are associated with the risk for diseases and resistance to xenobiotic agents. Rare polymorphisms in the coding regions of GCLC are more clearly causative of functional deficits in GSH synthesis with the predicted negative effect on health.

Finally, the concentration of GSH and regulation of its synthesis decline with age in many tissues and cell types, suggesting a contribution to aging and the increased susceptibility of the aged to environmental stressors. Thus, the regulation of GSH synthesis remains a question for the ages.

REFERENCES

Almusafri, F., H. E. Elamin, T. E. Khalaf et al. 2015. Genetic variants of microsomal epoxide hydrolase and glutamate-cysteine ligase in COPD. *Free Radic Biol Med* 12:588–95.

Almusafri, F., H. E. Elamin, T. E. Khalaf et al. 2017. Clinical and molecular characterization of 6 children with glutamate-cysteine ligase deficiency causing hemolytic anemia. *Blood Cells Mol Dis* 65:73–7.

Alzoghaibi, M. A. 2013. Concepts of oxidative stress and antioxidant defense in Crohn's disease. *World J Gastroenterol* 19 (39):6540–7.

Amir Aslani, B., and S. Ghobadi. 2016. Studies on oxidants and antioxidants with a brief glance at their relevance to the immune system. *Life Sci* 146:163–73.

Anderson, M. E., and A. Meister. 1983. Transport and direct utilization of gamma-glutamylcyst(e)ine for glutathione synthesis. *Proc Natl Acad Sci U S A* 80 (3):707–11.

Arisawa, S., K. Ishida, N. Kameyama et al. 2009. Ursodeoxycholic acid induces glutathione synthesis through activation of PI3K/Akt pathway in HepG2 cells. *Biochem Pharmacol* 77 (5):858–66.

Banerjee, R., and C. G. Zou. 2005. Redox regulation and reaction mechanism of human cystathionine-beta-synthase: A PLP-dependent hemesensor protein. *Arch Biochem Biophys* 433 (1):144–56.

Barcelos, G. R., D. Grotto, K. C. de Marco et al. 2013. Polymorphisms in glutathione-related genes modify mercury concentrations and antioxidant status in subjects environmentally exposed to methylmercury. *Sci Total Environ* 463–464:319–25.

Bekris, L. M., C. Shephard, M. Janer et al. 2007. Glutamate cysteine ligase catalytic subunit promoter polymorphisms and associations with type 1 diabetes age-at-onset and GAD65 autoantibody levels. *Exp Clin Endocrinol Diabetes* 115 (4):221–8.

Bekris, L. M., H. M. Viernes, F. M. Farin et al. 2006. Chronic beryllium disease and glutathione biosynthesis genes. *J Occup Environ Med* 48 (6):599–606.

Berk, M., S. Johansson, N. R. Wray et al. 2011. Glutamate cysteine ligase (GCL) and self reported depression: An association study from the HUNT. *J Affect Disord* 131 (1–3):207–13.

Beutler, E., T. Gelbart, T. Kondo, and A. T. Matsunaga. 1999. The molecular basis of a case of gamma-glutamylcysteine synthetase deficiency. *Blood* 94 (8):2890–4.

Beutler, E., R. Moroose, L. Kramer et al. 1990. Gamma-glutamylcysteine synthetase deficiency and hemolytic anemia. *Blood* 75 (1):271–3.

Butticaz, C., R. Gysin, M. Cuenod, and K. Q. Do. 2011. Interaction of GAG trinucleotide repeat and C-129T polymorphisms impairs expression of the glutamate-cysteine ligase catalytic subunit gene. *Free Radic Biol Med* 50 (5):617–23.

Campolo, J., S. Penco, E. Bianchi et al. 2007. Glutamate-cysteine ligase polymorphism, hypertension, and male sex are associated with cardiovascular events. Biochemical and genetic characterization of Italian subpopulation. *Am Heart J* 154 (6):1123–9.

Chappell, S., L. Daly, K. Morgan et al. 2008. Genetic variants of microsomal epoxide hydrolase and glutamate-cysteine ligase in COPD. *Eur Respir J* 32 (4):931–7.

Chen, Z. H., Y. Saito, Y. Yoshida et al. 2008. Regulation of GCL activity and cellular glutathione through inhibition of ERK phosphorylation. *Biofactors* 33 (1):1–11.

Chen, Y., H. G. Shertzer, S. N. Schneider et al. 2005. Glutamate cysteine ligase catalysis: Dependence on ATP and modifier subunit for regulation of tissue glutathione levels. *J Biol Chem* 280 (40):33766–74.

Cheng, L., J. Liu, B. Li et al. 2016. Cigarette smoke-induced hypermethylation of the GCLC gene is associated with COPD. *Chest* 149 (2):474–82.

Choi, J., R. M. Liu, R. K. Kundu et al. 2000. Molecular mechanism of decreased glutathione content in human immunodeficiency virus type 1 Tat-transgenic mice. *J Biol Chem* 275 (5):3693–8.

Ciofu, O., and J. Lykkesfeldt. 2014. Antioxidant supplementation for lung disease in cystic fibrosis. *Cochrane Database Syst Rev* (8):CD007020.

Custodio, H. M., K. Broberg, M. Wennberg et al. 2004. Polymorphisms in glutathione-related genes affect methylmercury retention. *Arch Environ Health* 59 (11):588–95.

Dalton, T. P., M. Z. Dieter, Y. Yang et al. 2000. Knockout of the mouse glutamate cysteine ligase catalytic subunit (Gclc) gene: Embryonic lethal when homozygous, and proposed model for moderate glutathione deficiency when heterozygous. *Biochem Biophys Res Commun* 279 (2):324–9.

Davis, J. M., and R. L. Auten. 2010. Maturation of the antioxidant system and the effects on preterm birth. *Semin Fetal Neonatal Med* 15 (4):191–5.

De, A. K., and R. Darad. 1991. Age-associated changes in antioxidants and antioxidative enzymes in rats. *Mech Ageing Dev* 59 (1–2):123–8.

de Andrade, K. Q., F. A. Moura, J. M. dos Santos et al. 2015. Oxidative stress and inflammation in hepatic diseases: Therapeutic possibilities of N-acetylcysteine. *Int J Mol Sci* 16 (12):30269–308.

de Lima Marson, F. A., C. S. Bertuzzo, R. Secolin et al. 2013. Genetic interaction of GSH metabolic pathway genes in cystic fibrosis. *BMC Med Genet* 14:60.

Deneke, S. M., and B. L. Fanburg. 1989. Regulation of cellular glutathione. *Am J Physiol* 257 (4 Pt 1):L163–73.

Dickinson, D. A., K. E. Iles, N. Watanabe et al. 2002. 4-hydroxynonenal induces glutamate cysteine ligase through JNK in HBE1 cells. *Free Radic Biol Med* 33 (7):974.

Dickinson, D. A., K. E. Iles, H. Zhang et al. 2003. Curcumin alters EpRE and AP-1 binding complexes and elevates glutamate-cysteine ligase gene expression. *FASEB J* 17 (3):473–5.

Eaton, D. L., and D. M. Hamel. 1994. Increase in gamma-glutamylcysteine synthetase activity as a mechanism for butylated hydroxyanisole-mediated elevation of hepatic glutathione. *Toxicol Appl Pharmacol* 126 (1):145–9.

Engstrom, K. S., M. Wennberg, U. Stromberg et al. 2011. Evaluation of the impact of genetic polymorphisms in glutathione-related genes on the association between methylmercury or n-3 polyunsaturated long chain fatty acids and risk of myocardial infarction: A case-control study. *Environ Health* 10:33.

Fagerberg, L., B. M. Hallstrom, P. Oksvold et al. 2014. Analysis of the human tissue-specific expression by genome-wide integration of transcriptomics and antibody-based proteomics. *Mol Cell Proteomics* 13 (2):397–406.

Fahey, R. C. 2013. Glutathione analogs in prokaryotes. *Biochim Biophys Acta* 1830 (5):3182–98.

Favilli, F., T. Iantomasi, P. Marraccini et al. 1994. Relationship between age and GSH metabolism in synaptosomes of rat cerebral cortex. *Neurobiol Aging* 15 (4):429–33.

Fontenot, A. P., M. T. Falta, J. W. Kappler et al. 2016. Beryllium-induced hypersensitivity: Genetic susceptibility and neoantigen generation. *J Immunol* 196 (1):22–7.

Franklin, C. C., M. E. Rosenfeld-Franklin, C. White et al. 2003. TGFbeta1-induced suppression of glutathione antioxidant defenses in hepatocytes: Caspase-dependent post-translational and caspase-independent transcriptional regulatory mechanisms. *FASEB J* 17 (11):1535–7.

Fu, Y., S. Zheng, S. C. Lu, and A. Chen. 2008. Epigallocatechin-3-gallate inhibits growth of activated hepatic stellate cells by enhancing the capacity of glutathione synthesis. *Mol Pharmacol* 73 (5):1465–73.

Fullerton, J. M., Y. Tiwari, G. Agahi et al. 2010. Assessing oxidative pathway genes as risk factors for bipolar disorder. *Bipolar Disord* 12 (5):550–6.

Garcia-Gimenez, J. L., C. Roma-Mateo, G. Perez-Machado et al. 2017. Role of glutathione in the regulation of epigenetic mechanisms in disease. *Free Radic Biol Med* 112:36–48.

Godwin, A. K., A. Meister, P. J. O'Dwyer et al. 1992. High resistance to cisplatin in human ovarian cancer cell lines is associated with marked increase of glutathione synthesis. *Proc Natl Acad Sci U S A* 89 (7):3070–4.

Gould, N. S., E. Min, S. Gauthier et al. 2010. Aging adversely affects the cigarette smoke-induced glutathione adaptive response in the lung. *Am J Respir Crit Care Med* 182 (9):1114–22.

Gu, F., V. Chauhan, and A. Chauhan. 2015. Glutathione redox imbalance in brain disorders. *Curr Opin Clin Nutr Metab Care* 18 (1):89–95.

Gysin, R., R. Kraftsik, O. Boulat et al. 2011. Genetic dysregulation of glutathione synthesis predicts alteration of plasma thiol redox status in schizophrenia. *Antioxid Redox Signal* 15 (7):2003–10.

Gysin, R., I. M. Riederer, M. Cuenod et al. 2009. Skin fibroblast model to study an impaired glutathione synthesis: Consequences of a genetic polymorphism on the proteome. *Brain Res Bull* 79 (1):46–52.

Ha, K. N., Y. Chen, J. Cai, and P. Sternberg, Jr. 2006. Increased glutathione synthesis through an ARE-Nrf2-dependent pathway by zinc in the RPE: Implication for protection against oxidative stress. *Invest Ophthalmol Vis Sci* 47 (6):2709–15.

Han, K. H., N. Hashimoto, and M. Fukushima. 2016. Relationships among alcoholic liver disease, antioxidants, and antioxidant enzymes. *World J Gastroenterol* 22 (1):37–49.

Hanzawa, R., T. Ohnuma, Y. Nagai et al. 2011. No association between glutathione-synthesis-related genes and Japanese schizophrenia. *Psychiatry Clin Neurosci* 65 (1):39–46.

Harari, R., F. Harari, L. Gerhardsson et al. 2012. Exposure and toxic effects of elemental mercury in gold-mining activities in Ecuador. *Toxicol Lett* 213 (1):75–82.

Harman, D. 1956. Aging: A theory based on free radical and radiation chemistry. *J Gerontol* 11 (3):298–300.

Hashemi, M., H. Hoseini, P. Yaghmaei et al. 2011. Association of polymorphisms in glutamate-cysteine ligase catalytic subunit and microsomal triglyceride transfer protein genes with nonalcoholic fatty liver disease. *DNA Cell Biol* 30 (8):569–75.

Hazelton, G. A., and C. A. Lang. 1980. Glutathione contents of tissues in the aging mouse. *Biochem J* 188 (1):25–30.

Hirono, A., H. Iyori, I. Sekine et al. 1996. Three cases of hereditary nonspherocytic hemolytic anemia associated with red blood cell glutathione deficiency. *Blood* 87 (5):2071–4.

Hobbs, C. A., M. A. Cleves, S. L. Macleod et al. 2014. Conotruncal heart defects and common variants in maternal and fetal genes in folate, homocysteine, and transsulfuration pathways. *Birth Defects Res A Clin Mol Teratol* 100 (2):116–26.

Hu, G., W. Yao, Y. Zhou et al. 2008. Meta- and pooled analyses of the effect of glutathione S-transferase M1 and T1 deficiency on chronic obstructive pulmonary disease. *Int J Tuberc Lung Dis* 12 (12):1474–81.

Huang, C. S., M. E. Anderson, and A. Meister. 1993a. Amino acid sequence and function of the light subunit of rat kidney gamma-glutamylcysteine synthetase. *J Biol Chem* 268 (27):20578–83.

Huang, C. S., M. E. Anderson, and A. Meister. 1993b. The function of the light subunit of g-glutamylcysteine synthetase (rat kidney). *FASEB J* 7:A1102.

Huang, C. S., L. S. Chang, M. E. Anderson, and A. Meister. 1993c. Catalytic and regulatory properties of the heavy subunit of rat kidney gamma-glutamylcysteine synthetase. *J Biol Chem* 268 (26):19675–80.

Jardine, H., W. MacNee, K. Donaldson, and I. Rahman. 2002. Molecular mechanism of transforming growth factor (TGF)-beta1-induced glutathione depletion in alveolar epithelial cells. Involvement of AP-1/ARE and Fra-1. *J Biol Chem* 277 (24):21158–66.

Jiang, L., N. Kon, T. Li et al. 2015. Ferroptosis as a p53-mediated activity during tumour suppression. *Nature* 520 (7545):57–62.

Johnson, W. M., A. L. Wilson-Delfosse, and J. J. Mieyal. 2012. Dysregulation of glutathione homeostasis in neurodegenerative diseases. *Nutrients* 4 (10):1399–440.

Jonsson, L. S., B. A. Jonsson, A. Axmon et al. 2008. Influence of glutathione-related genes on symptoms and immunologic markers among vulcanization workers in the southern Sweden rubber industries. *Int Arch Occup Environ Health* 81 (7):913–19.

Jung, Y. S. 2015. Metabolism of sulfur-containing amino acids in the liver: A link between hepatic injury and recovery. *Biol Pharm Bull* 38 (7):971–4.

Katakami, N., H. Kaneto, T. A. Matsuoka et al. 2010. Accumulation of gene polymorphisms related to oxidative stress is associated with myocardial infarction in Japanese type 2 diabetic patients. *Atherosclerosis* 212 (2):534–8.

Katakami, N., H. Kaneto, T. A. Matsuoka et al. 2014. Accumulation of oxidative stress-related gene polymorphisms and the risk of coronary heart disease events in patients with type 2 diabetes--an 8-year prospective study. *Atherosclerosis* 235 (2):408–14.

Kay, H. Y., J. Won Yang, T. H. Kim et al. 2010. Ajoene, a stable garlic by-product, has an antioxidant effect through Nrf2-mediated glutamate-cysteine ligase induction in HepG2 cells and primary hepatocytes. *J Nutr* 140 (7):1211–19.

Khadzhieva, M. B., N. N. Lutcenko, I. V. Volodin et al. 2014. Association of oxidative stress-related genes with idiopathic recurrent miscarriage. *Free Radic Res* 48 (5):534–41.

Kim, S. K., K. J. Woodcroft, S. S. Khodadadeh, and R. F. Novak. 2004. Insulin signaling regulates gamma-glutamylcysteine ligase catalytic subunit expression in primary cultured rat hepatocytes. *J Pharmacol Exp Ther* 311 (1):99–108.

Ko, K., H. Yang, M. Noureddin et al. 2008. Changes in S-adenosylmethionine and GSH homeostasis during endotoxemia in mice. *Lab Invest* 88 (10):1121–9.

Koide, S., K. Kugiyama, S. Sugiyama et al. 2003. Association of polymorphism in glutamate-cysteine ligase catalytic subunit gene with coronary vasomotor dysfunction and myocardial infarction. *J Am Coll Cardiol* 41 (4):539–45.

Kondo, T., K. Yoshida, Y. Urata et al. 1993. gamma-Glutamylcysteine synthetase and active transport of glutathione S-conjugate are responsive to heat shock in K562 erythroid cells. *J Biol Chem* 268 (27):20366–72.

Konrad, P. N., F. Richards, 2nd, W. N. Valentine, and D. E. Paglia. 1972. g-Glutamyl-cysteine synthetase deficiency. A cause of hereditary hemolytic anemia. *N Engl J Med* 286 (11):557–61.

Koralkova, P., W. W. van Solinge, and R. van Wijk. 2014. Rare hereditary red blood cell enzymopathies associated with hemolytic anemia—Pathophysiology, clinical aspects, and laboratory diagnosis. *Int J Lab Hematol* 36 (3):388–97.

Koutros, S., G. Andreotti, S. I. Berndt et al. 2011. Xenobiotic-metabolizing gene variants, pesticide use, and the risk of prostate cancer. *Pharmacogenet Genomics* 21 (10):615–23.

Krzywanski, D. M., D. A. Dickinson, K. E. Iles et al. 2004. Variable regulation of glutamate cysteine ligase subunit proteins affects glutathione biosynthesis in response to oxidative stress. *Arch Biochem Biophys* 423 (1):116–25.

Kumar, D., J. C. Lim, and P. J. Donaldson. 2013. A link between maternal malnutrition and depletion of glutathione in the developing lens: A possible explanation for idiopathic childhood cataract? *Clin Exp Optom* 96 (6):523–8.

Kuo, M. T., J. Bao, M. Furuichi et al. 1998. Frequent coexpression of MRP/GS-X pump and gamma-glutamylcysteine synthetase mRNA in drug-resistant cells, untreated tumor cells, and normal mouse tissues. *Biochem Pharmacol* 55 (5):605–15.

Lavoie, S., M. Berger, M. Schlogelhofer et al. 2017. Erythrocyte glutathione levels as long-term predictor of transition to psychosis. *Transl Psychiatry* 7 (3):e1064.

Lee, J. I., J. Kang, and M. H. Stipanuk. 2006. Differential regulation of glutamate-cysteine ligase subunit expression and increased holoenzyme formation in response to cysteine deprivation. *Biochem J* 393 (Pt 1):181–90.

Lee, T. D., H. Yang, J. Whang, and S. C. Lu. 2005. Cloning and characterization of the human glutathione synthetase 5'-flanking region. *Biochem J* 390 (Pt 2):521–8.

Levonen, A. L., D. A. Dickinson, D. R. Moellering et al. 2001. Biphasic effects of 15-deoxy-delta(12,14)-prostaglandin J(2) on glutathione induction and apoptosis in human endothelial cells. *Arterioscler Thromb Vasc Biol* 21 (11):1846–51.

Li, M., Z. Zhang, J. Yuan et al. 2014. Altered glutamate cysteine ligase expression and activity in renal cell carcinoma. *Biomed Rep* 2 (6):831–4.

Lin, W., L. Y. Tang, Y. L. Cen et al. 2013. Interaction of body mass index and a polymorphism in gene of catalytic subunit of glutamate-cysteine ligase on breast cancer risk among Chinese women. *Zhonghua Liu Xing Bing Xue Za Zhi* 34 (11):1115–19.

Liu, R., and J. Choi. 2000. Age-associated decline in gamma-glutamylcysteine synthetase gene expression in rats. *Free Radic Biol Med* 28 (4):566–74.

Liu, R. M., and D. A. Dickinson. 2003. Decreased synthetic capacity underlies the age-associated decline in glutathione content in Fisher 344 rats. *Antioxid Redox Signal* 5 (5):529–36.

Liu, R. M., L. Gao, J. Choi, and H. J. Forman. 1998. Gamma-glutamylcysteine synthetase: mRNA stabilization and independent subunit transcription by 4-hydroxy-2-nonenal. *Am J Physiol* 275 (5 Pt 1):L861–9.

Liu, R. M., H. Hu, T. W. Robison, and H. J. Forman. 1996. Increased gamma-glutamylcysteine synthetase and gamma-glutamyl transpeptidase activities enhance resistance of rat lung epithelial L2 cells to quinone toxicity. *Am J Respir Cell Mol Biol* 14 (2):192–7.

Liu, S., B. Li, Y. Zhou et al. 2007. Genetic analysis of CC16, OGG1 and GCLC polymorphisms and susceptibility to COPD. *Respirology* 12 (1):29–33.

Liu, R. M., P. K. Vayalil, C. Ballinger et al. 2012. Transforming growth factor beta suppresses glutamate-cysteine ligase gene expression and induces oxidative stress in a lung fibrosis model. *Free Radic Biol Med* 53 (3):554–63.

Lu, S. C. 2009. Regulation of glutathione synthesis. *Mol Aspects Med* 30 (1–2):42–59.

Lu, S. C. 2013. Glutathione synthesis. *Biochim Biophys Acta* 1830 (5):3143–53.

Lu, S. C., J. L. Ge, J. Kuhlenkamp, and N. Kaplowitz. 1992. Insulin and glucocorticoid dependence of hepatic gamma-glutamylcysteine synthetase and glutathione synthesis in the rat. Studies in cultured hepatocytes and in vivo. *J Clin Invest* 90 (2):524–32.

Lu, S. C., Z. Z. Huang, J. M. Yang, and H. Tsukamoto. 1999. Effect of ethanol and high-fat feeding on hepatic gamma-glutamylcysteine synthetase subunit expression in the rat. *Hepatology* 30 (1):209–14.

Lu, S. C., J. M. Mato, C. Espinosa-Diez, and S. Lamas. 2016. MicroRNA-mediated regulation of glutathione and methionine metabolism and its relevance for liver disease. *Free Radic Biol Med* 100:66–72.

Ma, J., D. M. Li, R. Zhang et al. 2010. Genetic analysis of glutamate cysteine ligase modifier (GCLM) gene and schizophrenia in Han Chinese. *Schizophr Res* 119 (1–3):273–4.

Man, B. L., L. Baum, Y. P. Fu et al. 2010. Genetic polymorphisms of Chinese patients with ischemic stroke and concurrent stenoses of extracranial and intracranial vessels. *J Clin Neurosci* 17 (10):1244–7.

Manu Pereira, M., T. Gelbart, E. Ristoff et al. 2007. Chronic non-spherocytic hemolytic anemia associated with severe neurological disease due to gamma-glutamylcysteine synthetase deficiency in a patient of Moroccan origin. *Haematologica* 92 (11):e102–5.

Marson, F. A., C. S. Bertuzzo, A. F. Ribeiro, and J. D. Ribeiro. 2014. Polymorphisms in the glutathione pathway modulate cystic fibrosis severity: A cross-sectional study. *BMC Med Genet* 15:27.

Matera, M. G., L. Calzetta, and M. Cazzola. 2016. Oxidation pathway and exacerbations in COPD: The role of NAC. *Expert Rev Respir Med* 10 (1):89–97.

May, M. J., T. Vernoux, C. Leaver et al. 1998. Glutathione homeostasis in plants : Implications for environmental sensing and plant development. *J Exp Bot* 49:649–67.

Mayer, A., and N. Hamzeh. 2015. Beryllium and other metal-induced lung disease. *Curr Opin Pulm Med* 21 (2):178–84.

McBean, G. J., M. Aslan, H. R. Griffiths, and R. C. Torrao. 2015. Thiol redox homeostasis in neurodegenerative disease. *Redox Biol* 5:186–94.

McCarty, M. F., and J. J. DiNicolantonio. 2015. An increased need for dietary cysteine in support of glutathione synthesis may underlie the increased risk for mortality associated with low protein intake in the elderly. *Age (Dordr)* 37 (5):96.

McConnachie, L. A., I. Mohar, F. N. Hudson et al. 2007. Glutamate cysteine ligase modifier subunit deficiency and gender as determinants of acetaminophen-induced hepatotoxicity in mice. *Toxicol Sci* 99 (2):628–36.

McKone, E. F., J. Shao, D. D. Frangolias et al. 2006. Variants in the glutamate-cysteine-ligase gene are associated with cystic fibrosis lung disease. *Am J Respir Crit Care Med* 174 (4):415–19.

Meister, A. 1981. On the cycle of glutathione metabolism and transport. *Curr Top Cell Regul* 21–58.

Meister, A. 1988. Glutathione metabolism and its selective modification. *J Biol Chem* 263 (33):17205–8.

Mistry, R. K., and A. C. Brewer. 2017. Redox-dependent regulation of sulfur metabolism in biomolecules: Implications for cardiovascular health. *Antioxid Redox Signal.* doi: 10.1089/ars.2017.7224. [Epub ahead of print].

Mordukhovich, I., J. Lepeule, B. A. Coull et al. 2015. The effect of oxidative stress polymorphisms on the association between long-term black carbon exposure and lung function among elderly men. *Thorax* 70 (2):133–7.

Muehlhause, A., S. Kropf, and A. Gardemann. 2007. C-588T polymorphism of the human glutamate-cysteine ligase modifier subunit gene is not associated with the risk and extent of ischemic heart disease in a German cohort. *Clin Chem Lab Med* 45 (10):1416–18.

Mulcahy, R. T., S. Untawale, and J. J. Gipp. 1994. Transcriptional up-regulation of gamma-glutamylcysteine synthetase gene expression in melphalan-resistant human prostate carcinoma cells. *Mol Pharmacol* 46 (5):909–14.

Mulcahy, R. T., M. A. Wartman, H. H. Bailey, and J. J. Gipp. 1997. Constitutive and beta-naphthoflavone-induced expression of the human gamma-glutamylcysteine synthetase heavy subunit gene is regulated by a distal antioxidant response element/TRE sequence. *J Biol Chem* 272 (11):7445–54.

Murphy, D. D., E. C. Reddy, N. Moran, and S. O'Neill. 2014. Regulation of platelet activity in a changing redox environment. *Antioxid Redox Signal* 20 (13):2074–89.

Nakamura, S., K. Kugiyama, S. Sugiyama et al. 2002. Polymorphism in the 5'-flanking region of human glutamate-cysteine ligase modifier subunit gene is associated with myocardial infarction. *Circulation* 105 (25):2968–73.

Nakamura, S., S. Sugiyama, D. Fujioka et al. 2003. Polymorphism in glutamate-cysteine ligase modifier subunit gene is associated with impairment of nitric oxide-mediated coronary vasomotor function. *Circulation* 108 (12):1425–7.

Nakata, K., M. Kawase, S. Ogino et al. 1996. Effects of age on levels of cysteine, glutathione and related enzyme activities in livers of mice and rats and an attempt to replenish hepatic glutathione level of mouse with cysteine derivatives. *Mech Ageing Dev* 90 (3):195–207.

Ndrepepa, G., and A. Kastrati. 2016. Gamma-glutamyl transferase and cardiovascular disease. *Ann Transl Med* 4 (24):481.

Nichenametla, S. N., J. E. Muscat, J. G. Liao et al. 2013. A functional trinucleotide repeat polymorphism in the 5'-untranslated region of the glutathione biosynthetic gene GCLC is associated with increased risk for lung and aerodigestive tract cancers. *Mol Carcinog* 52 (10):791–9.

Nichenametla, S. N., P. Lazarus, J. P. Richie, Jr. 2011.A GAG trinucleotide-repeat polymorphism in the gene for glutathione biosynthetic enzyme, GCLC, affects gene expression through translation. *FASEB J.* 25 (7):2180–7.

Noctor, G., A. C. M. Arisi, L. Jouanin et al. 1998. Glutathione: Biosynthesis, metabolism and relationship to stress tolerance explored in transformed plants *J Exp Bot* 49:623–47.

Oguri, T., Y. Fujiwara, T. Isobe et al. 1998. Expression of gamma-glutamylcysteine synthetase (gamma-GCS) and multidrug resistance-associated protein (MRP), but not human canalicular multispecific organic anion transporter (cMOAT), genes correlates with exposure of human lung cancers to platinum drugs. *Br J Cancer* 77 (7):1089–96.

Oliveira, C. P., J. T. Stefano, A. M. Cavaleiro et al. 2010. Association of polymorphisms of glutamate-cystein ligase and microsomal triglyceride transfer protein genes in non-alcoholic fatty liver disease. *J Gastroenterol Hepatol* 25 (2):357–61.

Orlowski, M., and A. Meister. 1970. The gamma-glutamyl cycle: A possible transport system for amino acids. *Proc Natl Acad Sci U S A* 67 (3):1248–55.

Paul, B. D., J. I. Sbodio, R. Xu et al. 2014. Cystathionine gamma-lyase deficiency mediates neurodegeneration in Huntington's disease. *Nature* 509 (7498):96–100.

Payabvash, S., M. H. Ghahremani, A. Goliaei et al. 2006. Nitric oxide modulates glutathione synthesis during endotoxemia. *Free Radic Biol Med* 41 (12):1817–28.

Polonikov, A. V., V. P. Ivanov, M. A. Solodilova et al. 2007. The relationship between polymorphisms in the glutamate cysteine ligase gene and asthma susceptibility. *Respir Med* 101 (11):2422–4.

Prins, H. K., M. Oort, C. Zurcher, and T. Beckers. 1966. Congenital nonspherocytic hemolytic anemia, associated with glutathione deficiency of the erythrocytes. Hematologic, biochemical and genetic studies. *Blood* 27 (2):145–66.

Rabbani, N., M. Xue, and P. J. Thornalley. 2016. Methylglyoxal-induced dicarbonyl stress in aging and disease: First steps towards glyoxalase 1-based treatments. *Clin Sci (Lond)* 130 (19):1677–96.

Ravindranath, V., B. R. Shivakumar, and H. K. Anandatheerthavarada. 1989. Low glutathione levels in brain regions of aged rats. *Neurosci Lett* 101 (2):187–90.

Richman, P. G., and A. Meister. 1975. Regulation of gamma-glutamyl-cysteine synthetase by nonallosteric feedback inhibition by glutathione. *J Biol Chem* 250 (4):1422–6.

Ristoff, E., C. Augustson, J. Geissler et al. 2000. A missense mutation in the heavy subunit of gamma-glutamylcysteine synthetase gene causes hemolytic anemia. *Blood* 95 (7):2193–6.

Rizvi, S. I., and P. K. Maurya. 2007. Markers of oxidative stress in erythrocytes during aging in humans. *Ann N Y Acad Sci* 1100:373–82.

Salnikova, L. E., T. V. Smelaya, A. M. Golubev et al. 2013. CYP1A1, GCLC, AGT, AGTR1 gene-gene interactions in community-acquired pneumonia pulmonary complications. *Mol Biol Rep* 40 (11):6163–76.

Santus, P., A. Corsico, P. Solidoro et al. 2014. Oxidative stress and respiratory system: Pharmacological and clinical reappraisal of N-acetylcysteine. *COPD* 11 (6):705–17.

Sanz, N., C. Diez-Fernandez, A. Alvarez, and M. Cascales. 1997. Age-dependent modifications in rat hepatocyte antioxidant defense systems. *J Hepatol* 27 (3):525–34.

Sasaki, T., M. Senda, S. Kim et al. 2001. Age-related changes of glutathione content, glucose transport and metabolism, and mitochondrial electron transfer function in mouse brain. *Nucl Med Biol* 28 (1):25–31.

Seelig, G. F., and A. Meister. 1985. Gamma-glutamylcysteine synthetase from erythrocytes. *Methods Enzymol* 113:390–2.

Seelig, G. F., R. P. Simondsen, and A. Meister. 1984. Reversible dissociation of gamma-glutamylcysteine synthetase into two subunits. *J Biol Chem* 259 (15):9345–7.

Sekhar, R. V., S. G. Patel, A. P. Guthikonda et al. 2011. Deficient synthesis of glutathione underlies oxidative stress in aging and can be corrected by dietary cysteine and glycine supplementation. *Am J Clin Nutr* 94 (3):847–53.

Shi, M. M., A. Kugelman, T. Iwamoto et al. 1994. Quinone-induced oxidative stress elevates glutathione and induces gamma-glutamylcysteine synthetase activity in rat lung epithelial L2 cells. *J Biol Chem* 269 (42):26512–17.

Shi, Z. Z., J. Osei-Frimpong, G. Kala et al. 2000. Glutathione synthesis is essential for mouse development but not for cell growth in culture. *Proc Natl Acad Sci U S A* 97 (10):5101–6.

Shih, A. Y., D. A. Johnson, G. Wong et al. 2003. Coordinate regulation of glutathione biosynthesis and release by Nrf2-expressing glia potently protects neurons from oxidative stress. *J Neurosci* 23 (8):3394–406.

Short, J. D., K. Downs, S. Tavakoli, and R. Asmis. 2016. Protein thiol redox signaling in monocytes and macrophages. *Antioxid Redox Signal* 25 (15):816–35.

Siedlinski, M., D. S. Postma, C. C. van Diemen et al. 2008. Lung function loss, smoking, vitamin C intake, and polymorphisms of the glutamate-cysteine ligase genes. *Am J Respir Crit Care Med* 178 (1):13–19.

Skvortsova, L., A. Perfelyeva, E. Khussainova et al. 2017. Association of GCLM -588C/T and GCLC -129T/C promoter polymorphisms of genes coding the subunits of glutamate cysteine ligase with ischemic heart disease development in Kazakhstan population. *Dis Markers* 2017:4209257.

Song, I. S., S. Tatebe, W. Dai, and M. T. Kuo. 2005. Delayed mechanism for induction of gamma-glutamylcysteine synthetase heavy subunit mRNA stability by oxidative stress involving p38 mitogen-activated protein kinase signaling. *J Biol Chem* 280 (31):28230–40.

Stepniewska, J., E. Golembiewska, B. Dolegowska et al. 2015. Oxidative stress and antioxidative enzyme activities in chronic kidney disease and different types of renal replacement therapy. *Curr Protein Pept Sci* 16 (3):243–8.

Stipanuk, M. H. 2004. Sulfur amino acid metabolism: Pathways for production and removal of homocysteine and cysteine. *Annu Rev Nutr* 24:539–77.

Stipanuk, M. H., R. M. Coloso, R. A. Garcia, and M. F. Banks. 1992. Cysteine concentration regulates cysteine metabolism to glutathione, sulfate and taurine in rat hepatocytes. *J Nutr* 122 (3):420–7.

Sun, W. M., Z. Z. Huang, and S. C. Lu. 1996. Regulation of gamma-glutamylcysteine synthetase by protein phosphorylation. *Biochem J* 320 (Pt 1):321–8.

Tang, W., A. R. Bentley, S. B. Kritchevsky et al. 2013. Genetic variation in antioxidant enzymes, cigarette smoking, and longitudinal change in lung function. *Free Radic Biol Med* 63:304–12.

Tang, X., M. A. Cleves, T. G. Nick et al. 2015. Obstructive heart defects associated with candidate genes, maternal obesity, and folic acid supplementation. *Am J Med Genet A* 167 (6):1231–42.

Tang, X., T. G. Nick, M. A. Cleves et al. 2014. Maternal obesity and tobacco use modify the impact of genetic variants on the occurrence of conotruncal heart defects. *PLoS One* 9 (9):e108903.

Teramoto, S., Y. Fukuchi, Y. Uejima et al. 1994. Age-related changes in the antioxidant screen of the distal lung in mice. *Lung* 172 (4):223–30.

Thompson, J. A., C. C. White, D. P. Cox et al. 2009. Distinct Nrf1/2-independent mechanisms mediate As 3+-induced glutamate-cysteine ligase subunit gene expression in murine hepatocytes. *Free Radic Biol Med* 46 (12):1614–25.

Tiitto, L. H., M. J. Peltoniemi, R. L. Kaarteenaho-Wiik et al. 2004. Cell-specific regulation of gamma-glutamylcysteine synthetase in human interstitial lung diseases. *Hum Pathol* 35 (7):832–9.

Tomasi, M. L., M. Ryoo, H. Yang et al. 2014. Molecular mechanisms of lipopolysaccharide-mediated inhibition of glutathione synthesis in mice. *Free Radic Biol Med* 68:148–58.

Toroser, D., and R. S. Sohal. 2007. Age-associated perturbations in glutathione synthesis in mouse liver. *Biochem J* 405 (3):583–9.

Tosic, M., J. Ott, S. Barral et al. 2006. Schizophrenia and oxidative stress: Glutamate cysteine ligase modifier as a susceptibility gene. *Am J Hum Genet* 79 (3):586–92.

Tu, Z., and M. W. Anders. 1998. Up-regulation of glutamate-cysteine ligase gene expression by butylated hydroxytoluene is mediated by transcription factor AP-1. *Biochem Biophys Res Commun* 244 (3):801–5.

Urata, Y., S. Honma, S. Goto et al. 1999. Melatonin induces gamma-glutamylcysteine synthetase mediated by activator protein-1 in human vascular endothelial cells. *Free Radic Biol Med* 27 (7–8):838–47.

Ursini, F., and M. Maiori. 2010. Redox pioneer: Professor Leopold Flohé. *Antioxid Redox Signal* 13 (10):1617–22.

Vieira, S. M., M. B. Monteiro, T. Marques et al. 2011. Association of genetic variants in the promoter region of genes encoding p22phox (CYBA) and glutamate cysteine ligase catalytic subunit (GCLC) and renal disease in patients with type 1 diabetes mellitus. *BMC Med Genet* 12:129.

Walsh, A. C., J. A. Feulner, and A. Reilly. 2001. Evidence for functionally significant polymorphism of human glutamate cysteine ligase catalytic subunit: Association with glutathione levels and drug resistance in the National Cancer Institute tumor cell line panel. *Toxicol Sci* 61 (2):218–23.

Walsh, A. C., W. Li, D. R. Rosen, and D. A. Lawrence. 1996. Genetic mapping of GLCLC, the human gene encoding the catalytic subunit of gamma-glutamyl-cysteine synthetase, to chromosome band 6p12 and characterization of a polymorphic trinucleotide repeat within its 5' untranslated region. *Cytogenet Cell Genet* 75 (1):14–16.

Wang, D., A. Curtis, A. C. Papp et al. 2012. Polymorphism in glutamate cysteine ligase catalytic subunit (GCLC) is associated with sulfamethoxazole-induced hypersensitivity in HIV/AIDS patients. *BMC Med Genomics* 5:32.

Wang, H., H. Liu, and R. M. Liu. 2003. Gender difference in glutathione metabolism during aging in mice. *Exp Gerontol* 38 (5):507–17.

Warnatz, H. J., D. Schmidt, T. Manke et al. 2011. The BTB and CNC homology 1 (BACH1) target genes are involved in the oxidative stress response and in control of the cell cycle. *J Biol Chem* 286 (26):23521–32.

Watanabe, Y., R. A. Cohen, and R. Matsui. 2016. Redox regulation of ischemic angiogenesis-another aspect of reactive oxygen species. *Circ J* 80 (6):1278–84.

Watanabe, Y., K. Watanabe, T. Kobayashi et al. 2013. Chronic depletion of glutathione exacerbates ventricular remodelling and dysfunction in the pressure-overloaded heart. *Cardiovasc Res* 97 (2):282–92.

Wells, P. G., G. P. McCallum, C. S. Chen et al. 2009. Oxidative stress in developmental origins of disease: Teratogenesis, neurodevelopmental deficits, and cancer. *Toxicol Sci* 108 (1):4–18.

Wild, A. C., J. J. Gipp, and T. Mulcahy. 1998. Overlapping antioxidant response element and PMA response element sequences mediate basal and beta-naphthoflavone-induced expression of the human gamma-glutamylcysteine synthetase catalytic subunit gene. *Biochem J* 332 (Pt 2):373–81.

Willis, A. S., M. L. Freeman, S. R. Summar et al. 2003. Ethnic diversity in a critical gene responsible for glutathione synthesis. *Free Radic Biol Med* 34 (1):72–6.

Woods, J. S., H. A. Davis, and R. P. Baer. 1992. Enhancement of gamma-glutamylcysteine synthetase mRNA in rat kidney by methyl mercury. *Arch Biochem Biophys* 296 (1):350–3.

Wu, A.-L. L., and W. S. S. Moye-Rowley. 1994. GSH1, which encodes gamma-glutamylcysteine synthetase, is a target gene for yAP-1 transcriptional regulation. *Mol Cell Biol* 14:5832–9.

Xin, L., R. Mekle, M. Fournier et al. 2016. Genetic polymorphism associated prefrontal glutathione and its coupling with brain glutamate and peripheral redox status in early psychosis. *Schizophr Bull* 42 (5):1185–96.

Yang, Y. C., C. K. Lii, A. H. Lin et al. 2011. Induction of glutathione synthesis and heme oxygenase 1 by the flavonoids butein and phloretin is mediated through the ERK/Nrf2 pathway and protects against oxidative stress. *Free Radic Biol Med* 51 (11):2073–81.

Yang, H., N. Magilnick, C. Lee et al. 2005a. Nrf1 and Nrf2 regulate rat glutamate-cysteine ligase catalytic subunit transcription indirectly via NF-kappaB and AP-1. *Mol Cell Biol* 25 (14):5933–46.

Yang, H., N. Magilnick, X. Ou, and S. C. Lu. 2005b. Tumour necrosis factor alpha induces co-ordinated activation of rat GSH synthetic enzymes via nuclear factor kappaB and activator protein-1. *Biochem J* 391 (Pt 2):399–408.

Yang, H., Y. Zeng, T. D. Lee et al. 2002. Role of AP-1 in the coordinate induction of rat glutamate-cysteine ligase and glutathione synthetase by tert-butylhydroquinone. *J Biol Chem* 277 (38):35232–9.

Yue, F., Y. Cheng, A. Breschi et al. 2014. A comparative encyclopedia of DNA elements in the mouse genome. *Nature* 515 (7527):355–64.

Yuniastuti, A., R. Susanti, and D. Mustikaningtyas. 2017. Polymorphism of glutamate-cysteine ligase subunit catalytic (GCLC) gene in pulmonary tuberculosis patients. *Pak J Biol Sci* 20 (8):397–402.

Zhang, H., N. Court, and H. J. Forman. 2007. Submicromolar concentrations of 4-hydroxynonenal induce glutamate cysteine ligase expression in HBE1 cells. *Redox Rep* 12 (1):101–6.

Zhang, H., K. J. A. Davies, and H. J. Forman. 2015. Oxidative stress response and Nrf2 signaling in aging. *Free Radic Biol Med* 88 (Pt B):314–36.

Zhang, H., and H. J. Forman. 2010. Reexamination of the electrophile response element sequences and context reveals a lack of consensus in gene function. *Biochim Biophys Acta* 1799 (7):496–501.

Zhang, H., H. Liu, K. J. Davies et al. 2012. Nrf2-regulated phase II enzymes are induced by chronic ambient nanoparticle exposure in young mice with age-related impairments. *Free Radic Biol Med* 52 (9):2038–46.

Zhang, H., L. Zhou, J. Yuen et al. 2017. Delayed Nrf2-regulated antioxidant gene induction in response to silica nanoparticles. *Free Radic Biol Med* 108:311–19.

Zipper, L. M., and R. T. Mulcahy. 2000. Inhibition of ERK and p38 MAP kinases inhibits binding of Nrf2 and induction of GCS genes. *Biochem Biophys Res Commun* 278 (2):484–92.

Zuo, H. P., W. J. Xu, M. Luo et al. 2007. The glutamate-cysteine ligase catalytic subunit gene C-129T and modifier subunit gene G-23T polymorphisms and risk for coronary diseases. *Zhonghua Xin Xue Guan Bing Za Zhi* 35 (7):637–40.

2 Renal Glutathione Transport Systems

Roles in Redox Homeostasis, Cytoprotection, and Bioactivation

Lawrence H. Lash
Wayne State University School of Medicine

CONTENTS

2.1 INTRODUCTION

The established dogma before about 1980 was that GSH and GSH *S*-conjugates were not transported into mammalian cells, based primarily on studies in hepatocytes, but also on erroneous conclusions from studies in other tissues, including the kidneys (Meister, 1976; Griffith et al., 1979; Anderson et al., 1980; Anderson and Meister, 1983). For the kidneys, however, *in vivo* studies in rats and *in vitro* studies in isolated perfused rat kidneys demonstrated that glomerular filtration and degradation of GSH could not account for the renal extraction of the tripeptide (McIntyre and Curthoys, 1980; Ormstad et al., 1982; Orrenius et al., 1983). In fact, the extent of involvement of basolateral membrane transport in the renal extraction of GSH from the circulation was greater than that due to glomerular filtration, highlighting the importance of plasma membrane transport in overall renal handling of GSH.

Whereas transport of GSH across the basolateral plasma membrane typically involves movement of GSH from the plasma and renal interstitial space into the proximal tubular cell, GSH is also transported across the brush-border or luminal plasma membrane. In this case, by contrast, the predominant direction of trans-port under physiological conditions is the efflux of intracellular GSH into the tubular lumen. Here, transport serves to deliver the tripeptide to the active site of γ-glutamyltransferase (GGT) (Griffith et al., 1979).

A critical determinant of the disposition and transport of GSH is that the mole-cule has a net negative charge at physiological pH. The importance of this in relation to membrane transport is that membrane potentials exist across both the cellular plasma and intracellular organelle membranes. The biological significance of this is further highlighted by considering the status and regulation of the mitochondrial GSH pool. In renal proximal tubular cells and other cell types, including hepatocytes, distinct subcellular compartments exist. Although GSH synthesis appears to occur exclusively in the cytoplasm (Griffith and Meister, 1985; Olafsdottir et al., 1988; McKernan et al., 1991), the presence of the negatively charged GSH molecule in the mitochondrial matrix and the membrane potential across the mitochondrial inner membrane (highly negative inside) indicate that a regulated and energy-dependent transport process must exist to account for the mitochondrial pool.

This chapter will review the current state of knowledge regarding membrane transport processes for GSH and GSH S-conjugates in mammalian kidney. Although there is a fairly large literature and many years of studies on these processes, only limited background information is presented for context. The reader can refer to any of several reviews that detail this background in more depth (e.g., Ballatori et al., 2009; Lash, 2005, 2006, 2009, 2011, 2012, 2015). The primary focus will be on the more recent studies that have identified the carrier proteins responsible, helped to elucidate the physiological roles of each transporter, and assessed the influence of disease states in the function and regulation of these transporters, and the potential for these transporters to be targets for novel therapeutic approaches.

2.2   RENAL PLASMA MEMBRANE GSH TRANSPORT

2.2.1   Physiological Evidence for Basolateral Plasma Membrane Transport

The isopeptide bond of the GSH molecule is a critical feature of the molecule that makes it resistant to degradation by the myriad proteases that exist in plasma. Rather, only the enzyme GGT is capable of hydrolyzing this bond between the γ-carboxyl group of the glutamyl residue and the α-amino group of the cysteinyl residue. Although GGT is expressed in membranes of several tissues, it is an ectoenzyme (i.e., its active site is extracellular), whose expression is highest in the kidneys, where it is found predominantly in the brush-border plasma membrane of proximal tubular cells (Glossman and Neville, 1972). Accordingly, the GSH molecule can circulate through the blood as an intact tripeptide. The liver has the highest activity of GSH synthesis in the body, is very efficient at exporting the molecule into the bile and blood, and is thus the primary source of plasma GSH (Bartoli and Sies, 1978; Wendel and Cikryt, 1980). Although the biliary epithelium has GGT activity and degrades GSH, that released from the liver by efflux across the sinusoidal plasma membrane is not degraded and is delivered by the blood to the kidneys.

Around 1980, it was observed that GSH was present in plasma and that its concentration in blood vessels leaving and entering certain tissues differed. This observation also served as the basis of the concept of interorgan GSH transport, in which tissues with high GSH biosynthetic capacity, primarily the liver, supplied the plasma with GSH, whereas tissues with high GSH degradative capacity, such as the kidneys and small intestine, extracted GSH (Hahn et al., 1978; Griffith and Meister, 1979a; Häberle et al., 1979; Anderson and Meister, 1980; McIntyre and Curthoys, 1980; Wendel and Cikryt, 1980). The lower GSH concentration in blood vessels leaving a tissue compared to that entering the tissue indicates extraction of the GSH molecule by that tissue. Because of the high activities of GGT and dipeptidase (DP) in those tissues that seem to extract GSH, particularly the kidneys, and the dogma that GSH was not transported into cells, it was presumed that this extraction occurred by degradation of GSH to its constituent amino acids, uptake of the amino acids, and resynthesis of GSH within the cell. Furthermore, reported evidence of GGT expression in other subcellular locations besides the luminal membrane led some investigators to conclude that GSH was readily degraded in both luminal and renal interstitial fluids (Dass et al., 1981; Abbott et al., 1984; Inoue et al., 1986).

Despite the high GGT activity in renal proximal tubular brush-border membranes and the suggestion that some GGT activity was also present elsewhere in the kidneys, the observation that GSH concentration in renal arterial plasma was about fourfold higher than that in renal vein plasma, careful *in vivo* studies in rats and *in vitro* studies in isolated perfused rat kidneys using inhibitors of GGT with appropriate time courses, demonstrated that the basolateral extraction process was largely independent of GGT activity (Fonteles et al., 1976; Ormstad et al., 1982; Rankin and Curthoys, 1982; Rankin et al., 1985). As illustrated in Figure 2.1, renal extraction of GSH occurs by two mechanisms: a luminal process involving glomerular filtration and GGT- and DP-catalyzed degradation, and a basolateral process involving membrane transport.

2.2.2 IDENTIFICATION OF CARRIERS FOR RENAL PLASMA MEMBRANE GSH TRANSPORT

A characteristic feature of the renal proximal tubules is the presence of a large and diverse array of plasma membrane transporters. With the realization that most of the renal GSH extraction occurs by a basolateral transport mechanism, the first approach that was used to identify the specific membrane carriers involved in the uptake of the tripeptide was to study the transport in suspensions of isolated basolateral plasma membrane vesicles (Lash and Jones, 1983, 1984). A key advantage

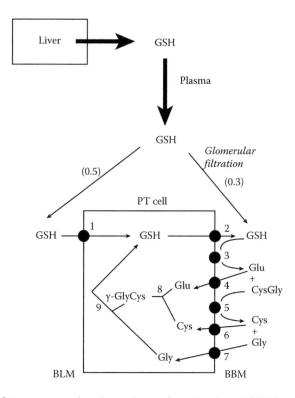

FIGURE 2.1 Interorgan translocation pathways for extraction of GSH from plasma by the kidneys. The scheme summarizes the hepatorenal translocation of GSH and its extraction by the renal proximal tubular (PT) cells. The liver efficiently pumps GSH into the plasma. Due to the very high blood flow rate relative to tissue weight and the existence of active filtration and transport processes, the kidneys are the primary organ responsible for overall turnover of GSH. During a single pass through the kidneys, 80% of plasma GSH is extracted—30% via glomerular filtration and 50% via basolateral plasma membrane transport. Uptake of GSH across basolateral plasma membrane (process 1). Efflux of intracellular GSH into tubular lumen for successive degradation (process 2) by GGT (process 3) and DP (process 5) activities, with an uptake of the constituent amino acids by transport across the brush-border plasma membrane (processes 4, 6, and 7). Once inside the PT cell, GSH can be resynthesized in two successive, ATP-dependent reactions (processes 8 and 9).

of this experimental model is that it allows measurement of transport in a highly controlled system without the potential complications of intracellular metabolism. For GSH, however, an additional complication is that degradation of GSH is catalyzed by a membrane-bound enzyme, GGT. Although GGT in the renal proximal tubule is predominantly, if not entirely, localized on the brush-border plasma membrane, as discussed previously, basolateral plasma membrane vesicle preparations typically exhibit a small degree of contamination with brush-border membranes (Lash and Jones, 1982). Due to the extremely high activity of GGT in brush-border membranes, particularly in rodents (Hinchman and Ballatori, 1990), even a very small degree of contamination of basolateral plasma membranes with brush-border

TABLE 2.1
Key Biochemical Properties of GSH Transport by Rat Renal Basolateral Plasma Membrane Vesicles

Parameter	Biochemical Property
Ion dependence	Coupled to transport of Na$^+$; 2 Na$^+$: GSH; net transfer of positive charge (+1)
Membrane potential dependence	Stimulated by negative membrane potential
Sensitivity to transport inhibitors	Inhibited by probenecid and dimethylsuccinate
Substrate specificity	Inhibited by γ-glutamyl amino acids; not inhibited by L-Glu, L-Cys, Gly, or L-Cys-Gly; GSSG also transported, but more slowly
Kinetics	$K_m = 3.0$ mM; $V_{max} = 19.5$ nmol/min per mg protein

Source: Summary of key properties of GSH transport from Lash and Jones (1983, 1984).

plasma membrane (e.g., <5%) can result in rapid and extensive degradation of GSH. Accordingly, transport must be measured in the presence of a GGT inhibitor.

Transport measurements in basolateral plasma membrane vesicles from rat kidney cortex demonstrated the process to be electrogenic, Na$^+$-dependent, and stimulated by a negative membrane potential. A summary of key biochemical properties observed in the basolateral membrane vesicles is shown in Table 2.1. Coupling to Na$^+$ ions with a stoichiometry of two Na$^+$ ions per GSH molecule and the stimulation by a negative membrane potential are consistent with the characteristics of energetics and distribution of GSH in the renal proximal tubule. These characteristics include the GSH molecule having a net charge of approximately -1 at the physiological pH of the plasma and renal interstitial space, a concentration gradient of GSH across the renal basolateral plasma membrane of ~500 (~10 μM in plasma and 5 mM in renal proximal tubule cytoplasm), and a typical membrane potential of -60 mV (negative inside).

Although analysis of kinetics in the basolateral membrane vesicles revealed only a single process, it is often difficult to discriminate among multiple transporters, particularly when the multiple carriers differ markedly in their kinetic parameters. Based on the multiplicity of membrane transporters for organic anions and amino acids and the analysis suggesting the function of both a Na$^+$-dependent and a Na$^+$-independent process, the next step in identifying potential membrane carrier proteins responsible for GSH transport across the basolateral plasma membrane was to correlate energetics, substrate specificity, and inhibitor sensitivity with known carriers. As summarized in the scheme in Figure 2.2, these properties implicated the potential function of either of the major organic anion transporters, organic anion transporter 1 (Oat1; *Slc22a6*) and organic anion transporter 3 (Oat3; *Slc22a8*), for the Na$^+$-independent transport and the sodium dicarboxylate carrier-3 (NaC3; *Slc13a3*) for the Na$^+$-coupled transport. Oat1 and Oat3 catalyze the facilitated uptake of organic anions in exchange for 2-oxoglutarate (2-OG) and are the major Oats expressed in the basolateral plasma membrane of renal proximal tubules, whereas NaC3 catalyzes the Na$^+$-coupled uptake of dicarboxylates such as succinate and involves the cotransport of dicarboxylates with two Na$^+$ ions.

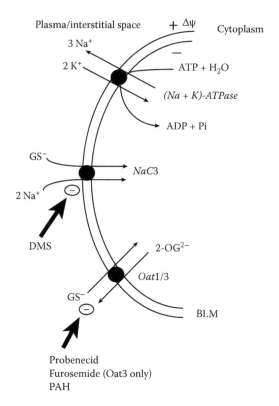

FIGURE 2.2 Established and presumed transport pathways for GSH across the renal basolateral plasma membrane. The scheme shows the two potential transport processes that mediate GSH transport across the basolateral plasma membrane (BLM) into the renal proximal tubular cell. To highlight the energetics of the processes, the $(Na^+ + K^+)$-ATPase generates the Na^+ ion gradient across the membrane, and the membrane potential $(\Delta\Psi)$ exists across the membrane. Putative or known transporters for GSH include the NaC3 (*Slc13a3*) and the Oat1/3 (*Slc22a6/8*). NaC3 mediates the cotransport of GSH with Na^+ ions, whereas Oat1/3 mediates the exchange of GSH with 2-OG^{2-}. Direct evidence supporting the function of a specific carrier on the basolateral plasma membrane in GSH uptake only exists for Oat3. Consistent with functions of these carriers, GSH uptake is partially inhibited by dimethylsuccinate (DMS; NaC3-specific inhibitor), probenecid (Oat1/3 inhibitor), furosemide (Oat3-selective inhibitor), or *p*-aminohippurate (PAH; Oat1/3 substrate and thus competitive inhibitor).

Although the function of NaC3 in the Na^+-coupled uptake of GSH is presumed, based on the transport properties in renal proximal tubular cells and basolateral plasma membrane vesicles, no direct evidence is available to support this. For rat kidney Oat3, however, direct evidence has been obtained (Lash et al., 2007). Bacterially expressed and reconstituted rat Oat3 as well as NRK-52E cells transfected with rat Oat3 cDNA both transported GSH with properties consistent with its function *in vivo*. This included stimulation of exchange transport of *p*-aminohippurate and 2-OG and inhibition by probenecid and furosemide. As for Oat1, again there is no direct evidence for its role in GSH transport, only a presumption based on energetics and substrate specificity. However, Hagos et al. (2013) expressed human OAT1 and OAT3 in HEK293 cells and concluded that neither carrier was involved in mediated GSH uptake. Factors contributing to the discrepancy may include species differences (i.e., rat vs. human) and differences in experimental models used.

2.2.3 PHYSIOLOGICAL AND TOXICOLOGICAL ROLES OF RENAL PLASMA MEMBRANE GSH TRANSPORT

2.2.3.1 Cytoprotection and Redox Homeostasis

Although glomerular filtration delivers significant amounts of GSH to the renal proximal tubular lumen, about 60% of the total GSH extracted by the kidneys is through the basolateral membrane route. Further realization that it is removal of circulating GSH by the kidneys that is primarily responsible for overall turnover of GSH in the body (Fonteles et al., 1976; Häberle et al., 1979) highlights the obvious functional importance of GSH transport across the basolateral plasma membrane in maintenance of GSH homeostasis. Although it has been shown that inhibition of GGT results in a profound glutathionuria (Griffith and Meister, 1979b; Anderson and Meister, 1986), thus disrupting GSH homeostasis, clearly the ability of the kidneys to remove most of the circulating GSH is initially dependent on active transporters on the renal basolateral plasma membrane. Analysis of GSH homeostasis on a cellular level showed that under certain conditions, such as oxidant exposure (Visarius et al., 1996; Lash et al., 1998; Lash and Putt, 1999), recovery of cellular GSH concentration can be derived much more rapidly and completely by transport of exogenous GSH than by intracellular resynthesis from precursors.

Thus, from the previous discussion, it can be concluded that two key functions of basolateral plasma membrane transport in kidneys are to (1) replenish and maintain intracellular concentrations of GSH, primarily in renal proximal tubular cells, and (2) participate in turnover of GSH.

As the predominant intracellular, nonprotein thiol in virtually all cells, maintenance of adequate supplies of intracellular GSH, through either *de novo* or resynthesis from precursor amino acids or from extracellular compartments via transport, is critical to cellular redox balance and prevention of oxidative injury from toxicants or reactive oxygen species. Accordingly, inhibition or enhancement of basolateral membrane GSH uptake in isolated renal cell preparations has been directly associated with exacerbation of or protection from, respectively, a large array of toxic agents and pathological conditions that have oxidative injury (or so-called oxidative stress) as a key mechanistic component. Examples include protection of isolated renal cells from *tert*-butyl hydroperoxide (Hagen et al., 1988; Lash and Tokarz, 1990; Lash et al., 1993; Visarius et al., 1996), the redox-cycling quinone menadione and hydrogen peroxide (Lash and Tokarz, 1990), the redox imbalance due to hypoxia (Lash et al., 1993), the redox imbalance due to chemical inhibition of mitochondrial respiration (Lash et al., 1996), and cytotoxicity due to S-(1,2-dichlorovinyl)-L-cysteine (DCVC), which is the penultimate nephrotoxic metabolite of the environmental contaminant and known human carcinogen trichloroethylene (Lash et al., 2002a).

By functioning to deliver GSH to intracellular sites and counteract oxidative injury, basolateral membrane transport thus supports what one might call the "classic" functions of GSH as a reductant and nucleophile.

2.2.3.2 Bioactivation of Nephrotoxic S-Conjugates and Prodrugs

Despite being classically known as an antioxidant and cytoprotective agent, for some classes of chemicals, conjugation with GSH is actually a bioactivation reaction by which further metabolism eventually yields a reactive and cytotoxic sulfur-containing metabolite (Anders et al., 1988; Lash et al., 1988; Lash, 2018). Although most research with regard to nephrotoxic GSH S-conjugates has focused on metabolism, interactions of the reactive metabolites with cellular macromolecules (i.e., DNA, protein, and lipid), and effects on renal cellular function and renal membrane transport processes also play an important part in the overall disposition and nephrotoxicity of these chemicals.

The predominant function of the liver with respect to GSH conjugates is similar to that for synthesis of GSH, in that the liver is the primary site of their formation, but the organ is highly efficient at mediating the efflux of the product into either the bile or the plasma. Hence, GSH conjugates released from the liver into the plasma are extracted by the kidneys in the same manner as is GSH. Studies in isolated rat proximal tubular cells showed that the GSH conjugate of trichloroethylene, S-(1,2-dichlorovinyl)glutathione (DCVG), is transported across the basolateral plasma membrane by Na^+-dependent and Na^+-independent carriers that appear to be the same as those that transport GSH (Lash and Jones, 1985). As the majority of the renal extraction of DCVG is basolateral rather than by glomerular filtration, these transport processes are critical determinants of the delivery of DCVG to sites of further metabolism and subsequent nephrotoxicity (Lash et al., 2014). Inhibition of transport with probenecid results in protection of the renal proximal tubular cells from DCVC-induced injury (Lash and Anders, 1986). Similar considerations exist for GSH S-conjugates of other halogenated hydrocarbons (Cichocki et al., 2016; Lash, 2018).

A different type of role for GSH conjugate transport has been demonstrated, in which a GSH conjugate is synthesized as a prodrug and transport functions to deliver it to further sites of metabolism so that it can be converted to its active form. The GSH S-conjugate of the chemotherapeutic drug 6-mercaptopurine, S-(6-purinyl)glutathione, was transported into rat renal proximal tubular cells similar to DCVG and GSH (Lash et al., 1997). Here, by contrast, the role of transport is not for turnover or delivery to sites of metabolism that result in cytotoxicity but for delivery to sites of further metabolism to generate the pharmacologically active drug inside the cell where it can exert its therapeutic effects.

2.3 RENAL MITOCHONDRIAL GSH TRANSPORT

2.3.1 IDENTIFICATION OF CARRIERS

Three sets of facts or observations led to the hypothesis that in many cell types, such as hepatocytes or renal proximal tubules, a distinctly regulated pool of GSH that is determined by transport from the cytoplasm exists within the mitochondria. These key facts or observations are as follows: (1) Studies in hepatocytes and proximal tubular cells indicate that most, if not all, intracellular synthesis of GSH occurs in the cytoplasm with no detectable synthesis in the mitochondria (Griffith and Meister, 1985; McKernan et al., 1991); (2) analysis of the turnover of cytoplasmic and mitochondrial GSH pools in hepatocytes (Meredith and Reed, 1982) and renal proximal tubules (Schnellmann and Mandel, 1985) showed the pool in the cytoplasm to have a half-life on the order of ~2 h, whereas that in the mitochondria was ~30 h; (3) the cytoplasmic and mitochondrial GSH pools exhibit differential susceptibility to oxidants and various cytotoxic chemicals; and (4) GSH is a charged molecule with a net negative charge at typical cytoplasmic pH. Considering the membrane potential across the mitochondrial inner membrane, the only way for GSH to get into the mitochondria is by facilitated or active transport.

To identify potential carrier proteins that could mediate GSH transport in renal mitochondria, a similar strategy as used to identify the renal basolateral membrane carriers for GSH was used, namely, to examine transport energetics, substrate specificity, and sensitivity to inhibition by known, selective inhibitors of specific carrier proteins (McKernan et al., 1991; Chen and Lash, 1998; Chen et al., 2000). Mitochondria are known to possess a family of carriers (primarily of the *Slc25a* gene family) that catalyze the uptake of various mono-, di-, and tricarboxylates from the cytoplasm into the mitochondrial matrix (Palmieri et al., 1996; Kaplan, 2001). The relatively broad substrate specificity of these carriers suggests several as potential carriers that are responsible for catalyzing the uptake of GSH into renal mitochondria.

Based on the studies in suspensions of freshly isolated renal cortical mitochondria from rats (Chen and Lash, 1998; McKernan et al., 1991) and partially purified and reconstituted mitochondrial inner membrane carriers from rat kidney (Chen et al., 2000), a dependence on phosphate ions, a lack of dependence on pH or membrane potential, and partial inhibition by both butylmalonate and phenylsuccinate led us to identify the two carriers as being responsible for mitochondrial GSH uptake, the dicarboxylate carrier (DIC; *Slc25a10*) and the OGC (*Slc25a11*). As shown in Figure 2.3, the DIC is selectively inhibited by butylmalonate and catalyzes an electroneutral exchange of GSH with inorganic phosphate; the OGC is selectively inhibited by phenylsuccinate and catalyzes the electroneutral exchange of GSH with other dicarboxylates such as 2-OG.

The DIC and OGC, which of course play critical roles in the transport of citric acid cycle intermediates, are present in the mitochondria from most tissues. Accordingly, the DIC in the mitochondria from the liver (Zhong et al., 2008) and brain (Kamga et al., 2010) and the OGC in the mitochondria from the liver (Coll et al., 2003; Zhong et al., 2008; Baulies et al., 2018), brain (Wilkins et al., 2012, 2014), and colonic epithelial cells (Circu et al., 2009) have been shown to be involved in GSH transport. Although these two transporters are present in the mitochondria from many tissues, tissue-specific differences in their relative expression as well as differential expression of other transporters exist, which result in tissue-dependent differences in GSH transport (Zhong et al., 2008; Baulies et al., 2018). For example, whereas the DIC and OGC are estimated to account for nearly all the measureable GSH transport in renal mitochondria (Chen and Lash, 1998; Chen et al., 2000), a clear role

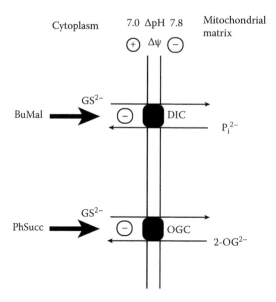

FIGURE 2.3 Properties of renal mitochondrial GSH uptake. The scheme highlights the basic factors of energetics (pH gradient [ΔpH], membrane potential [ΔΨ]), substrate specificity, and susceptibility to carrier-selective inhibitors, butylmalonate (BuMal) for the DIC and phenylsuccinate (PhSucc) for the OGC. The DIC, which exchanges various dicarboxylates such as L-malate for inorganic phosphate (P_i^{2-}), and the OGC, which exchanges 2-OG for various dicarboxylates, also can use GSH in the di-anionic form as a substrate.

for other, still unidentified carrier(s) has been found for liver mitochondria (Zhong et al., 2008). Additionally, whereas substrate specificity studies excluded a role for the tricarboxylate carrier for GSH transport in renal mitochondria (Chen and Lash, 1998), the tricarboxylate carrier does function in GSH transport in brain mitochondria (Wadey et al., 2009).

2.3.2 ROLE IN GSH AND REDOX HOMEOSTASIS

The importance of understanding GSH transport is that the mitochondria are the primary site of oxygen consumption in aerobic cells, which results in the potential for generation of significant amounts of reactive oxygen species under certain toxic or pathological states. The function of these organelles is also critically dependent on the preservation of a strong reducing environment. Maintenance of adequate concentrations of GSH within the mitochondria is, therefore, required for mitochondrial function, and this is largely determined by the function of the DIC and OGC (Passarella et al., 2003; Ribas et al., 2014).

2.3.3 MODULATION OF MITOCHONDRIAL GSH TRANSPORT IN DISEASE

Disturbances in mitochondrial GSH transport function have been associated with several diseases and toxic chemical exposures (Huizing et al., 1998; Lash, 2012,

2015). In kidney mitochondria, diabetic nephropathy is associated with alterations in mitochondrial redox status and GSH transport function (Putt et al., 2012). Similar alterations in mitochondrial GSH transport and GSH redox status are observed in renal proximal tubular cells and renal cortical mitochondria isolated from hypertrophied kidneys of uni-nephrectomized rats (Benipal and Lash, 2011, 2013). Although underlying molecular mechanisms of renal cellular hypertrophy certainly differ from those associated with diabetic nephropathy, both processes involve oxidative stress and mitochondrial changes that can be adaptive but can progress to dysfunction. In both states, overexpression of one or the other mitochondrial GSH transporter results in decreases in oxidative injury, preservation of mitochondrial function, and reversion of the hypertrophied state. Oxidative injury in NRK-52E cells caused by either *tert*-butyl hydroperoxide or DCVC exposure was attenuated by overexpression of either the DIC or the OGC (Lash et al., 2002b; Xu et al., 2006). This approach has also been applied to other tissues, such as the brain (Kamga et al., 2010; Wilkins et al., 2014) and the intestine (Circu et al., 2009). These findings support the potential use of overexpression of the DIC and/or OGC as a therapeutic approach to treat both toxic chemical exposures and a broad range of diseases or pathological states.

2.4 SUMMARY AND CONCLUSIONS

This chapter has summarized some key aspects of GSH and GSH *S*-conjugate transport across renal membranes. The focus was on two processes, transport from plasma and interstitial fluid into the proximal tubular cell across the basolateral membrane and transport from cytoplasm into the mitochondrial matrix. Although some historical perspective was provided, particularly for basolateral plasma membrane transport, the primary areas of focus were on physiological function, role in response to nephrotoxic chemicals or in chronic diseases, and finally, potential modulation of these transporters as a therapeutic approach.

The kidneys play unique roles in the processing of GSH and GSH *S*-conjugates by virtue of the high activity of GGT on the brush-border plasma membrane of the proximal tubules and the presence of a diverse array of plasma membrane transporters on both brush-border and basolateral plasma membranes that can mediate both uptake and efflux. Despite the high rates of glomerular filtration and high activity of GGT, uptake of GSH by carriers on the basolateral plasma membrane accounts for ~60% of the extraction of GSH by the kidneys. Studies characterizing transport energetics, substrate specificity, and sensitivity to known, carrier-selective inhibitors using isolated proximal tubular cells, isolated plasma membrane vesicles, and bacterially expressed and reconstituted carrier proteins showed that in the rat kidney, Oat3 was identified as one of the carriers that can catalyze the uptake of GSH, and Oat1 and NaC3 may also function in GSH uptake. Besides extraction of plasma GSH, basolateral membrane transport of GSH *S*-conjugates of nephrotoxicants plays an important role in the delivery of these conjugates to further sites of metabolism, which results in generation of reactive and toxic species. Thus, besides the role of these transporters in cytoprotection and maintenance of redox status, they may also be a key step in the bioactivation of certain nephrotoxic chemicals.

GSH transport across the mitochondrial inner membrane is also critical in maintaining cellular redox homeostasis. The critical nature of this transport lies in the fact that GSH concentrations in the mitochondrial matrix are similarly high as those in the cytoplasm, yet GSH synthesis is believed to occur solely in the cytoplasm. The charged and amphiphilic nature of the GSH molecule and the membrane potential across the mitochondrial inner membrane indicate that specific, carrier-mediated processes must exist for GSH to cross that membrane. Similar strategies that were used to identify plasma membrane GSH carriers showed that two anion carriers from the *Slc25a* gene family, the DIC and OGC, account for most of the mitochondrial GSH uptake in rat kidneys. The function of these carriers in the mitochondria from other tissues, including the brain, liver, and large intestine, has also been confirmed.

By identifying the carriers that can transport GSH and GSH *S*-conjugates, potential therapeutic targets for prevention of nephrotoxicity from exposure to toxic drugs and environmental contaminants and amelioration of redox imbalances associated with several diseases and pathological states may be developed. Some studies in which carrier proteins have been overexpressed are described that provide a basis for future development of novel therapeutic regimens.

REFERENCES

Abbott, W. A., R. J. Bridges, and A. Meister. 1984. Extracellular metabolism of glutathione accounts for its disappearance from the basolateral circulation of the kidneys. *J Biol Chem* 259:15393–400.

Anders, M. W., L. H. Lash, W. Dekant et al. 1988. Biosynthesis and metabolism of glutathione conjugates to toxic forms. *CRC Crit Rev Toxicol* 18:311–41.

Anderson, M. E., R. J. Bridges, and A. Meister. 1980. Direct evidence for inter-organ transport of glutathione and that the non-filtration mechanism for glutathione utilization involves γ-glutamyl transpeptidase. *Biochem Biophys Res Commun* 96:848–53.

Anderson, M. E., and A. Meister. 1983. Transport and direct utilization of γ-glutamylcyst(e)ine for glutathione synthesis. *Proc Natl Acad Sci USA* 80:707–11.

Anderson, M. E., and A. Meister. 1986. Inhibition of γ-glutamyl transpeptidase and induction of glutathionuria by γ-glutamyl amino acids. *Proc Natl Acad Sci USA* 83:5029–32.

Anderson, M. E., A. Meister. 1980. Dynamic state of glutathione in blood plasma. *J Biol Chem* 255 (20):9530–3

Ballatori, N., S. M. Krance, S. Notenboom et al. 2009. Glutathione dysregulation and the etiology and progression of human diseases. *Biol Chem* 390:191–214.

Bartoli, G. M., and H. Sies. 1978. Reduced and oxidized glutathione efflux from liver. *FEBS Lett* 86:89–91.

Baulies, A., J. Montero, N. Matías et al. 2018. The 2-oxoglutarate carrier promotes liver cancer by sustaining mitochondrial GSH despite cholesterol loading. *Redox Biol* 14:164–77.

Benipal, B., and L. H. Lash. 2011. Influence of renal compensatory hypertrophy on mitochondrial energetics and redox status. *Biochem Pharmacol* 81:295–303.

Benipal, B., and L. H. Lash. 2013. Modulation of mitochondrial glutathione status and cellular energetics in primary cultures of proximal tubular cells from remnant kidney of uninephrectomized rats. *Biochem Pharmacol* 85:1379–88.

Chen, Z., and L. H. Lash. 1998. Evidence for mitochondrial uptake of glutathione by dicarboxylate and 2-oxoglutarate carriers. *J Pharmacol Exp Ther* 285:608–18.

Chen, Z., D. A. Putt, and L. H. Lash. 2000. Enrichment and functional reconstitution of glutathione transport activity from rabbit kidney mitochondria: Further evidence for the role of the dicarboxylate and 2-oxoglutarate carriers in mitochondrial glutathione transport. *Arch Biochem Biophys* 373:193–202.

Chikhi, N., N. Holic, G. Guellaen et al. 1999. Gamma-glutamyl transpeptidase gene organization and expression: A comparative analysis in rat, mouse, pig and human species. *Comp Biochem Physiol* 122:367–80.

Cichocki, J. A., K. Z. Guyton, N. Guha et al. 2016. Target organ metabolism, toxicity, and mechanisms of trichloroethylene and perchloroethylene: Key similarities, differences, and data gaps. *J Pharmacol Exp Ther* 359:110–23.

Circu, M. L., M. P. Moyer, L. Harrison et al. 2009. Contribution of glutathione status to oxidant-induced mitochondrial DNA damage in colonic epithelial cells. *Free Radic Biol Med* 47:1190–98.

Coll, O., C. Garcia-Ruiz, N. Kaplowitz et al. 2003. Sensitivity of the 2-oxoglutarate carrier to alcohol intake contributes to mitochondrial glutathione depletion. *Hepatology* 38:692–702.

Dass, P. D., R. P. Misra, and T. C. Welbourne. 1981. Presence of γ-glutamyltransferase in the renal microvascular compartment. *Can J Biochem* 59:383–6.

Fernandez-Checa, J., and N. Kaplowitz. 2005. Hepatic mitochondrial glutathione: Transport and role in disease and toxicity. *Toxicol Appl Pharmacol* 204:263–73.

Fonteles, M. C., D. J. Pillion, A. H. Jeske et al. 1976. Extraction of glutathione by the isolated perfused rabbit kidney. *J Surg Res* 21:169–74.

Glossman, H., and D. M. Neville Jr. 1972. γ-Glutamyltransferase in kidney brush border membranes. *FEBS Lett* 19:340–4.

Griffith, O. W., and A. Meister.1979a. Glutathione: Interorgan translocation, turnover, and metabolism. *Proc Natl Acad Sci USA* 76:5606–10.

Griffith, O. W., and A. Meister. 1979b. Translocation of intracellular glutathione to membrane-bound γ-glutamyl transpeptidase as a discrete step in the γ-glutamyl cycle: Glutathionuria after inhibition of transpeptidase. *Proc Natl Acad Sci USA* 76:268–72.

Griffith, O. W., R. J. Bridges, and A. Meister. 1979. Transport of γ-glutamyl amino acids: Role of glutathione and γ-glutamyl transpeptidase. *Proc Natl Acad Sci USA* 76:6319–22.

Griffith, O. W., and A. Meister. 1985. Origin and turnover of mitochondrial glutathione. *Proc Natl Acad Sci USA* 82:4668–72.

Häberle, D., A. Wähllander, and H. Sies. 1979. Assessment of the kidney function in maintenance of plasma glutathione concentration and redox state in anaesthetized rats. *FEBS Lett* 108:335–40.

Hagen, T. M., T. Y. Aw, and D. P. Jones. 1988. Glutathione uptake and protection against oxidative injury in isolated kidney cells. *Kidney Int* 34:74–81.

Hagos, Y., G. Burckhardt, and B. C. Burckhardt. 2013. Human organic anion transporter OAT1 is not responsible for glutathione transport but mediates transport of glutamate derivatives. *Am J Physiol* 304:F403–9.

Hahn, R., A. Wendel, and L. Flohé. 1978. The fate of extracellular glutathione in the rat. *Biochim Biophys Acta* 539:324–37.

Hinchman, C. A., and N. Ballatori. 1990. Glutathione-degrading capacities of liver and kidney in different species. *Biochem Pharmacol* 40:1131–5.

Huizing, M., W. Ruitenbeek, L. P. van den Heuvel et al. 1998. Human mitochondrial transmembrane metabolite carriers: Tissue distribution and its implication for mitochondrial disorders. *J Bioenerg Biomembr* 30:277–84.

Inoue, M., S. Shinozuka, and Y. Mori. 1986. Mechanism of renal peritubular extraction of plasma glutathione: The catalytic activity of contralumenal γ-glutamyltransferase is prerequisite to the apparent peritubular extraction of plasma glutathione. *Eur J Biochem* 157:605–9.

Kamga, C. K., S. X. Zhang, and Y. Wang. 2010. Dicarboxylate carrier-mediated glutathione transport is essential for reactive oxygen species homeostasis and normal respiration in rat brain mitochondria. *Am J Physiol* 299:C497–C505.

Kaplan, R. S. 2001. Structure and function of mitochondrial anion transport proteins. *J Membrane Biol* 179:165–83.

Lash, L. H. 2005. Role of glutathione transport processes in kidney function. *Toxicol Appl Pharmacol* 204:329–42.

Lash, L. H. 2006. Mitochondrial glutathione transport: Physiological, pathological and toxicological implications. *Chem Biol Interact* 163:54–67.

Lash, L. H. 2009. Renal glutathione transport: Identification of carriers, physiological functions, and controversies. *BioFactors* 35:500–8.

Lash, L. H. 2011. Renal membrane transport of glutathione in toxicology and disease. *Vet Pathol* 48:408–19.

Lash, L. H. 2012. Mitochondrial glutathione in toxicology and disease of the kidneys. *Toxicology Res* 1:39–46.

Lash, L. H. 2015. Mitochondrial glutathione in diabetic nephropathy. *J Clin Med* 4:1428–47.

Lash, L. H. 2018. Halogenated hydrocarbons. In *Comprehensive Toxicology*, edited by C. A. McQueen, 3rd ed., vol. 14, 380–409. Oxford: Elsevier Ltd.

Lash, L. H., and M. W. Anders. 1986. Cytotoxicity of *S*-(1,2-dichlorovinyl)glutathione and *S*-(1,2-dichlorovinyl)-L-cysteine in isolated rat kidney cells. *J Biol Chem* 261:13076–81.

Lash. L. H., W. A. Chiu, K. Z. Guyton et al. 2014. Trichloroethylene biotransformation and its role in mutagenicity, carcinogenicity and target organ toxicity. *Mutat Res Rev* 762:22–36.

Lash, L. H., and D. P. Jones. 1982. Localization of the membrane-associated thiol oxidase of rat kidney to the basal-lateral plasma membrane. *Biochem J* 203:371–6.

Lash, L. H., and D. P. Jones. 1983. Transport of glutathione by renal basal-lateral membrane vesicles. *Biochem Biophys Res Commun* 112:55–60.

Lash, L. H., and D. P. Jones. 1984. Renal glutathione transport: Characteristics of the sodium-dependent system in the basal-lateral membrane. *J Biol Chem* 259:14508–14.

Lash, L. H., and D. P. Jones. 1985. Uptake of the glutathione conjugate *S*-(1,2-dichlorovinyl)-glutathione by renal basal-lateral membrane vesicles and isolated kidney cells. *Mol Pharmacol* 28:278–82.

Lash, L. H., D. P. Jones, and M. W. Anders. 1988. Glutathione homeostasis and glutathione *S*-conjugate toxicity in kidney. *Rev Biochem Toxicol* 9:29–67.

Lash, L. H., and D. A. Putt. 1999. Renal cellular transport of exogenous glutathione: Heterogeneity at physiological and pharmacological concentrations. *Biochem Pharmacol* 58:897–907.

Lash, L. H., D. A. Putt, S. E. Hueni et al. 2002a. Cellular energetics and glutathione status in NRK-52E cells: Toxicological implications. *Biochem Pharmacol* 64:1533–46.

Lash, L. H., D. A. Putt, and L. H. Matherly. 2002b. Protection of NRK-52E cells, a rat renal proximal tubular cell line, from chemical induced apoptosis by overexpression of a mitochondrial glutathione transporter. *J Pharmacol Exp Ther* 303:476–86.

Lash, L. H., D. A. Putt, F. Xu et al. 2007. Role of rat organic anion transporter 3 (Oat3) in the renal basolateral transport of glutathione. *Chem Bio Interact* 170:124–34.

Lash, L. H., A. Shivnani, J. Mai et al. 1997. Renal cellular transport, metabolism and cytotoxicity of *S*-(6-purinyl)glutathione, a prodrug of 6-mercaptopurine, and analogues. *Biochem Pharmacol* 54:1341–9.

Lash, L. H., and J. J. Tokarz. 1990. Oxidative stress in isolated rat renal proximal and distal tubular cells. *Am J Physiol* 259:F338–47.

Lash, L. H., J. J. Tokarz, Z. Chen et al. 1996. ATP depletion by iodoacetate and cyanide in renal distal tubular cells. *J Pharmacol Exp Ther* 276:194–205.

Lash, L. H., J. J. Tokarz, E. B. Woods et al. 1993. Hypoxia and oxygen dependence of cytotoxicity in renal proximal tubular and distal tubular cells. *Biochem Pharmacol* 45:191–200.

Lash, L. H., T. M. Visarius, J. M. Sall et al. 1998. Cellular and subcellular heterogeneity of glutathione metabolism and transport in rat kidney cells. *Toxicology* 130:1–15.

McIntyre, T. M., and N. P. Curthoys. 1980. The interorgan metabolism of glutathione. *Int J Biochem* 12:545–51.

McKernan, T. B., E. B. Woods, and L. H. Lash. 1991. Uptake of glutathione by renal cortical mitochondria. *Arch Biochem Biophys* 288:653–63.

Meister, A. 1976. Glutathione and related γ-glutamyl compounds: Biosynthesis and utilization. *Annu Rev Biochem* 45:559–604.

Meredith, M. J., and D. J. Reed. 1982. Status of the mitochondrial pool of glutathione in the isolated hepatocyte. *J Biol Chem* 257:3747–53.

Olafsdottir, K., G. A. Pascoe, and D. J. Reed. 1988. Mitochondrial glutathione status during Ca^{2+} ionophore-induced injury to isolated hepatocytes. *Arch. Biochem Biophys* 263:226–35.

Ormstad, K., T. Lastböm, and S. Orrenius. 1982. Evidence for different localization of glutathione oxidase and γ-glutamyltransferase activities during extracellular glutathione metabolism in isolated perfused kidney. *Biochim Biophys Acta* 700:148–53.

Orrenius, S., K. Ormstad, T. Thor et al. 1983. Turnover and functions of glutathione studied with isolated hepatic and renal cells. *Fed Proc* 42:3177–88.

Palmieri, F., F. Bisaccia, L. Capobianco et al. 1996. Mitochondrial metabolite transporters. *Biochim Biophys Acta* 1275:127–32.

Passarella, S., A. Atlante, D. Valenti, and L. de Bari. 2003. The role of mitochondrial transport in energy metabolism. *Mitochondrion* 2:319–43.

Putt, D. A., Q. Zhong, and L. H. Lash. 2012. Adaptive changes in renal mitochondrial redox status in diabetic nephropathy. *Toxicol Appl Pharmacol* 258:188–98.

Rankin, B. B., and N. P. Curthoys. 1982. Evidence for renal paratubular transport of glutathione. *FEBS Lett* 147:193–6.

Rankin, B. B., W. Wells, and N. P. Curthoys. 1985. Rat renal peritubular transport and metabolism of plasma [^{35}S]glutathione. *Am J Physiol* 249:F198–204.

Ribas, V., C. Garcia-Ruiz, and J. C. Fernandez-Checa. 2014. Glutathione and mitochondria. *Front Pharmacol* 5:151.

Schnellmann, R. G., and L. J. Mandel. 1985. Intracellular compartmentation of glutathione in rabbit renal proximal tubules. *Biochem Biophys Res Commun* 133:1001–5.

Visarius, T. M., D. A. Putt, J. M. Schare et al. 1996. Pathways of glutathione metabolism and transport in isolated proximal tubular cells from rat kidney. *Biochem Pharmacol* 52:259–72.

Wadey, A. L., H. Muyderman, P. T. Kwek et al. 2009. Mitochondrial glutathione uptake: Characterization in isolated brain mitochondria and astrocytes in culture. *J Neurochem* 109(Suppl. 1):101–8.

Wendel, A., and P. Cikryt. 1980. The level and half-life of glutathione in human plasma. *FEBS Lett* 120:209–11.

Wilkins, H. M., K. Marquardt, L. H. Lash et al. 2012. Bcl-2 is a novel interacting partner for the 2-oxoglutarate carrier and a key regulator of mitochondrial glutathione. *Free Radic Biol Med* 52:410–19.

Wilkins, H. M., S. Brock, J. J. Gray et al. 2014. Stable over-expression of the 2-oxoglutarate carrier enhances neuronal cell resistance to oxidative stress via Bcl-2-dependent mitochondrial GSH transport. *J Neurochem* 130:75–86.

Xu, F., D. A. Putt, L. H. Matherly et al. 2006. Modulation of expression of rat mitochondrial 2-oxoglutarate carrier in NRK-52E cells alters mitochondrial transport and accumulation of glutathione and susceptibility to chemically induced apoptosis. *J Pharmacol Exp Ther* 316:1175–86.

Zhong, Q., D. A. Putt, F. Xu et al. 2008. Hepatic mitochondrial transport of glutathione: Studies in isolated rat liver mitochondria and H4IIE rat hepatoma cells. *Arch Biochem Biophys* 474:119–27.

Part II

Glutathione-Dependent Hydroperoxide Metabolism

3 The Catalytic Mechanism of Glutathione Peroxidases

Laura Orian, Giorgio Cozza, Matilde Maiorino,
Stefano Toppo, and Fulvio Ursini
Università degli Studi di Padova

CONTENTS

3.1 INTRODUCTION

Glutathione (GSH) peroxidases (GPx) are the members of a protein family which is widespread in the whole tree of life, whose definitive recognition did not take place until the second half of the 20th century. Before then, the term "peroxidase" (EC 1.11.1.-) was synonymous with heme-containing catalysts using hydroperoxide as an oxidant (Flohé and Ursini, 2008). The typical redox-sensitive spectral absorbance of their heme prosthetic group indeed made enzymological studies easy. For this reason, the unexpected discovery of non-heme peroxidases encountered some skepticism in the scientific community. The first experimental evidence of a peroxidase missing the colored prosthetic group, obtained by Gordon C. Mills (1957), took several years, indeed, before its acknowledgment. Its identification as the first GPx and the first selenoprotein, now known as GPx1, was achieved in 1973 by the groups of Flohé (1973) and Hoekstra (Rotruck et al., 1973).

The class of thiol peroxidases is split into GPx and peroxiredoxins. Both share the folding architecture known as thioredoxin fold (Martin, 1995), whose minimal

DOI: 10.1201/9781351261760-5

information is given by a structural motif consisting of a central core of a four-stranded β-sheet and three surrounding α-helices. The basic shared mechanism of the reaction catalyzed is the reduction of hydroperoxides by thiols, where the redox transitions are faster than the formation of an enzyme–substrate complex and, consequently, K_M and V_{max} are infinite (Flohé et al., 1972, 2011).

The second discovered GPx is "phospholipid hydroperoxide GPx" (Ursini et al., 1985), now also known as GPx4, from the name assigned to the corresponding gene. GPx4 extends the activity of GPx1 to hydroperoxy groups of complex lipids in biomembranes. At a first glance, this specific function descends from the monomeric nature of GPx4 that lacks the oligomerization interfaces, which in the tetrameric GPx1 hamper the accessibility of bulky substrates such as membrane phospholipid hydroperoxides (Flohé et al., 2011). Moreover, a specific interaction of a cationic area of the enzyme with a polar head of phospholipids had been detected in GPx4 but not GPx1 (Cozza et al., 2017).

Presently, eight GPx homologs are known in mammals. These belong to a vast family of paralogs spread over all domains of life (Toppo et al., 2008). Most of the mammalian GPx are selenoenzymes, namely, GPx1, GPx2, GPx3, GPx4, and, depending on species, GPx6. By contrast, GPx5, GPx7, and GPx8 are homologs whose catalytic Sec residue is replaced by Cys. The redox center of these enzymes encompasses three more conserved residues, which is considered as the minimal requirement for the efficiency of the catalytic center: Gln 81, Trp 136, and Asn 137 (human cytosolic GPX4 numbering) (Tosatto et al., 2008). The only exception is GPx8, the peroxidase activity of which is still elusive, in which a Ser substitutes for Gln 81 (Bosello-Travain et al., 2015).

GPx are also characterized by their structural organization: GPx1, GPx2, GPx3, GPx5, and GPx6 are homo-tetrameric, whereas GPx4, GPx7, and GPx8 are monomeric (Toppo et al., 2008).

Monomeric GPx containing a catalytic Cys (CysGPx) are also the vast majority of GPx in invertebrate species. Thereof, it had been questioned whether these Cys paralogs could be efficient peroxidases, because a drop in the catalytic activity by orders of magnitude was observed in recombinant mammalian GPx4, in which the catalytic Sec was replaced by Cys (Maiorino et al., 1995). More recently, however, kinetic analyses of naturally occurring CysGPx revealed surprisingly high catalytic activities. This unexpected efficiency remained unexplained until the physiological reducing substrate was unraveled. During the catalytic cycle, these invertebrate monomeric CysGPx undergo ample conformational changes that allow the interaction with reducing substrates of the family of "redoxins." This was observed for the first time in the CysGPx of *Plasmodium falciparum* (Sztajer et al., 2001), and the mechanism was disclosed in the enzyme of *Drosophila melanogaster* (Maiorino et al., 2007). This suggested a paradigm, where CysGPx mimics a peroxiredoxin working with two redox centers, that is, a peroxidatic Cys and a resolving Cys (Hofmann et al., 2002; Schlecker et al., 2007; Fourquet et al., 2008). In this respect, the mammalian CysGPx, that is, GPx5, GPx7, and possibly GPx8, are notable exception lacking a defined resolving Cys and using, besides GSH, the redoxin-like protein disulfide isomerase (PDI) as the reducing substrate (Bosello-Travain et al., 2013).

In this chapter, we will focus on the steps of the catalytic cycle GPx, having a closer look at the undisclosed chemistry at the quantum mechanics and the molecular dynamics (MD) level of theory.

3.2   GPX KINETICS

The basic reaction catalyzed by GPx is as follows (Figure 3.1):

$$ROOH + 2GSH \rightarrow GSSG + ROH + H_2O$$

A hydroperoxide (ROOH) is reduced to alcohol (ROH) at the expense of two molecules of GSH, forming its disulfide (GSSG).

The kinetic mechanism corresponds to the mechanism IV delineated by Dalziel that describes a ping-pong reaction without substrate saturation (Dalziel, 1957). The kinetic behavior observed in steady-state kinetic analyses for GPx1 and GPx4, where V_{max} and Michaelis constants are infinite (Flohé et al., 1972; Günzler et al., 1972; Ursini et al., 1985), indeed provided an early proof that the mechanism IV, as defined by Dalziel, existed in real life. The mechanism implies that the catalytic cycle is split into two main independent events that theoretically reach infinite velocities at infinite concentrations of the reaction partners. These events encompass the oxidation of the reduced enzyme by ROOH and the reduction of the oxidized enzyme by GSH.

The Dalziel equation is as follows:

$$[E_0]/v_0 = \Phi_1/[ROOH] + \Phi_2/[GSH]$$

where Φ_1 and Φ_2 are the reciprocal apparent rate constants for the net forward oxidative and reductive parts of the catalytic cycle, respectively:

$$[E_0]/v_0 = 1/k'_{+1} \cdot [ROOH] + 1/k'_{+2} \cdot [GSH]$$

However, as shown in Figure 3.1, GPx catalysis involves at least six forward and five reverse steps. This raises the question how the experimentally accessible coefficients k'_{+1} and k'_{+2} translate into the microscopic rate constants $k_{\pm 1-6}$. Indeed, from kinetic studies, it is possible to infer the relationships among these parameters.

Ground-state (reduced GPx) "E" is apparently oxidized by a hydroperoxide substrate without forming a canonical enzyme–substrate complex. Accordingly, the apparent rate constants for this step, k'_{+1}, are near 10^8 M^{-1} s^{-1}, among the fastest ever determined for bimolecular enzymatic reactions. Since the oxidation of the catalytic center, forming F (Figure 3.1), is not reversible, it follows that the microscopic k_{+1} is identical to the experimentally accessible k'_{+1}.

The reductive part is more complex. The oxidized enzyme F reacts with the first molecule of GSH forming the glutathionylated intermediate G at the active site, as validated by tandem mass spectrometry analysis on GPx4 (Mauri et al., 2003).

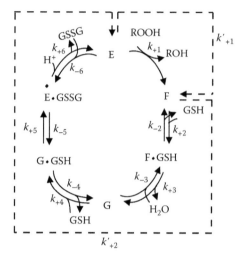

FIGURE 3.1 General scheme of the canonical catalytic cycle of GPx. Adapted from Toppo et al. (2009). Schematic presentation of partial reactions and microscopic rate constants k; E (enzyme ground state), F (first oxidized intermediate), and G (half reduced intermediate).

Therefore, the selenyl-sulfide bridge of G is reduced by the second GSH molecule releasing GSSG and recovering the reduced enzyme.

The observation that there is no saturation by GSH implies that none of the substrate-independent steps, that is, the decay of the complexes (F·GSH) or (G·GSH), can be rate limiting. In other words, rate constants should have the following dependencies: $k_{+2} \ll k_{+3}$ and $k_{+4} \ll k_{+5}$. Accordingly, the experimentally accessible apparent rate constant k'_{+2} approximates either $(k_{+2} - k_{-2})$ or $(k_{+4} - k_{-4})$. As summarized in Table 3.1, this is the rate-limiting phase of the catalytic cycle.

Ping-pong kinetics were also reported for the CysGPx of *P. falciparum* and *Trypanosoma brucei* (Alphey et al., 2008), *Leishmania major* (König and Fairlamb, 2007), and *D. melanogaster* (Maiorino et al., 2007), but, like the great majority of the other invertebrate GPx, these enzymes are not real glutathione peroxidases because they mimic 2-Cys peroxiredoxins and employ either thioredoxin or other redoxins carrying the reactive C-X-X-C motif as the reducing substrate. Even the mammalian monomeric CysGPx7 prefers PDI over GSH as a reductant, although, differently from the CysGPx of invertebrates, it works with a 1-Cys peroxiredoxin-like mechanism (Bosello-Travain et al., 2013; Maiorino et al., 2015).

The extremely high catalytic efficiency stimulated studies at the analytical and computational level. Moreover, although the oxidation of the catalytic Se or S by ROOH is seemingly taking place by a common mechanism (nucleophilic displacement), not only the reason underlying the high rate of the reaction but also the different catalytic efficiency of sulfur and selenium have been a matter of debate. One aspect accounting for the different catalytic efficiency was typically ascribed to the different pK_a of Sec (5.2) versus Cys (8.4), although this explanation does not satisfy at all (Toppo et al., 2009). Indeed, all the amino, imino, and amide groups

TABLE 3.1
Selected Kinetic Parameters of Mammalian GPx

Enzyme substrates	k_{+1} (M^{-1} s^{-1})	k'_{+2} (M^{-1} s^{-1})	Reference
Bovine GPx1[a]			
H$_2$O$_2$/GSH[b]	4.5×10^7	2.0×10^5	Flohé et al., 1972
H$_2$O$_2$/GSH[c]	1.5×10^7	4.2×10^5	Günzler et al., 1972
Et-OOH/GSH[c]	7.8×10^6	4.2×10^5	
C-OOH/GSH[c]	3.2×10^6	4.2×10^5	
t-b-OOH/GSH[c]	1.9×10^6	4.2×10^5	
Peroxynitrite	8.0×10^6	n.d.	Briviba et al., 1998
Human GPX1			
H$_2$O$_2$/GSH	4.1×10^7	2.3×10^5	Takebe et al., 2002
t-b-OOH/GSH	4.2×10^6	2.3×10^5	
Hamster GPx1			
H$_2$O$_2$/GSH[d]	5.0×10^7	n.d.	Chaudiere and Tappel,
p-M-OOH/GSH[d]	2.0×10^7	n.d.	1983
L-OOH/GSH[d]	4.0×10^7	n.d.	
Human GPX3			
H$_2$O$_2$/GSH	4.0×10^7	7.9×10^4	Takebe et al., 2002
t-b-OOH/GSH	2.3×10^6	7.9×10^4	
PC-OOH/GSH	3.4×10^5	7.9×10^4	
Porcine GPx4			
PC-OOH/GSH	1.4×10^7	1.2×10^5	Ursini et al. 1985
L-OOH/GSH	3.0×10^7	7.2×10^4	
H$_2$O$_2$/GSH	3.0×10^6	1.3×10^5	
C-OOH/GSH	1.8×10^6	6.7×10^4	
t-b-OOH/GSH	1.2×10^6	5.3×10^3	
Porcine GPx4 U46C			
PC-OOH/GSH	5.0×10^4	25	Maiorino et al., 1995
Human GPX4			
PC-OOH/GSH	1.5×10^7	5.7×10^4	Takebe et al., 2002
Mouse GPx7			
PC-OOH/GSH	9.5×10^3	12.6	Bosello-Travain et al., 2013
PC-OOH/HsPDI	4.9×10^3	3.5×10^3	

Source: Adapted and updated from Toppo et al. (2009).

n.d., not determined; H$_2$O$_2$, hydrogen peroxide; GSH, glutathione; Et-OOH, ethyl hydroperoxide; C-OOH, cumene hydroperoxide; L-OOH, linoleic acid hydroperoxide; PC-OOH, phosphatidylcholine hydroperoxide; p-M-OOH, p-menthane hydroperoxide; t-b-OOH, t-butyl hydroperoxide; HsPDI, human protein disulfide isomerase.

[a] Rate constants of tetrameric enzymes were recalculated per subunit.

[b] pH 7.0.

[c] pH 6.7.

[d] pH 7.6.

structurally coordinated into a hydrogen bond network to the catalytic Se or S contribute to the different efficiency, as it emerged from quantum mechanics studies.

3.3 QUANTUM CHEMISTRY MECHANISTIC INVESTIGATIONS

To date hardware technology and computational methodologies are powerful enough to investigate *in silico* chemical reaction mechanisms with pretty good accuracy. Although structural properties, including conformational analysis and docking, can be entirely treated at the classic level, that is, using the equations of Newtonian mechanics, chemical reactivity can be investigated only using the computationally expensive quantum chemistry methods. So, with increasing system size, strategies are required to model bond disruption and formation. Among quantum molecular methodologies, the density functional theory (DFT) approach allows a solution of the electronic Schrödinger equation referring to the electron density. This drastically reduces the number of coordinates, which rapidly increase with the total number of the electrons (Parr and Weitao, 1994). A largely adopted strategy is to extract a cluster of residues, that is, those forming the catalytic site, and perform the mechanistic investigation on this sizably small system. As an alternative, the study can be initially tackled with classical simulations (MD) and the mechanistic investigation carried out using a hybrid scheme, that is, describing the catalytic site at the DFT level and the remaining protein with molecular mechanics (Chung et al., 2015). This approach offers the advantage of retaining the whole system. In this case, the availability of the force fields becomes limiting, that is, the classical Hamiltonians required for simulations, which are not available for all residues, fragments, and atoms. Enzymes are too large to be tackled entirely at the DFT level, and simulations of selenoproteins have so far not been straightforward due to the lack of accurate force fields for Sec. The case of GPx is emblematic, because Se is crucial for the efficiency and rigorous comparison with S challenging. The pioneering computational mechanistic work by Morokuma and coworkers focused on a simplified model cluster based on human GPX3 described at the PCM-B3LYP/6–311+G(d,p)//B3LYP/6–31G(d) level and on the whole enzyme using a hybrid quantum mechanics/molecular mechanics scheme (Prabhakar et al., 2008). In this study, apart from Sec, the active site contained Gly 50, Leu 51, Tyr 48, Gln 83, and Trp 157, the latter modeled as formamide and indole, respectively, as well as one to two water molecules (numbers refer to the human GPX3 sequence). For the oxidation (see Figure 3.1), a two-step as well as a direct mechanism has been proposed, the former implying the transfer of the selenol proton to Gln 83 followed by the reduction of the peroxide with formation of F.

A mechanistic investigation elucidating the different role of Se and S in GPx4 was later carried out on a cluster of six conserved amino acids (Cys/Sec 46, Gly 47, Gln 81, Trp 136, Asn 137, and Phe 138) with the addition of a seventh amino acid, Lys 48, in GPx4, which, being a nonconserved residue, was replaced by a Gly. This system was entirely modeled at the quantum chemistry level without any constraint (Orian et al., 2015). This revealed an identical mechanism for Sec and Cys, but the different energetics (barriers and stability of intermediates) is consistent with the notion that the reaction path is faster when Sec is the redox center.

These conclusions have been confirmed in a new theoretical study combining classic MD simulations with DFT mechanistic studies (Bortoli et al., 2017). The same cluster extracted from an MD simulation was used, maintaining the backbone frozen for the quantum chemical calculations. This was possible having an accurate parametrization of a flexible force field for Sec, permitting the in-depth analysis of the role of a single atom (Se rather than S). The essential features of GPx mechanism, as presented by Orian et al. (2015), have been confidently assessed.

From a chemical perspective, we can state that the GPx mechanism entirely relies on nucleophilic substitutions in which a chalcogenide attacks a dichalcogenide substrate, that is, selenol on hydrogen peroxide (step I), GSH on selenenic acid (step II), and GSH on selenenyl sulfide (step III) (Figure 3.1).

A fundamental issue in this class of reactions is to identify the nucleophile. It is well known that nucleophilicity is a periodic table property and increases along a group. Based on this, considering the selenol/thiol as the nucleophile and the peroxide as the substrate, a SecGPx is expected to be more efficient than a CysGPx. However, also other factors such as charge, solvent, and steric hindrance have an impact on nucleophilicity. Deprotonation of the thiol/selenol reverts the trend along the chalcogen group, but this holds true only in gas phase, where thiolate is a better nucleophile than selenolate. Conversely, in polar media like in the catalytic pocket of GPx, selenolate is more nucleophilic than thiolate, the negative charge being more diffuse. This enhanced nucleophilicity in part explains why the presence of Se rather than S in the first catalytic reaction is advantageous. These aspects are unraveled in a recent quantum mechanical study on reactions between a methylchalcogenolate and a dimethyldichalcogenide substrate (Bortoli et al., 2016).

In the reductive part of the catalytic cycle (reactions II and III of Scheme 3.1), GSH is the nucleophile and the chemical issue is whether the attack occurs at X_1 or O and at X_1 or S in the steps II and III of the scheme, respectively. An answer comes in part from the results obtained with model reactions (Cardey and Enescu, 2005), and we can draw the conclusions as follows: (i) in step II, the attack of S occurs at X_1, protonated HO^- being an excellent leaving group, and (ii) in step III, the attack of S occurs at S, because the attack at Se in model compounds would lead to the formation of a trinuclear intermediate also in a polar medium (addition–elimination mechanism) (Bortoli et al., 2016), although there is no evidence for the formation of a trichalcogenide in the enzyme, seemingly prevented by steric hindrance. This latter issue can be invoked also to explain why in the Cys enzyme ($X_1 = S$) the attack of GSH leads to the release of GSSG rather than to a scrambling reaction leading to the regeneration of the initial thiol form.

In the first step of the catalytic cycle, when ROOH is reduced, steric effects are approximately identical in the CysGPx and in the SecGPx, where the actual nucleophilicity stands out. Relying on model systems, accurate calculations of the energetics

$$E\text{-}X_1H + H\text{-}O\text{-}O\text{-}H \rightarrow E\text{-}X_1\text{-}OH + H_2O \quad \text{(I)}$$

$$GSH + E\text{-}X_1\text{-}OH \rightarrow E\text{-}X_1\text{-}SG + H_2O \quad \text{(II)}$$

$$GSH + E\text{-}X_1\text{-}SG \rightarrow E\text{-}X_1H + GS\text{-}SG \quad \text{(III)}$$

SCHEME 3.1 S_N2 reactions in the three-step mechanism of GPx. $X_1 = S$, Se.

of the reduction of H_2O_2 by a methylthiolate/selenolate do not predict significant differences between their kinetics, that is, the calculated Gibbs free activation energies are 20.7 and 19.2 kcal mol^{-1} in gas phase and 18.4 and 16.5 kcal mol^{-1} in water, respectively (Cardey and Enescu, 2005). When considering the deprotonated Cys and Sec (Cardey and Enescu, 2007), the calculated Gibbs free activation energies depend on the molecular conformation and on the presence of hydrogen bonds between the amino acid and the substrate. Again, no appreciable differences are computed for free Cys and Sec (e.g., for the conformation with the lowest activation energies of 16.7 and 14.9 kcal mol^{-1} computed in gas phase and 18.6 and 17.0 kcal mol^{-1} in water, respectively). These results suggest that specific molecular interactions amplify the intrinsic differences between S and Se while reducing a hydroperoxide in the active site of the enzyme.

The computational mechanistic studies on GPx4 (Orian et al., 2015, Bortoli et al., 2017) converge to the notion that the reduction of H_2O_2 takes place in two steps: first, deprotonation of Cys/Sec 46 occurs and a suitable proton acceptor is the indol nitrogen of Trp 136. This is a long-range proton transfer involving H_2O_2, facing the Cys/Sec 46 residue, and a water molecule, located between the peroxide and Trp 136; the Gibbs free activation energies are 27.1 and 21.6 kcal mol^{-1}, respectively (Bortoli et al., 2017). This difference is ascribed mainly to a larger strain in the transition state of the Cys cluster, due to stronger variation in the S-H/Se-H bond length (0.66 and 0.47 Å, respectively). Therefore, the Gibbs free activation energy is mostly related to the nature of the chalcogen. The charge-separated intermediates are in both cases energetically destabilized with respect to the initial adducts, that is, by 18.3 and 11.4 kcal mol^{-1}.

One could argue whether H_2O_2 is actually involved in the deprotonation process, because it is well known that in physiological conditions, Cys/Sec is already deprotonated (Flohé and Günzler, 1974; Schlecker et al., 2007). By replacing H_2O_2 with a water molecule in the GPx cluster, the energetics changes are negligible: the charge-separated products remain destabilized by 15.7 and 15.1 kcal mol^{-1}, respectively. Water molecules stabilize transition states and charge-separated species, because the hydrogen bond network reduces the strain. As during MD simulations water enters and leaves continuously the catalytic pocket (Bortoli et al., 2017), we may safely conclude that in this pocket the dissociation of both Cys and Sec is facilitated.

In the presence of hydroperoxide substrate, the oxidation of the chalcogen is extremely fast, because the anion forms of Cys/Sec are strong nucleophiles that attack and break the O–O bond. Moreover, the leaving basic group OH$^-$ is easily neutralized by the proton returning from Trp 136, and the reaction rate is further deeply accelerated. This is seemingly why no appreciable activation energy for this reaction has been computed for both CysGPx and SecGPx (Orian et al., 2015; Bortoli et al., 2017).

The nucleophilic attack of the thiolate/selenolate on H_2O_2 and the proton returning from the protonated Trp 136 are concerted events, and this makes the whole process barrierless, forming water, the ideal leaving group. Thus, it is not just the dissociation of the chalcogenol, but the dissociation *and* the parking of the dissociated proton in a strategic position of the reaction center that account for the observed extremely high reaction rate.

Yet, this evidence cannot provide justification about the advantage of Se vs S in the oxidative step. In this respect, specific molecular interactions in the catalytic pocket must also be taken into account. In initial adducts and charge-separated species, the number of hydrogen bonds in CysGPx or SecGPx ranges from four to six (excluding those involved in the bond network through which the proton transfer to Trp 136 occurs) and is systematically greater in the SecGPx intermediates. Notably, Gln 81 and Asn 137 and, somewhat less, Phe 138 are involved. That these individually weak interactions are responsible of the overall energy stabilization of the system has been unambiguously and quantitatively demonstrated (Bortoli et al., 2017), where removal of Gln 81, Trp 136, Asn 137, and Phe 138 raises the energy of the initial adduct and almost doubles the activation energy for the formation of the charge-separated species. Since the performance of the catalytic cycle in terms of turnover frequency is an integrated function of the whole cycle, the lowering of the energy intermediates and the transition states of the SecGPx vs CysGPx, albeit weak, leads to the marked advantageous performance of the former. Conversely, the reactivity difference is modest when comparing selenides to sulfides in molecular systems where the reactivity of the chalcogen is not tuned by intermolecular interactions.

Another remarkable difference between Se or S catalysis descends from the reactivity of the oxidized species (F in Figure 3.1). When GSH is limiting, the selenenic acid moiety generates indeed an eight-membered ring by forming a bond with the nitrogen of the peptide bond two amino acids downstream the Sec (Orian et al., 2015). Instead, when S is the redox moiety, the sulfenic acid is prone to further oxidation to sulfinic and sulfonic acids. This is expected to have a major physiological relevance because the cyclic complex can be recovered by GSH, whereas the over-oxidized S leads to an irreversible inactivation.

3.4 MOLECULAR MECHANICS OF THE REDUCTIVE PHASE OF THE GPX CATALYTIC CYCLE

Although no tridimensional structure of GPx in complex with GSH is currently available, important indications of the reductive phase of the catalytic cycle of GPx have been obtained through molecular modeling approaches (Aumann et al., 1997; Flohé et al., 2011; Bosello-Travain et al., 2013; Cozza et al., 2017). In particular, molecular docking technique has been successfully applied to determine the binding motif of the first GSH on GPx1, GPx4, and GPx7, exploiting three different semiflexible algorithms (MOE-Dock, GOLD, and Glide) (Flohé et al., 2011; Bosello-Travain et al., 2013). Crystal structures of GPx1, GPx4, and GPx7, retrieved from Protein Data Bank (PDB codes: 2F8A, 2OBI, and 2P31), have been processed in order to remove water molecules and unwanted ligands. Protein structures were further prepared by adding hydrogen atoms using standard geometries (Molecular Operating Environment, MOE) and by minimizing contacts between hydrogens (AMBER99 force field), as described in detail in the works of Bosello-Travain et al. (2013) and Flohé et al. (2011). GSH structure was built and minimized using MOE builder tool (MMFF94x force field); charges were calculated using quantum mechanical molecular electrostatic potential methodology (Bosello-Travain et al., 2013). This docking procedure exploits a docking box or a docking sphere (according to the docking algorithm) centered at residues 46, 47, and 57 for GPx4, GPx1, and GPx7, respectively (Bosello-Travain et al., 2013).

In the case of GPx1 and GPx4, GSH in the catalytic site is bound by electrostatic interactions, being the docking pose almost superimposable. Indeed, the carboxylic functions of GSH are electrostatically linked to conserved and positively charged residues, Arg 52 and Arg 180 in the case of human GPX1 and Lys 48 and Lys 135 in the case of human GPX4. In both cases, the S of GSH points toward the Se at the active site (Figure 3.2A and B) (Flohé et al., 2011). A different docking approach, used for analyzing the interaction between GSH and bovine GPx1, showed a different orientation involving the electrostatic interaction with Arg 57 and Arg 84 (Aumann et al., 1997). Apparently, this could imply different options for GSH positioning, permitted by the rotational freedom of the molecule in the active site. Moreover, we can assume that a flexible molecule, like GSH, requires time to adopt the most suitable conformation(s) and orientation(s) toward the active site.

The case of GPx7 is quite intriguing. Although the molecular docking pose suggests that the GSH binding motif is quite similar to those retrieved for GPx1 and GPx4 (Figure 3.2A and B) (Bosello-Travain et al., 2013), relevant differences also emerge. In particular, the electrostatic network between the carboxylic functions of GSH and Arg 180/Arg 52 (GPx1) or Lys 135/Lys 48 (GPx4) is replaced by hydrogen bonding/charge–dipole interactions with His 63 and Thr 162 (Figure 3.2C). Moreover, hydrophobic interactions provided by Pro 161 and Phe 59 contribute to the strength of GSH interaction (Figure 3.2C). However, GPx7 prefers PDI as the substrate, the oxidation of which is more than 2 orders of magnitude faster than that of GSH (Bosello-Travain et al., 2013).

An *in silico* study to clarify the molecular mechanism of interaction of the second GSH was carried out on GPx4 (Figure 3.2) (Cozza et al., 2017). The interaction of GPx4 with our *in silico* membrane system was analyzed by molecular docking and MD simulation. The outcome of this study reveals that, during the normal catalytic cycle, that is, in the presence of lipid hydroperoxides and GSH, GPx4 "jumps" onto the membrane, being bound to the double layer during the hydroperoxide reduction phase and released during the GSH-dependent regeneration of the catalyst. The positively charged area on the surface of GPx4 docks the enzyme to the membrane via electrostatic contacts with the polar head of phospholipids, as is also supported by surface plasmon resonance (SPR) evidence. By generating an *in silico* unit cell containing 200 phospholipid molecules containing 50% stearyl-arachidonyl phosphatidyl choline and 50% tetra linoleyl-cardiolipin (TLCL), it could be demonstrated that the hydroperoxide group (13-OOH TLCL and 9-OOH TLCL) becomes water exposed after 130 ns of MD simulation. Molecular docking analysis for the TLCL–GPx4 complex together with MD simulation of GPx4 in the membrane surface (400 ns) indicated that Arg 152, Lys 125, and Lys 135 are responsible for the interaction with the polar head of CL, addressing the hydroperoxide close to the catalytic center (Figure 3.2D). Moreover, the analysis of the CL–GPx4 complex, immersed into the membrane model, suggested that the first GSH is not able to access the active site. However, MD simulations (500 ns) demonstrate that the redox shift occurring on the enzyme in the first phase of the catalytic cycle (Cys/Sec 46-OH and phospholipid-OH) primes a repulsion that facilitates the access of GSH. The covalent binding of GSH to the redox

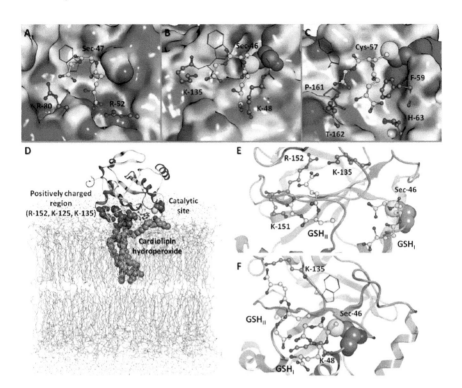

FIGURE 3.2 Molecular mechanics simulations of GSH–GPx complexes. (A)–(C) Molecular docking of the first GSH (green) bound to GPx1 (A), GPx4 (B), and GPx7 (C), respectively. The Connolly electrostatic charge distribution surfaces are also shown (red negative and blue positive charges). (D) Molecular docking and MD simulation of GPx4–13-OOH TLCL complex immersed in a membrane system. (E) Molecular docking of the second GSH (GSH_{II}) toward GPx4 and interaction model between the first GSH (GSH_I) and GSH_{II} after MD simulation (F). The most important residues involved are highlighted as balls and sticks.

center (Mauri et al., 2003; Flohé et al., 2011) further undocks the enzyme from the membrane surface and fully exposes the catalytic center to the solvent. The second GSH can now interact with the enzyme. Exploiting a Site Finder approach, a surface interaction zone was identified, encompassing the same residues implicated in the binding of phospholipid polar heads (Arg 152, Lys 125, and Lys 135; Figure 3.2E). This competition of GSH for the same docking site complies with the SPR evidence that GSH decreases the affinity of GPx4 for phospholipid bilayers. This study was extended by means of MD simulations to clarify how the two GSH molecules located in GPx4 at 7 Å distance could evolve toward a more productive interaction, eventually leading to the formation of GSSG. The computational approach, indeed, suggests that the second GSH "flows" toward the first one after 130 ns of MD simulation, reaching a 3 Å distance (Figure 3.2F) (Cozza et al., 2017). The second GSH now, through its carboxymethylamino group, shares with the first GSH an equidistant Lys 48 interaction, whereas the other carboxyl group is electrostatically linked to Lys 135

(Figure 3.2F). These data are compatible with the formation and release of GSSG and the regeneration of the reduced enzyme.

3.5 CONCLUSION

The mechanism of the reactions at the active site of GPx accounting for the extremely high kinetic efficiency of the catalysis had so far remained elusive. More recently, however, the whole catalytic cycle has been dissected into microscopic steps by means of quantum chemistry mechanistic investigations and the interactions between GSH and enzyme unraveled by molecular mechanics. Notably, the combination of these approaches has been particularly fruitful in studying GPx, quantum chemical being the only efficient approach to describe reactions when bond formation and breakage occurs, and MD giving the structural insight required to fully elucidate interaction pattern and compatibility.

ACKNOWLEDGMENTS

This work was supported by Human Frontier Science Program, Grant RGP0013/2014 to FU; by the University of Padova, Progetti di Ricerca di Ateneo, CPDA151400/15 to MM, and Progetto Giovani Ricercatori to GC; and by the CINECA ISCRA Grant REBEL to LO.

REFERENCES

Alphey, M. S., J. König, and A. H. Fairlamb. 2008. Structural and mechanistic insights into type II trypanosomatid tryparedoxin-dependent peroxidases. *Biochem J* 414:375–81.

Aumann, K. D., N. Bedorf, R. Brigelius-Flohé, D. Schomburg, and L. Flohé. 1997. Glutathione peroxidase revisited. Simulation of the catalytic cycle by computer-assisted molecular modelling. *Biomed Environ Sci* 10:136–55.

Bortoli, M., M. Torsello, F. M. Bickelhaupt, and L. Orian. 2017. Role of the chalcogen (S, Se, Te) in the oxidation mechanism of the glutathione peroxidase active site. *ChemPhysChem* 18:2990–8.

Bortoli, M., L. P. Wolters, L. Orian, and F. M. Bickelhaupt. 2016. Addition-elimination or nucleophilic substitution? Understanding the energy profiles for the reaction of chalcogenolates with dichalcogenides. *J Chem Theory Comput* 12:2752–61.

Bosello-Travain, V., M. Conrad, G. Cozza et al. 2013. Protein disulfide isomerase and glutathione are alternative substrates in the one Cys catalytic cycle of glutathione peroxidase 7. *Biochim Biophys Acta* 1830:3846–57.

Bosello-Travain, V., H. J. Forman, A. Roveri et al. 2015. Glutathione peroxidase 8 is transcriptionally regulated by HIFalpha and modulates growth factor signaling in Hela cells. *Free Radic Biol Med* 81:58–68.

Briviba, K., R. Kissner, W. H. Koppenol, and H. Sies. 1998. Kinetic study of the reaction of glutathione peroxidase with peroxynitrite. *Chem Res Toxicol* 11:1398–401.

Cardey, B., and M. Enescu. 2005. A computational study of thiolate and selenolate oxidation by hydrogen peroxide. *ChemPhysChem* 6:1175–80.

Cardey, B., and M. Enescu. 2007. Selenocysteine versus cysteine reactivity: A theoretical study of their oxidation by hydrogen peroxide. *J Phys Chem A* 111:673–8.

Chaudiere, J., and A. L. Tappel. 1983. Purification and characterization of selenium-glutathione peroxidase from hamster liver. *Arch Biochem Biophys* 226:448–57.

Chung, L. W., W. M. Sameera, R. Ramozzi, et al. 2015. The ONIOM method and its applications. *Chem Rev* 115:5678–796.

Cozza, G., M. Rossetto, V. Bosello-Travain, et al. 2017. Glutathione peroxidase 4-catalyzed reduction of lipid hydroperoxides in membranes: The polar head of membrane phospholipids binds the enzyme and addresses the fatty acid hydroperoxide group toward the redox center. *Free Radic Biol Med* 112:1–11.

Dalziel, K. 1957. Initial steady state velocities in the evaluation of enzyme–coenzyme–substrate reaction mechanisms. *Acta Chemica Scandinavica* 11:1706–23.

Flohé, L., and W. A. Günzler. 1974. Glutathione peroxidase. In *Glutathione*, edited by L. Flohé, H. C. Benöhr, H. Sies, H. D. Waller, and A. Wendel, 132–45. Stuttgart, Germany: Georg Thieme Verlag.

Flohé, L., W. A. Günzler, and H. H. Schock. 1973. Glutathione peroxidase: A selenoenzyme. *FEBS Lett* 32:132–4.

Flohé, L., G. Loschen, W. A. Günzler, and E. Eichele. 1972. Glutathione peroxidase V. The kinetic mechanism. *Hoppe Seylers Z Physiol Chem* 353:987–99.

Flohé, L., S. Toppo, G. Cozza, and F. Ursini. 2011. A comparison of thiol peroxidase mechanisms. *Antioxid Redox Signal* 15:763–80.

Flohé, L., and F. Ursini. 2008. Peroxidase: A term of many meanings. *Antioxid Redox Signal* 10:1485–90.

Fourquet, S., M. E. Huang, B. D'Autreaux, and M. B. Toledano. 2008. The dual functions of thiol-based peroxidases in H_2O_2 scavenging and signaling. *Antioxid Redox Signal* 10:1565–76.

Günzler, W. A., H. Vergin, I. Müller, and L. Flohé. 1972. Glutathione peroxidase VI: The reaction of glutahione peroxidase with various hydroperoxides. *Hoppe Seylers Z Physiol Chem* 353:1001–4.

Hofmann, B., Hecht, H.-J., and Flohé, L. 2002. Peroxiredoxins. *Biol Chem* 383:347–364.

König, J., and A. H. Fairlamb. 2007. A comparative study of type I and type II tryparedoxin peroxidases in *Leishmania major*. *FEBS J* 274:5643–58.

Maiorino, M., K. D. Aumann, R. Brigelius-Flohé et al. 1995. Probing the presumed catalytic triad of selenium-containing peroxidases by mutational analysis of phospholipid hydroperoxide glutathione peroxidase (PHGPx). *Biol Chem Hoppe Seyler* 376:651–60.

Maiorino, M., V. Bosello-Travain, G. Cozza et al. 2015. Understanding mammalian glutathione peroxidase 7 in the light of its homologs. *Free Radic Biol Med* 83:352–60.

Maiorino, M., F. Ursini, V. Bosello et al. 2007. The thioredoxin specificity of drosophila GPx: A paradigm for a peroxiredoxin-like mechanism of many glutathione peroxidases. *J Mol Biol* 365:1033–46.

Martin, J. L. 1995. Thioredoxin—A fold for all reasons. *Structure* 3:245–50.

Mauri, P., L. Benazzi, L. Flohé et al. 2003. Versatility of selenium catalysis in PHGPx unraveled by LC/ESI-MS/MS. *Biol Chem* 384:575–88.

Mills, G. C. 1957. Hemoglobin catabolism. I. Glutathione peroxidase, an erythrocyte enzyme which protects hemoglobin from oxidative breakdown. *J Biol Chem* 229:189–97.

Orian, L., P. Mauri, A. Roveri, et al. 2015. Selenocysteine oxidation in glutathione peroxidase catalysis: An MS-supported quantum mechanics study. *Free Radic Biol Med* 87:1–14.

Parr, R. G., and Y. Weitao. 1994. *Density-Funcional Theory of Atoms and Molecules, International Series of Monographs on Chemistry (Book 16)*. New York: Oxford University Press.

Prabhakar, R., K. Morokuma, and D. G. Musaev. 2008. Computational insights into the structural properties and catalytic functions of selenoprotein glutathione peroxidase (GPx). In Computational *Modeling* for *Homogeneous* and *Enzymatic Catalysis*: A *Knowledge*-base for *Designing Efficient Catalysts*, edited by K. Morokuma and D. G. Musaev. Weinheim, New York, Chichester, Brisbane, Singapore, Toronto: Wiley-VCH.

Rotruck, J. T., A. L. Pope, H. E. Ganther, et al. 1973. Selenium: Biochemical role as a component of glutathione peroxidase. *Science* 179:588–90.

Schlecker, T., M. A. Comini, J. Melchers, T. Ruppert, and R. L. Krauth-Siegel. 2007. Catalytic
mechanism of the glutathione peroxidase-type tryparedoxin peroxidase of trypanosoma brucei. *Biochem J* 405:445–54.

Sztajer, H., B. Gamain, K. D. Aumann et al. 2001. The putative glutathione peroxidase gene of
plasmodium falciparum codes for a thioredoxin peroxidase. *J Biol Chem* 276:7397–403.

Takebe, G., J. Yarimizu, Y. Saito, et al. 2002. A comparative study on the hydroperoxide and
thiol specificity of the glutathione peroxidase family and selenoprotein P. *J Biol Chem*
277:41254–8.

Toppo, S., L. Flohé, F. Ursini, S. Vanin, and M. Maiorino. 2009. Catalytic mechanisms and
specificities of glutathione peroxidases: Variations of a basic scheme. *Biochim Biophys
Acta* 1790:1486–500.

Toppo, S., S. Vanin, V. Bosello, and S. C. Tosatto. 2008. Evolutionary and structural insights
into the multifaceted glutathione peroxidase (GPx) superfamily. *Antioxid Redox Signal*
10:1501–14.

Tosatto, S. C., V. Bosello, F. Fogolari et al. 2008. The catalytic site of glutathione peroxidases.
Antioxid Redox Signal 10:1515–26.

Ursini, F., M. Maiorino, and C. Gregolin. 1985. The selenoenzyme phospholipid hydroperoxide glutathione peroxidase. *Biochim Biophys Acta* 839:62–70.

4 GPx1-Dependent Regulatory Processes in Health and Disease

Diane E. Handy and Joseph Loscalzo
Harvard Medical School

CONTENTS

4.1   INTRODUCTION

Glutathione (GSH) peroxidase 1 (GPx1) is the first member of the GPx family discovered (Mills, 1957) and is one of the selenium-dependent enzymes of this family of antioxidant enzymes (Flohé et al., 1973). In mammals, GPx1 is ubiquitously expressed and, in many cell types, is highly abundant [reviewed in the work of Lubos et al. (2011b)]; it plays an essential role in regulating the balance between oxidant and reductant stresses via the enzymatic reduction of hydrogen peroxide, soluble lipid hydroperoxides, and peroxynitrite (Flohé, 2010), and can also influence the cellular redox state by its obligate oxidation of GSH in the process. Moreover, by reducing hydroperoxides, GPx affects the generation of other reactive oxygen species (ROS) that are derived from hydroperoxides. As has been shown in many studies, GPx1-dependent regulation of ROS modulates the cellular activity by regulating redox-dependent protein functions that control receptor-mediated cell signaling, nuclear factor activation, and apoptosis (Figure 4.1). This chapter highlights the fundamental cellular processes that are regulated by GPx1 in mammalian cells and discusses how the actions of GPx1 on the cellular redox state regulate disease processes.

DOI: 10.1201/9781351261760-6

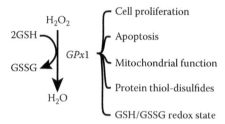

FIGURE 4.1 The effects of GPx1 and hydrogen peroxide on normal cellular functions. Through its enzymatic actions in reducing hydrogen peroxide levels in cells, GPx1 regulates many important cellular functions as illustrated here.

4.2 REGULATION OF ROS-DEPENDENT CELLULAR PROCESSES BY GPX1

Hydrogen peroxide is an effective signaling molecule owing, in part, to its stability and to its ability to pass through membranes via either passive diffusion or passage through membrane pores, such as aquaporin (Bienert et al., 2007). Importantly, hydrogen peroxide can cause a multitude of redox-dependent posttranslational protein modifications to regulate a variety of cell functions, including proliferation and apoptosis. GPx1 is present in the cytoplasm and mitochondria, where it facilitates the enzymatic reduction of hydrogen peroxide to modulate intracellular as well as extracellular hydrogen peroxide fluxes. Paradoxically, the effects of GPx1 on basic cell functions, such as cellular proliferation, differ in different cell types. These differences may be due, in part, to the unique complement of antioxidants and oxidants expressed in various cells; nonetheless, disruption of the balance between reductive and oxidative potentials, in either direction, can have deleterious biological consequences.

The necessary balance between cellular oxidant-generating and antioxidant systems limits excess accumulation of harmful ROS while allowing a low-level flux of oxidants that are essential for normal ROS-dependent processes. Thus, although antioxidant enzymes, such as GPx1, protect against ROS-induced oxidation of proteins, lipids, and nucleic acids, ROS generation is an integral part of growth factor-mediated signal transduction. In particular, growth factor-mediated signal transduction is regulated, in part, by hydrogen peroxide via the reversible oxidation and subsequent inactivation of redox-active protein tyrosine phosphatases to promote the activity of protein kinases that are effectors of receptor-mediated signaling [first shown for GPx1 by Kretz-Remy et al. (1996) and reviewed by Burgoyne et al. (2013)].

We have studied the impact of GPx1 on these ROS-dependent pathways and have found that a modest twofold upregulation of GPx1 is sufficient to decrease cellular accumulation of oxidants resulting in a decrease of growth factor-dependent or hydrogen peroxide-mediated activation of the downstream effector kinase, Akt (Handy et al., 2009). Furthermore, the suppressive effect of GPx1 overexpression on signal transduction can be mimicked by overexpression of catalase, confirming that hydrogen peroxide is necessary for these signaling events. *In vivo, Gpx1$^{-/-}$* mice fed a high fat diet show enhanced insulin-mediated Akt phosphorylation in skeletal muscle that involves inactivation of the phosphatase and tensin homolog deleted on chromosome 10 (PTEN) by oxidation of a redox-active cysteine (Loh et al., 2009). In our cell culture studies, we also found that knockdown of GPx1 enhanced Akt phosphorylation in response to growth factor-mediated signaling; however, GPx1 expression did not alter PTEN oxidation. Rather, we linked GPx1-mediated regulation of signal transduction with alterations of mitochondrial oxidants and global protein thiol–disulfide bond formation (Handy et al., 2009). Thus, the resulting reductant stress caused by excess GPx1 decreased ROS to decrease disulfide bond formation, attenuating mitochondrial function and suppressing cell proliferation.

Excess oxidants may promote cell growth, in part, by augmenting hydrogen peroxide-mediated activation of survival pathways, such as Akt and ERK1/2; however, in some cell types, GPx1 deficiency attenuates cell growth and promotes cell death and apoptosis in response to noxious stimuli. Thus, isolated neurons, aortic smooth muscle cells, astrocytes, and fibroblasts from *Gpx1$^{-/-}$* mice were found to be more susceptible to apoptotic stimuli, such as hydrogen peroxide, than the corresponding wild-type cells (De Haan et al., 2004; Van Remmen et al., 2004; Taylor et al., 2005; Liddell et al., 2006a). In some cells, enhanced sensitivity to apoptosis may stem, in part, from an attenuation of Akt-pro-survival signaling (De Haan et al., 1998; Taylor et al., 2005), which enhances the effects of excess ROS on apoptotic pathways. In the absence of GPx1, some cells may also be more sensitive to ROS-mediated cell death and apoptosis because the remaining cellular antioxidants are not sufficient to prevent the adverse oxidative modification of macromolecules or the activation of redox-sensitive apoptotic signaling.

Excess ROS can stimulate pro-apoptotic signaling by many mechanisms, including redox-dependent mechanisms that stimulate the stress-activated MAP kinase JNK (Son et al., 2013). Thus, in unstressed cells, ASK1, one of the upstream kinase activators of JNK, is maintained in an inactive state by the binding of thioredoxin (Trx), a redox-sensitive protein. Hydrogen peroxide-mediated oxidation of Trx dissociates it from ASK1, resulting in the activation of ASK1 and causing downstream phosphorylation and activation of JNK. JNK activation, in turn, can further increase the production of cellular ROS and initiate the caspase cascade to promote cell death. Similarly, excess ROS or caspase activation promotes mitochondrial-mediated apoptosis that involves the translocation of apoptogens, such as apoptosis-inducing factor and cytochrome *c*, from the mitochondria (Zemlyak et al., 2009). Importantly, GPx1 and other antioxidants can attenuate these ROS-mediated events, inhibiting JNK activation, the activation of the caspase cascade, and the mitochondrial release of apoptogens [these pathways are reviewed in the work of Lubos et al. (2011b)].

Depending on the cell type, increased expression of GPx1 can be beneficial in the context of excess oxidants, allowing for cell survival. Upregulation of GPx1 is the basis for the protective effect of glial cell line-derived neurotrophic factor against 6-hydroxydopamine and hydrogen peroxide-induced toxicity in PC12 cells, a pheochromocytoma cell line (Gharib et al., 2013). Similarly, in a variety of cell types grown in culture (Lu et al., 2012; Colle et al., 2013; Colle et al., 2016) as well as in *in vivo* systems discussed in more detail next, excess GPx1 prevents oxidant-related cell death in response to toxins, ischemia–reperfusion, and inflammation-induced injury. By contrast, in the context of hepatic acetaminophen toxicity, mice overexpressing GPx1 are more sensitive to acetaminophen than wild-type mice, whereas *Gpx1*−/− mice are more resistant to liver damage and death (Lei and Cheng, 2005). Enhanced sensitivity to acetaminophen may be due, in part, to the depletion of GSH in the livers of overexpressing GPx1 mice exposed to this hepatic toxin. Interestingly, under many oxidant stress conditions, the upregulation of GPx1 gene expression is observed at the same time as the redox-dependent activation of the Nrf2 transcription factor, which increases the expression of genes encoding the enzymes involved in GSH biosynthesis and recycling of oxidized GSH. The resulting increase in GSH substrate along with the GPx1 enzyme may allow for a more efficient detoxification of oxidants.

4.3 REGULATION OF HEALTH AND DISEASE BY GPX1

As mentioned previously, upregulation of GPx1 is protective against oxidant-generating toxins, but upregulation of GPx1 may also contribute to survival of cancer cells, and excess GPx1 can promote reductant stress by eliminating necessary hydrogen peroxide to enhance the development of cardiovascular diseases. GPx1 can also affect insulin sensitivity and diabetes mellitus (these effects are specifically addressed in Chapter 5). GPx1 deficiency has been shown to cause pro-atherogenic activation of vascular endothelial cells to augment the development of atherosclerosis. GPx1 deficiency also promotes ischemia–reperfusion injury, decreases endothelial regeneration and angiogenesis, and promotes endothelial dysfunction. In this section, we examine more specifically how the basic effects of GPx1 on the cellular redox state may contribute to normal and disease states (Figure 4.2).

4.3.1 ENDOTHELIAL DYSFUNCTION AND VASCULAR TONE

Endothelial cells lining blood vessels play an important role in maintaining vascular integrity and vascular tone. The production of nitric oxide (NO) by endothelial nitric oxide synthase (eNOS) in endothelial cells causes vascular relaxation of the underlying vascular smooth muscle cells by stimulating the generation of cyclic guanosine monophosphate (cGMP) through the enzymatic activation of soluble guanylyl cyclase. In addition, NO is antithrombotic and anti-inflammatory, and its release from endothelial cells limits the proliferation of smooth muscle cells to prevent vascular remodeling. Endothelial dysfunction is characterized by a decrease in normal endothelium-dependent vascular relaxation and is a symptom of decreased availability of bioactive NO. We linked the deficiency of GPx1 with endothelial dysfunction via ROS-dependent mechanisms. Thus, mesenteric arterioles from *Gpx1* homozygous knockout mice (*Gpx1*−/−) as well as *Gpx1* heterozygous knockout mice (*Gpx1*+/−) have attenuated responses to vasoactive substances, such as beta-methacholine and bradykinin, that stimulate endothelium-dependent NO production (Forgione et al., 2002a,b).

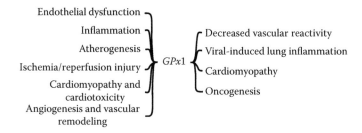

FIGURE 4.2   The effects of GPx1 and hydrogen peroxide on pathobiology. GPx1 is protective against oxidant stress to prevent dysfunction and the development of diseases, such as those listed on the left side. GPx1, however, can be deleterious and lead to the development of dysfunctional or disease states, such as those listed on the right. Note that excess GPx1 can reduce hydrogen-peroxide-mediated vasodilation responses; however, GPx1 can also prevent endothelial dysfunction by preserving nitric oxide-mediated vascular function. In addition, cardiomyopathy in Keshan disease is caused by a lack of GPx1, whereas cardiomyopathy caused by reductant stress is promoted by excess GPx1.

By contrast, a nitric oxide donor elicits normal endothelium-independent vascular relaxation in these vessels. These findings suggest that the vascular dysfunction caused by GPx1 deficiency originates from a defective endothelium. GPx1-deficient mice had elevated concentrations of plasma and aortic $iPF_{2\alpha}$-III isoprostanes and elevated aortic 3-nitrotyrosine, consistent with oxidant stress.

The importance of oxidant stress to the dysfunction of the endothelium is exemplified by the ability of antioxidants to restore normal vascular responsiveness in $Gpx1^{-/-}$ mice (Forgione et al., 2002b). Furthermore, in hyperhomocysteinemic mice, we found that an acquired deficiency of vascular GPx1 caused by excess homocysteine similarly led to endothelial dysfunction and elevated oxidant stress (Eberhardt et al., 2000). Consistent with a role for GPx1 deficiency in endothelial dysfunction in hyperhomocysteinemia, transgenic overexpression of GPx1 attenuated the endothelial dysfunction caused by hyperhomocysteinemia (Weiss et al., 2001). Similarly, in endothelial cells in culture, exposure to homocysteine decreased cellular GPx1 activity as well as the release of NO. Overexpression of GPx1 was sufficient to restore normal NO release in homocysteine-treated endothelial cells, confirming the protective effect of GPx1 against endothelial dysfunction. In addition, angiotensin II (AII)-mediated vascular dysfunction in coronary arteries was increased in $Gpx1^{+/-}$ mice compared to wild-type mice, and overexpression of catalase or GPx1 decreased vascular dysfunction caused by AII administration (Chrissobolis et al., 2008).

In aging mice, GPx1 deficiency was found to affect aortic function by both endothelium-dependent and endothelium-independent mechanisms, implicating smooth muscle cell dysfunction along with endothelial dysfunction in the abnormal vasodilatory responses (Oelze et al., 2014). This study, however, only found consistent evidence for smooth muscle cell dysfunction in aging (12-month-old) mice. Mechanistically, this study suggested a role of eNOS uncoupling in the pathogenesis of vascular dysfunction caused by GPx1 deficiency. Oxidant stress has been shown to promote eNOS uncoupling, resulting in an enzyme that produces superoxide instead of NO. Thus, in $Gpx1^{-/-}$ mice, there was an age-dependent increase in eNOS phosphorylation at Thr495, as well as eNOS S-glutathionylation. These posttranslational eNOS modifications have been shown to be promoted by excess hydrogen peroxide and to be associated with eNOS uncoupling.

In addition to NO, vascular tone can also be regulated by other endothelium-derived factors, including hydrogen peroxide (Ellinsworth et al., 2016). Thus, at low micromolar concentrations, hydrogen peroxide mediates endothelium-dependent vasodilation in a variety of vascular beds. More studies are needed to dissect the precise contribution of these pathways to vascular tone, but these pathways can be stimulated by stretch or arachidonic acid and are sensitive to inactivation by excess GPx1 or catalase (Modrick et al., 2009; Ellinsworth et al., 2016).

4.3.2 ATHEROGENESIS AND INFLAMMATION

In human subjects with coronary heart disease, GPx1 activity was found to be protective against future cardiovascular events in prospective and case–control studies (Blankenberg et al., 2003; Schnabel et al., 2005; Espinola-Klein et al., 2007; Flores-Mateo et al., 2009). In addition, an inverse correlation was reported between erythrocyte GPx1 activity and the extent of atherosclerosis (Espinola-Klein et al., 2007), and the combination of lower levels of GPx1 activity and hyperhomocysteinemia promoted cardiovascular risk (Schnabel et al., 2005). Furthermore, in human coronary atherectomy samples, GPx1 transcript levels are decreased in atherosclerotic vessels (Ali et al., 2014). In the apolipoprotein-deficient mouse ($Apoe^{-/-}$) fed a Western diet, knockout of GPx1 augmented lesion development compared to $Apoe^{-/-}$ mice with intact GPx1 (Torzewski et al., 2007). Similarly, the double knockout $Apoe^{-/-}/Gpx1^{-/-}$ had increased lesion development when stressed with streptozotocin to induce diabetes compared to diabetic $Apoe^{-/-}$ mice (Lewis et al., 2007). In each of these atherogenic model systems, lack of GPx1 increased inflammation and oxidant stress. Furthermore, ebselen, a GPx1 mimic, was shown to reduce atherogenesis in diabetic $Apoe^{-/-}$ mice, suggesting an important role for oxidant stress in atherosclerotic lesion development (Chew et al., 2010) and indicating that GPx1 is atheroprotective.

To understand better the mechanisms by which GPx1 regulates endothelial cell function to promote atherogenesis, we tested whether knockdown of GPx1 augmented inflammatory activation in human endothelial cells. Our studies showed that loss of GPx1 alone increased pro-inflammatory changes in human endothelial cells, increasing the expression of adhesion molecules, such as vascular cell adhesion molecule 1 (VCAM-1) and intercellular adhesion molecule 1 (Lubos et al., 2010, 2011a). Loss of GPx1 altered the expression of other inflammatory mediators, including CD14, a component of the innate immune system that acts as a cofactor for recognition of endotoxin by toll-like receptor 4 (Lubos et al., 2011a). Consequently, lipopolysaccharide-induced upregulation of adhesion molecules was enhanced by GPx1 deficiency in these cells. Similarly, TNF-α-induced activation was also enhanced by GPx1 deficiency, increasing ROS to promote the activation of JNK and ERK1/2 MAP kinases, to prolong the activation of nuclear factor (NF)-κB, and to increase adhesion molecule expression (Lubos et al., 2010). Consistent with a role for ROS in fostering these inflammatory responses, overexpression of GPx1 or antioxidant treatments attenuated inflammatory activation (Lubos et al., 2010, 2011a). Additional studies showed a similar pattern of augmented TNF-α-induced activation in primary aortic endothelial cells isolated from GPx1$^{-/-}$ mice (Sharma et al., 2016). Furthermore, the augmented expression of VCAM caused by GPx1 deficiency in mice increased the binding of leukocytes to endothelial cells in culture and in *ex vivo* aortas, processes that could be eliminated by ebselen treatment. Together, these findings support the notion that excess ROS generated in GPx1 deficiency can contribute to a pro-inflammatory environment to promote atherogenesis. Macrophages from *Gpx1*$^{-/-}$ mice were also found to proliferate more readily in response to macrophage colony-stimulating growth factor or oxidized low-density lipoprotein than those from wild-type mice, suggesting another mechanism by which GPx1 deficiency may contribute to atherogenesis (Cheng et al., 2013). Our recent findings illustrate that a modest suppression of GPx1 activity by less than 50% in human endothelial cells is sufficient to augment cellular ROS, increase adhesion molecule expression, and enhance leukocyte binding in endothelial cells, highlighting the importance of GPx1 in maintaining cellular homeostasis (Barroso et al., 2014).

In some models of inflammation-induced tissue injury, GPx1 has been found to be protective. Thus, in studies of ROS-induced lung injury from cigarette smoke, infiltrates of neutrophils and macrophages were increased in lungs of *Gpx1*$^{-/-}$ mice compared to wild-type mice, whereas ebselen administration limited lung inflammation (Duong et al., 2010). Influenza A virus-induced lung inflammation was also enhanced in *Gpx1*$^{-/-}$ mice, suggesting that GPx1 decreases influenza-induced lung inflammation (Yatmaz et al., 2013). By contrast, GPx1 contributes to inflammatory pathways that mediate disease in other model systems by fostering the production of inflammatory mediators. Thus, in a model of allergen-induced asthma, GPx1 deficiency attenuated inflammation by altering T-cell proliferation and differentiation in a ROS-dependent manner (Won et al., 2010). Similarly, ROS-dependent suppression of T-cell signaling in *Gpx1*$^{-/-}$ mice decreased inflammation and reduced liver injury in the concanavalin A model of hepatitis (Lee et al., 2016).

4.3.3 REGULATION OF CARDIAC DYSFUNCTION BY GPx1

Keshan disease, a dilated cardiomyopathy endemic to regions of China with low selenium concentrations in the soil, has long been attributed to the decreased expression of antioxidant selenoproteins, such as GPx1 and thioredoxin reductase 1 (Lei et al., 2009). Additional inflammatory stress caused by virulent strains of the coxsackie virus may also contribute to the development of Keshan disease (Peng et al., 2000). Nonetheless, recent studies report that the myocardium of Keshan disease patients has elevated levels of 8-hydroxy-2-deoxyguanosine, a marker of oxidant stress, along with decreased expression of GPx1 and thioredoxin reductase 1 compared to control myocardium (Pei et al., 2013), providing additional evidence for oxidant stress as well as decreased myocardial selenoprotein expression in the pathogenesis of Keshan disease (Loscalzo, 2014).

Mouse models confirm that GPx1 is essential for cardiac protection against oxidant stress. Deficiency of GPx1 in *Gpx1*$^{+/-}$ as well as *Gpx1*$^{-/-}$ mice results in significant increases in ventricular dysfunction in response to ROS-dependent cardiac ischemia–reperfusion injury compared to wild-type mice (Forgione et al., 2002a; Lim et al., 2009). Interestingly, male *Gpx1*$^{-/-}$ mice show greater contractile and diastolic dysfunction compared to female *Gpx1*$^{-/-}$ mice, which showed no dysfunction following ischemia–reperfusion injury (Lim et al., 2009). We attributed the extra protection in female hearts to preservation of ascorbate redox status and increased nitrate-to-nitrite conversion, suggesting the presence of additional sex-specific mechanisms that enhance the production of nitric oxide to protect the heart in these mice. Interestingly, GPx1 appears to be the crucial and greatest determinant of GSH utilization in hypoxia and ischemia (Fu et al., 2001; Liddell et al., 2006).

GPx1 is thought to protect cardiac mitochondria against damage during the reoxygenation that follows ischemia to limit the production of mitochondrial ROS and reduce oxidant damage to mitochondrial DNA (Thu et al., 2010). Similarly, excess GPx1 protects against doxorubicin-mediated cardiac toxicity by decreasing mitochondrial production of oxidants to preserve cardiac function, whereas *Gpx1*$^{-/-}$ mice are more susceptible to doxorubicin-induced apoptosis and cardiac dysfunction (Wang et al., 2002; Xiong et al., 2006). *Gpx1*$^{-/-}$ mice are also more susceptible to AII-induced cardiac hypertrophy, resulting in increased left ventricular mass, increased myocyte cross-sectional area, and intraventricular septum thickening (Ardanaz et al., 2010). These structural changes correlated with increased ventricular dysfunction in AII-treated *Gpx1*$^{-/-}$ mice compared to AII-treated wild-type mice.

In contrast to the protective roles of GPx1 in the heart described previously, other studies implicate the antioxidant function of GPx1 in cardiomyopathies caused by reductant stress. In particular, mutations in the small heat shock protein αB-crystallin (CryAB), as well as mutations in desmin, the intermediate myofibrillar protein that CryAB chaperones, lead to protein misfolding, accumulation of cytoplasmic aggregates, and cardiomyopathy in human patients and in mouse models (Wang et al., 2001; Rajasekaran et al., 2007). In the mouse, the CryAB mutation was found to upregulate GPx-1 and other antioxidant enzymes by mechanisms involving Nrf2 activation to promote reductant stress and subsequent cardiac dysfunction. In the case of the CryAB mutant, cardiac dysfunction could be alleviated by introducing a deficiency of Nrf2 or G6PD (Rajasekaran et al., 2007; Kannan et al., 2013), an important source for NADPH required for reductive recycling of GSSG and other antioxidant systems. Similar findings suggest that an upregulation of heat shock protein 27 increases cardiac GPx1 and decreases ROS to cause hypertrophy and cardiac dysfunction (Zhang et al., 2010). In the context of heat shock protein 27-induced cardiomyopathy, the deleterious consequences on cardiac hypertrophy and function were attenuated by inhibition of GPx1.

4.3.4 ISCHEMIA AND VASCULAR INJURY

Consistent with the protective effect of GPx1 in cardiac ischemia–reperfusion injury, GPx1 is also protective against ischemia–reperfusion injury in the brain. Thus,

Gpx1⁻/⁻ mice are more susceptible to injury in the middle cerebral artery occlusion model of stroke than wild-type mice. Brain injury is augmented in *Gpx1⁻/⁻* mice, in part, by the upregulation of NF-κB, as injury can be attenuated with ebselen or inhibitors of NF-κB (Crack et al., 2006). Furthermore, GPx1-overexpressing mice are more protected against ischemia–reperfusion brain injury than wild-type controls (Weisbrot-Lefkowitz et al., 1998).

Gpx1⁻/⁻ mice also have decreased angiogenic potential as they fail to revascularize muscle tissue adequately in the hind limb ischemia model (Galasso et al., 2006). Our findings suggested that the loss of GPx1 alters the survival of endothelial progenitor cells (EPCs). Neither *in vivo* ischemia nor injections of vascular endothelial growth factor were capable of stimulating EPC expansion in *Gpx1⁻/⁻* mice. Furthermore, cultured GPx1-deficient murine EPCs were more sensitive to hydrogen peroxide-mediated apoptosis compared to wild-type EPCs. Other studies confirm that GPx1 deficiency reduces endothelial regeneration (Ali et al., 2014). Thus, in balloon-induced vascular injury, lack of GPx1 decreased regeneration of the vascular endothelium in *Apoe⁻/⁻/Gpx1⁻/⁻* mice compared to *Apoe⁻/⁻* mice. Interestingly, in this model, vascular injury resulted in increased expression of the enzymes that regulate GSH biosynthesis in addition to an increase in GSH metabolizing enzymes, consistent with an oxidant-mediated Nrf2 activation, although Nrf2 was not assessed in this study. Nonetheless, the apparent excess ROS in GPx1 deficiency caused an upregulation of antioxidant genes to result in reductive stress. Loss of oxidants, in turn, promoted oxidant stress. Thus, the pathogenic phenotype of vascular remodeling was caused, in part, by the *S*-glutathionylation and inactivation of SHP-2 phosphatases that resulted in the activation of Ros1, a protein tyrosine kinase, to promote smooth muscle cell proliferation (Ali et al., 2014).

4.3.5 Oncogenesis and Cancer Cell Survival

Alterations in cellular metabolism can lead to excess ROS production in cancer cells. Similar to its protective actions in normal cells, upregulation of GPx1 in response to excess ROS enhances cell survival in cancer cells (Jin et al., 2015; Marengo et al., 2016; Zhang et al., 2018). In addition, excess GPx1 has been shown to promote the invasion and migration of cancer cells. In a number of cancers, including laryngeal squamous cell carcinoma and oral squamous cell carcinoma, high expression of GPx1 correlates with poor prognosis (Lee et al., 2017; Zhang et al., 2018). Genetic polymorphisms of GPx1 may also contribute to increased risk of breast, lung, bladder, and other cancers in patients (Marengo et al., 2016). This effect may be due to the differential partitioning of polymorphic forms of the GPx1 protein between mitochondrial and cytoplasmic compartments to alter the cellular redox state. These polymorphic GPx1 proteins altered cellular levels of signaling molecules, such as NF-κB and phosphorylated Akt (Bera et al., 2014; Ekoue et al., 2017). Nonetheless, in many cancers, no association has been found between these polymorphisms and cancer risk, and in some cancers, excess GPx1 may act as a tumor suppressor, improving patient survival (Zhang et al., 2018).

4.4   SUMMARY

The balance between oxidant production and oxidant removal can be readily disrupted by alterations in either direction to create either oxidant or reductant stress (Handy and Loscalzo, 2017). As a ubiquitously expressed antioxidant enzyme, GPx1 plays an important role in regulating cellular levels of hydrogen peroxide, an important signaling molecule that regulates cellular redox function. Thus, our understanding of the role of GPx1 has grown to include not only its protective actions that eliminate harmful oxidants but also its role in mediating reductant stress. Thus, through its basic function of reducing cellular hydroperoxides, GPx1 regulates cell signaling, mitochondrial activity and metabolism, cellular proliferation, as well as cell death and apoptosis. Ongoing and future studies will, no doubt, provide additional insights into the mechanisms by which GPx1 regulates essential biological and disease processes.

REFERENCES

Ali, Z. A., V. de Jesus Perez, K. Yuan et al. 2014. Oxido-reductive regulation of vascular remodeling by receptor tyrosine kinase ROS1. *J Clin Invest* 124 (12):5159–74.

Ardanaz, N., X. P. Yang, M. E. Cifuentes et al. 2010. Lack of glutathione peroxidase 1 accelerates cardiac-specific hypertrophy and dysfunction in angiotensin II hypertension. *Hypertension* 55 (1):116–23.

Barroso, M., C. Florindo, H. Kalwa et al. 2014. Inhibition of cellular methyltransferases promotes endothelial cell activation by suppressing glutathione peroxidase 1 protein expression. *J Biol Chem* 289 (22):15350–62.

Bera, S., F. Weinberg, D. N. Ekoue et al. 2014. Natural allelic variations in glutathione peroxidase-1 affect its subcellular localization and function. *Cancer Res* 74 (18):5118–26.

Bienert, G. P., A. L. Moller, K. A. Kristiansen et al. 2007. Specific aquaporins facilitate the diffusion of hydrogen peroxide across membranes. *J Biol Chem* 282 (2):1183–92.

Blankenberg, S., H. J. Rupprecht, C. Bickel et al. 2003. Glutathione peroxidase 1 activity and cardiovascular events in patients with coronary artery disease. *N Engl J Med* 349 (17):1605–13.

Burgoyne, J. R., S. Oka, N. Ale-Agha, and P. Eaton. 2013. Hydrogen peroxide sensing and signaling by protein kinases in the cardiovascular system. *Antioxid Redox Signal* 18 (9):1042–52.

Cheng, F., M. Torzewski, A. Degreif et al. 2013. Impact of glutathione peroxidase-1 deficiency on macrophage foam cell formation and proliferation: Implications for atherogenesis. *PLoS One* 8 (8):e72063.

Chew, P., D. Y. Yuen, N. Stefanovic et al. 2010. Antiatherosclerotic and renoprotective effects of ebselen in the diabetic apolipoprotein E/GPx1-double knockout mouse. *Diabetes* 59 (12):3198–207.

Chrissobolis, S., S. P. Didion, D. A. Kinzenbaw et al. 2008. Glutathione peroxidase-1 plays a major role in protecting against angiotensin II-induced vascular dysfunction. *Hypertension* 51 (4):872–7.

Colle, D., D. B. Santos, J. M. Hartwig et al. 2016. Succinobucol, a lipid-lowering drug, Protects against 3-nitropropionic acid-induced mitochondrial dysfunction and oxidative stress in SH-SY5Y cells via upregulation of glutathione levels and glutamate cysteine ligase activity. *Mol Neurobiol* 53 (2):1280–95.

Colle, D., D. B. Santos, E. L. Moreira et al. 2013. Probucol increases striatal glutathione peroxidase activity and protects against 3-nitropropionic acid-induced pro-oxidative damage in rats. *PLoS One* 8 (6):e67658.

Crack, P. J., J. M. Taylor, U. Ali et al. 2006. Potential contribution of NF-kappaB in neuronal cell death in the glutathione peroxidase-1 knockout mouse in response to ischemia-reperfusion injury. *Stroke* 37 (6):1533–8.

de Haan, J. B., C. Bladier, P. Griffiths et al. 1998. Mice with a homozygous null mutation for the most abundant glutathione peroxidase, Gpx1, show increased susceptibility to the oxidative stress-inducing agents paraquat and hydrogen peroxide. *J Biol Chem* 273 (35):22528–36.

de Haan, J. B., C. Bladier, M. Lotfi-Miri et al. 2004. Fibroblasts derived from Gpx1 knockout mice display senescent-like features and are susceptible to H2O2-mediated cell death. *Free Radic Biol Med* 36 (1):53–64.

Duong, C., H. J. Seow, S. Bozinovski et al. 2010. Glutathione peroxidase-1 protects against cigarette smoke-induced lung inflammation in mice. *Am J Physiol Lung Cell Mol Physiol* 299 (3):L425–33.

Eberhardt, R. T., M. A. Forgione, A. Cap et al. 2000. Endothelial dysfunction in a murine model of mild hyperhomocyst(e)inemia. *J Clin Invest* 106 (4):483–91.

Ekoue, D. N., C. He, A. M. Diamond, and M. G. Bonini. 2017. Manganese superoxide dismutase and glutathione peroxidase-1 contribute to the rise and fall of mitochondrial reactive oxygen species which drive oncogenesis. *Biochim Biophys Acta* 1858 (8):628–32.

Ellinsworth, D. C., S. L. Sandow, N. Shukla et al. 2016. Endothelium-derived hyperpolarization and coronary vasodilation: Diverse and integrated roles of epoxyeicosatrienoic acids, hydrogen peroxide, and gap junctions. *Microcirculation* 23 (1):15–32.

Espinola-Klein, C., H. J. Rupprecht, C. Bickel et al. 2007. Glutathione peroxidase-1 activity, atherosclerotic burden, and cardiovascular prognosis. *Am J Cardiol* 99 (6):808–12.

Flohé, L. 2010. Changing paradigms in thiology from antioxidant defense toward redox regulation. *Methods Enzymol* 473:1–39.

Flohé, L., W. A. Günzler, and H. H. Schock. 1973. Glutathione peroxidase: A selenoenzyme. *FEBS Lett* 32 (1):132–4.

Flores-Mateo, G., P. Carrillo-Santisteve, R. Elosua et al. 2009. Antioxidant enzyme activity and coronary heart disease: Meta-analyses of observational studies. *Am J Epidemiol* 170 (2):135–47.

Forgione, M. A., A. Cap, R. Liao et al. 2002a. Heterozygous cellular glutathione peroxidase deficiency in the mouse: Abnormalities in vascular and cardiac function and structure. *Circulation* 106 (9):1154–8.

Forgione, M. A., N. Weiss, S. Heydrick et al. 2002b. Cellular glutathione peroxidase deficiency and endothelial dysfunction. *Am J Physiol Heart Circ Physiol* 282 (4):H1255–61.

Fu, Y., H. Sies, and X. G. Lei. 2001. Opposite roles of selenium-dependent glutathione peroxidase-1 in superoxide generator diquat- and peroxynitrite-induced apoptosis and signaling. *J Biol Chem* 276 (46):43004–9.

Galasso, G., S. Schiekofer, K. Sato et al. 2006. Impaired angiogenesis in glutathione peroxidase-1-deficient mice is associated with endothelial progenitor cell dysfunction. *Circ Res* 98 (2):254–61.

Gharib, E., M. Gardaneh, and S. Shojaei. 2013. Upregulation of glutathione peroxidase-1 expression and activity by glial cell line-derived neurotrophic factor promotes high-level protection of PC12 cells against 6-hydroxydopamine and hydrogen peroxide toxicities. *Rejuvenation Res* 16 (3):185–99.

Handy, D. E., and J. Loscalzo. 2017. Responses to reductive stress in the cardiovascular system. *Free Radic Biol Med* 109:114–24.

Handy, D. E., E. Lubos, Y. Yang et al. 2009. Glutathione peroxidase-1 regulates mitochondrial function to modulate redox-dependent cellular responses. *J Biol Chem* 284 (18):11913–21.

Jin, L., D. Li, G. N. Alesi et al. 2015. Glutamate dehydrogenase 1 signals through antioxidant glutathione peroxidase 1 to regulate redox homeostasis and tumor growth. *Cancer Cell* 27 (2):257–70.

Kannan, S., V. R. Muthusamy, K. J. Whitehead et al. 2013. Nrf2 deficiency prevents reductive stress-induced hypertrophic cardiomyopathy. *Cardiovasc Res* 100 (1):63–73.

Kretz-Remy, C., P. Mehlen, M. E. Mirault, and A. P. Arrigo. 1996. Inhibition of I kappa B-alpha phosphorylation and degradation and subsequent NF-kappa B activation by glutathione peroxidase overexpression. *J Cell Biol* 133 (5):1083–93.

Lee, J. R., J. L. Roh, S. M. Lee et al. 2017. Overexpression of glutathione peroxidase 1 predicts poor prognosis in oral squamous cell carcinoma. *J Cancer Res Clin Oncol* 143 (11):2257–65.

Lee, D. H., D. J. Son, M. H. Park et al. 2016. Glutathione peroxidase 1 deficiency attenuates concanavalin A-induced hepatic injury by modulation of T-cell activation. *Cell Death Dis* 7:e2208.

Lei, X. G., and W. H. Cheng. 2005. New roles for an old selenoenzyme: Evidence from glutathione peroxidase-1 null and overexpressing mice. *J Nutr* 135 (10):2295–8.

Lei, C., X. Niu, J. Wei et al. 2009. Interaction of glutathione peroxidase-1 and selenium in endemic dilated cardiomyopathy. *Clin Chim Acta* 399 (1–2):102–8.

Lewis, P., N. Stefanovic, J. Pete et al. 2007. Lack of the antioxidant enzyme glutathione peroxidase-1 accelerates atherosclerosis in diabetic apolipoprotein E-deficient mice. *Circulation* 115 (16):2178–87.

Liddell, J. R., R. Dringen, P. J. Crack, and S. R. Robinson. 2006a. Glutathione peroxidase 1 and a high cellular glutathione concentration are essential for effective organic hydroperoxide detoxification in astrocytes. *Glia* 54 (8):873–9.

Liddell, J. R., H. H. Hoepken, P. J. Crack et al. 2006. Glutathione peroxidase 1 and glutathione are required to protect mouse astrocytes from iron-mediated hydrogen peroxide toxicity. *J Neurosci Res* 84 (3):578–86.

Lim, C. C., N. S. Bryan, M. Jain et al. 2009. Glutathione peroxidase deficiency exacerbates ischemia-reperfusion injury in male but not female myocardium: Insights into antioxidant compensatory mechanisms. *Am J Physiol Heart Circ Physiol* 297 (6):H2144–53.

Loh, K., H. Deng, A. Fukushima et al. 2009. Reactive oxygen species enhance insulin sensitivity. *Cell Metab* 10 (4):260–72.

Loscalzo, J. 2014. Keshan disease, selenium deficiency, and the selenoproteome. *N Engl J Med* 370 (18):1756–60.

Lu, W., Z. Chen, H. Zhang et al. 2012. ZNF143 transcription factor mediates cell survival through upregulation of the GPX1 activity in the mitochondrial respiratory dysfunction. *Cell Death Dis* 3:e422.

Lubos, E., N. J. Kelly, S. R. Oldebeken et al. 2011a. Glutathione peroxidase-1 deficiency augments proinflammatory cytokine-induced redox signaling and human endothelial cell activation. *J Biol Chem* 286 (41):35407–17.

Lubos, E., J. Loscalzo, and D. E. Handy. 2011b. Glutathione peroxidase-1 in health and disease: From molecular mechanisms to therapeutic opportunities. *Antioxid Redox Signal* 15 (7):1957–97.

Lubos, E., C. E. Mahoney, J. A. Leopold et al. 2010. Glutathione peroxidase-1 modulates lipopolysaccharide-induced adhesion molecule expression in endothelial cells by altering CD14 expression. *FASEB J* 24 (7):2525–32.

Marengo, B., M. Nitti, A. L. Furfaro et al. 2016. Redox homeostasis and cellular antioxidant systems: Crucial players in cancer growth and therapy. *Oxid Med Cell Longev* 2016:6235641.

Mills, G. C. 1957. Hemoglobin catabolism. I. Glutathione peroxidase, an erythrocyte enzyme which protects hemoglobin from oxidative breakdown. *J Biol Chem* 229 (1):189–97.

Modrick, M. L., S. P. Didion, C. M. Lynch et al. 2009. Role of hydrogen peroxide and the impact of glutathione peroxidase-1 in regulation of cerebral vascular tone. *J Cereb Blood Flow Metab* 29 (6):1130–7.

Oelze, M., S. Kroller-Schon, S. Steven et al. 2014. Glutathione peroxidase-1 deficiency potentiates dysregulatory modifications of endothelial nitric oxide synthase and vascular dysfunction in aging. *Hypertension* 63 (2):390–6.

Pei, J., W. Fu, L. Yang et al. 2013. Oxidative stress is involved in the pathogenesis of Keshan disease (an endemic dilated cardiomyopathy) in China. *Oxid Med Cell Longev* 2013:474203.

Peng, T., Y. Li, Y. Yang et al. 2000. Characterization of enterovirus isolates from patients with heart muscle disease in a selenium-deficient area of China. *J Clin Microbiol* 38 (10):3538–43.

Rajasekaran, N. S., P. Connell, E. S. Christians et al. 2007. Human alpha B-crystallin mutation causes oxido-reductive stress and protein aggregation cardiomyopathy in mice. *Cell* 130 (3):427–39.

Schnabel, R., K. J. Lackner, H. J. Rupprecht et al. 2005. Glutathione peroxidase-1 and homocysteine for cardiovascular risk prediction: Results from the AtheroGene study. *J Am Coll Cardiol* 45 (10):1631–7.

Sharma, A., D. Yuen, O. Huet et al. 2016. Lack of glutathione peroxidase-1 facilitates a pro-inflammatory and activated vascular endothelium. *Vascul Pharmacol* 79:32–42.

Son, Y., S. Kim, H. T. Chung, and H. O. Pae. 2013. Reactive oxygen species in the activation of MAP kinases. *Methods Enzymol* 528:27–48.

Taylor, J. M., U. Ali, R. C. Iannello et al. 2005. Diminished Akt phosphorylation in neurons lacking glutathione peroxidase-1 (Gpx1) leads to increased susceptibility to oxidative stress-induced cell death. *J Neurochem* 92 (2):283–93.

Thu, V. T., H. K. Kim, S. H. Ha et al. 2010. Glutathione peroxidase 1 protects mitochondria against hypoxia/reoxygenation damage in mouse hearts. *Pflugers Arch* 460 (1):55–68.

Torzewski, M., V. Ochsenhirt, A. L. Kleschyov et al. 2007. Deficiency of glutathione peroxidase-1 accelerates the progression of atherosclerosis in apolipoprotein E-deficient mice. *Arterioscler Thromb Vasc Biol* 27 (4):850–7.

Van Remmen, H., W. Qi, M. Sabia et al. 2004. Multiple deficiencies in antioxidant enzymes in mice result in a compound increase in sensitivity to oxidative stress. *Free Radic Biol Med* 36 (12):1625–34.

Wang, S., S. Kotamraju, E. Konorev et al. 2002. Activation of nuclear factor-kappaB during doxorubicin-induced apoptosis in endothelial cells and myocytes is pro-apoptotic: The role of hydrogen peroxide. *Biochem J* 367 (Pt 3):729–40.

Wang, X., H. Osinska, R. Klevitsky et al. 2001. Expression of R120G-alphaB-crystallin causes aberrant desmin and alphaB-crystallin aggregation and cardiomyopathy in mice. *Circ Res* 89 (1):84–91.

Weisbrot-Lefkowitz, M., K. Reuhl, B. Perry et al. 1998. Overexpression of human glutathione peroxidase protects transgenic mice against focal cerebral ischemia/reperfusion damage. *Brain Res Mol Brain Res* 53 (1–2):333–8.

Weiss, N., Y. Y. Zhang, S. Heydrick et al. 2001. Overexpression of cellular glutathione peroxidase rescues homocyst(e)ine-induced endothelial dysfunction. *Proc Natl Acad Sci U S A* 98 (22):12503–8.

Won, H. Y., J. H. Sohn, H. J. Min et al. 2010. Glutathione peroxidase 1 deficiency attenuates allergen-induced airway inflammation by suppressing Th2 and Th17 cell development. *Antioxid Redox Signal* 13 (5):575–87.

Xiong, Y., X. Liu, C. P. Lee et al. 2006. Attenuation of doxorubicin-induced contractile and mitochondrial dysfunction in mouse heart by cellular glutathione peroxidase. *Free Radic Biol Med* 41 (1):46–55.

Yatmaz, S., H. J. Seow, R. C. Gualano et al. 2013. Glutathione peroxidase-1 reduces influenza A virus-induced lung inflammation. *Am J Respir Cell Mol Biol* 48 (1):17–26.

Zemlyak, I., S. M. Brooke, M. H. Singh, and R. M. Sapolsky. 2009. Effects of overexpression of antioxidants on the release of cytochrome c and apoptosis-inducing factor in the model of ischemia. *Neurosci Lett* 453 (3):182–5.

Zhang, X., X. Min, C. Li et al. 2010. Involvement of reductive stress in the cardiomyopathy in transgenic mice with cardiac-specific overexpression of heat shock protein 27. *Hypertension* 55 (6):1412–17.

Zhang, Q., H. Xu, Y. You et al. 2018. High Gpx1 expression predicts poor survival in laryngeal squamous cell carcinoma. *Auris Nasus Larynx* 45 (1):13–9.

5 Glutathione Peroxidase 1 as a Modulator of Insulin Production and Signaling
Implications for Its Dual Role in Diabetes

Holger Steinbrenner and Lars-Oliver Klotz
Friedrich Schiller University Jena

CONTENTS

5.1 INTRODUCTION

The family of glutathione peroxidase (GPx) enzymes, catalyzing the glutathione-dependent reduction of hydrogen peroxide (H_2O_2) and organic hydroperoxides to water and alcohols, respectively, comprises eight mammalian isoforms. Five of these are selenoproteins in humans (Brigelius-Flohé and Maiorino, 2013; Steinbrenner et al., 2016). Indeed, the GPx isoform 1 (GPx1) was the first mammalian selenoprotein to be identified in 1973 (Flohé et al., 1973; Rotruck et al., 1973). Human GPX1 occurs as a homotetramer, with a highly reactive selenocysteine moiety at the active site of its monomers of ~22 kDa molecular mass each (Lubos et al., 2011; Brigelius-Flohé and Maiorino, 2013). It has been hypothesized that this tetrameric structure might exclude complex hydroperoxides from the active site of the enzyme, thus confining the group of GPx1 substrates to H_2O_2 and soluble low-molecular-weight hydroperoxides (Brigelius-Flohé and Maiorino, 2013). Based on its substrate profile,

DOI: 10.1201/9781351261760-7

its highly abundant ubiquitous expression and its location in both the cytosol and the mitochondria, GPx1 has long been considered as an important intracellular antioxidant enzyme (Lubos et al., 2011; Brigelius-Flohé and Maiorino, 2013). However, this had been somewhat questioned due to its low ranking in the so-called hierarchy of selenoproteins, meaning that cellular GPx1 mRNA and protein levels diminish very rapidly and strongly in response to dietary selenium (Se) deficiency (Lubos et al., 2011; Brigelius-Flohé and Maiorino, 2013). Studies on GPx1 knockout (KO) mice revealed that GPx1 is the primary antioxidant selenoenzyme under conditions of severe acute oxidative stress, whereas it appears to be dispensable during embryonic development and under homeostatic (physiological) conditions (Ho et al., 1997; Cheng et al., 1998).

Oxidative stress has been implicated in the pathogenesis of a number of chronic diseases, including diabetes mellitus and other metabolic disorders. Chronic nutrient overload is thought to provoke an increase in mitochondrial production of H_2O_2 and its release into the cytosol, eventually resulting in disruption of insulin signaling and in insulin resistance (Fisher-Wellman and Neufer, 2012). On the other hand, it became increasingly recognized during the past few decades that H_2O_2—at low levels—has a crucial physiological function, serving as a second messenger molecule within the signal transduction cascades of several growth factors and hormones including insulin (Fisher-Wellman and Neufer, 2012; Steinbrenner, 2013). A dual role of reactive oxygen species (ROS) has also been discussed with respect to pancreatic insulin production and secretion: Insulin secretion appears to be stimulated by low (physiological) levels of H_2O_2 (Pi et al., 2007), whereas oxidative stress may promote damage and dysfunction of the insulin-producing pancreatic beta cells. The latter phenomena represent common features of both type 1 and type 2 diabetes (Steinbrenner, 2013; Schwartz et al., 2017). Antioxidant enzymes, including peroxidases such as GPx1, catalase, and peroxiredoxins, have been studied with regard to their role in the regulation of the insulin-controlled energy metabolism and in pathogenesis and prevention of diabetes mellitus, sometimes with surprising and paradoxical outcomes (Lei and Vatamaniuk, 2011; Lei et al., 2016). This chapter will deal in particular with GPx1 as a modulator of both insulin production and signaling, and it will discuss potential implications for its dual and somewhat paradoxical roles in diabetes. As GPx1 is a selenoenzyme whose activity depends in large part on Se availability, this topic is also related to the puzzling and still insufficiently understood role of Se in diabetes.

5.2   THE PUZZLING LINK BETWEEN SELENIUM AND DIABETES, AND A POTENTIAL ROLE OF GPX1 THEREIN

Given the incorporation of Se, in the form of selenocysteine, into antioxidant selenoenzymes such as GPx1 and the involvement of oxidative stress in the pathogenesis of diabetes and its comorbidities, antidiabetic effects of dietary Se supplementation would be expected. And indeed, antidiabetic as well as insulin-mimetic properties of Se compounds, such as selenate, have been reported from *in vitro* and animal experiments, even though high and sometimes nonphysiological Se doses have been used in these early studies (Ezaki, 1990; Steinbrenner et al., 2011). By contrast, a secondary analysis of data from the U.S. Nutritional Prevention of Cancer (NPC) study, which was published in 2007, initiated a yet ongoing discussion on the safety of dietary Se supplements and on a previously unexpected role of Se as a potential risk factor for diabetes. The authors reported that NPC participants with high baseline plasma Se levels already in the beginning of the trial had a significantly higher risk to develop type 2 diabetes when supplemented with 200 µg Se/day for 12 years than persons from the placebo group (Stranges et al., 2007). Numerous reports on the relationship between Se and type 2 diabetes risk have been published since then; however, the currently available epidemiological data from both cross-sectional and intervention studies are still insufficient and inconsistent, as discussed in more detail elsewhere (Rayman and Stranges, 2013; Ogawa-Wong et al., 2016; Steinbrenner et al., 2016). This situation is illustrated by two recent studies, yielding conflicting results: Although plasma Se levels were positively associated with incidence and prevalence of diabetes in a Spanish cohort (Galan-Chilet et al., 2017), higher dietary Se intake (up to 1.6 µg/kg per day) was associated with lower insulin resistance in a population from Newfoundland (Wang et al., 2017). Also, the slightly increased diabetes risk in the Se group of the large U.S. Selenium and Vitamin E Cancer Prevention Trial was statistically not significant (Lippman et al., 2009). By contrast, data from animal and *in vitro* studies largely suggest that high (supra-nutritional) Se intake and/or elevated expression/activity of selenoproteins (above physiological levels) may promote the occurrence of diabetes features such as insulin resistance, hyperinsulinemia, and hyperglycemia (Steinbrenner, 2013; Ogawa-Wong et al., 2016; Steinbrenner et al., 2016). It is important to note that Se-induced biological responses may not be mediated solely through selenoproteins and may be influenced not only by the dose but also by the chemical form of the applied Se compound (e.g., selenate, selenite, or selenomethionine) (Pinto et al., 2011; Lennicke et al., 2017).

Several selenoproteins have been linked to prevention, pathogenesis, and/or progression of diabetes and related comorbidities such as obesity and metabolic syndrome; among those, the roles of GPx1, selenoprotein P, and thioredoxin reductases are characterized best (Steinbrenner, 2013; Ogawa-Wong et al., 2016; Tinkov et al., 2018). A role of GPx1 as modulator of insulin production and signaling is well established from studies in transgenic animals; however, the significance of GPx1 for fine-tuning of human energy metabolism under physiological and pathophysiological (diabetic) conditions might be limited for several reasons:

1. Although GPx1 expression is suppressed in cells kept in Se-deficient culture medium and in animals fed a Se-deficient diet, it has been found to be saturated already at adequate dietary Se levels in two major insulin target tissues, liver and white adipose tissue (WAT), of both rodent and pig animal models. Only in skeletal muscle, the third major insulin target tissue, did dietary Se supply above adequate levels result in further enhanced GPx1 expression and activity (Barnes et al., 2009; Pinto et al., 2012). Even though dietary Se intake in many European countries is somewhat lower than required for saturation of the plasma GPx isoenzyme GPx3, overt Se deficiency is very rare in humans (Steinbrenner and Brigelius-Flohé, 2015).

Thus, it appears questionable whether Se supplementation will result in more than a slight increase in GPx1 activity, if any, in humans on a regular diet.

2. GPx1 is not the only and quantitatively not the most important enzyme for the removal of intracellular H_2O_2. Catalase and several peroxiredoxins also contribute. The latter are particularly well suited for their roles both in antioxidant defense and in the regulation of redox signaling, as they efficiently reduce H_2O_2, are ubiquitously and abundantly expressed, and are found in all cellular compartments (Chae et al., 1999; Cox et al., 2009; Rhee and Kil, 2017). It has been estimated that peroxiredoxin 3 eliminates ~90% of mitochondrial H_2O_2 (Cox et al., 2009). Transient inactivation of the fraction of cytosolic peroxiredoxin 1 that is localized near the plasma membrane has been shown to allow for an accumulation of H_2O_2 for signaling purposes upon growth factor stimulation (Woo et al., 2010); whether this also holds true for insulin signaling has yet to be explored.

3. GPx1 has been reported to be downregulated and/or inactivated under conditions associated with diabetes, such as oxidative stress and hyperglycemia: High glucose concentrations (20 mM) suppressed GPx1 mRNA and protein levels (Wu et al., 2017). H_2O_2 (at a high dose of 1 mM) and dicarbonyl compounds (e.g., methylglyoxal, which is capable of non-enzymatic glycation reactions) decreased the enzymatic activity of GPx1 (Miyamoto et al., 2003; Cho et al., 2010). Interestingly, this is counteracted by an adaptive response, in which the resulting elevated intracellular H_2O_2 levels may cause the upregulation of GPx1 biosynthesis (Miyamoto et al., 2003; Touat-Hamici et al., 2014).

Notwithstanding the above-discussed limitations, reduction of H_2O_2 provides an obvious mechanistic rationale for the capacity of GPx1 to modulate cellular redox homeostasis; this will be delineated with respect to insulin production and signaling in Sections 5.3 and 5.4.

5.3 MODULATION OF PANCREATIC INSULIN BIOSYNTHESIS AND SECRETION BY GPX1

Insulin is produced and secreted solely by beta cells in the Langerhans islets of the pancreas. Therefore, pancreatic beta cells are crucial for the regulation of systemic energy metabolism and the maintenance of blood glucose homeostasis. With the exception of peroxiredoxins (Bast et al., 2002), beta cells are poorly endowed with common antioxidant enzymes such as SOD2, catalase, and also GPx1 (Tiedge et al., 1997). Expression and specific activity of GPx1 in beta cells accounts for only about 5% of the respective levels found in the liver, and they are not elevated upon exposure to stressful conditions (Tiedge et al., 1997). The low specific activities of antioxidant enzymes are thought to render beta cells exceptionally prone to cytotoxic effects of oxidative stress, as they occur during the development of diabetes or as they are experimentally induced in animal models by diabetogenic compounds (Tiedge et al., 1997). Physiologically, the low activity of antioxidant enzymes in beta cells might set the threshold low for signaling actions of ROS derived from carbohydrate metabolism. Intracellular H_2O_2 levels in isolated islets and beta-cell lines were reported to be elevated in response to glucose treatment; this increase was required for glucose-induced insulin secretion (GSIS) (Pi et al., 2007). By contrast, higher-than-physiological concentrations of exogenously added H_2O_2 disrupted GSIS (Maechler et al., 1999).

Early analyses of GPx1 KO mouse models did not reveal any obvious signs of diabetes or an elevated risk to develop diabetes (Ho et al., 1997; Cheng et al., 1998; Esposito et al., 2000), which supported the idea that loss of GPx1 activity can be compensated by other H_2O_2-reducing enzymes under physiological (homeostatic) conditions. However, a more targeted search for pancreas-specific alterations due to global disruption of the *GPx1* gene uncovered a mild pancreatitis, a decrease in beta-cell mass, and a decreased capacity to reduce ROS generated upon exposure to high glucose or xanthine/xanthine oxidase. GSIS was slightly impaired but not suppressed, and fasting plasma insulin levels were 50% lower in the GPx1 KO mice compared to their wild-type littermates (Wang et al., 2011). However, these slight modifications of islet morphology and function did not result in a statistically significant increase in fasting blood glucose concentrations or in impaired glucose tolerance, and thus were not sufficient to induce overt diabetes (Wang et al., 2011). The authors of this study also examined Cu, Zn-SOD (SOD1) KO, and GPx1/SOD1 double-KO mice: Both models displayed a phenotype similar to GPx1 KO mice, but more severe, with completely suppressed GSIS and mild hyperglycemia (Wang et al., 2011). Ebselen, a synthetic seleno-organic GPx mimic (Steinbrenner et al., 2016), enhanced GSIS in islets isolated from the wild-type as well as from the three KO mice, with the most pronounced improvement in GPx1 KO animals (Wang et al., 2014). When GPx1 KO mice were exposed to metabolic stress by feeding them an obesogenic high-fat diet, they developed more severe symptoms of a defective pancreatic insulin production/secretion than their wild-type littermates. GSIS as well as plasma insulin levels in both the fed and the fasted states were strongly decreased, and plasma glucose levels in the fed state were slightly enhanced (Merry et al., 2014).

The study of transgenic mice overexpressing GPx1 yielded, in part, expected outcomes, but in part also surprising ones. Apparently paradoxical phenomena were observed: specific overexpression of GPx1 in pancreatic beta cells protected mice from beta-cell loss and ameliorated the hyperglycemia induced by treatment with the diabetogenic beta-cell toxin streptozotocin (Harmon et al., 2009). In *db/db* mice, a rodent model of type 2 diabetes, beta-cell-specific overexpression of GPx1 also ameliorated hyperglycemia, prevented loss in beta-cell mass during the development of diabetes, and increased the insulin content in islets (Harmon et al., 2009). Similar results were obtained by using a pharmacological approach: treatment with the GPx mimic ebselen protected the beta cells of Zucker diabetic fatty (ZDF) rats from oxidative stress-induced apoptotic cell death, resulting in elevated beta-cell mass, higher plasma insulin levels, and lower plasma glucose levels, compared to controls (Mahadevan et al., 2013). In contrast to these overall beneficial effects, mice with global overexpression of GPx1 were unexpectedly found to develop a type 2 diabetes-like phenotype at older age, characterized by obesity, insulin resistance, hyperglycemia, and hyperinsulinemia (McClung et al., 2004). Although the authors initially hypothesized that the globally increased GPx1 expression above physiological levels might interfere with insulin signaling through an over-scavenging of ROS (McClung et al., 2004), their follow-up study identified seemingly beneficial changes in the pancreatic beta cells as the primary reason for the paradoxical development of diabetes in the GPx1-overexpressing mice. When these mice were kept under dietary restriction, the occurrence of all diabetes features was prevented with the sole exception of hyperinsulinemia (Wang et al., 2008). This is reminiscent of observations from several animal studies, reporting hyperinsulinemia and disturbance of insulin-regulated energy metabolism in response to supra-nutritional Se intake (Steinbrenner, 2013). Moreover, selenite has been shown to stimulate insulin production and secretion from beta cells and islets cultured *in vitro*, probably through upregulation of GPx1 (Campbell et al., 2008).

Oxidative stress, elicited by prolonged hyperglycemia or H_2O_2 treatment, has been reported to induce sequential inactivation (by nuclear exclusion) and/or downregulation of a specific subset of four beta-cell-enriched transcription factors: MafA, MafB, homeobox protein Nkx-6.1, and pancreas/duodenum homeobox protein 1 (Pdx1). The observations were made in cultured beta-cell lines as well as in isolated islets from type 2 diabetes patients and diabetic *db/db* mice (Guo et al., 2013). The four transcription factors participate in the regulation of beta-cell proliferation and differentiation as well as in the transcriptional control of insulin production, and are thus required for the maintenance of both beta-cell mass and function. Protein levels and/or transcriptional activity of these factors appear to be elevated when the activity of GPx1 in beta cells is elevated. This likely explains the increases in beta-cell mass, islet insulin content, and plasma insulin levels observed in the above-referenced studies with transgenic mice overexpressing GPx1: Beta-cell-specific overexpression of GPx1 restored nuclear MafA and Nkx-6.1 levels, and rescued beta-cell function in *db/db* mice (Harmon et al., 2009; Guo et al., 2013). Compared to their wild-type littermates, mice with global overexpression of GPx1 showed elevated Pdx1 levels in the islets, which was associated with a twofold higher beta-cell mass and a 40% higher insulin content (Wang et al., 2008). Treatment of ZDF rats with the GPx mimic ebselen maintained the intranuclear localization of MafA and Pdx1 within beta cells (Mahadevan et al., 2013). Conversely, broilers fed a Se-deficient diet developed pancreatic atrophy and hypoinsulinemia associated with suppressed GPx1 activity and a decrease in Pdx1 mRNA levels in the pancreas (Xu et al., 2017), and GPx1 KO mice fed an obesogenic diet showed hypoinsulinemia as well as low pancreatic insulin content and Pdx1 expression (Merry et al., 2014).

5.4 MODULATION OF INSULIN SIGNALING IN THE MAJOR INSULIN TARGET TISSUES BY GPX1

It has been known for decades that H_2O_2 may mimic insulin-induced glucose uptake in adipocytes (Czech et al., 1974). Later on, insulin was found to stimulate intracellular H_2O_2 generation through activation of NADPH oxidase; H_2O_2, in turn, augments insulin signaling by transiently oxidizing and thus inactivating counter-regulatory phosphatases (Mahadev et al., 2001). As reviewed in detail elsewhere (Fisher-Wellman and Neufer, 2012), overall phosphatase activity was estimated to be approximately tenfold higher than kinase activity in resting cells, keeping the major components of the insulin signaling cascade in a dephosphory-lated state. In line with this notion, the insulin-induced increase in H_2O_2 concentrations was interpreted as a "dimming signal" for phosphatases, thereby facilitating ensuing phosphorylation events and the respective physiological responses (Fisher-Wellman and Neufer, 2012). This provides a mechanistic rationale for the concept that GPx1 (as well as other H_2O_2-reducing enzymes) may interfere with insulin signaling and with insulin-regulated energy metabolism (Steinbrenner, 2013). GPx1 is expressed in each of the three major insulin target tissues, with the highest GPx1 protein levels and specific activities measured in the liver, compared to moderate levels in WAT and relatively low levels in skeletal muscle (Barnes et al., 2009; Pinto et al., 2012). Under Se-replete conditions, ~50% of the selenium in mouse liver is bound in the form of GPx1 (Hill et al., 2012). GPx1 mRNA and protein levels as well as GPx1 activity are rapidly and strongly suppressed in response to a Se-deficient diet (Lubos et al., 2011; Brigelius-Flohé and Maiorino, 2013). Of note, GPx1 activity in the liver and WAT is saturated already at adequate dietary Se supply, whereas GPx1 activity in skeletal muscle can be further enhanced even at supra-nutritional Se intake (Barnes et al., 2009; Pinto et al., 2012). This pattern suggests that the observed alterations in energy metabolism and the occurrence of some diabetes-like features in animals fed diets supplying supra-nutritional Se doses are most likely not entirely due to high GPx1 activity (Steinbrenner, 2013). On the other hand, several studies with GPx1 KO mice have provided compelling evidence for a physiological role of GPx1 in curtailing insulin signaling in skeletal muscle and the liver.

Mice with global GPx1 deficiency did not develop insulin resistance in skeletal muscle and the liver when fed an obesogenic high-fat diet, despite an increase in intracellular ROS levels (Loh et al., 2009; Merry et al., 2014). In contrast to GPx1 wild-type mice, whose skeletal muscle became insulin resistant due to the high-fat diet, the muscle tissue of GPx1 KO mice remained insulin sensitive. Insulin-induced phosphorylation (activation) of Akt and the "Akt substrate of 160 kDa" (AS160), which regulates the translocation of the glucose transporter GLUT4 to the plasma membrane and hence myocyte glucose uptake, was more pronounced in muscle tissue as well as in cultivated differentiated muscle cells (myotubes) lacking GPx1, compared to wild-type controls. Oxidation (and inactivation) of PTEN, a phosphatidylinositol-3,4,5-trisphosphate 3-phosphatase and dual-specificity protein phosphatase, was enhanced in myocytes lacking GPx1. As PTEN counteracts insulin signaling, its inactivating oxidation would result in a stronger stimulation of Akt, which is downstream of PTEN, upon insulin treatment. The authors concluded that higher intracellular H_2O_2 concentrations in the skeletal muscle of GPx1 KO mice may, through enhancing oxidation (inactivation) of PTEN, result in better insulin sensitivity (Loh et al., 2009). A follow-up study reported that an obesogenic diet induced less liver steatosis and damage in GPx1 KO mice than in their wild-type littermates; these beneficial effects of GPx1 deficiency were associated with better hepatic insulin sensitivity (Merry et al., 2014). More recently, hepatocyte-specific GPx1-deficient (HGKO) mice were generated by the same group (Merry et al., 2016). HGKO mice showed improved insulin sensitivity in the liver and a better control of glucose and lipid metabolism than wild-type mice, under homeostatic as well as pathological conditions. Hepatic insulin signaling was enhanced, and the key enzymes of gluconeogenesis were downregulated in the HGKO mice. These effects were associated with increased oxidation of several phosphatases, including tyrosine-protein phosphatase non-receptor type 2, protein tyrosine phosphatase 1B, and small heterodimer partner 1, known to attenuate insulin signaling. Moreover, HGKO mice were protected from the development of nonalcoholic steatohepatitis (Merry et al., 2016).

Analogous studies in transgenic mouse models in order to analyze the role of GPx1 in insulin signaling in WAT have not been performed yet. However, an *in vitro* study in cultured mature adipocytes may point to a more important function of GPx1 in counteracting oxidative stress than in redox signaling in the WAT. In contrast to the overall beneficial effects of GPx1 deficiency on insulin sensitivity of the liver and skeletal muscle, pharmacological inhibition of GPx1 activity and small interfering RNA-mediated downregulation of GPx1 resulted in impaired insulin signaling in the adipocytes (Kobayashi et al., 2009). Moreover, GPx1 activity was decreased in the WAT of diabetic *db/db* mice (Kobayashi et al., 2009).

5.5 SINGLE-NUCLEOTIDE POLYMORPHISMS WITHIN THE HUMAN *GPX1* GENE AND DIABETES RISK

As the susceptibility to diabetes has a genetic component and diabetes has been linked to oxidative stress, potential associations between single-nucleotide polymorphisms (SNPs) within the genes of human antioxidant enzymes including GPx1 and the risk to develop diabetes and related comorbidities have been explored in a number of case–control studies (Crawford et al., 2012). Among the 46 SNPs identified within the human *GPX1* gene (Crawford et al., 2012), two have been studied with respect to diabetes:

1. The SNP rs3448 represents a C→T mutation located within the 3′ untranslated region (3′-UTR) of the human *GPX1* gene (Crawford et al., 2012). The less frequent T allele of rs3448 was recently found to be associated with kidney complications (e.g., incidences of microalbuminuria, renal events, end-stage renal disease) in patients with type 1 diabetes (Mohammedi et al., 2016).

2. The SNP rs1050450 has attracted the most scientific attention. It appears to affect both the GPx1 activity and the systemic Se status. It is a C→T mutation located within the coding sequence of the human *GPX1* gene, which results in an amino acid exchange (Pro→Leu) close to the C-terminus of the GPx1 protein (Crawford et al., 2012). Although the UniProt database (www.uniprot.org/uniprot/P07203) lists this Pro→Leu exchange at position 200 of the GPx1 protein, other resources and studies refer to positions 197 or 198 (Crawford et al., 2012). Compared to the more frequent C allele of rs1050450, the T allele has been reported to be associated with lower GPx1 activity in erythrocytes. However, it should be noted that only the homozygous T/T genotype showed a statistically significant decrease in GPx1 activity, whereas the slight difference between the C/C and C/T

genotypes did not reach statistical significance (Ravn-Haren et al., 2006). Individuals with the T/T genotype showed significantly lower basal plasma Se levels and increased urinary Se excretion in response to dietary Se supplementation, compared to persons with the C/C genotype (Combs et al., 2012). Four case–control studies reported that carriers of the T allele had a higher risk to develop diabetes or diabetes-related comorbidities. In a Turkish study, new-onset diabetes after renal transplantation was more frequent in patients with the T/T genotype, compared to patients with the C/C or C/T genotype (Yalin et al., 2017). In two further studies from the United Kingdom and Poland, the T allele was associated with increased occurrence of peripheral neuropathy in diabetes patients (Tang et al., 2012; Buraczynska et al., 2017). In Japanese men, but not in women, the T allele was associated with higher prevalence of the metabolic syndrome (Kuzuya et al., 2008).

5.6 CONCLUDING REMARKS

ROS, and particularly H_2O_2, have a dual role in physiology and pathophysiology, serving as second messengers in redox signaling at low physiological levels but causing damage to cellular biomolecules and cell death at excess levels (Kehrer and Klotz, 2015; Sies et al., 2017). The discrepancy between beneficial actions of H_2O_2 in physiological insulin signaling and its involvement in the dysregulation of insulin signaling and production during the development of diabetes and diabetes-related comorbidities makes a good example for this apparent paradox (Fisher-Wellman and Neufer, 2012). The dual role of H_2O_2 appears to be reflected by the dual role of hydroperoxide-reducing antioxidant enzymes including GPx1 in the fine-tuning of insulin signaling and production, and sometimes the paradoxical effects of their knockdown or overexpression in transgenic animal models (Lei and Vatamaniuk, 2011; Steinbrenner, 2013; Lei et al., 2016). GPx1 has been shown to support maintenance and function of the insulin-producing pancreatic beta cells, through the enhancement of protein levels and/or the activity of four beta-cell-enriched transcription factors, MafA, MafB, Nkx-6.1, and Pdx1. However, such seemingly beneficial effects of high GPx1 activity may result in hyperinsulinemia and the paradoxical occurrence of a diabetes-like phenotype, as observed in mice with global GPx1 overexpression. On the other hand, GPx1-deficient mice did not become diabetic, despite slightly lowered plasma insulin levels. GPx1 deficiency actually improved insulin sensitivity in skeletal muscle and the liver, probably due to an increased insulin-induced intracellular H_2O_2 level, which is required for transient oxidation (inactivation) of counter-regulatory phosphatases within the insulin signaling cascade (Figure 5.1). Taken together, GPx1 participates in the control of cellular redox homeostasis and may act as an antidiabetic, mainly through protection of pancreatic beta cells from oxidative damage. Nevertheless, the partially counterintuitive findings in GPx1-overexpressing and GPx1-deficient mice argue for the importance of balancing GPx1 activity to avoid undesirable pro-diabetic and insulin-antagonistic effects.

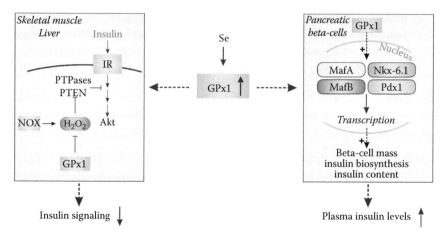

FIGURE 5.1 Modulation of insulin production in pancreatic beta cells as well as insulin signaling in the liver and skeletal muscle by GPx1. The biosynthesis of GPx1 requires selenium. GPx1 supports maintenance and function of the insulin-producing pancreatic beta cells, through the enhancement of protein levels and/or the activity of beta-cell-enriched transcription factors, MafA, MafB, Nkx-6.1, and Pdx1. This may result in increased beta-cell mass and insulin secretion. On the other hand, GPx1 reduces H_2O_2 that is required for transient oxidation (inactivation) of counter-regulatory phosphatases (PTPases, PTEN) within the insulin signaling cascade. This may decrease insulin sensitivity of the liver and skeletal muscle.

REFERENCES

Barnes, K. M., J. K. Evenson, A. M. Raines, and R. A. Sunde. 2009. Transcript analysis of the selenoproteome indicates that dietary selenium requirements of rats based on selenium-regulated selenoprotein mRNA levels are uniformly less than those based on glutathione peroxidase activity. *J Nutr* 139 (2):199–206.

Bast, A., G. Wolf, I. Oberbaumer, and R. Walther. 2002. Oxidative and nitrosative stress induces peroxiredoxins in pancreatic beta cells. *Diabetologia* 45 (6):867–76.

Brigelius-Flohé, R., and M. Maiori. 2013. Glutathione peroxidases. *Biochim Biophys Acta* 1830 (5):3289–303.

Buraczynska, M., K. Buraczynska, M. Dragan, and A. Ksiazek. 2017. Pro198Leu polymorphism in the glutathione peroxidase 1 gene contributes to diabetic peripheral neuropathy in type 2 diabetes patients. *Neuromolecular Med* 19 (1):147–53.

Campbell, S. C., A. Aldibbiat, C. E. Marriott et al. 2008. Selenium stimulates pancreatic beta-cell gene expression and enhances islet function. *FEBS Lett* 582 (15):2333–7.

Chae, H. Z., H. J. Kim, S. W. Kang, and S. G. Rhee. 1999. Characterization of three isoforms of mammalian peroxiredoxin that reduce peroxides in the presence of thioredoxin. *Diabetes Res Clin Pract* 45 (2–3):101–12.

Cheng, W. H., Y. S. Ho, B. A. Valentine et al. 1998. Cellular glutathione peroxidase is the mediator of body selenium to protect against paraquat lethality in transgenic mice. *J Nutr* 128 (7):1070–6.

Cho, C. S., S. Lee, G. T. Lee et al. 2010. Irreversible inactivation of glutathione peroxidase 1 and reversible inactivation of peroxiredoxin II by H2O2 in red blood cells. *Antioxid Redox Signal* 12 (11):1235–46.

Combs, G. F., Jr., M. I. Jackson, J. C. Watts et al. 2012. Differential responses to selenomethionine supplementation by sex and genotype in healthy adults. *Br J Nutr* 107 (10):1514–25.

Cox, A. G., C. C. Winterbourn, and M. B. Hampton. 2009. Mitochondrial peroxiredoxin involvement in antioxidant defence and redox signalling. *Biochem J* 425 (2):313–25.

Crawford, A., R. G. Fassett, D. P. Geraghty et al. 2012. Relationships between single nucleotide polymorphisms of antioxidant enzymes and disease. *Gene* 501 (2):89–103.

Czech, M. P., J. C. Lawrence, Jr., and W. S. Lynn. 1974. Hexose transport in isolated brown fat cells. A model system for investigating insulin action on membrane transport. *J Biol Chem* 249 (17):5421–7.

Esposito, L. A., J. E. Kokoszka, K. G. Waymire et al. 2000. Mitochondrial oxidative stress in mice lacking the glutathione peroxidase-1 gene. *Free Radic Biol Med* 28 (5):754–66.

Ezaki, O. 1990. The insulin-like effects of selenate in rat adipocytes. *J Biol Chem* 265 (2):1124–8.

Fisher-Wellman, K. H., and P. D. Neufer. 2012. Linking mitochondrial bioenergetics to insulin resistance via redox biology. *Trends Endocrinol Metab* 23 (3):142–53.

Flohé, L., W. A. Günzler, and H. H. Schock. 1973. Glutathione peroxidase: A selenoenzyme. *FEBS Lett* 32 (1):132–4.

Galan-Chilet, I., M. Grau-Perez, G. De Marco et al. 2017. A gene-environment interaction analysis of plasma selenium with prevalent and incident diabetes: The Hortega study. *Redox Biol* 12:798–805.

Guo, S., C. Dai, M. Guo et al. 2013. Inactivation of specific beta cell transcription factors in type 2 diabetes. *J Clin Invest* 123 (8):3305–16.

Harmon, J. S., M. Bogdani, S. D. Parazzoli et al. 2009. Beta-cell-specific overexpression of glutathione peroxidase preserves intranuclear MafA and reverses diabetes in db/db mice. *Endocrinology* 150 (11):4855–62.

Hill, K. E., S. Wu, A. K. Motley et al. 2012. Production of selenoprotein P (Sepp1) by hepatocytes is central to selenium homeostasis. *J Biol Chem* 287 (48):40414–24.

Ho, Y. S., J. L. Magnenat, R. T. Bronson et al. 1997. Mice deficient in cellular glutathione peroxidase develop normally and show no increased sensitivity to hyperoxia. *J Biol Chem* 272 (26):16644–51.

Kehrer, J. P., and L. O. Klotz. 2015. Free radicals and related reactive species as mediators of tissue injury and disease: Implications for health. *Crit Rev Toxicol* 45 (9):765–98.

Kobayashi, H., M. Matsuda, A. Fukuhara et al. 2009. Dysregulated glutathione metabolism links to impaired insulin action in adipocytes. *Am J Physiol Endocrinol Metab* 296 (6):E1326–34.

Kuzuya, M., F. Ando, A. Iguchi, and H. Shimokata. 2008. Glutathione peroxidase 1 Pro198Leu variant contributes to the metabolic syndrome in men in a large Japanese cohort. *Am J Clin Nutr* 87 (6):1939–44.

Lei, X. G., and M. Z. Vatamaniuk. 2011. Two tales of antioxidant enzymes on beta cells and diabetes. *Antioxid Redox Signal* 14 (3):489–503.

Lei, X. G., J. H. Zhu, W. H. Cheng et al. 2016. Paradoxical roles of antioxidant enzymes: Basic mechanisms and health implications. *Physiol Rev* 96 (1):307–64.

Lennicke, C., J. Rahn, A. P. Kipp et al. 2017. Individual effects of different selenocompounds on the hepatic proteome and energy metabolism of mice. *Biochim Biophys Acta* 1861 (1 Pt A):3323–34.

Lippman, S. M., E. A. Klein, P. J. Goodman et al. 2009. Effect of selenium and vitamin E on risk of prostate cancer and other cancers: The Selenium and Vitamin E Cancer Prevention Trial (SELECT). *JAMA* 301 (1):39–51.

Loh, K., H. Deng, A. Fukushima et al. 2009. Reactive oxygen species enhance insulin sensitivity. *Cell Metab* 10 (4):260–72.

Lubos, E., J. Loscalzo, and D. E. Handy. 2011. Glutathione peroxidase-1 in health and disease: From molecular mechanisms to therapeutic opportunities. *Antioxid Redox Signal* 15 (7):1957–97.

Maechler, P., L. Jornot, and C. B. Wollheim. 1999. Hydrogen peroxide alters mitochondrial activation and insulin secretion in pancreatic beta cells. *J Biol Chem* 274 (39):27905–13.

Mahadev, K., A. Zilbering, L. Zhu, and B. J. Goldstein. 2001. Insulin-stimulated hydrogen peroxide reversibly inhibits protein-tyrosine phosphatase 1b in vivo and enhances the early insulin action cascade. *J Biol Chem* 276 (24):21938–42.

Mahadevan, J., S. Parazzoli, E. Oseid et al. 2013. Ebselen treatment prevents islet apoptosis, maintains intranuclear Pdx-1 and MafA levels, and preserves beta-cell mass and function in ZDF rats. *Diabetes* 62 (10):3582–8.

McClung, J. P., C. A. Roneker, W. Mu et al. 2004. Development of insulin resistance and obesity in mice overexpressing cellular glutathione peroxidase. *Proc Natl Acad Sci U S A* 101 (24):8852–7.

Merry, T. L., M. Tran, G. T. Dodd et al. 2016. Hepatocyte glutathione peroxidase-1 deficiency improves hepatic glucose metabolism and decreases steatohepatitis in mice. *Diabetologia* 59 (12):2632–44.

Merry, T. L., M. Tran, M. Stathopoulos et al. 2014. High-fat-fed obese glutathione peroxidase 1-deficient mice exhibit defective insulin secretion but protection from hepatic steatosis and liver damage. *Antioxid Redox Signal* 20 (14):2114–29.

Miyamoto, Y., Y. H. Koh, Y. S. Park et al. 2003. Oxidative stress caused by inactivation of glutathione peroxidase and adaptive responses. *Biol Chem* 384 (4):567–74.

Mohammedi, K., T. A. Patente, N. Bellili-Munoz et al. 2016. Glutathione peroxidase-1 gene (GPX1) variants, oxidative stress and risk of kidney complications in people with type 1 diabetes. *Metabolism* 65 (2):12–19.

Ogawa-Wong, A. N., M. J. Berry, and L. A. Seale. 2016. Selenium and metabolic disorders: An emphasis on type 2 diabetes risk. *Nutrients* 8 (2):80.

Pi, J., Y. Bai, Q. Zhang et al. 2007. Reactive oxygen species as a signal in glucose-stimulated insulin secretion. *Diabetes* 56 (7):1783–91.

Pinto, A., D. T. Juniper, M. Sanil et al. 2012. Supranutritional selenium induces alterations in molecular targets related to energy metabolism in skeletal muscle and visceral adipose tissue of pigs. *J Inorg Biochem* 114:47–54.

Pinto, A., B. Speckmann, M. Heisler et al. 2011. Delaying of insulin signal transduction in skeletal muscle cells by selenium compounds. *J Inorg Biochem* 105 (6):812–20.

Ravn-Haren, G., A. Olsen, A. Tjonneland et al. 2006. Associations between GPX1 Pro198Leu polymorphism, erythrocyte GPX activity, alcohol consumption and breast cancer risk in a prospective cohort study. *Carcinogenesis* 27 (4):820–5.

Rayman, M. P., and S. Stranges. 2013. Epidemiology of selenium and type 2 diabetes: Can we make sense of it? *Free Radic Biol Med* 65:1557–64.

Rhee, S. G., and I. S. Kil. 2017. Multiple functions and regulation of mammalian peroxiredoxins. *Annu Rev Biochem* 86:749–75.

Rotruck, J. T., A. L. Pope, H. E. Ganther et al. 1973. Selenium: Biochemical role as a component of glutathione peroxidase. *Science* 179 (4073):588–90.

Schwartz, S. S., S. Epstein, B. E. Corkey et al. 2017. A unified pathophysiological construct of diabetes and its complications. *Trends Endocrinol Metab* 28 (9):645–55.

Sies, H., C. Berndt, and D. P. Jones. 2017. Oxidative Stress. *Annu Rev Biochem* 86:715–48.

Steinbrenner, H. 2013. Interference of selenium and selenoproteins with the insulin-regulated carbohydrate and lipid metabolism. *Free Radic Biol Med* 65:1538–47.

Steinbrenner, H., and R. Brigelius-Flohé. 2015. The essential trace element selenium: Requirements for selenium intake in health and disease. *Aktuelle Ernahrungsmedizin* 40 (6):368–78.

Steinbrenner, H., B. Speckmann, and L. O. Klotz. 2016. Selenoproteins: Antioxidant selenoenzymes and beyond. *Arch Biochem Biophys* 595:113–19.

Steinbrenner, H., B. Speckmann, A. Pinto, and H. Sies. 2011. High selenium intake and increased diabetes risk: Experimental evidence for interplay between selenium and carbohydrate metabolism. *J Clin Biochem Nutr* 48 (1):40–5.

Stranges, S., J. R. Marshall, R. Natarajan et al. 2007. Effects of long-term selenium supplementation on the incidence of type 2 diabetes: A randomized trial. *Ann Intern Med* 147 (4):217–23.

Tang, T. S., S. L. Prior, K. W. Li et al. 2012. Association between the rs1050450 glutathione peroxidase-1 (C > T) gene variant and peripheral neuropathy in two independent samples of subjects with diabetes mellitus. *Nutr Metab Cardiovasc Dis* 22 (5):417–25.

Tiedge, M., S. Lortz, J. Drinkgern, and S. Lenzen. 1997. Relation between antioxidant enzyme gene expression and antioxidative defense status of insulin-producing cells. *Diabetes* 46 (11):1733–42.

Tinkov, A. A., G. Bjorklund, A. V. Skalny et al. 2018. The role of the thioredoxin/thioredoxin reductase system in the metabolic syndrome: Towards a possible prognostic marker? *Cell Mol Life Sci* 75(9):1567–86

Touat-Hamici, Z., Y. Legrain, A. L. Bulteau, and L. Chavatte. 2014. Selective up-regulation of human selenoproteins in response to oxidative stress. *J Biol Chem* 289 (21):14750–61.

Wang, X., M. Z. Vatamaniuk, C. A. Roneker et al. 2011. Knockouts of SOD1 and GPX1 exert different impacts on murine islet function and pancreatic integrity. *Antioxid Redox Signal* 14 (3):391–401.

Wang, X. D., M. Z. Vatamaniuk, S. K. Wang et al. 2008. Molecular mechanisms for hyperinsulinaemia induced by overproduction of selenium-dependent glutathione peroxidase-1 in mice. *Diabetologia* 51 (8):1515–24.

Wang, X., J. W. Yun, and X. G. Lei. 2014. Glutathione peroxidase mimic ebselen improves glucose-stimulated insulin secretion in murine islets. *Antioxid Redox Signal* 20 (2):191–203.

Wang, Y., M. Lin, X. Gao et al. 2017. High dietary selenium intake is associated with less insulin resistance in the Newfoundland population. *PLoS One* 12 (4):e0174149.

Woo, H. A., S. H. Yim, D. H. Shin et al. 2010. Inactivation of peroxiredoxin I by phosphorylation allows localized $H(2)O(2)$ accumulation for cell signaling. *Cell* 140 (4):517–28.

Wu, Y., S. Lee, S. Bobadilla et al. 2017. High glucose-induced p53 phosphorylation contributes to impairment of endothelial antioxidant system. *Biochim Biophys Acta* 1863 (9):2355–62.

Xu, J., L. Wang, J. Tang et al. 2017. Pancreatic atrophy caused by dietary selenium deficiency induces hypoinsulinemic hyperglycemia via global down-regulation of selenoprotein encoding genes in broilers. *PLoS One* 12 (8):e0182079.

Yalin, G. Y., S. Akgul, S. Tanrikulu et al. 2017. Evaluation of glutathione peroxidase and KCNJ11 gene polymorphisms in patients with new onset diabetes mellitus after renal transplantation. *Exp Clin Endocrinol Diabetes* 125 (6):408–13.

6 GPx2
Role in Physiology and Carcinogenesis

Anna P. Kipp
Friedrich Schiller University Jena

CONTENTS

6.1 INTRODUCTION

Glutathione peroxidase 2 (GPx2) has first been described in 1993 as an enzyme of the gastrointestinal tract (Chu et al., 1993). It was therefore called gastrointestinal glutathione peroxidase (GI-GPx). However, it is now clear that GPx2 is expressed not only in the intestinal epithelium but also in almost all epithelia, including the lung, breast, gall bladder, and urinary bladder (Chu et al., 1993; Florian et al., 2001; Cho et al., 2002). Accordingly, a better name for GPx2 would be epithelium-specific GPx. At the transcript level, GPx2 expression has been further detected in keratinocytes of the skin (Schäfer et al., 2012) as well as in neuronal cells of the pituitary (Zhou et al., 2009). In two mouse models of depression, GPx2 expression is downregulated in the frontal cortex. In addition, GPx2 expression is further upregulated in inflamed epithelia, for example, during colitis (Florian et al., 2010), or in the chronically inflamed lung of smokers (Singh et al., 2006). During colorectal carcinogenesis, GPx2 is upregulated in all tumor stages in comparison with nontransformed tissue: adenomas (Mörk et al., 2000), carcinomas (Murawaki et al., 2008), and moderately differentiated tumors (Florian et al., 2001; Banning

DOI: 10.1201/9781351261760-8

et al., 2008a). GPx2 expression is enhanced in squamous cell carcinomas (SCCs) of the skin (Serewko et al., 2002; Walshe et al., 2007), adenocarcinomas of the lung (Wönckhaus et al., 2006), ductal mammary carcinomas (Naiki-Ito et al., 2007), hepatocellular carcinomas (Suzuki et al., 2013), and prostate cancer (Naiki et al., 2014). This tumor-specific expression pattern is unique for GPx2, as all other members of the GPx family are unaffected or even downregulated during carcinogenesis [overview in the work of Kipp (2017)].

As described for other members of the GPx family, GPx2 is supposed to reduce hydroperoxides by using glutathione as a reducing agent (Brigelius-Flohé and Maiorino, 2013). A preference for organic fatty acid hydroperoxides has been deduced from cell culture experiments (Wingler et al., 2000), but so far no kinetic data are available for GPx2, which has been neither purified nor produced as a recombinant protein. Thus, it can be only speculated about its exact function. GPx2 belongs to the selenoprotein subfamily, meaning that its translation depends on the availability of selenium. When selenium becomes limited, cells prioritize the translation of selenoproteins following a so-called hierarchy. GPx2 ranks relatively high in this hierarchy and is synthetized at the expense of other selenoproteins, for example, GPx1 (Wingler et al., 1999). This indicates that GPx2 is of importance for the survival of cells. The expression of GPx2 is regulated by several transcription factors which are involved in mediating proliferation and differentiation, such as the nuclear factor erythroid-derived 2 like 2 (Nrf2) (Banning et al., 2005), homeobox protein Nkx-3.1 (Ouyang et al., 2005), β-catenin/TCF (Kipp et al., 2007), DNp63 (Yan and Chen, 2006), and the STAT family (Hiller et al., 2015). Research progress has been made in recent years toward understanding how GPx2 affects proliferation and differentiation processes in both healthy epithelia but also during inflammation and carcinogenesis. For several types of cancer, evidence is increasing that GPx2 could be a prognostic marker for tumor recurrence and thus survival probability of patients. Whether this is linked to the selenium status of those patients is not clear yet. These novel aspects about the physiological and pathophysiological roles of GPx2 will be described and discussed in this chapter.

6.2 GPX2 IN HEALTHY ORGANISMS

6.2.1 Embryonic Development and Pluripotent Stem Cells

During early embryonic development of mice, GPx2 mRNA is mainly located in extra-embryonic tissues including placenta. At embryonic days 9.5–12.5, GPx2 mRNA is abundantly expressed in nervous tissues such as the telencephalon, mesencephalon, and dorsal neural tube. After embryonic day 13.5, GPx2 mRNA levels are high in the cerebral cortex, meta-nephric corpuscle, pancreatic ducts, surface epithelia of the skin, inner ear, and nasal conchae, gastrointestinal tract, liver, urinary bladder, airway passages of the lung, and whisker follicles (Baek et al., 2011). As described previously, this expression pattern is more or less retained in adult organisms. Even though GPx2 is abundantly expressed during embryonic development, it does not appear to be essential for this process as constitutive GPx2 knockout mice do neither have any obvious phenotype nor impaired survival rates (Esworthy et al., 2000).

In a genetic model of impaired nephrogenesis, several genes including GPx2 were identified to be differentially expressed on E 15.5, and thus might act as crucial players for proper nephron development (Herlan et al., 2015). During late embryogenesis (E 13.5–18.5), the mouse colon develops from a pseudostratified, undifferentiated endoderm to a single-layered epithelium, which is accompanied by the upregulation of GPx2 together with GPx3 mRNA (Park et al., 2005). This process is critical, because during this time frame the adult stem and progenitor cells develop defining the regenerative capacity of the adult epithelium (Gregorieff and Clevers, 2005). Also in the adult intestine, GPx2 is maintained at the crypt base where stem cells and proliferating cells are located (see Section 6.2.3). Even though most of the results on GPx2 expression during embryogenesis are only based on mRNA data, it provides some evidence for an involvement of GPx2, which however needs to be confirmed on the protein level and further studied mechanistically.

The generation of induced pluripotent human stem cells (iPSCs) from adult somatic cells is an important approach in stem cell biology. iPSCs share the features of self-renewal and pluripotency with embryonic stem cells (ESCs) but also sets of expressed genes and proteins to maintain those features (Takahashi et al., 2007; Yu et al., 2007; Park et al., 2008). iPSCs as well as ESCs depend on efficient mechanisms to protect their DNA from damage. One approach is the upregulation of antioxidant proteins, with GPx2 being the most prominent one (1000-fold increased mRNA in iPSCs compared to fibroblasts) (Dannenmann et al., 2015). Genomic integrity of iPSCs was substantially impaired by a knockdown of GPx2. Interestingly, GPx2 expression is further enhanced in iPSCs derived from aged donor mice (A-iPSC) in comparison with those of young donors (Skamagki et al., 2017). In consequence, an imbalance of the cellular redox status is established with excess reduction of hydro-peroxides resulting in reduced apoptosis induction and permit of survival of cells with genomic instability. The pluripotency-inducing transcription factor ZSCAN10 (Yu et al., 2009) was shown to normalize (meaning reduce) GPx2 levels of A-iPSCs indirectly via enhancing the RNA exosome complex. This complex removes aberrantly accumulated RNA transcripts but is also important to target specific mRNAs to degradation, for example, during embryonic development (Wan et al., 2012). By contrast, disrupting the exosome complex by short hairpin RNA (shRNA)-mediated knockdown against one of the main exosome subunits resulted in a dramatic increase in GPx2 gene expression (Skamagki et al., 2017). Thus, GPx2 can be considered as an important factor involved in stem cell maintenance.

6.2.2 THE LUNG

The airway epithelium serves as the first line of defense against a variety of insults, including respiratory toxicants or hyperoxia. One of the major transcription factors activated under such circumstances is Nrf2. Under basal conditions, Nrf2 is kept in the cytosol and constantly degraded via the proteasome. The pathway becomes activated by thiol modification of the scavenger protein Keap1, which continues to bind Nrf2 but impairs its degradation. Thus, newly synthesized Nrf2 can enter the nucleus and activate the transcription of a multitude of target genes [overview in the work of Hayes and Dinkova-Kostova (2014)]. Its target genes include detoxifying enzymes such as glutathione *S*-transferases, enzymes related to glutathione synthesis such as glutamate-cysteine ligase, and antioxidant enzymes such as peroxiredoxin 1 or sulfiredoxin 1. Also, GPx2 is a target gene of the Nrf2 pathway with a functional antioxidant response element within the human GPX2 promoter for binding of Nrf2 (Banning et al., 2005). *In vivo*, basal GPx2 expression is strongly downregulated in Nrf2 knockout mice both in the lung (Singh et al., 2006) and in the intestine (Hiller et al., 2015). By contrast, it is upregulated in response to Nrf2 activators such as sulforaphane both in the intestine (Krehl et al., 2012; Lippmann et al., 2014) and in the lung (Ano et al., 2017). In addition, all other GPx isoforms are not directly part of the adaptive Nrf2-regulated gene signature. Based on this, GPx2 can be considered as the stress-inducible GPx isoform accounting for an important part of the pulmonary antioxidant defense system.

Indeed, GPx2 is strongly induced in the lungs in response to hyperoxia together with further Nrf2-regulated genes (Cho et al., 2002, 2012), which is not observed in neonatal Nrf2 knockout mice at P1–P4 (Cho et al., 2012). Furthermore, the lung injury in response to hyperoxia indicated by the number of neutrophiles in bronchoalveolar lavage is significantly stronger in GPx2 knockout mice in comparison with the wild type. By contrast, loss of the ubiquitously and constitutively expressed GPx1 does not affect the survival rates of mice in response to hyperoxia (Ho et al., 1997). Also in a murine model of respiratory syncytial virus disease, Nrf2 is involved in the antiviral defense accompanied by an Nrf2-mediated upregulation of GPx2 (Cho et al., 2009).

The strong interrelation between Nrf2 and GPx2 in the lung epithelium was further studied using additional models of lung injury: GPx2 was strongly induced in response to cigarette smoking (Harvey et al., 2007), which was recently confirmed by the experimental usage of electric cigarettes as well (Solleti et al., 2017). Also, the inhalation exposure of novel nanoparticles resulted in inflammation of the respective murine lungs and Nrf2-induced upregulation of GPx2 expression (Rouse et al., 2008). The lung tissue of patients exposed to sulfur mustard was characterized by high GPx2 levels as part of an activated Nrf2 response (Tahmasbpour et al., 2016). The same effect was observed after chlorine exposure of mice (Ano et al., 2017). Deletion of Nrf2 impaired GPx2 upregulation by chlorine and prolonged the concomitant inflammation and airway hyperresponsiveness. In a murine model of radiation-induced lung disease, the duration of asymptomatic survival time before the development of pneumonitis and fibrosis significantly correlated with total pulmonary GPx activity. However, out of the four studied GPx isoforms, only the basal GPx2 expression levels were predictive for the tissue injury response in the post-irradiation phase (Kunwar and Haston, 2014). In a mouse model of allergic airway inflammation, GPx2 knockout mice developed a more severe inflammation and stronger infiltration of immune cells into the lung epithelium after challenging with antigen (Dittrich et al., 2010). Thus, it can be concluded that GPx2 is involved in the Nrf2-mediated regeneration of the lung epithelium in response to damage. These repair processes need to be tightly controlled. To maintain airway health is the main duty of resident airway basal stem cells (ABSCs). Changes of the ABSC redox status activate Nrf2, which activates the Notch pathway to stimulate the self-renewal of ABSCs on the one hand and upregulates the antioxidant enzymes presumably including GPx2 on the other hand to counter-regulate and switch off the redox signal (Paul et al., 2014).

6.2.3 THE INTESTINE

The small intestine represents the primary defense against everything ingested by food. Initially, it was suggested that GPx2 acts as a barrier against hydroperoxides produced during food digestion (Chu et al., 1993; Wingler et al., 2000). In response to the n-3 polyunsaturated fatty acid (PUFA) oxidation product 4-hydroxy-2-hexenal and the n-6 PUFA oxidation product 4-hydroxy-2-nonenal, GPx2 mRNA levels were upregulated both in the small intestine and in CaCo2 cells (Awada et al., 2012). This upregulation was also detectable after feeding a general high-fat diet (Awada et al., 2013). During fasting, GPx2 mRNA expression was reduced in the intestine (Jonas et al., 2000). Thus, GPx2 expression levels can indeed be adapted to food intake.

GPx2 is not uniformly expressed in the intestinal epithelium but rather shows a gradient with high levels at the crypt base, which decreases when moving upward the crypt. Interestingly, GPx1 shows the completely opposite localization being highly expressed in differentiated cells and low in proliferating cells (Florian et al., 2010). GPx2 knockout mice do not have an obvious phenotype, but GPx1/GPx2 double knockout mice spontaneously develop colitis (Esworthy et al., 2001) and cancer (Chu et al., 2004). Effects can be rescued by one wild-type allele of GPx2 but not of GPx1 (Esworthy et al., 2005), indicating that GPx2 is more important for the maintenance of the intestinal homeostasis than GPx1. The severity of these symptoms depends on the microbiotic milieu of the intestine. Colitis is more severe under the conventional housing conditions than under specific pathogen-free or even germ-free conditions (Esworthy et al., 2003). In comparison with germ-free mice, GPx2 expression was upregulated after colonization with bacteria, whereas GPx1 expression was unaffected (Esworthy et al., 2003). Recently, it has been shown that GPx2 is specifically enhanced by the colonization of piglets with *Lactobacillus reuteri* (Zhang et al., 2017). Thus, the microbiota itself appear to specifically modulate GPx2 expression.

Even though there are no massive, life-threatening effects detectable upon loss of GPx2, the intestinal epithelium shows some major differences (Figure 6.1). The intestinal epithelium of GPx2 knockout mice is characterized by ectopic apoptotic cells at the crypt base (Florian et al., 2010) (Figure 6.1, green cells). It is not entirely clarified yet which cell types are dying, but it appears to involve stem and progenitor cells as well as Paneth cells [or Reg4-positive deep crypt secretory cells (Sasaki et al., 2016) in the colon, respectively]. The mRNA expression of the stem cell marker Lgr5 was reduced in the colon of GPx2 knockout mice, whereas mRNA levels of the Paneth cell marker lysozyme were unaffected (Lennicke et al., 2018). This suggests lower stem cell numbers upon loss of GPx2. At the same time, there are more proliferating, PCNA-positive cells per crypt, which might be an attempt to compensate for the loss of cells at the crypt base (Müller et al., 2013). Colonic proteome profiles recently revealed that the calcium-activated chloride channel regulator 1 (Clca1), which is specifically expressed in mucin-producing Goblet cells, was significantly downregulated in GPx2 knockout mice (Lennicke et al., 2018). Unexpectedly, the number of Goblet cells was strongly increased and not decreased in GPx2 knockout mice. This indicates that loss of GPx2 supports the cell fate decision toward

Goblet cells, but these Goblet cells appear to be less capable of producing Clca1. In addition, these Clca1-negative Goblet cells are located in the lower third of the crypt, a region where normally no Goblet cells are found. The mRNA expression of chromogranin A (Chga), a marker for enteroendocrine cells, was reduced, indicating that Goblet cell numbers are increased at the expense of enteroendocrine cells. Cell fate decisions toward the different lineages are mainly under the control of the Notch signaling pathway (Vandussen et al., 2012). In the GPx2 knockout epithelium, the direct Notch target gene Hes1 was downregulated, whereas Atoh1, which is repressed by Hes1, was upregulated (Lennicke et al., 2018). Thus, alterations in the Notch signaling pathway appear to be involved in modulating changes in the proliferation to differentiation transition. Furthermore, GPx2 is expressed in Paneth cells (Florian et al., 2001) (Figure 6.1, white cells), which are interspaced between intestinal stem cells (ISCs; Figure 6.1) and express Notch ligands on their surfaces which interact with the respective Notch receptors on the next ISC (Sato et al., 2010). Thus, impaired Notch signaling by loss of GPx2 could destroy the functionality of the stem cell niche and thus could contribute to cell death at the crypt base. As already described for lung stem cells (see Section 6.2.2), the Notch pathway is regulated by the cellular redox status (Coant et al., 2010), indicating that GPx2 is a putative modulator.

In addition, loss of GPx2 results in the development of a low-grade intestinal inflammation as shown by higher numbers of mobile F4/80-positive intra-epithelial myeloid cells (Müller et al., 2013) (Figure 6.1, violet cells). In addition, local tumor necrosis factor-α levels were enhanced, which could also contribute to the induction of apoptosis. By contrast, a reduced barrier function due to cell death could increase the contact between the microbiota and the host resulting in inflammation. To clearly decipher the cause and consequence, inducible GPx2 knockout mice would be a suitable model. Overall, major differences have been observed in the intestinal epithelium upon loss of GPx2, which affect the intestinal homoeostasis and integrity.

To better understand the results obtained from constitutive GPx2 knockout mice, it needs to be noted that the total GPx activity against H_2O_2 of the intestinal epithelium

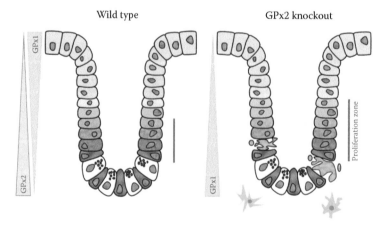

FIGURE 6.1   Modulation of the intestinal epithelium by loss of GPx2. For details, see the text.

is even increased upon loss of GPx2 in comparison with wild-type mice instead of reduced. This can be attributed to GPx1, which becomes aberrantly expressed at the crypt base when GPx2 is lost in addition to its usual localization in differentiated cells (Florian et al., 2010). Thus, the effects of GPx2 knockout either can be directly attributed to the reduced levels of GPx2 or are an indirect result of the aberrantly high GPx1 levels or a combination of both. One approach to shed further light on this complex relationship is the analysis of GPx2 knockout mice with different selenium statuses. As explained before, a limited selenium supply strongly impairs GPx1 expression but only mildly reduces GPx2. Thus, under suboptimal selenium conditions, the upregulation of GPx1 in GPx2 knockout mice can be more or less prohibited. If the effects of the knockout persist under both selenium suboptimal and adequate conditions, it appears to be a direct GPx2 effect. The increased number of apoptotic cells (Florian et al., 2010) was detectable independently of the selenium status, as was the increase in PCNA-positive and F4/80-positive cells (Müller et al., 2013). Also, the expression of Clca1, Lgr5, and Chga was reduced in both selenium groups, whereas the increase in the Goblet cell number was only detectable under suboptimal selenium conditions (Lennicke et al., 2018). Interestingly, the knockout of the intestinal NADPH oxidase Nox1 shows the same effect as observed upon loss of GPx2: impaired Notch signaling and higher Goblet cell numbers (Coant et al., 2010). In both knockouts, the redox tone appears to be too low (either by loss of the H_2O_2-producing enzyme or by locally higher H_2O_2 reduction by GPx1 at the crypt base) to activate Notch and maintain the balance of cell fate decisions. By contrast, Goblet cell numbers were reduced in GPx1/GPx2 double knockout mice, indicating that loss of both isoenzymes has the opposite effect most probably resulting from higher H_2O_2 levels (Chu et al., 2017).

To identify the source of hydroperoxides that are usually reduced by GPx1 and GPx2 in the intestine, triple knockout mice were generated that lack a further gene, in this case for either Nox1 or Duox2, in addition to GPx1 and GPx2. Both enzymes are the members of the family of NADPH oxidases, which are transmembrane proteins known to produce H_2O_2. Indeed, loss of Nox1 rescued the colitis symptoms and apoptosis induction described in the GPx1/GPx2 double knockout epithelium (Esworthy et al., 2014). The knockout of Duox2 only partially rescued the inflammatory phenotype but not the induction of apoptosis (Chu et al., 2017). Thus, there exists a critical interrelationship between NADPH oxidases and GPx1 and GPx2 to balance the redox status of the stem cell compartment to fine-tune redox signaling as discussed for the Notch pathway.

6.3   GPX2 DURING CARCINOGENESIS

6.3.1   GPx2 DURING INFLAMMATION AND TISSUE REGENERATION

The GPx2 expression is enhanced under conditions of inflammation, for example, in the intestine of patients with inflammatory bowel disease (Florian et al., 2001; Te Velde et al., 2008). In addition, GPx2 is upregulated in several murine models such as allergic airway inflammation (Dittrich et al., 2010) or dextran sodium sulfate (DSS)-induced colitis (Kaushal et al., 2014; Hiller et al., 2015) and in genetic

models such as mucin 2 (Yang et al., 2008) and Reg2 knockout mice after infection
with *Helicobacter hepaticus* (Mangerich et al., 2012). Following the course of DSS-
induced colitis, it became obvious that GPx2 upregulation is a biphasic effect. GPx2
is increased during the acute phase but also during the resolution of inflammation
and tissue regeneration. During the latter process, GPx2 is highly induced in
regenerating crypts that are located next to ulcerations (Hiller et al., 2015). These
crypts have a high proliferation rate which might be maintained by GPx2. Similarly,
GPx2 is induced in ulcerated tissue of the stomach 7 days after injury in comparison with the uninjured tissue. In aged mice, impaired wound healing can be rescued by transplanting organoids containing stem cells and high GPx2 expression
(Engevik et al., 2016). Inflammatory mediators including the eicosanoids are essential to balance tissue regeneration and to shut off proliferation when wound healing
is achieved (Fullerton and Gilroy, 2016). The rate-limiting enzyme for prostaglandin
synthesis is the cyclooxygenase 2 (COX-2). In the colorectal cancer cell line HT-29,
loss of GPx2 expression results in an overactivation of COX-2 (Banning et al., 2008a).
Thus, GPx2 appears to be a switch to limit prostaglandin synthesis. Comparing the
severity of lung inflammation in Nrf2 knockout and wild-type mice after treatment
with chlorine gas revealed no difference in the acute phase but an impaired epithelial regeneration in Nrf2 knockout mice. As GPx2 is one of the Nrf2 target genes
strongly induced by chlorine treatment, it is obviously involved in modulating the
resolution of airway inflammation (Ano et al., 2017). In the intestine, GPx2 upregulation during inflammation was independent of Nrf2 but appears to involve STAT
transcription factors and β-catenin (Hiller et al., 2015). Both DSS-induced colitis
and allergic airway inflammation were more severe in GPx2 knockout than in wild-
type mice (Dittrich et al., 2010; Krehl et al., 2012).

Chronic inflammation substantially contributes to the proliferation of established
cancer cells and, thus, is part of a tumor-promoting microenvironment (Hanahan
and Weinberg, 2011). In the gastric mucosa, chronic inflammation results in parietal cell loss and disruption of tissue homeostasis followed by transdifferentiation of
chief cells into spasmolytic polypeptide-expressing metaplasia (SPEM). During this
process, GPx2 was strongly upregulated (Petersen et al., 2014). In a mouse model of
colitis-driven colorectal carcinogenesis, GPx2 knockout mice developed more adenomas in comparison with wild-type mice (Krehl et al., 2012). This indicates that
GPx2 is able to limit inflammation, which can also slow down inflammation-driven
carcinogenesis.

6.3.2 GPx2 during Tumor Development

Tumorigenesis is a long-lasting, multifactorial process. In recent years, it became
more and more clear that the role of GPx2 during carcinogenesis changes depending on the stage of tumor development. The initiation of tumor cells is induced by
DNA damage acquired most probably in adult stem cells which stays unrepaired
and results in the establishment of a mutation with functional consequences for the
cell. The reduction of hydroperoxides by GPx2 in the stem cell compartment of the
airway and intestinal epithelium can reduce DNA damage as already described for
iPSCs (see Section 6.2.1). Also in established colorectal cancer cells, GPx2 protects

from DNA damage, for example, induced by hypoxia. In addition, hypoxia-induced effects resulting in impaired proliferation and enhanced apoptosis are strongly blocked in cells overexpressing GPx2 (Jongen et al., 2017). Thus, changes of the redox status also affect the pathways switching from proliferation to cell death. In the intestinal epithelium, GPx2 protects cells of the crypt base from apoptotic cell death. However, such protection would support tumor development, when tumor cells acquire these survival properties. In a model of colorectal carcinogenesis induced by azoxymethane (AOM), GPx2 knockout mice developed fewer preneoplastic lesions and adenomas than wild-type mice. Shortly after AOM injection, the number of apoptotic cells was higher in GPx2 knockout mice, indicating that the damaged cells were more efficiently eliminated (Müller et al., 2013). Thus, upregulation of GPx2 might be an approach to prevent the initiation of tumor cells but might also interfere with the efficient elimination of established tumor cells. Recently, it has been shown that lung SCC can be suppressed by treatment with digitoxin (Huang et al., 2017). Mechanistically, digitoxin activated the transcription factor YAP which decreased DNp63. GPx2 is a target gene of DNp63 (Yan and Chen, 2006) and was concomitantly downregulated. The increased redox status upon GPx2 inhibition apparently induced apoptosis of SCC.

Besides limiting cell death, GPx2 also enhances proliferation, which is beneficial for the maintenance of tissue homeostasis and regeneration but detrimental in cancer cells. Knocking down GPx2 expression indeed inhibited proliferation of several tumor cells derived from the breast (Naiki-Ito et al., 2007), liver (Suzuki et al., 2013), prostate (Naiki et al., 2014), and colon (Banning et al., 2008b; Emmink et al., 2014). In castration-resistant prostate cancer cells, loss of GPx2 induced a cyclin B1-dependent G2/M arrest and an inhibited tumor growth, when transplanted into castrated mice (Naiki et al., 2014). In line with this, the capability of colorectal cancer cells with low GPx2 expression was reduced to develop colonies or tumors in a xenograft model. Tumor tissue with low GPx2 expression mainly consisted of slow-growing, undifferentiated, stem cell-like cells, whereas GPx2 overexpression stimulated multi-lineage differentiation, proliferation, and tumor growth (Emmink et al., 2014). Thus, GPx2 protects from DNA damage and tumor initiation but clearly is a pro-survival factor in established tumor cells. The widely used chemotherapeutic drug 5-fluorouracil was recently shown to enhance the translation of a set of mRNAs including GPx2 (Bash-Imam et al., 2017). In light of the discussed effects, further upregulation of GPx2 is an undesired side effect of this drug treatment.

6.3.3   GPx2 as a Prognostic Marker?

The upregulation in many tumor types and the obvious growth supporting properties of GPx2 are the reasons for an increasing number of studies analyzing the value of GPx2 as a prognostic marker for the survival of patients. Low GPx2 expression in biopsies of prostate cancer was a marker for a significantly longer prostate-specific antigen recurrence-free and overall survival time of patients in comparison with those with high tumor-resident GPx2 expression (Naiki et al., 2014). Also high-risk hepatocellular carcinoma patients with high GPx2 expression in the tumor tissue had a poor prognosis (Liu et al., 2017a). In gastric tumors, GPx2 expression was upregulated but even further enhanced in lymph node metastasis. Again, high GPx2 expression was correlated with a worse overall survival (Liu et al., 2017b). In patients with the aggressive colorectal cancer subtype 3, high tumor-resident GPx2 expression was also a predictor for a shorter relapse-free survival time (Emmink et al., 2014). The main reason for a shorter relapse-free survival time after resection of the tumor is early metastasis formation. As described before (see Section 6.3.2), patient-derived colorectal cancer cells with a GPx2 knockdown have a stem cell-like character which would imply a higher capability to form metastasis. But unexpectedly, low GPx2 levels abolished metastasis formation of colorectal cancer cells in mouse liver (Emmink et al., 2014). Also hepatocellular carcinoma cells with low GPx2 expression were unable to form metastatic nodules in the lungs when injected into nude mice (Suzuki et al., 2013). The most likely explanation is that GPx2 knockdown cells are more prone to anoikis, a form of cell death induced by the detachment from the previous matrix, for example, the tumor, which is accompanied by an increase in H_2O_2 levels. Thus, high tumor-resident GPx2 levels appear to support the survival of floating tumor cells resulting in a higher probability for metastasis formation.

In esophageal SCC, GPx2 was also upregulated in comparison with the surrounding tissue. However, in this case, high GPx2 expression was a marker for a better prognosis (Lei et al., 2016). Also in urothelial carcinoma, high tumor-resident GPx2 expression was associated with longer survival. Interestingly, even though GPx2 levels were upregulated in tumors, they were strongly downregulated in metastasis (Chang et al., 2015). This indicates that the role of GPx2 and its putative value as a prognostic marker clearly depends on the tumor's tissue of origin, the mutations acquired in the respective tumor, and the tumor grade.

6.4 CONCLUSION

Major processes such as apoptosis, proliferation, and differentiation but also inflammation and regeneration are not only affected by GPx2 in healthy tissues but also modulated during tumor development. Thus, the same function of GPx2 might have different effects in terms of health outcome under either physiological or pathophysiological conditions. Also during carcinogenesis, both protective and detrimental effects of GPx2 have been observed depending on the model, the tumor stage, the tumor environment, and the involvement of inflammation. Further studies are needed to decipher the effect of upregulating GPx2 more precisely and, thus, to evaluate its value as a prognostic marker at both mRNA and protein levels.

ACKNOWLEDGMENTS

The author is grateful to Stefan Holtz for his support to generate Figure 6.1.

REFERENCES

Ano, S., A. Panariti, B. Allard et al. 2017. Inflammation and airway hyperresponsiveness after chlorine exposure are prolonged by Nrf2 deficiency in mice. *Free Radic Biol Med* 102:1–15.

Awada, M., A. Meynier, C. O. Soulage et al. 2013. n-3 PUFA added to high-fat diets affect differently adiposity and inflammation when carried by phospholipids or triacylglycerols in mice. *Nutr Metab (Lond)* 10 (1):23.

Awada, M., C. O. Soulage, A. Meynier et al. 2012. Dietary oxidized n-3 PUFA induce oxidative stress and inflammation: Role of intestinal absorption of 4-HHE and reactivity in intestinal cells. *J Lipid Res* 53 (10):2069–80.

Baek, I. J., J. M. Yon, S. R. Lee et al. 2011. Differential expression of gastrointestinal glutathione peroxidase (GI-GPx) gene during mouse organogenesis. *Anat Histol Embryol* 40 (3):210–18.

Banning, A., S. Deubel, D. Kluth et al. 2005. The GI-GPx gene is a target for Nrf2. *Mol Cell Biol* 25 (12):4914–23.

Banning, A., S. Florian, S. Deubel et al. 2008a. GPx2 counteracts PGE2 production by dampening COX-2 and mPGES-1 expression in human colon cancer cells. *Antioxid Redox Signal* 10 (9):1491–1500.

Banning, A., A. Kipp, S. Schmitmeier et al. 2008b. Glutathione peroxidase 2 inhibits cyclooxygenase-2-mediated migration and invasion of HT-29 adenocarcinoma cells but supports their growth as tumors in nude mice. *Cancer Res* 68 (23):9746–53.

Bash-Imam, Z., G. Thérizols, A. Vincent et al. 2017. Translational reprogramming of colorectal cancer cells induced by 5-fluorouracil through a miRNA-dependent mechanism. *Oncotarget* 8 (28):46219–33.

Brigelius-Flohé, R., and M. Maiorino. 2013. Glutathione peroxidases. *Biochim Biophys Acta* 1830 (5):3289–303.

Chang, I. W., V. C. Lin, C. H. Hung et al. 2015. GPX2 underexpression indicates poor prognosis in patients with urothelial carcinomas of the upper urinary tract and urinary bladder. *World J Urol* 33 (11):1777–89.

Cho, H. Y., F. Imani, L. Miller-DeGraff et al. 2009. Antiviral activity of Nrf2 in a murine model of respiratory syncytial virus disease. *Am J Respir Crit Care Med* 179 (2):138–50.

Cho, H. Y., A. E. Jedlicka, S. P. Reddy et al. 2002. Role of NRF2 in protection against hyperoxic lung injury in mice. *Am J Respir Cell Mol Biol* 26 (2):175–82.

Cho, H. Y., B. van Houten, X. Wang et al. 2012. Targeted deletion of nrf2 impairs lung development and oxidant injury in neonatal mice. *Antioxid Redox Signal* 17 (8):1066–82.

Chu, F. F., J. H. Doroshow, and R. S. Esworthy. 1993. Expression, characterization, and tissue distribution of a new cellular selenium-dependent glutathione peroxidase, GSHPx-GI. *J Biol Chem* 268 (4):2571–6.

Chu, F. F., R. S. Esworthy, P. G. Chu et al. 2004. Bacteria-induced intestinal cancer in mice with disrupted Gpx1 and Gpx2 genes. *Cancer Res* 64 (3):962–8.

Chu, F. F., R. S. Esworthy, J. H. Doroshow et al. 2017. Deficiency in Duox2 activity alleviates ileitis in GPx1- and GPx2-knockout mice without affecting apoptosis incidence in the crypt epithelium. *Redox Biol* 11:144–56.

Coant, N., S. Ben Mkaddem, E. Pedruzzi et al. 2010. NADPH oxidase 1 modulates WNT and NOTCH1 signaling to control the fate of proliferative progenitor cells in the colon. *Mol Cell Biol* 30 (11):2636–50.

Dannenmann, B., S. Lehle, D. G. Hildebrand et al. 2015. High glutathione and glutathione peroxidase-2 levels mediate cell-type-specific DNA damage protection in human induced pluripotent stem cells. *Stem Cell Rep* 4 (5):886–98.

Dittrich, A. M., H. A. Meyer, M. Krokowski et al. 2010. Glutathione peroxidase-2 protects from allergen-induced airway inflammation in mice. *Eur Respir J* 35 (5):1148–54.

Emmink, B. L., J. Laoukili, A. P. Kipp et al. 2014. GPx2 suppression of H_2O_2 stress links the formation of differentiated tumor mass to metastatic capacity in colorectal cancer. *Cancer Res* 74 (22):6717–30.

Engevik, A. C., R. Feng, E. Choi et al. 2016. The development of spasmolytic Polypeptide/TFF2-expressing metaplasia (SPEM) during gastric repair is absent in the aged stomach. *Cell Mol Gastroenterol Hepatol* 2 (5):605–24.

Esworthy, R. S., R. Aranda, M. G. Martin et al. 2001. Mice with combined disruption of Gpx1 and Gpx2 genes have colitis. *Am J Physiol Gastrointest Liver Physiol* 281 (3):G848–55.

Esworthy, R. S., S. W. Binder, J. H. Doroshow, and F. F. Chu. 2003. Microflora trigger colitis in mice deficient in selenium-dependent glutathione peroxidase and induce Gpx2 gene expression. *Biol Chem* 384 (4):597–607.

Esworthy, R. S., B. W. Kim, J. Chow et al. 2014. Nox1 causes ileocolitis in mice deficient in glutathione peroxidase-1 and -2. *Free Radic Biol Med* 68:315–25.

Esworthy, R. S., J. R. Mann, M. Sam, and F. F. Chu. 2000. Low glutathione peroxidase activity in Gpx1 knockout mice protects jejunum crypts from gamma-irradiation damage. *Am J Physiol Gastrointest Liver Physiol* 279 (2):G426–36.

Esworthy, R. S., L. Yang, P. H. Frankel, and F. F. Chu. 2005. Epithelium-specific glutathione peroxidase, Gpx2, is involved in the prevention of intestinal inflammation in selenium-deficient mice. *J Nutr* 135 (4):740–5.

Florian, S., S. Krehl, M. Loewinger et al. 2010. Loss of GPx2 increases apoptosis, mitosis, and GPx1 expression in the intestine of mice. *Free Radic Biol Med* 49 (11):1694–702.

Florian, S., K. Wingler, K. Schmehl et al. 2001. Cellular and subcellular localization of gastrointestinal glutathione peroxidase in normal and malignant human intestinal tissue. *Free Radic Res* 35 (6):655–63.

Fullerton, J. N., and D. W. Gilroy. 2016. Resolution of inflammation: A new therapeutic frontier. *Nat Rev Drug Discov* 15 (8):551–67.

Gregorieff, A., and H. Clevers. 2005. Wnt signaling in the intestinal epithelium: From endoderm to cancer. *Genes Dev* 19 (8):877–90.

Hanahan, D., and R. A. Weinberg. 2011. Hallmarks of cancer: The next generation. *Cell* 144 (5):646–74.

Harvey, B. G., A. Heguy, P. L. Leopold et al. 2007. Modification of gene expression of the small airway epithelium in response to cigarette smoking. *J Mol Med (Berl)* 85 (1):39–53.

Hayes, J. D., and A. T. Dinkova-Kostova. 2014. The Nrf2 regulatory network provides an interface between redox and intermediary metabolism. *Trends Biochem Sci* 39 (4):199–218.

Herlan, L., A. Schulz, L. Schulte et al. 2015. Novel candidate genes for impaired nephron development in a rat model with inherited nephron deficit and albuminuria. *Clin Exp Pharmacol Physiol* 42 (10):1051–8.

Hiller, F., K. Besselt, S. Deubel et al. 2015. GPx2 induction is mediated through STAT transcription factors during acute colitis. *Inflamm Bowel Dis* 21 (9):2078–89.

Ho, Y. S., J. L. Magnenat, R. T. Bronson et al. 1997. Mice deficient in cellular glutathione peroxidase develop normally and show no increased sensitivity to hyperoxia. *J Biol Chem* 272 (26):16644–51.

Huang, H., W. Zhang, Y. Pan et al. 2017. YAP suppresses lung squamous cell carcinoma progression via deregulation of the DNp63-GPX2 axis and ROS accumulation. *Cancer Res* 77 (21):5769–81.

Jonas, C. R., C. L. Farrell, S. Scully et al. 2000. Enteral nutrition and keratinocyte growth factor regulate expression of glutathione-related enzyme messenger RNAs in rat intestine. *JPEN J Parenter Enteral Nutr* 24 (2):67–75.

Jongen, M. J., L. M. van der Waals, K. Trumpi et al. 2017. Downregulation of DNA repair proteins and increased DNA damage in hypoxic colon cancer cells is a therapeutically exploitable vulnerability. *Oncotarget* 8 (49):86296–311.

Kaushal, N., A. K. Kudva, A. D. Patterson et al. 2014. Crucial role of macrophage selenoproteins in experimental colitis. *J Immunol* 193 (7):3683–92.

Kipp, A. P. 2017. Selenium-dependent glutathione peroxidases during tumor development. *Adv Cancer Res* 136:109–38.

Kipp, A., A. Banning, and R. Brigelius-Flohé. 2007. Activation of the glutathione peroxidase 2 (GPx2) promoter by beta-catenin. *Biol Chem* 388 (10):1027–33.

Krehl, S., M. Loewinger, S. Florian et al. 2012. Glutathione peroxidase-2 and selenium decreased inflammation and tumors in a mouse model of inflammation-associated carcinogenesis whereas sulforaphane effects differed with selenium supply. *Carcinogenesis* 33 (3):620–8.

Kunwar, A., and C. K. Haston. 2014. Basal levels of glutathione peroxidase correlate with onset of radiation induced lung disease in inbred mouse strains. *Am J Physiol Lung Cell Mol Physiol* 307 (8):L597–L604.

Lei, Z., D. Tian, C. Zhang et al. 2016. Clinicopathological and prognostic significance of GPX2 protein expression in esophageal squamous cell carcinoma. *BMC Cancer* 16:410.

Lennicke, C., J. Rahn, C. Wickenhauser et al. 2018. Loss of epithelium-specific GPx2 results in aberrant cell fate decisions during intestinal differentiation. *Oncotarget* 9:539–52.

Lippmann, D., C. Lehmann, S. Florian et al. 2014. Glucosinolates from pak choi and broccoli induce enzymes and inhibit inflammation and colon cancer differently. *Food Funct* 5 (6):1073–81.

Liu, T., X. F. Kan, C. Ma et al. 2017a. GPX2 overexpression indicates poor prognosis in patients with hepatocellular carcinoma. *Tumour Biol* 39 (6):1–10.

Liu, D., L. Sun, J. Tong et al. 2017b. Prognostic significance of glutathione peroxidase 2 in gastric carcinoma. *Tumour Biol* 39 (6):1–9.

Mangerich, A., C. G. Knutson, N. M. Parry et al. 2012. Infection-induced colitis in mice causes dynamic and tissue-specific changes in stress response and DNA damage leading to colon cancer. *Proc Natl Acad Sci USA* 109 (27):E1820–9.

Mörk, H., O. H. al-Taie, K. Bahr et al. 2000. Inverse mRNA expression of the selenocysteine-containing proteins GI-GPx and SeP in colorectal adenomas compared with adjacent normal mucosa. *Nutr Cancer* 37 (1):108–16.

Müller, M. F., S. Florian, S. Pommer et al. 2013. Deletion of glutathione peroxidase-2 inhibits azoxymethane-induced colon cancer development. *PLoS One* 8 (8):e72055.

Murawaki, Y., H. Tsuchiya, T. Kanbe et al. 2008. Aberrant expression of selenoproteins in the progression of colorectal cancer. *Cancer Lett* 259 (2):218–30.

Naiki, T., A. Naiki-Ito, M. Asamoto et al. 2014. GPX2 overexpression is involved in cell proliferation and prognosis of castration-resistant prostate cancer. *Carcinogenesis* 35 (9):1962–7.

Naiki-Ito, A., M. Asamoto, N. Hokaiwado et al. 2007. Gpx2 is an overexpressed gene in rat breast cancers induced by three different chemical carcinogens. *Cancer Res* 67 (23):11353–8.

Ouyang, X., T. L. DeWeese, W. G. Nelson, and C. Abate-Shen. 2005. Loss-of-function of Nkx3.1 promotes increased oxidative damage in prostate carcinogenesis. *Cancer Res* 65 (15):6773–9.

Park, Y. K., J. L. Franklin, S. H. Settle et al. 2005. Gene expression profile analysis of mouse colon embryonic development. *Genesis* 41 (1):1–12.

Park, I. H., R. Zhao, J. A. West et al. 2008. Reprogramming of human somatic cells to pluripotency with defined factors. *Nature* 451 (7175):141–6.

Paul, M. K., B. Bisht, D. O. Darmawan et al. 2014. Dynamic changes in intracellular ROS levels regulate airway basal stem cell homeostasis through Nrf2-dependent Notch signaling. *Cell Stem Cell* 15 (2):199–214.

Petersen, C. P., V. G. Weis, K. T. Nam et al. 2014. Macrophages promote progression of spasmolytic polypeptide-expressing metaplasia after acute loss of parietal cells. *Gastroenterology* 146 (7):1727–38.

Rouse, R. L., G. Murphy, M. J. Boudreaux et al. 2008. Soot nanoparticles promote biotransformation, oxidative stress, and inflammation in murine lungs. *Am J Respir Cell Mol Biol* 39 (2):198–207.

Sasaki, N., N. Sachs, K. Wiebrands et al. 2016. Reg4+ deep crypt secretory cells function as epithelial niche for Lgr5+ stem cells in colon. *Proc Natl Acad Sci USA* 113 (37):E5399–407.

Sato, T., J. H. van Es, H. J. Snippert et al. 2010. Paneth cells constitute the niche for Lgr5 stem cells in intestinal crypts. *Nature* 469 (7330):415–18.

Schäfer, M., H. Farwanah, A. H. Willrodt et al. 2012. Nrf2 links epidermal barrier function with antioxidant defense. *EMBO Mol Med* 4 (5):364–79.

Serewko, M. M., C. Popa, A. L. Dahler et al. 2002. Alterations in gene expression and activity during squamous cell carcinoma development. *Cancer Res* 62 (13):3759–65.

Singh, A., T. Rangasamy, R. K. Thimmulappa et al. 2006. Glutathione peroxidase 2, the major cigarette smoke-inducible isoform of GPX in lungs, is regulated by Nrf2. *Am J Respir Cell Mol Biol* 35 (6):639–50.

Skamagki, M., C. Zhang, C. A. Ross et al. 2017. RNA exosome complex-mediated control of redox status in pluripotent stem cells. *Stem Cell Rep* 9 (4):1053–61.

Solleti, S. K., S. Bhattacharya, A. Ahmad et al. 2017. MicroRNA expression profiling defines the impact of electronic cigarettes on human airway epithelial cells. *Sci Rep* 7 (1):1081.

Suzuki, S., P. Pitchakarn, K. Ogawa et al. 2013. Expression of glutathione peroxidase 2 is associated with not only early hepatocarcinogenesis but also late stage metastasis. *Toxicology* 311 (3):115–23.

Tahmasbpour, E., M. Ghanei, A. Qazvini et al. 2016. Gene expression profile of oxidative stress and antioxidant defense in lung tissue of patients exposed to sulfur mustard. *Mutat Res Genet Toxicol Environ Mutagen* 800–801:12–21.

Takahashi, K., K. Tanabe, M. Ohnuki et al. 2007. Induction of pluripotent stem cells from adult human fibroblasts by defined factors. *Cell* 131 (5):861–72.

Te Velde, A. A., I. Pronk, F. de Kort, and P. C. Stokkers. 2008. Glutathione peroxidase 2 and aquaporin 8 as new markers for colonic inflammation in experimental colitis and inflammatory bowel diseases: An important role for H_2O_2? *Eur J Gastroenterol Hepatol* 20 (6):555–60.

VanDussen, K. L., A. J. Carulli, T. M. Keeley et al. 2012. Notch signaling modulates proliferation and differentiation of intestinal crypt base columnar stem cells. *Development* 139 (3):488–97.

Walshe, J., M. M. Serewko-Auret, N. Teakle et al. 2007. Inactivation of glutathione peroxidase activity contributes to UV-induced squamous cell carcinoma formation. *Cancer Res* 67 (10):4751–8.

Wan, J., M. Yourshaw, H. Mamsa et al. 2012. Mutations in the RNA exosome component gene EXOSC3 cause pontocerebellar hypoplasia and spinal motor neuron degeneration. *Nat Genet* 44 (6):704–8.

Wingler, K., M. Böcher, L. Flohé et al. 1999. mRNA stability and selenocysteine insertion sequence efficiency rank gastrointestinal glutathione peroxidase high in the hierarchy of selenoproteins. *Eur J Biochem* 259 (1–2):149–57.

Wingler, K., C. Müller, K. Schmehl et al. 2000. Gastrointestinal glutathione peroxidase prevents transport of lipid hydroperoxides in CaCo-2 cells. *Gastroenterology* 119 (2):420–30.

Wönckhaus, M., L. Klein-Hitpass, U. Grepmeier et al. 2006. Smoking and cancer-related gene expression in bronchial epithelium and non-small-cell lung cancers. *J Pathol* 210 (2):192–204.

Yan, W., and X. Chen. 2006. GPX2, a direct target of p63, inhibits oxidative stress-induced apoptosis in a p53-dependent manner. *J Biol Chem* 281 (12):7856–62.

Yang, K., N. V. Popova, W. C. Yang et al. 2008. Interaction of Muc2 and Apc on Wnt signaling and in intestinal tumorigenesis: Potential role of chronic inflammation. *Cancer Res* 68 (18):7313–22.

Yu, H. B., G. Kunarso, F. H. Hong, and L. W. Stanton. 2009. Zfp206, Oct4, and Sox2 are integrated components of a transcriptional regulatory network in embryonic stem cells. *J Biol Chem* 284 (45):31327–35.

Yu, J., M. A. Vodyanik, K. Smuga-Otto et al. 2007. Induced pluripotent stem cell lines derived from human somatic cells. *Science* 318 (5858):1917–20.

Zhang, D., T. Shang, Y. Huang et al. 2017. Gene expression profile changes in the jejunum of weaned piglets after oral administration of Lactobacillus or an antibiotic. *Sci Rep* 7 (1):15816.

Zhou, J. C., H. Zhao, J. G. Li et al. 2009. Selenoprotein gene expression in thyroid and pituitary of young pigs is not affected by dietary selenium deficiency or excess. *J Nutr* 139 (6):1061–6.

7 GPx4
From Prevention of Lipid Peroxidation to Spermatogenesis and Back

Matilde Maiorino, Antonella Roveri, and Fulvio Ursini
Università degli Studi di Padova

CONTENTS

7.1 FOR INTRODUCTION: THE HISTORY OF PHGPX AND LIPID PEROXIDATION

Lipid peroxidation is a transition metal-dependent, oxidative degradation of polyunsaturated fatty acids of membrane phospholipids. Comprehensive available reviews (Cheng and Li, 2007; Yin et al., 2011; Davies and Guo, 2014) underscore the notion that phospholipid hydroperoxides (PLOOH) are both the primary product of lipid peroxidation and the most prominent initiating species. Minute amounts of PLOOH, necessary to start lipid peroxidation, are continuously generated in a biological environment. The methylene bond between two nonconjugated double bonds in the fatty acid chain esterified in a phospholipid is, in fact, prone to the abstraction of a hydrogen atom, thus forming a carbon-centered radical. Free radicals competent for this reaction, the hydroperoxyl (HOO$^\bullet$) and hydroxyl (HO$^\bullet$) radicals, are continuously produced by metabolic oxygen activation. The reversible oxygen addition to the carbon-centered radical in an unsaturated chain of a phospholipid produces a phospholipid hydroperoxyl radical (PLOO$^\bullet$), and a hydrogen transfer finally stabilizes this species forming a PLOOH. Available evidence on lipid peroxidation supports the notion that propagation is rather limited, since PLOO$^\bullet$ is prone to undergo a termination reaction. However, from PLOOH, in the presence of a redox-active metal such as iron, the highly reactive phospholipid alkoxy radical (PLO˙) is produced, which primes new oxidations. The most relevant initiator of iron-dependent lipid peroxidation is therefore a PLOOH. The final products of iron-dependent lipid peroxidation are a large series of electrophiles generated from decomposition of PLOOH. It is therefore not surprising that it is the reduction of PLOOH that accounts for the most efficient antiperoxidant mechanism known, irrespective of the mechanism of PLOOH generation.

In 1960, the occurrence of lipid peroxidation was observed in tissues of vitamin E-deficient animals (Zalkin and Tappel, 1960) and in subcellular fractions incubated in the presence of oxygen and hematin (Tappel and Zalkin, 1960), ferrous ions (Ottolenghi, 1959; Hunter et al., 1963), or ascorbate (Ottolenghi, 1959; Thiele and Huff, 1960). Shortly later, an enzymatic form of lipid peroxidation was described, sparked by the activity of microsomal rat liver NADPH P_{450} oxidoreductase, reduced pyridine nucleotide, and Fe^{3+} chelates of ADP or ATP (Hochstein and Ernster, 1963). The observation that iron-dependent lipid peroxidation is primed by physiological compounds at physiological concentrations contributed to the perception that the process could have a physiological relevance, envisaging biomedical implications (Hochstein and Ernster, 1964; Ernster et al., 1982). This motivated the search of the antioxidant defense system(s) and expanded the field of free radicals in biology and medicine, on the assumption that all these processes must proceed exclusively by free radical reactions. This came out not to be true.

In 1976, McCay first demonstrated that reduced glutathione (GSH), when incubated with dialyzed rat liver cytosol, prevents NADPH-induced microsomal lipid peroxidation. This effect was attributed to a "thermolabile factor," tentatively identified as glutathione peroxidase (McCay et al., 1976). Glutathione peroxidase (known today as the tetrameric GPx1, E.C. 1.11.1.9) was already known for almost 20 years as the enzyme that, by catalyzing the reduction of H_2O_2 by GSH, protects hemoglobin from oxidative breakdown (Mills, 1957). Yet, the real nature of the "thermolabile factor" remained controversial. Burk proposed that protection could be due, at least in part, to a glutathione transferase (E.C. 2.5.1.18) (Burk et al., 1980), while Gibson showed that enzymatic lipid peroxidation was not prevented by an enriched preparation of glutathione transferase. These authors argued that, instead, a still unidentified protein acting on intact PLOOH could be the actual antiperoxidant factor (Gibson et al., 1980).

In the meanwhile, in our laboratory at the "Istituto di Chimica Biologica" in Padova, while studying the regulation of HMG-CoA reductase using microsomes as the enzyme source, we had to deal with the problem of lipid peroxidation. Thus, we resorted to learn more about the mechanism of the antioxidant capacity of GSH and the "thermolabile factor" by purifying it. Looking for a simple and quick lipid peroxidation test for assaying the protective activity present in the chromatographic fractions, we came across the experiments of Kaschnitz and Hatefi (1975), showing that lipid peroxidation could be induced by hematin in a suspension of phospholipids containing polyunsaturated fatty acids and traces of PLOOH. These could be easily obtained by spontaneous autoxidation, i.e., leaving the phospholipid suspension at +4°C for some days. However, GSH alone inhibited hematin-driven peroxidation (Kaschnitz and Hatefi, 1975), and thus this compound was not suitable for assaying the activity of the antiperoxidant factor. Exploring other iron complexes inducing lipid peroxidation regardless of the presence of GSH, we identified Fe^{3+}-triethylenetetramine (TETA), a complex previously described by Wang (1955) as a catalase mimic. Fe-TETA induced a fast lipid peroxidation in autoxidized phospholipid suspensions by a mechanism that was assumed to be PLOOH-dependent, similar to that of hematin, with the advantage that GSH was not inhibitory in the absence of the "thermolabile factor" (Maiorino et al., 1980).

This approach was pivotal to its purification. This revealed a monomeric protein, which was devoid of any GSH transferase activity and accounted for the activity of the cytosol in the McCay protection assay (Ursini et al., 1982). We showed that the purified enzyme catalyzed a GSH oxidation coupled to a 2:1 stoichiometric reduction of the PLOOH contained in the phospholipid suspension. We also provided an mass spectrometry (MS) evidence that phospholipid alcohols (PLOH) were produced in the enzymatic reaction (Ursini et al., 1982; Daolio et al., 1983). The molecular weight of the protein, as detected by size exclusion chromatography and polyacrylamide gel electrophoresis, was approximately 20 kDa, quite different from the previously known homo-tetrameric GPx1. Finally, by proton induced X-ray fluorescence, we demonstrated the presence of selenium. The second glutathione peroxidase came to light and, due to the peroxidase activity on PLOOH, we named it "phospholipid hydroperoxide glutathione peroxidase" (PHGPx) (E.C. 1.11.1.12) (Ursini et al., 1985). Today, PHGPx, also known as GPx4, is the product of the *GPx4* gene (Gladyshev et al., 2016). Enzymological and structural characterization followed in a span of a decade.

7.2 PHGPX: KINETIC CHARACTERIZATION AND SUBSTRATE SPECIFICITY

Steady-state kinetic analysis of the reaction indicated that PHGPx adopts a uni–ter ping-pong mechanism, similar to that of GPx1 (Ursini et al., 1985; Maiorino et al., 1990; Ursini et al., 1995). The reaction was assumed encompassing two ordered independent chemical events: the catalytic selenol (–SeH) is first oxidized by the hydroperoxide, and then reduced back in two steps by two molecules of GSH. Kinetic data were compatible with a mechanism where the –SeH of catalytic selenocysteine residue is oxidized by the hydroperoxide to selenenic acid (–SeOH) while the corresponding alcohol and water are produced, then glutathionylated by the first molecule of GSH (-Se-SG), and eventually reduced back by the second molecule of GSH, whereby oxidized glutathione (GSSG) is released and the ground state enzyme is regenerated.

As already observed for GPx1 (Flohé et al., 1972), in the reaction catalyzed by PHGPx, the values of V_{max} and K_m are infinite, since the redox transitions are faster than the formation of the enzyme–substrate complexes. In other words, the kinetics do not comply with the Michaelis–Menten model. PHGPx acts according to the concept of "Zwischenstoff-Katalyse" (catalysis by intermediate formation) (Flohé, 2011) and does not display any saturation pattern. Therefore, any variation in substrate concentration invariably results in a corresponding variation in the enzymatic rate, and thus, the Michaelis–Menten parameters K_m and V_{max} are meaningless.

The rate constants used for describing kinetic features of PHGPx were deduced by applying a simplified version of the Dalziel equation [see Ursini et al. (1995) and Toppo et al. (2009) for a review]. In pig heart PHGPx, the rate constant for the oxidizing step (k_{+1}), measured in the presence of PLOOH substrate in mixed micellar form with Triton X-100 is 1.4×10^7 M^{-1} s^{-1}, which is among the fastest ones measured for bimolecular enzymatic reactions. A recent computational approach by Density Functional Theory Quantum Mechanics (DFT-QM) [(Orian et al., 2015), reviewed in Chapter 3] indicated that such a fast rate is obtained by the formation of a charge-separated intermediate. This represents the clue of the catalysis of PHGPx and of the other selenium-dependent glutathione peroxidases, where the selenol undergoes deprotonation via long-range proton transfer involving the selenol proton, two water molecules, and the indole group of the Trp present in the catalytic tetrad. Upon binding of the peroxide substrate, the products of the oxidative part of the catalytic cycle (i.e., the –SeOH, PLOH, and H_2O) are generated in a barrierless reaction, since the proton, when shifting back, hits the hydroperoxide group. This primes the nucleophilic substitution, whereby the hydroperoxide is reduced and the selenium is oxidized (see also Chapter 3).

Since the reductive part of the catalytic cycle is comparatively slow ($k'_{+2} = 1.2 \times 10^5$ $M^{-1}s^{-1}$ for the pig heart PHGPx) and the regeneration of reduced glutathione is typically even slower, the availability of GSH at the catalytic center emerges as a major controller of PHGPx turnover. This, therefore, is expected to play a physiological role in respect to either activation of lipid peroxidation or oxidation of alternative thiol substrates.

Unlike GPx1, PHGPx has a broad specificity for both the hydroperoxide and the reducing substrates. All of the hydroperoxides so far assayed are reduced by PHGPx, from H_2O_2 and free fatty acid hydroperoxides, which are also the substrates for GPx1, to the hydroperoxide derivatives of phospholipids, cholesterol, or cholesterol esters inserted in membranes or lipoproteins, which are reduced only by PHGPx (Thomas et al., 1990a,b; Maiorino et al., 1991). This lack of specificity is apparently due to the fact that only the monomeric structure allows the access of large substrates to the active site (Ursini et al., 1995; Maiorino et al., 2015). As for the reducing substrate, PHGPx accepts different thiols such as the monothiol mercaptoethanol or dithiols such as dithiothreitol or dithioerythritol, which indeed react even faster than GSH (Roveri et al., 1994). In a physiological context, this is comparable with the reactivity of PHGPx on adjacent cysteine residues in specific proteins (Maiorino et al., 2005) (see below).

The reduction of hydroperoxides in membranes by PHGPx is a multistage process (Cozza et al., 2017). First, PHGPx binds to the lipid monolayer by specific electrostatic interactions between a cationic area adjacent to the redox center and the phospholipid polar head. This addresses the esterified fatty acid hydroperoxide, which floats and is stabilized on the surface of the membrane, to the active site of the enzyme. The redox reaction between the –SeH and the PLOOH facilitates the access of GSH, which reacts with the –SeOH forming –Se–SG. The second GSH interacts with the catalytic center competing with the same amino acids docking the enzyme to the membrane and brings the catalytic cycle to completion. Consequently, the reduced enzyme is therefore released from the membrane, ready for the next catalytic cycle (see also Chapter 3).

7.3   PHGPX: SEQUENCING, CLONING, AND MRNA EXPRESSION

The issue of the primary structure of PHGPx was initially attempted on the enzyme purified from pig heart by the laborious Edman degradation, the best technique available at the time. Although this approach yielded no more than partial information on the structure, it allowed to conceive degenerated primers that were used successfully to fish, by polymerase chain reaction, the corresponding mRNA from a pig heart cDNA library. The combination of these two techniques provided an almost complete information on the sequence, while just a small stretch at the N-terminus was missing (Schuckelt et al., 1991). Interestingly, the peptides containing the selenium could not be identified by direct sequencing, while the presence of selenocysteine was just suggested by a tiny peak in the amino acid analysis, compatible with the presence of carboxy-methyl-selenocysteine. The presence of selenocysteine was eventually proven by the identification of the in-frame TGA codon (Schuckelt et al., 1991). Some years before indeed, it had been discovered, first in GPx1 (Chambers et al., 1986) and later in other selenocysteine-containing proteins of mammalian (Takahashi et al., 1990) and bacterial origin (Stadtman, 1987; Leinfelder et al., 1988), that selenocysteine is inserted co-translationally by an in-frame TGA, previously known as a termination codon only. The complete coding region of PHGPx was obtained 3 years later, when the gene structure was elucidated. These results established for PHGPx a maximum length of 170 amino acids (Brigelius-Flohé et al., 1994).

Not surprisingly, some parts of the PHGPx primary structure proved to be similar to GPx1. The most homologous part involved sequences building up the active site, adding strong support to the notion that the catalytic mechanism of PHGPx is identical to that of GPx1 (see earlier and Chapter 3). Divergent strings of amino acids included large gaps in the PHGPx sequence, in regions corresponding to the subunit interaction sites in bovine GPx1, thus giving an account for the monomeric nature of PHGPx (Brigelius-Flohé et al., 1994). Primary structure elucidation allowed the modeling of PHGPx on the scaffold of the known crystal structure of the homo-tetrameric GPx1. This yielded the first model of three-dimensional structure of PHGPx (Ursini et al., 1995) that was largely confirmed by the PHGPx crystal structure resolved years later on a Cys variant of PHGPx (Scheerer et al., 2007).

These studies established that the residues involved in GSH binding in GPx1 are mutated or deleted in PHGPx, shading doubts on whether PHGPx could be strictly considered a *glutathione* peroxidase (Brigelius-Flohé et al., 1994; Ursini et al., 1995). It was later shown that GSH actually docks to PHGPx by different amino acids that, worth of note, are the same involved also in electrostatic interaction with the polar head of membrane phospholipids (Flohé et al., 2011; Bosello-Travain et al., 2013; Cozza et al., 2017).

The overall homology of pig PHGPx with GPx1 of different species is approximately 30%, indicating that the two peroxidases are phylogenetically related (Schuckelt et al., 1991; Brigelius-Flohé et al., 1994; Toppo et al., 2008). To date, mammalian glutathione peroxidase homologs encompass eight phylogenetically related proteins (Brigelius-Flohé and Maiorino, 2013).

The gene encoding PHGPx, *GPx4*, has been initially described as a complex 7 exons-containing gene, yielding a product of 170 amino acids (Brigelius-Flohé et al., 1994). Later it emerged that *GPx4* produces different transcripts mainly expressed in a tissue-dependent manner. Rat testis contains a cDNA encoding for the additional 27 amino acids at the N terminus of the mitochondrial localization sequence (Pushpa-Rekha et al., 1995). This longer transcript, usually referred to as the mitochondrial *GPx4* (m-*GPx4*), also carries an upstream translational start. The smaller transcript of somatic tissues, therefore, was identified as the "nonmitochondrial" *GPx4* transcript (Pushpa-Rekha et al., 1995), i.e., the cytosolic *GPx4* (c-*GPx4*) transcript yielding the product previously isolated, purified, and sequenced (Schuckelt et al., 1991; Brigelius-Flohé et al., 1994). Few years later, a third mRNA product arising from an alternative exon inside the first intron of *GPx4* was discovered. This yields a 32-kDa protein containing an N-terminal nuclear localization sequence (referred to as nuclear PHGPx, the product of the n-*GPx4* transcript). This protein is predominantly expressed in the testis (Pfeifer et al., 2001), although all the three transcripts coexist, in different proportions, in somatic and germ cell (Maiorino et al., 2003b; Liang et al., 2009). In the mature protein, the mitochondrial but not the nuclear targeting sequence is cleaved off, so that both the m- and the c-*GPx4* transcripts yield the identical mature PHGPx (Arai et al., 1996).

7.4 PHGPX AND SPERMATOGENESIS

In 1992, we observed a large PHGPx activity in rat testis, which, differently from other tissues, was higher in the nuclei and in the mitochondrial fraction than in the cytosol (Roveri et al., 1992). Immunohistochemical examination of the testicular tissue revealed that the majority of PHGPx is located in maturating spermatogenic cells, where the enzyme is located in the peripheral part of the cytoplasm, while a relatively larger portion is associated with the membranes of the nuclei and the mitochondria. These results established, for the first time, that male germ cells are a preferred site of PHGPx expression, with a distribution pattern quite different from somatic tissues. A high level of PHGPx expression in the male germ cells was also observed by *in situ* hybridization (Maiorino et al., 1998). Yet, the signaling pathway leading to high transcriptional activation is still unknown, while just the direct role of testosterone was ruled out. Testosterone, in fact, induces testicular spermatogenesis and thus stimulates the proliferation of the germ layer where PHGPx is located, mainly in spermatocytes and spermatids (Maiorino et al., 1998; Haraguchi et al., 2003; Puglisi et al., 2003).

At that time, a critical role of selenium in male fertility had been known for almost 20 years. From selenium deprivation studies, indeed, it was known that second-generation selenium deficiency results in reduced fertility in mouse and loss of fertility in rat (McCoy and Weswig, 1969). The seminiferous tubule from the Se-deficient animals contains fewer elongating spermatids and spermatozoa. Furthermore, when spermatogenesis proceeds, microscopic lesions at the level of the mid piece of maturating spermatozoa become apparent. An abnormal shape and arrangement of the mitochondria in the mid piece sheath, resulted in bends between the principal piece and the mid piece (i.e., a "kink morphology") (McCoy and Weswig, 1969;

Wu et al., 1979; Wallace et al., 1983). The discovery of PHGPx in testis germ cells paved the way to the understanding of a major molecular mechanism underlying the phenotype produced by Se deficiency.

The evidence that PHGPx is a major protein of mitochondrial capsule of spermatozoa, i.e., the outermost layer of the external mitochondrial membrane (Ursini et al., 1999), was a milestone achievement. However, the enzyme was inactive and only high concentrations of thiols (i.e., 0.1 M of 2-mercaptoethanol or dithiotreitol), in the presence of guanidine, could regenerate the activity. Interestingly, the increase of PHGPx activity by this solubilizing procedure (herein comprehensively referred to as "rescuing") was not observed in spermatogenic cells from testicular tubules, showing a high activity without any "rescuing" procedure. These results suggested a functional shift of PHGPx during spermatogenesis. This selenoenzyme, in fact, is transformed from a soluble, active peroxidase in immature germ cells to an enzymatically inactive protein in spermatozoa, where, apparently, it gains a structural role. We proposed that this moonlighting is triggered by a shortage of GSH taking place in late spermatogenesis, when also a massive protein disulfide formation takes place (Shalgi et al., 1989; Bauché et al., 1994; Fisher and Aitken, 1997). As a consequence of GSH depletion, PHGPx, due to its broad thiol specificity (Roveri et al., 1994), can use the capsular protein thiols as alternative substrates and produces protein disulfides and mixed Se-disulfides instead of GSSG. In doing this, the enzyme remains self-incorporated in the capsule. This functional shift occurs in step 19 spermatids, when PHGPx migrates from the matrix of the mitochondria to the outermost membrane region (Haraguchi et al., 2003). The PHGPx enzymatic activity is actually "rescued" *in vitro* when these mixed disulfides and Se-disulfides are reduced. Altogether, these observations point for the specific functional role for the production of a hydroperoxide in maturating germ cells and epididymal spermatozoa (Fisher and Aitken, 1997), which results in massive GSH depletion.

In spite of this remarkable increase of knowledge, the fine structure of the sperm mitochondrial capsules still remains largely unknown. Some studies attempted to define the protein substrate(s) interacting with PHGPx indicated as a likely candidate the "sperm mitochondrion associated cysteine-rich protein (SMCP)," the protein that was initially erroneously declared to be the selenoprotein in the capsule (Pallini and Bacci, 1979). Indeed, peptides containing the typical motifs of SMCP, i.e., adjacent cysteine residues (Cys-Cys), are oxidized by PHGPx. Since a disulfide between Cys-Cys generates a large conformational redox switch, it has been speculated that the cystine residues produced by PHGPx on SMCP are a transient intermediate that, by producing the conformational switch, primes the complex oxidative polymerization involving forthcoming thiol-disulfide exchanges (Maiorino et al., 2005).

In these studies, however, the individual role of the products of the three *GPx4* transcripts was not specifically addressed. Moreover, after it had become evident that deletion of *GPx4* yields a lethal phenotype (Imai et al., 2003; Yant et al., 2003), it became a real challenge to unravel which of the three *GPx4* transcripts is endowed of the vital function. This question had been elegantly answered by inverse genetic approaches that remarkably disclosed the multifaceted physiological roles of the PHGPx reaction.

At first, the expression of either n-*GPx4* or m-*GPx4* mRNA was disrupted in the mice by whole body knocked out (KO) (Conrad et al., 2005; Schneider et al., 2009). From these experiments, it has been concluded that PHGPx from n-*GPx4* is neither vital nor crucial to male fertility. It just stabilizes nuclear chromatin in maturating spermatozoa; n-*GPx4*$^{-/-}$ mice exhibiting defective chromatin condensation only in the sperm isolated from the caput epididymis (Conrad et al., 2005). Consistently, PHGPx in the nuclei had been identified as a chromatin-bound enzyme (Godeas et al., 1996) where it exhibits a protamine thiol peroxidase activity (Godeas et al., 1997). PHGPx expressed from the mitochondrial transcript is not vital either. However, the male offspring is infertile, and sperm shows the typical kink morphology observed under severe Se deficiency. Notably, despite the abnormal spermatozoa causing infertility in m-*GPx4*$^{-/-}$, the male germinal epithelium is not morphologically altered (Schneider et al., 2009).

Eventually, the experiments of Liang et al. clearly established that PHGPx from c-*GPx4* accounts for survival, while confirming that the deficiency of PHGPx from m-*GPx4* just results in the kink morphology of spermatozoa. *GPx4*$^{-/-}$ mice, indeed, could be rescued by PHGPx from the c-*GPx4* but not by the m-*GPx4* transcript, indicating that embryogenesis is supported only by the "cytosolic" form of enzyme (Liang et al., 2009). As expected, the male offspring of mice expressing c-*GPx4* over a *GPx4*$^{-/-}$ background, and thus lacking the product from the m-*GPx4* transcript, is infertile. The spermatozoa exhibit the typical kink morphology also described in m-*GPx4*$^{-/-}$ mice (Schneider et al., 2009), while testis and epididymis do not show any relevant histological alteration (Liang et al., 2009). All together, these studies show that it is the PHGPx from the m-*GPx4* transcript that plays the structural role in sperm, while the PHGPx from the c-*GPx4* one supports survival of the whole embryo and normal cell growth in testis germinal epithelium.

The double, vital and structural, role of PHGPx in testicular cells was further confirmed by Imai (Imai et al., 2009), showing that targeted deletion of the entire *GPx4* in spermatocytes but not in spermatogonia (i.e., in the haploid but not diploid stem cells of the testis) or somatic cells induces death of the haploid spermatocytes where *GPx4* is knocked out. Furthermore, the few spermatozoa derived from these spermatocytes show the typical kink morphology reported for *GPx4*$^{-/-}$ male mice expressing c-*GPx4* only (Liang et al., 2009) or for m-*GPx4* whole-body KO mice (Schneider et al., 2009).

The burst of PHGPx expression during spermatogenesis in rodents was also described in humans (Maiorino et al., 1998; Imai et al., 2001). As in rodents, PHGPx in human sperm is largely inactive if not previously "rescued" (Foresta et al., 2002). A lower "rescued" PHGPx activity in human spermatozoa is associated with infertility, irrespective of the etiology. PHGPx content/activity in spermatozoa is positively related to typical parameters of sperm quality, such as motility, morphology, and viability, although the correlation with the latter is less straightforward. Thus, PHGPx appears as a global marker of human fertility (Imai et al., 2001; Foresta et al., 2002). On the other hand, although screened on a limited number of infertile males, gene analysis failed to demonstrate that *GPx-4* polymorphism may account for the described correlation between PHGPx activity and the sperm and parameters of fertility (Maiorino et al., 2003a).

In summary, PHGPx has a well-documented role in male fertility, and a decreased activity gives an account for the lesions observed in the germinal epithelium and spermatozoa of rodents under severe Se deficiency (McCoy and Weswig, 1969; Wu et al., 1979; Wallace et al., 1983).

7.5   BACK TO LIPID PEROXIDATION

Using inducible deletion of PHGPx in mouse embryonic fibroblasts (mouse embryonic fibroblast, MEFs), and C11 BODIPY, a probe for detecting membrane peroxidation in living cells, it has been shown that cell death due to *GPx4* disappearance is preceded by an increased C11 BODIPY fluorescence, indicating membrane peroxidation (Seiler et al., 2008). This evidence that the control of lipid peroxidation is a critical element for cell survival, revived the whole field, which before had largely been confined to toxicology and experimental pathology (Plaa and Witschi, 1976).

In the meanwhile, while searching for new anticancer drugs by high-throughput screening, two nonstructurally related compounds, namely, erastin and RSL3, have been identified as selectively lethal to cell lines containing an oncogenic RAS mutant. In the studied cells, the RAS-RAF-MEK signaling pathway was constitutively activated and, for this reason, the compounds were named "RAS-selective lethal (RSL)" (Dolma et al., 2003; Yang and Stockwell, 2008).

The cell death induced by erastin was associated with an increase of oxidants as detected by dichlorofluorescein and BODIPY fluorescence. This suggested that lipid peroxidation could be involved in the mechanism of erastin-mediated cell death. This type of cell death was prevented by iron chelation and genetic inhibition of cellular iron uptake (Yang and Stockwell, 2008; Dixon et al., 2012) and, at least partially, by the NOX inhibitor diphenylene iodonium (Dixon et al., 2012). Moreover, RAS-mutated tumor cells contained both (1) an increased iron content (Yang and Stockwell, 2008) and (2) upregulated superoxide-producing NADPH oxidases (NOX1–5, DUOX1,2) (Shinohara et al., 2010). Taken together, these observations suggested that NOX-derived superoxide and iron are involved in erastin-mediated cell death.

Notably, the nuclei of erastin-treated cells did not display the typical morphological pattern of apoptosis. Major changes in mitochondrial morphology were instead observed, pointing to a dysfunction of mitochondria as a critical element of this form of regulated cell death. Mitochondrial dysfunction, on the other hand, is not detectable when cell death is primed by agonists known to specifically induce apoptosis, necrosis, or autophagy (Yagoda et al., 2007; Dixon et al., 2012). Further, classic features of apoptosis, such as mitochondrial cytochrome c release, caspase activation, and chromatin fragmentation, were not detected (Dolma et al., 2003; Yagoda et al., 2007; Yang and Stockwell, 2008). Thus, in order to describe the phenotype of regulated cell death (RCD) by the RSL compounds, which depends on the presence of iron, the term "ferroptosis" was introduced in 2012 (Dixon et al., 2012).

Direct targets of erastin were first identified as the voltage-dependent anion channels 2 and 3 (VDAC2/3), but overexpression studies demonstrated that these channels are necessary but not sufficient to induce cell death by this compound (Yagoda et al., 2007). Later, the system x_c^- emerged as the most relevant erastin target (Dixon et al., 2012). The system x_c^- is a cystine/glutamate antiporter composed of a light-chain subunit (xCT, SLC7A11) and a heavy-chain subunit (CD98hc, SLC3A2). The system imports cystine to cells, where it is reduced to cysteine and feeds GSH biosynthesis (Lo et al., 2008). Thus, erastin apparently induces a deep decrease of cellular GSH, and this primes the particular form of iron-dependent and lipid peroxidation-mediated cell death.

Repression of expression of the x_c^- subunit SLC7A11 is one of the mechanisms by which p53 suppresses tumorigenesis (Jiang et al., 2015). This link expands the concept of ferroptosis to the physiological control of cell survival. Notably, also the toxicity of glutamate to neuronal cells was interpreted as the outcome of a functional x_c^- inhibition (Tan et al., 2001).

RSL3, the other RSL compound identified, similar to erastin induces ferroptosis, with a downstream phenotype impinging on mitochondria. Differently from erastin, however, RSL3 action appears not to be confined to cells expressing mutant RAS and has an effect neither on VDACs nor on the x_c^- system (Yang and Stockwell, 2008). Instead, RSL3 inactivated PHGPx (Yang et al., 2014) by targeting the selenocysteine at the active site (Yang et al., 2016).

Thus, the investigation of the RSL compounds disclosed the common cell death pattern of ferroptosis: decreased PHGPx turnover due to GSH shortage or direct PHGPx inhibition. The two distinct mechanisms converge on iron-dependent lipid peroxidation, which executes ferroptotic cell death. Accordingly, the Nomenclature Committee on Cell Death, 2015 filed ferroptosis as an "iron-dependent form of regulated cell death, under the control of glutathione peroxidase 4" (Galluzzi et al., 2014).

PHGPx is therefore considered as a promising target for induction of ferroptosis in cancer chemotherapy. The concept has been validated by the observation that *GPx4* expression is critical for survival of chemoresistant mesenchymal epithelial tumor cells (Viswanathan et al., 2017) and that drug-tolerant persister cancer cells are vulnerable to PHGPx inhibition (Hangauer et al., 2017). Despite the obvious relevance of the issue of PHGPx activity in supporting the life of cancer cells, the function of this enzyme as regulators of cell life and death is also been discussed for seemingly unrelated pathophysiological conditions. Indeed, in the brain from Alzheimer's patient, *GPx4* expression is downregulated (Yoo et al., 2010) and neuron-specific PHGPx depletion causes neurodegeneration *in vivo* and *ex vivo* (Seiler et al., 2008). Furthermore, inducible total body inactivation of *GPx4* in adult mice activates ferroptosis in kidney tubular cells (Friedmann Angeli et al., 2014), and conditional silencing in the immune system prevents immunity (Matsushita et al., 2015). All this evidence converges to the notion that iron-dependent lipid peroxidation is a physiological controller of cell death, and PHGPx is an appealing molecular target for diverse medical intervention strategies. It is to be inhibited for cancer treatment, where the goal is cell death, and its activity should be nutritionally optimized in degenerative diseases, where the goal is cell survival.

Apart from the key players dampening ferroptosis, PHGPx and GSH, also the nature of the fatty acids in the membranes plays an important role. This has been elegantly confirmed in cells while searching for agonists rescuing the death phenotype of *GPx4* silencing. In this study, an acyl-CoA synthetase long-chain family member 4 (ACSL4), which quite specifically activates polyunsaturated fatty acids for the insertion into phospholipids, was identified. Cells carrying the double KO *GPx4/ACSL4* were found to be resistant to ferroptosis (Doll et al., 2017). In agreement, α6β4 integrin promotes resistance to ferroptosis by activating Src and STAT3 which suppresses the expression of ACSL4 and, in consequence, decreases the content of pro-ferroptotic fatty acids esterified in the membrane phospholipids (Brown et al., 2017).

In summary, we learned from *in vitro* and *in vivo* studies that missing PHGPx activity yields iron-dependent lipid peroxidation of membrane containing polyunsaturated fatty acids, and this evolves into a novel form of RCD, the ferroptosis. The remarkable feature of ferroptosis is that, unlike the other RCD pathways, it does not require any specific agonist, such as FAS ligand or TNF. It is sufficient that the critical mechanism for the inhibition of lipid peroxidation becomes limiting. Inversely, the PHGPx/GSH system minimizes cell death and allows proliferation. This general concept elegantly fits the recent observation that ATF4, a mediator of the metabolic and oxidative homeostasis, promotes glioma cell growth by transcriptional targeting of the xCT system, thus increasing cellular GSH content (Chen et al., 2017).

Yet, many crucial questions remain still unresolved:

Vitamin E and free radical scavengers such as ferrostatin and liproxstatin were reported to inhibit ferroptosis in the absence of PHGPx activity. This has been attributed to their peroxidation chain breaking activity (Dixon et al., 2012; Friedmann Angeli et al., 2014). However, the reduction of PLOO$^\bullet$ yields PLOOH, which continuously generates new chain reactions in the presence of iron. Thus, scavenging of lipid hydroperoxyl radical cannot likely be the mechanism of inhibition of ferroptosis. An interaction of vitamin E with the species distinct from the PLOO$^\bullet$ should be investigated.

How is PHGPx turnover regulated *in vivo*?

How is the PHGPx substrate PLOOH formed in cells? Is superoxide produced by mitochondria or NADPH oxidases the chain-initiating species or do we have to consider an initial involvement of lipoxygenases? Indeed, a lipoxygenase could act on intact phospholipids. In favor of this, it is known for years that vitamin E, which rescues cell death by GPx4 depletion, dampens lipoxygenase activity (Panganamala and Cornwell, 1982), although the *in vivo* relevance of this vitamin E action remains controversial. Reverse genetic experiments so far failed to provide a nonambiguous answer about the effect of the hottest candidate for ferroptosis induction, the product of 12/15 lipoxygenase. While in cells 12/15 LOX deletion rescued the lethal phenotype of *GPx4* KO (Seiler et al., 2008), this was not confirmed in whole mice (Friedmann Angeli et al., 2014; Brutsch et al., 2015). Is any other lipoxygenase involved?

Are only, or just mainly, the mitochondrial membranes the target of lipid peroxidation in ferroptosis? Indeed the high affinity of PHGPx for cardiolipin polar heads (Cozza et al., 2017) supports the direct involvement of mitochondrial membranes in ferroptosis. Also the association of PHGPx with the inner mitochondrial membranes in wild-type mice or in mice overexpressing c-*GPx4* over a *GPx4*$^{-/-}$ background (Liang et al., 2009) points into this direction.

How is the availability of iron of the labile cytosolic pool modulated?

Is there any specific physiological species competent for the inactivation of PHGPx acting similar to the candidate drug RSL3?

We are confident that answers to these questions will contribute to the description of an intriguing scenario where oxygen, GSH, PHGPx, and vitamin E are the pivotal regulators of both life and death of aerobic cells. Needless to say, biomedicine and pharmacology is expecting a remarkable gain of information, which will pave the way to innovative therapeutic approaches.

ACKNOWLEDGMENTS

This work was supported by Human Frontier Science Program, Grant RGP0013/2014 to FU, and by the University of Padova, Progetti di Ricerca di Ateneo, CPDA151400/15 to MM.

REFERENCES

Arai, M., H. Imai, D. Sumi et al. 1996. Import into mitochondria of phospholipid hydroperoxide glutathione peroxidase requires a leader sequence. *Biochem Biophys Res Commun* 227:433–9.

Bauché, F., M. H. Fouchard, and B. Jégou. 1994. Antioxidant system in rat testicular cells. *FEBS Lett* 349:392–6.

Bosello-Travain, V., M. Conrad, G. Cozza et al. 2013. Protein disulfide isomerase and glutathione are alternative substrates in the one cys catalytic cycle of glutathione peroxidase 7. *Biochim Biophys Acta* 1830:3846–57.

Brigelius-Flohé, R., K. D. Aumann, H. Blöcker et al. 1994. Phospholipid- hydroperoxide glutathione peroxidase. Genomic DNA, cDNA, and deduced amino acid sequence. *J Biol Chem* 269:7342–8.

Brigelius-Flohé, R., and M. Maiorino. 2013. Glutathione peroxidases. *Biochim Biophys Acta* 1830:3289–303.

Brown, C. W., J. J. Amante, H. L. Goel, and A. M. Mercurio. 2017. The α6β4 integrin promotes resistance to ferroptosis. *J Cell Biol* 216:4287–4297

Brutsch, S. H., C. C. Wang, L. Li et al. 2015. Expression of inactive glutathione peroxidase 4 leads to embryonic lethality, and inactivation of the alox15 gene does not rescue such knock-in mice. *Antioxid Redox Signal* 22:281–93.

Burk, R. F., M. J. Trumble, and R. A. Lawrence. 1980. Rat hepatic cytosolic glutathione-dependent enzyme protection against lipid peroxidation in the NADPH-microsomal lipid peroxidation system. *Biochim Biophys Acta* 618:35–41.

Chambers, I., J. Frampton, P. Goldfarb et al. 1986. The structure of the mouse glutathione peroxidase gene: The selenocysteine in the active site is encoded by the 'termination' codon, TGA. *EMBO J* 5:1221–7.

Chen, D., Z. Fan, M. Rauh et al. 2017. Atf4 promotes angiogenesis and neuronal cell death and confers ferroptosis in a xCT-dependent manner. *Oncogene* 36:5593–608.

Cheng, Z. Y., and Y. Z. Li. 2007. What is responsible for the initiating chemistry of iron-mediated lipid peroxidation: An update. *Chemical Reviews* 107:748–66.

Conrad, M., S. G. Moreno, F. Sinowatz et al. 2005. The nuclear form of phospholipid hydroperoxide glutathione peroxidase is a protein thiol peroxidase contributing to sperm chromatin stability. *Mol Cell Biol* 25:7637–44.

Cozza, G., M. Rossetto, V. Bosello-Travain et al. 2017. Glutathione peroxidase 4-catalyzed reduction of lipid hydroperoxides in membranes: The polar head of membrane phospholipids binds the enzyme and addresses the fatty acid hydroperoxide group toward the redox center. *Free Radic Biol Med* 112:1–11.

Daolio, S., P. Traldi, F. Ursini, M. Maiorino, and C. Gregolin. 1983. Evidence of peroxidase activity in the peroxidation inhibiting protein on dilinoleyl phosphatidylcholine hydroperoxide as obtained in direct electron impact conditions. *Biomed Mass Spectrom* 10:499–504.

Davies, S. S., and L. L. Guo. 2014. Lipid peroxidation generates biologically active phospholipids including oxidatively n-modified phospholipids. *Chem Phys Lipids* 181:1–33.

Dixon, S. J., K. M. Lemberg, M. R. Lamprecht et al. 2012. Ferroptosis: An iron- dependent form of nonapoptotic cell death. *Cell* 149:1060–72.

Doll, S., B. Proneth, Y. Y. Tyurina et al. 2017. Acsl4 dictates ferroptosis sensitivity by shaping cellular lipid composition. *Nat Chem Biol* 13:91–8.

Dolma, S., S. L. Lessnick, W. C. Hahn, and B. R. Stockwell. 2003. Identification of genotype-selective antitumor agents using synthetic lethal chemical screening in engineered human tumor cells. *Cancer Cell* 3:285–96.

Ernster, L., K. Nordenbrand, and S. Orrenius. 1982. Microsomal lipid peroxidation: Mechanism and some biomedical implications. In *Lipid Peroxides in Biology and Medicine*, edited by K. Yagi, 55–79. New York: Academic Press.

Fisher, H. M., and R. J. Aitken. 1997. Comparative analysis of the ability of precursor germ cells and epididymal spermatozoa to generate reactive oxygen metabolites. *J Exp Zool* 277:390–400.

Flohé, L. 2011. Glutathione peroxidases. In *Selenoproteins and Mimics*, edited by J. Liu, G. Luo and Y. Mu, 1–25. Berlin Heidelberg: Springer-Verlag.

Flohé, L., G. Loschen, W. A. Günzler, and E. Eichele. 1972. Glutathione peroxidase V. The kinetic mechanism. *Hoppe-Seyler Z Physiol Chem* 353:987–99.

Flohé, L., S. Toppo, G. Cozza, and F. Ursini. 2011. A comparison of thiol peroxidase mechanisms. *Antioxid Redox Signal* 15:763–80.

Foresta, C., L. Flohé, A. Garolla et al. 2002. Male fertility is linked to the selenoprotein phospholipid hydroperoxide glutathione peroxidase. *Biol Reprod* 67:967–71.

Friedmann Angeli, J. P., M. Schneider, B. Proneth et al. 2014. Inactivation of the ferroptosis regulator Gpx4 triggers acute renal failure in mice. *Nat Cell Biol* 16:1180–91.

Galluzzi, L., J. M. Bravo-San Pedro, I. Vitale et al. 2014. Essential versus accessory aspects of cell death: Recommendations of the NCCD 2015. *Cell Death Differ.* 22:58–73.

Gibson, D. D., K. R. Hornbrook, and P. B. McCay. 1980. Glutathione-dependent inhibition of lipid peroxidation by a soluble, heat-labile factor in animal tissues. *Biochim Biophys Acta* 620:572–82.

Gladyshev, V. N., E. S. Arner, M. J. Berry et al. 2016. Selenoprotein gene nomenclature. *J Biol Chem* 291:24036–40.

Godeas, C., F. Tramer, F. Micali et al. 1996. Phospholipid hydroperoxide glutathione peroxidase (PHGPx) in rat testis nuclei is bound to chromatin. *Biochem Mol Med* 59:118–24.

Godeas, C., F. Tramer, F. Micali et al. 1997. Distribution and possible novel role of phospholipid hydroperoxide glutathione peroxidase in rat epididymal spermatozoa. *Biol Reprod* 57:1502–8.

Hangauer, M. J., V. S. Viswanathan, M. J. Ryan et al. 2017. Drug-tolerant persister cancer cells are vulnerable to GPX4 inhibition. *Nature.* 551(7679):247–50.

Haraguchi, C. M., T. Mabuchi, S. Hirata et al. 2003. Spatiotemporal changes of levels of a moonlighting protein, phospholipid hydroperoxide glutathione peroxidase, in subcellular compartments during spermatogenesis in the rat testis. *Biol Reprod* 69:885–95.

Hochstein, P., and L. Ernster. 1963. ADP-activated lipid peroxidation coupled to the TPNH oxidase system of microsomes. *Biochem Biophys Res Commun* 12:388–94.

Hochstein, P., and L. Ernster. 1964. Microsomal peroxidation of lipids and its possible role in cell injury. In *Ciba Foundation Symposium on Cell Injury*, edited by A. V. S. deReuck and J. Knight, 123–34. London: Churchill.

Hunter, F. E., Jr., J. M. Gebicki, P. E. Hoffsten, J. Weinstein, and A. Scott. 1963. Swelling and lysis of rat liver mitochondria induced by ferrous ions. *J Biol Chem* 238:828–35.

Imai, H., N. Hakkaku, R. Iwamoto et al. 2009. Depletion of selenoprotein GPx4 in spermatocytes causes male infertility in mice. *J Biol Chem* 284:32522–32.

Imai, H., F. Hirao, T. Sakamoto et al. 2003. Early embryonic lethality caused by targeted disruption of the mouse PHGPx gene. *Biochem Biophys Res Commun* 305:278–86.

Imai, H., K. Suzuki, K. Ishizaka et al. 2001. Failure of the expression of phospholipid hydroperoxide glutathione peroxidase in the spermatozoa of human infertile males. *Biol Reprod* 64:674–83.

Jiang, L., N. Kon, T. Li et al. 2015. Ferroptosis as a p53-mediated activity during tumour suppression. *Nature* 520:57–62.

Kaschnitz, R. M., and Y. Hatefi. 1975. Lipid oxidation in biological membranes. Electron transfer proteins as initiators of lipid autoxidation. *Arch Biochem Biophys* 171:292–304.

Leinfelder, W., E. Zehelein, M. A. Mandrand-Berthelot, and A. Böck. 1988. Gene for a novel tRNA species that accepts l-serine and cotranslationally inserts selenocysteine. *Nature* 331:723–5.

Liang, H., S. E. Yoo, R. Na et al. 2009. Short form glutathione peroxidase 4 is the essential isoform required for survival and somatic mitochondrial functions. *J Biol Chem* 284:30836–44.

Lo, M., Y. Z. Wang, and P. W. Gout. 2008. The x(c)- cystine/glutamate antiporter: A potential target for therapy of cancer and other diseases. *J Cell Physiol* 215:593–602.

Maiorino, M., V. Bosello, F. Ursini et al. 2003a. Genetic variations of gpx-4 and male infertility in humans. *Biol Reprod* 68:1134–41.

Maiorino, M., V. Bosello-Travain, G. Cozza, et al. 2015. Understanding mammalian glutathione peroxidase 7 in the light of its homologs. *Free Radic Biol Med* 83:352–60.

Maiorino, M., C. Gregolin, and F. Ursini. 1990. Phospholipid hydroperoxide glutathione peroxidase. *Meth Enzymol* 186:448–57.

Maiorino, M., A. Roveri, L. Benazzi et al. 2005. Functional interaction of phospholipid hydroperoxide glutathione peroxidase with sperm mitochondrion- associated cysteine-rich protein discloses the adjacent cysteine motif as a new substrate of the selenoperoxidase. *J Biol Chem* 280:38395–402.

Maiorino, M., M. Scapin, F. Ursini et al. 2003b. Distinct promoters determine alternative transcription of gpx-4 into phospholipid-hydroperoxide glutathione peroxidase variants. *J Biol Chem* 278:34286–90.

Maiorino, M., J. P. Thomas, A. W. Girotti, and F. Ursini. 1991. Reactivity of phospholipid hydroperoxide glutathione peroxidase with membrane and lipoprotein lipid hydroperoxides. *Free Radic Res Commun* 12–13 (Pt 1):131–5.

Maiorino, M., F. Ursini, M. Valente, L. Ferri, and C. Gregolin. 1980. Attività dei complessi tra trietilenetetraamina e analoghi e Fe^{+++} nella perossidazione lipidica. *Atti dell'Istituto Veneto di Scienze, Lettere ed Arti Tomo* CXXXVIII:131–8.

Maiorino, M., J. B. Wissing, R. Brigelius-Flohé et al. 1998. Testosterone mediates expression of the selenoprotein PHGPx by induction of spermatogenesis and not by direct transcriptional gene activation. *FASEB J* 12:1359–70.

Matsushita, M., S. Freigang, C. Schneider et al. 2015. T cell lipid peroxidation induces ferroptosis and prevents immunity to infection. *J Exp Med* 212:555–68.

McCay, P. B., D. D. Gibson, K. L. Fong, and K. R. Hornbrook. 1976. Effect of glutathione peroxidase activity on lipid peroxidation in biological membranes. *Biochim Biophys Acta* 431:459–68.

McCoy, K. E., and P. H. Weswig. 1969. Some selenium responses in the rat not related to vitamin E. *J Nutr* 98:383–9.

Mills, G. C. 1957. Hemoglobin catabolism. I. Glutathione peroxidase, an erythrocyte enzyme which protects hemoglobin from oxidative breakdown. *J Biol Chem* 229:189–97.

Orian, L., P. Mauri, A. Roveri et al. 2015. Selenocysteine oxidation in glutathione peroxidase catalysis: An MS-supported quantum mechanics study. *Free Radic Biol Med* 87:1–14.

Ottolenghi, A. 1959. Interaction of ascorbic acid and mitochondrial lipids. *Arch Biochem Biophys* 79:355–63.

Pallini, V., and E. Bacci. 1979. Bull sperm selenium is bound to a structural protein of mitochondria. *J Submicrosc Cytol* 11:165–70.

Panganamala, R. V., and D. G. Cornwell. 1982. The effects of vitamin E on arachidonic acid metabolism. *Ann N Y Acad Sci* 393:376–91.

Pfeifer, H., M. Conrad, D. Roethlein et al. 2001. Identification of a specific sperm nuclei selenoenzyme necessary for protamine thiol cross-linking during sperm maturation. *FASEB J* 15:1236–8.

Plaa, G. L., and H. Witschi. 1976. Chemicals, drugs, and lipid peroxidation. *Annu Rev Pharmacol Toxicol* 16:125–41.

Puglisi, R., F. Tramer, E. Panfili et al. 2003. Differential splicing of the phospholipid hydroperoxide glutathione peroxidase gene in diploid and haploid male germ cells in the rat. *Biol Reprod* 68:405–11.

Pushpa-Rekha, T. R., A. L. Burdsall, L. M. Oleksa, G. M. Chisolm, and D. M. Driscoll. 1995. Rat phospholipid-hydroperoxide glutathione peroxidase. cDNA cloning and identification of multiple transcription and translation start sites. *J Biol Chem* 270:26993–9.

Roveri, A., A. Casasco, M. Maiorino et al. 1992. Phospholipid hydroperoxide glutathione peroxidase of rat testis. Gonadotropin dependence and immunocytochemical identification. *J Biol Chem* 267:6142–6.

Roveri, A., M. Maiorino, C. Nisii, and F. Ursini. 1994. Purification and characterization of phospholipid hydroperoxide glutathione peroxidase from rat testis mitochondrial membranes. *Biochim Biophys Acta* 1208:211–21.

Scheerer, P., A. Borchert, N. Krauss et al. 2007. Structural basis for catalytic activity and enzyme polymerization of phospholipid hydroperoxide glutathione peroxidase-4 (GPx4). *Biochemistry* 46:9041–9.

Schneider, M., H. Forster, A. Boersma et al. 2009. Mitochondrial glutathione peroxidase 4 disruption causes male infertility. *Faseb J* 23:3233–42.

Schuckelt, R., R. Brigelius-Flohé, M. Maiorino et al. 1991. Phospholipid hydroperoxide glutathione peroxidase is a selenoenzyme distinct from the classical glutathione peroxidase as evident from cDNA and amino acid sequencing. *Free Radic Res Commun* 14:343–61.

Seiler, A., M. Schneider, H. Forster et al. 2008. Glutathione peroxidase 4 senses and translates oxidative stress into 12/15-lipoxygenase dependent- and AIF-mediated cell death. *Cell Metab* 8:237–48.

Shalgi, R., J. Seligman, and N. S. Kosower. 1989. Dynamics of the thiol status of rat spermatozoa during maturation: Analysis with the fluorescent labeling agent monobromobimane. *Biol Reprod* 40:1037–45.

Shinohara, M., Y. Adachi, J. Mitsushita et al. 2010. Reactive oxygen generated by NADPH oxidase 1 (nox1) contributes to cell invasion by regulating matrix metalloprotease-9 production and cell migration. *J Biol Chem* 285:4481–8.

Stadtman, T. C. 1987. Specific occurrence of selenium in enzymes and amino acid tRNAs. *Faseb J* 1:375–9.

Takahashi, K., M. Akasaka, Y. Yamamoto et al. 1990. Primary structure of human plasma glutathione peroxidase deduced from cdna sequences. *J Biochem* 108:145–8.

Tan, S., D. Schubert, and P. Maher. 2001. Oxytosis: A novel form of programmed cell death. *Curr Top Med Chem* 1:497–506.

Tappel, A., and H. Zalkin. 1960. Inhibition of lipid peroxidation in microsomes by vitamin E. *Nature* 185:35.

Thiele, E. H., and J. W. Huff. 1960. Quantitative measurements of lipide peroxide formation by normal liver mitochondria under various conditions. *Arch Biochem Biophys* 88:203–7.

Thomas, J. P., P. G. Geiger, M. Maiorino, F. Ursini, and A. W. Girotti. 1990a. Enzymatic reduction of phospholipid and cholesterol hydroperoxides in artificial bilayers and lipoproteins. *Biochim Biophys Acta* 1045:252–60.

Thomas, J. P., M. Maiorino, F. Ursini, and A. W. Girotti. 1990b. Protective action of phospholipid hydroperoxide glutathione peroxidase against membrane-damaging lipid peroxidation. In situ reduction of phospholipid and cholesterol hydroperoxides. *J Biol Chem* 265:454–61.

Toppo, S., L. Flohé, F. Ursini, S. Vanin, and M. Maiorino. 2009. Catalytic mechanisms and specificities of glutathione peroxidases: Variations of a basic scheme. *Biochim Biophys Acta* 1790:1486–500.

Toppo, S., S. Vanin, V. Bosello, and S. C. E. Tosatto. 2008. Evolutionary and structural insights into the multifaceted glutathione peroxidase (GPx) superfamily. *Antioxid Redox Signal* 10:1501–13.

Ursini, F., S. Heim, M. Kiess et al. 1999. Dual function of the selenoprotein PHGPx during sperm maturation. *Science* 285:1393–6.

Ursini, F., M. Maiorino, R. Brigelius-Flohé et al. 1995. Diversity of glutathione peroxidases. *Meth Enzymol* 252:38–53.

Ursini, F., M. Maiorino, and C. Gregolin. 1985. The selenoenzyme phospholipid hydroperoxide glutathione peroxidase. *Biochim Biophys Acta* 839:62–70.

Ursini, F., M. Maiorino, M. Valente, L. Ferri, and C. Gregolin. 1982. Purification from pig liver of a protein which protects liposomes and biomembranes from peroxidative degradation and exhibits glutathione peroxidase activity on phosphatidylcholine hydroperoxides. *Biochim Biophys Acta* 710:197–211.

Viswanathan, V. S., M. J. Ryan, H. D. Dhruv et al. 2017. Dependency of a therapy- resistant state of cancer cells on a lipid peroxidase pathway. *Nature* 547:453–7.

Wallace, E., H. I. Calvin, and G. W. Cooper. 1983. Progressive defects observed in mouse sperm during the course of three generations of selenium deficiency. *Molecular Reproduction e Development* 7:377–87.

Wang, J. H. 1955. On the detailed mechanism of a new type of catalase-like action *J Amer Chem Soc* 77:4715–19.

Wu, A. S., J. E. Oldfield, L. R. Shull, and P. R. Cheeke. 1979. Specific effect of selenium deficiency on rat sperm. *Biol Reprod* 20:793–8.

Yagoda, N., M. von Rechenberg, E. Zaganjor et al. 2007. Ras-Raf-Mek- dependent oxidative cell death involving voltage-dependent anion channels. *Nature* 447:864–8.

Yang, W. S., K. J. Kim, M. M. Gaschler et al. 2016. Peroxidation of polyunsaturated fatty acids by lipoxygenases drives ferroptosis. *Proc Natl Acad Sci U S A* 113:E4966–75.

Yang, W. S., R. Sriramaratnam, M. E. Welsch et al. 2014. Regulation of ferroptotic cancer cell death by GPX4. *Cell* 156:317–31.

Yang, W. S., and B. R. Stockwell. 2008. Synthetic lethal screening identifies compounds activating iron-dependent, nonapoptotic cell death in oncogenic-RAS-harboring cancer cells. *Chem Biol* 15:234–45.

Yant, L. J., Q. Ran, L. Rao et al. 2003. The selenoprotein GPX4 is essential for mouse development and protects from radiation and oxidative damage insults. *Free Radic Biol Med* 34:496–502.

Yin, H. Y., L. B. Xu, and N. A. Porter. 2011. Free radical lipid peroxidation: Mechanisms and analysis. *Chemical Reviews* 111:5944–72.

Yoo, M. H., X. Gu, X. M. Xu et al. 2010. Delineating the role of glutathione peroxidase 4 in protecting cells against lipid hydroperoxide damage and in Alzheimer's disease. *Antioxid Redox Signal* 12:819–27.

Zalkin, H., and A. L. Tappel. 1960. Studies of the mechanism of vitamin e action. Iv. Lipide peroxidation in the vitamin e-deficient rabbit. *Arch Biochem Biophys* 88:113–17.

8 Thiols, Glutathione, GPx4, and Lipid Metabolism at the Crossroads of Cell Death and Survival

José Pedro Friedmann Angeli
University of Würzburg

Valerian E. Kagan
University of Pittsburgh

Marcus Conrad
Helmholtz Zentrum München

CONTENTS

8.1 INTRODUCTION

The role of glutathione (GSH) as a prosurvival molecule has been known for many decades. Landmark studies using animals deficient for γ-glutamylcysteine synthetase (γ-GCSc; also abbreviated GCL) have demonstrated the essential function of GSH for proper embryogenesis, tissue homeostasis, and cell viability (Dalton et al., 2000; Shi et al., 2000). These essential functions are supported by the inability of animals to synthesize GSH and, to survive past embryonic day 6.5 (E6.5) (Shi et al., 2000).

DOI: 10.1201/9781351261760-10

Additionally, cell models derived from these animals also fail to proliferate in cell culture unless exogenous GSH or other cysteine-donating compounds, such as N-acetylcysteine (NAC), are added (Shi et al., 2000). Yet, molecular and metabolic pathways involved in the loss of cell viability due to GSH depletion have remained obscure for many years. Nevertheless, the recognition that GSH deficiency leads to a defined form of cell death, dubbed ferroptosis (Dixon et al., 2012), provided renewed breath for the field, and is now an active and rapidly expanding topic of research, as it is proposed to be a main cellular event contributing to a wide array of patho(physio) logical conditions.

The aim of our present chapter is thus to provide a historical appraisal of the initial findings linking thiol metabolism with particular emphasis on its support to GSH biosynthesis, culminating in proper activity of what is now known to be the rate-limiting enzyme in providing the prosurvival function of GSH, glutathione peroxidase 4 (GPx4). Moreover, we highlight the relevance of these findings in the context of several pathologically relevant contexts.

8.2 CYSTEINE AND CELL SURVIVAL

Early landmark studies performed by Harry Eagle characterized the basic principles of what is known today to be the minimal requirement for cell growth in culture. His work identified a series of essential amino acids, namely arginine, glutamine, histidine, tyrosine, and cyst(e)ine to be an absolute requirement for cell proliferation and cell viability. In fact, all these amino acids have a direct impact on cell viability. Remarkably, glutamine and, in particular, cysteine were shown to affect cell viability at much faster rate, approximately 2–3 days (Eagle, 1955), than the others. Another important observation made by Eagle was that not all cells die in the absence of cysteine, thus suggesting that, at least in some cases, cysteine could be synthetized in the presence of defined precursors. In-depth studies of this phenomenon revealed that in some cells, when cyst(e)ine becomes limiting, they utilize methionine and glucose to replenish their cysteine pool (Eagle et al., 1961) (Figure 8.1). This was shown to proceed through metabolic intermediates consisting of homocysteine and serine deriving from methionine and glucose, respectively. This was the first indication for a potential enzyme-mediated demethylating activity acting on methionine to generate homocysteine. The activity was later demonstrated to be related to an active reverse transsulfuration pathway, which is responsible for the conversion of methionine to cysteine (Brosnan et al., 2006). This metabolic pathway is able to generate cysteine trough the activity of cystathionine β-synthase (CTH) and cystathionine γ-lyase (CBS) (Jhee et al., 2005). The combined activity of these enzymes transfers the thiol group from methionine to homocysteine, cystathionine and ultimately cysteine. Remarkably, he also detected that cells derived from patients diagnosed with cystathioninuria, which was later shown to be a deficiency in the transsulfuration pathway, were unable to survive in a cyst(e)ine-depleted medium containing methionine, cystathionine or homocystine as cysteine precursors. Despite this biosynthetic capacity, survival in a cysteine-free media is dependent on (a) the presence of trace amounts of preformed cyst(e)ine or homocyst(e)ine, or (b) the cell population density is increased to appropriate levels. Both features converge to the capacity of a cell

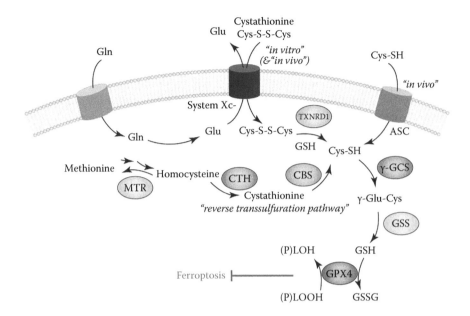

FIGURE 8.1 Thiol handling in mammalian cells. Several sources exist in cells to cope with the cells' demand for cysteine. While the neutral amino acid transporter ASC (alanine, serine, cysteine) is one of the key sources of cysteine in a whole organism, the cystine/glutamate antiporter system Xc- is the most relevant one in cell culture/organoid systems as the majority of cysteine is present in its oxidized form cystine in cell culture medium. CBH, cystathionine beta-synthase; CTH, cystathionase (cystathionine gamma-lyase); γ-GCS, γ-glutamylcysteine synthetase; GPx4, glutathione peroxidase 4; GSH, reduced glutathione; GSSG, oxidized glutathione; GSS, glutathione synthetase; TXNRD1, thioredoxin reductase 1; (P)LOOH, (phospho)lipid hydroperoxide.

line to maintain a critical steady state of a specific metabolite, since continuous synthesis will lead to increased secretion and diffusion of the amino acid produced. Conclusively, these early studies provided the basis of our current understanding of how a cell in culture sustains its cysteine pool.

Further characterization on the importance of cysteine for cell viability came from studies performed in Shiro Bannai's laboratory. For many years, his group worked on the characterization of amino acid transport systems across the plasma membrane, whereby particular interest was given to mechanisms that support cellular cystine uptake. Among the most important discoveries made by his group was the fact that cystine and glutamate functionally interact and share a similar transport system (Bannai et al., 1980) (Figure 8.1). This was reinforced by the finding that cystine uptake can be inhibited by high extracellular concentrations of glutamate in a competitive manner. Additionally, his group in a *tour-de-force* managed, by using expression of cDNA pools in *Xenopus* oocytes followed by subtractive analysis of these cDNA pools, to identify the two components of this transporter (i.e., system Xc-) required for cystine uptake activity (Sato et al., 1999). Sequence analysis revealed that one of the subunits is the 4F2 cell surface antigen

(4F2hc; SLC3A2) and the other one a novel protein conferring the specificity to the transporter, which the authors named xCT (SLC7A11). Further attesting for the importance of xCT in sustaining intracellular cysteine levels came from the recognition that excess extracellular glutamate, which inhibits system Xc-activity, leads to cysteine starvation and cell death. Unequivocal and important insights into the role of xCT in cysteine metabolism came from the generation of xCT knockout animals (Sato et al., 2005). xCT deficient animals unexpectedly presented no overt phenotype besides a decrease in intracellular GSH and an increase in cystine plasma levels. Nonetheless, embryonic fibroblasts derived from xCT deficient animals failed to grow under standard cell culture conditions (Sato et al., 2005). This reinforces the notion that amino acid uptake differs strongly depending on the redox environment (Conrad et al., 2012). Curiously enough, it was recently demonstrated that hepatocytes lacking xCT are able to survive and proliferate in cell culture (Lee et al., 2017). The viability of hepatocytes lacking xCT is accompanied by an increased consumption of methionine, a feature consistent with early observations showing that the transsulfuration pathway is able to sustain the cysteine pool.

Despite these early observations that cysteine deprivation causes loss of cell viability, little was known regarding the mechanism accounting for the cell death process. An initial characterization of the molecular events leading to cell death upon cysteine depletion were carried almost 40 years ago revealing two critical features: (1) Cysteine starvation is followed by a drastic and rapid decrease in GSH levels, and (2) cell viability could be maintained in the presence of the antioxidant α-tocopherol for days even when levels of GSH were almost undetectable (Bannai et al., 1977). Therefore, it was already hinted that loss of GSH was associated with the inability of the cell to prevent damage to lipids and associated cell death. Despite this initial discovery, the field remained largely rudimentary, and little mechanistic insights were gained over a long period of time.

In this context, it is important to highlight the work carried out by Schubert and Maher. By studying the mechanism of the toxicity caused by glutamate in neurons lacking N-methyl-D-aspartate receptors (Schubert et al., 1992), they provided evidence that glutamate-mediated inhibition of system Xc- leads to GSH depletion and cell death by a mechanism entailing lipid peroxidation and Ca^{2+} influx. Specifically, their work suggested that depletion of GSH is able to activate 12-lipoxygenase, and its metabolites stimulate the formation of soluble guanylate cyclase, which, in turn, is able to activate Ca^{2+} channels and ultimately cell death (Li et al., 1997a,b). This mode of cell death entailing glutamate-mediated GSH and lipid peroxidation-mediated cell death was later termed oxytosis (Tan et al., 2001).

8.3   GPX4 AS THE KEY ENZYME LINKING GSH TO CELL VIABILITY

At this point, the Bornkamm/Conrad and Stockwell groups have started to study two seemingly disparate phenomena. The Bornkamm and Conrad group were intrigued by a phenomenon observed in Burkitt's lymphoma (BL) cultures. BL is a B lymphocyte cancer, one of the most rapidly growing tumors in humans, characterized by deregulated c-myc activity. This cancer entity, when established from respective patients, shows a remarkable feature as it is unable to grow at lower cell densities without the support of an irradiated fibroblast feeder layer (Falk et al., 1993) (Figure 8.2). The survival factor was identified as cysteine (Brielmeier et al., 1998), but also copper chelating agents and low micromolar concentrations of sodium selenite (Conrad, unpublished observation) were able to rescue this phenotype (Falk et al., 1998). Cysteine secreted by feeder cells is taken up by BL cells in its reduced form to maintain the intracellular cysteine pool of BL because it is known that these cells have an extremely low xCT activity (Falk et al., 1993). In fact, forced expression of xCT in the BL cell line HH514 provided initial evidence that system Xc- maintains a cystine/cysteine redox cycle over the plasma membrane, which can act largely independent of intracellular GSH levels (Banjac et al., 2008). In fact, overexpression of xCT in cells lacking γ-GCS genetically even suppressed cell death induced by GSH deficiency, implying that cysteine is able to compensate for most of the effects conferred by intracellular GSH including proper functioning of the selenoenzyme GPx4 (Mandal et al., 2010). xCT overexpressing γ-GCS knockout cells could even support the growth of mock-transfected γ-GCS knockout cells, further supporting the idea that the feeder cells supply cocultured cells with sufficient cysteine, which can be easily taken up by the neutral amino acid transport system ASC (alanine, serine, and cysteine) (Figure 8.1). Moreover, cell death occurring in BL recapitulated the features observed previously for xCT knockout cultures and cells growing in cystine-free media. Additionally, the cell death pathway observed in these cell lines was independent of the classical antiapoptotic protein Bcl2 and other apoptosis-related factors (Falk et al., 1993).

In order to characterize factors that could support the growth of BL under cysteine-limiting conditions, the authors performed a cDNA expression approach that led to the discovery of GPx4 (previously referred to as phospholipid hydroperoxide glutathione peroxidase, PHGPx; EC 1.11.1. 12) as a highly cell-protective

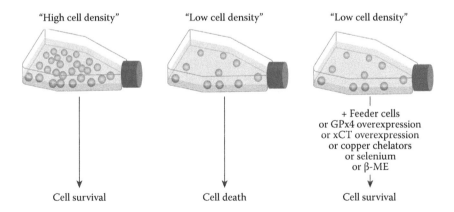

FIGURE 8.2 Growth behavior of Burkitt's lymphoma (BL) cells. Growth and survival of BL cells *in vitro* is directly related to cell densities. Growth at nonpermissive conditions can be compensated either by coculturing on an irradiated feeder layer, GPx4 overexpression, xCT overexpression, copper chelators, low nanomolar concentrations of sodium selenite or ß-mercaptoethanol (ß-ME).

protein (Brielmeier et al., 2001). GPx4 is a monomeric selenocysteine-containing glutathione peroxidase that had been described to be essential for the repair of oxidized phospholipids (Ursini et al., 1982). Also, the *in vivo* importance of selenium utilization in form of the 21st amino acid selenocysteine in GPx4 has been recently highlighted, when it was shown that it confers full protection against irreversible enzyme inactivation and associated cell death (Ingold et al., 2018).

The early finding that GPx4 suppresses lethality of BL cells at nonpermissive culture conditions provided further evidence for the importance of oxidatively induced damage to lipids in the onset of cell death caused by cysteine depletion (Figure 8.2). In fact, we and others initially considered this cell-protective function being an "antiapoptotic" activity [reviewed in Friedmann Angeli et al. (2016) and Conrad et al. (2007)]. However, in the meantime, it has been demonstrated in a number of studies that GPx4 is the key regulator of a recently recognized form of regulated necrotic cell death, called ferroptosis (see below). Motivated by this initial finding, the first conditional knockout mouse model for GPx4 was generated (Seiler et al., 2008). Important to say is that classical knockout mouse strains targeting GPx4 were generated earlier, unveiling the essentiality of GPx4 for early embryogenesis, as these animals fail to develop past embryonic day 7.5 (E7.5) (Imai et al., 2003; Yant et al., 2003). The utilization of a conditional mouse line allowed to explore the molecular underpinnings accounting for the loss of viability upon loss of GPx4 (Seiler et al., 2008). An initial assessment of the molecular events leading to loss of cell viability upon GPx4 recapitulated several features of the molecular events observed in cysteine-deprived cells, which are the rescue by α-tocopherol (vitamin E), and the lack of rescue of cell death by classical apoptosis inhibitors (Seiler et al., 2008). Moreover, at that time a potential contribution of 12/15-lipoxygenase (ALOX15) was proposed based on pharmacological inhibitors and the increased resistance of *Alox15*-deficient fibroblasts to GSH depletion. Nonetheless, as it will be detailed later, the picture seems to be rather complex.

Independently, the Stockwell group was applying high throughput screening campaigns with small molecules to target vulnerabilities in cancer cells carrying defined oncogenic mutations. Their initial efforts on the identification of small molecules targeting RAS mutants identified a class of small molecules that was able to efficiently kill these tumors by a mechanism involving the excessive production of reactive oxygen species (ROS) (Dolma et al., 2003). One of the initial hits was named erastin for "eradicator of RAS and ST-expressing cells." Further work from his group uncovered additional small molecules entailing a similar cell killing profile as the one observed for erastin (Yang et al., 2008). Subsequent work characterized the mode of action of erastin as it uncovered a critical and essential feature required for the cell death, which is the capacity of erastin to deplete the cellular GSH pool by a mechanism involving the direct inhibition of xCT (Dixon et al., 2012). Due to the uniqueness of this form of cell death and its intricate connections with iron metabolism, the authors coined the name "ferroptosis" (from the Latin word "ferrum" for iron and the Greek word "ptosis" for fall). This work also identified the first specific small molecules inhibitor of this pathway, called ferrostatin-1 (Dixon et al., 2012). Of particular interest was the further characterization of the compound named (1S, 3R)-RSL3 (RSL3, ras synthetic lethal 3) that similar to erastin triggered a cell death pathway that shared most of the biochemical and ultrastructural features, but unlike erastin did so without impacting the cysteine and the GSH pool (Yang et al., 2014). Using chemoproteomic approaches, the authors were able to identify GPx4 as the cellular target of RSL3 (Yang et al., 2014). Cell death induced by RSL3 was rescuable by ferroptosis inhibitors, and as we showed independently that they could also suppress the genetic loss of GPx4 (Friedmann Angeli et al., 2014). Therefore, work carried out by these two groups over the course of almost 15 years led to the identification of GPx4 as the enzyme responsible for the prosurvival function conferred by GSH (Stockwell et al., 2017). More importantly, they also established that this form of cell death can be elicited without a direct impact on the cysteine/GSH pool opening up a refreshed view onto mechanisms able to trigger ferroptosis (Seiler et al., 2008; Friedmann Angeli et al., 2014; Yang et al., 2014).

8.4   LIPID METABOLISM DETERMINES SENSITIVITY TO FERROPTOSIS

The identification of GPx4 as the limiting factor in the prosurvival function conferred by GSH in tissues and cells combined with the development of pharmacological and genetic tools provided the ideal framework to understand the factors dictating sensitivity to this form of cell death. It was already clear from work performed during the 1990s that lipid peroxidation was a critical feature of the cell death caused by GSH depletion. This was then further supported by the discovery of the critical role played by GPx4. Nonetheless, important questions were left unanswered including those concerning the nature of the lipids being oxidized, and which potential metabolic networks would dictate this sensitivity.

The fast developments in the field of global genetic screens accelerated the discovery of these factors. For instance, the Stockwell group using gene trap in haploid screens and the Conrad group using CRISPR/Cas9-based screens have uncovered an essential function of the enzyme acyl-CoA synthetase long chain family member 4 (ACSL4) in sensitizing cells to ferroptosis (Dixon et al., 2015; Doll et al., 2017). ACSL4 is an enzyme responsible for the ATP-dependent esterification of CoA to fatty acids and is thus responsible for their activation to enter catabolic and or anabolic pathways (Ellis et al., 2010). Specifically, ACSL4 has a marked preference for long and polyunsaturated fatty acids, such as arachidonic and adrenic acids. In accordance with this function, it was shown that absence/inactivation of ACSL4 leads to a dramatic change in a cell membrane fatty acid composition, switching longer and more unsaturated fatty acids for shorter and more saturated ones (Doll et al., 2017; Kagan et al., 2017). This change was then sufficient to provide an unforeseen resistance to cell death triggered by cysteine and GSH depletion or GPx4 inhibition (Doll et al., 2017). Interestingly, it was also proposed that lysophosphatidylcholine acyltransferase 3 (LPCAT3), one of the enzymes responsible for the reesterification of fatty acyl-CoA to anionic phospholipids, could also be involved (Dixon et al., 2015). Yet, this activity appears to be restricted to some cells (Doll et al., 2017), and more studies are certainly required to understand the importance of alternative pathways for reesterification of ferroptosis-sensitive fatty acid substrates. Moreover, with the advancement of analytic techniques the global analysis of oxidized phospholipids became possible, which then allowed the identification of the specific subset of lipids specifically oxidized during ferroptosis. The detailed investigation on the identities of these distinctively oxidized lipids during the ferroptotic process revealed that phosphatidylethanolamines (PE) containing arachidonic acid and its elongation product, adrenic acid, are the preferentially oxidized lipids in fibroblasts sensitive to ferroptosis (Kagan et al., 2017). Of note, supplementation with arachidonic acid strongly enhanced the sensitivity of cells to ferroptosis and, expectedly, stimulated the integration of arachidonic into phospholipids, including PE. On the other hand, these findings were also unexpected as previous studies have suggested a marked preference for GPx4 to protect cardiolipins from oxidative degradation *in vivo* and *in vitro* (Nomura et al., 2000; Friedmann Angeli et al., 2014). Even more surprisingly, cardiolipins, unique phospholipids of the inner mitochondrial membrane, were the only class of phospholipids entirely excluded from ferroptosis-associated oxidation reactions. This raised the question about the specific metabolic/enzymatic pathway(s) that could be activated by ferroptotic conditions. While PEs are still dominating phospholipids, whose oxidation is preventable by ferroptosis inhibitors (Doll et al., 2017; Kagan et al., 2017; Wenzel et al., 2017), further studies will be informative whether this is a general feature observed in all ferroptotic cells or it is restricted to specific cellular subtypes. Moreover, GPx4-catalyzed reduction of hydroperoxy-phospholipids to the respective alcohols prevents oxidative truncation and formation of diverse electrophilic products readily targeting multiple nucleophilic sites in cells. The significance of these very reactive intermediates versus more readily detectable and less reactive phospholipid oxidation products presents yet another important issue that remains to be resolved.

Another particular aspect of ferroptosis that remains controversial is whether ferroptosis-associated lipid peroxidation is a random free radical reaction or an enzymatic process. While effectiveness of diverse free radical scavengers seemingly support the first notion (Shah et al., 2017), the specificity of phospholipid products is more compatible with the concept of an enzymatic process and beseeches the identification of the enzymatic source(s) responsible for the oxidation of these lipid species. Initial studies have suggested an important role of lipoxygenases as culprits, but unequivocal (i.e., genetic) proof is still missing. Initial studies were performed relying on poorly characterized pharmacological inhibitors of these enzymes (Li et al., 1997b; Seiler et al., 2008), which perhaps had led to potentially misleading interpretations. This cautionary notice is particularly important *in lieu* of the data showing that the use of double knockout mice for *Gpx*4 and *Alox15* failed to show any improvement on the lethality caused by GPx4 loss (Friedmann Angeli et al., 2014; Brutsch et al., 2015). Nevertheless, it has also been proposed that a complex interplay between various lipoxygenases might exist, meaning one isoform could be active when the other one is absent. This was somehow corroborated by studies showing that knocking down several lipoxygenases presents an increased protection to ferroptosis compared to single lipoxygenase knockdown (Friedmann Angeli et al., 2014; Yang et al., 2016). Recently, the phosphatidylethanolamine binding protein-1 (PEBP1), a scaffold protein that interacts with Raf1 kinase and thereby inhibiting protein kinase signaling, was shown to be required for ferroptosis (Wenzel et al., 2017). Mechanistically, disruption of the interaction of

PEBP1 with Raf1 provides a platform for 15-lipoxygenase that stimulates the formation of oxidized PE species. This mechanism appears to be relevant for the pathology of kidney failure, brain trauma, and asthma.

Nevertheless, the proof of the unique role of lipoxygenase in ferroptosis is still missing. Other groups have advocated for an alternative mechanism that proposes the so-called "labile-iron-pool" being responsible for the lipid peroxidation process observed during ferroptosis. Many reports have suggested that a specific form of autophagy, known as ferritinophagy consisting of the lysosomal degradation of ferritin, stimulates ferroptosis (Gao et al., 2016; Hou et al., 2016). Increased flow of ferritin through the autophagic pathway would thus lead to a higher amount of free iron thereby increasing the labile iron pool. This, in turn, would facilitate Fenton-like chemistry with preexisting phospholipid hydroperoxides (Doll and Conrad, 2017).

8.5 PATHOLOGICAL RELEVANCE

Since its discovery, ferroptosis has been implicated in a wide array of disease contexts (Angeli et al., 2017; Stockwell et al., 2017). The first indications that ferroptosis could be involved in pathological setting came from the findings that *Gpx*4 conditional knockout animals presented delayed mortality (due to acute renal failure resulting from massive ferroptosis of kidney tubular cells), when treated with the ferroptosis inhibitor liproxstatin-1 (Friedmann Angeli et al., 2014). Therefore, it was rationalized that what was long known in cell culture conditions might indeed by relevant for the *in vivo* situation. With the development of ferroptosis inhibitors, shown to be active *in vivo*, more studies followed to implicate their contribution in other relevant settings. For instance, during ischemia/reperfusion in kidney a ferrostatin derivative was able to ameliorate all parameters of kidney damage (Linkermann et al., 2014). This finding further supported the relevance of the GPx4 inhibitory activity on the ferroptosis machinery in kidney tubular epithelial cells.

More recently, other studies have also started to implicate ferroptosis in a broad range of acute models of injury. For instance, during ischemic stroke, it was shown that Tau confers an important role in iron efflux, which is critical for the aggravation of neuronal damage if absent (Tuo et al., 2017). Additionally, this iron accumulation is critical for the aggravation of ferroptosis, which can be blunted by the intranasal administration of liproxstatin-1. Interestingly, it was also reported that in a model of intracerebral hemorrhage ferroptosis inhibitors were able to protect neurons by inhibiting hemoglobin-induced cell death (Li et al., 2017). This finding highlights the importance of heme proteins in propagating and aggravating the damage inflicted by preformed lipid hydroperoxides (Angeli et al., 2011). Additionally, the mechanisms proposed herein might also be relevant for damage prevention in tissues, such as the heart (Gao et al., 2015; Baba et al., 2017). In this context, it has recently been demonstrated that during cardiac infarction iron and excessive formation of ROS might be at the core of the cell death process leading to cardiomyocyte death. Curiously, in this model the mechanistic target of rapamycin (mTOR) is involved in suppressing ferroptosis at least in cardiomyocytes by a mechanism that might be related to ferritin recycling (Baba et al., 2017).

Despite the potential use of ferroptosis inhibitors to suppress tissue damage in a wide array of pathological settings (Angeli et al., 2017), in the last couple of years we have witnessed a steadily increasing relevance for the suppression of ferroptosis during cancer development. Initial insights into the importance to suppress ferroptosis in the process of carcinogenesis came with the finding that TP53 (p53) was able to sensitize cells to ferroptosis through a mechanism that involves the repression of xCT transcripts by p53 (Jiang et al., 2015). Potential *in vivo* relevance for this finding came from the recognition that mice lacking mouse double minute 2 homolog (MDM2), where p53 is constitutively active, leads to an early embryonic lethal phenotype that could be partially rescued by ferroptosis inhibitors (Jiang et al., 2015). This finding is interesting but fails to reconcile with the effects seen in xCT knockout animals, which are fully viable (Sato et al., 2005). Therefore, it might seem logical that overactivation of p53 would lead to a form of ferroptotic cell death that would account not only for its inhibitory activity on xCT. Additional findings supporting a role for p53 in ferroptosis came from studies showing that in mouse embryonic fibroblasts carrying a common African-descent variant form of p53 (S47) a marked resistance to ferroptosis was evidenced (Jennis et al., 2016). Moreover, the S47 variant was also shown to predispose to an increased incidence of cancers. Nonetheless, the effect of p53 on ferroptosis appears to be cell type-dependent and more work is definitely needed to fully understand its contribution to ferroptosis (Basu et al., 2016; Xie et al., 2017).

From these indications, it is provocative to propose that ferroptosis might be a roadblock from the conversion of a normal, nonmalignant state to a malignant cell. This notion might resemble what has been demonstrated recently in a model of neuronal reprograming (Gascon et al., 2016). Thereby, the direct conversion of a fully differentiated cell line (i.e., astrocytes) into another cell type (i.e., functional neurons) without the requirement to undergo an undifferentiated stem cell-like state requires the efficient suppression of ferroptosis (Gascon et al., 2016).

Interestingly, it has also been reported recently that during the progression of lung adenocarcinomas the tumor cells select for expression of a pathway dependent on mitochondrial cysteine desulfurase (NSF1) that protects cells from undergoing ferroptosis (Alvarez et al., 2017). Additional discoveries that demonstrate the importance of targeting ferroptosis in cancerous diseases came from the finding that several therapeutically refractory states are in fact sensitive to ferroptosis (Hangauer et al., 2017; Viswanathan et al., 2017). These intriguing findings provide compelling evidence that tumor cells refractory to current treatment paradigms may undergo a "dormant" state characteristic of a mesenchymal-like state. Thereby, these cells become refractory to standard treatments but for yet unknown reasons are exceedingly dependent of the GSH/GPx4 lipid peroxidation repair axis (Rennekamp, 2017). Despite these novel revelations, there are, however, currently no metabolically stable small molecule compounds at hand to efficiently target this axis in an *in vivo* setting. Yet, alternative ways to target ferroptosis in the context of malignancies *in vivo* have surfaced and include the use of nanoparticles and small molecules triggering ferroptosis by targeting noncanonical branches of the ferroptosis network (Kim et al., 2016; Mai et al., 2017; Ou et al., 2017).

8.6 CONCLUDING REMARKS

It has become apparent that incredible discoveries have been made since the early findings of Harry Eagle showing that cysteine deprivation inevitably causes cell death. Since still little is known regarding the importance of ferroptosis due to our limited understanding on its regulatory networks and also regarding reliable (bio) markers to detect this form of death, most of these studies need to be taken with care. The identification of pharmacological tools that specifically suppress ferroptosis *in vitro*, such as liproxstatins and ferrostatins (Conrad et al., 2016), offered us the opportunity to challenge pathological settings, yet also not without caveats. Since it was recently demonstrated that these molecules act through inhibition of lipid peroxidation through a radical trapping mechanism (Zilka et al., 2017), which involves an intermediary nitroxide, it is expected that these molecules could interfere in a more unspecific manner by impinging in other redox systems. Therefore, in order to better comprehend and exploit the translational potential of ferroptosis, it is mandatory to further develop better tools and biomarkers to probe and dissect this pathway (Stockwell et al., 2017), which will ultimately guide us to novel and unforeseen therapeutic strategies to treat what are now thought to be yet incurable disease conditions.

ACKNOWLEDGMENTS

This work was in part supported from the Deutsche Forschungsgemeinschaft (DFG; CO 291/2-3 and CO 291/5-1) to Marcus Conrad, and the Human Frontier Science Program (HFSP) RGP0013/2014 to Marcus Conrad and Valerian E. Kagan.

REFERENCES

Alvarez, S. W., V. O. Sviderskiy, E. M. Terzi et al. 2017. Nfs1 undergoes positive selection in lung tumours and protects cells from ferroptosis. *Nature* 551 (7682):639–43.

Angeli, J. P., C. C. Garcia, F. Sena et al. 2011. Lipid hydroperoxide-induced and hemoglobin-enhanced oxidative damage to colon cancer cells. *Free Radic Biol Med* 51 (2):503–15.

Angeli, J. P. F., R. Shah, D. A. Pratt, and M. Conrad. 2017. Ferroptosis inhibition: Mechanisms and opportunities. *Trends Pharmacol Sci* 38 (5):489–98.

Baba, Y., J. K. Higa, B. K. Shimada et al. 2017. Protective effects of the mechanistic target of rapamycin against excess iron and ferroptosis in cardiomyocytes. *Am J Physiol Heart Circ Physiol* Doi:10.1152/ajpheart.00452.2017.

Banjac, A., T. Perisic, H. Sato et al. 2008. The cystine/cysteine cycle: A redox cycle regulating susceptibility versus resistance to cell death. *Oncogene* 27 (11):1618–28.

Bannai, S., and E. Kitamura. 1980. Transport interaction of L-cystine and L-glutamate in human diploid fibroblasts in culture. *J Biol Chem* 255 (6):2372–6.

Bannai, S., H. Tsukeda, and H. Okumura. 1977. Effect of antioxidants on cultured human diploid fibroblasts exposed to cystine-free medium. *Biochem Biophys Res Commun* 74 (4):1582–8.

Basu, S., T. Barnoud, C. P. Kung et al. 2016. The African-specific S47 polymorphism of p53 alters chemosensitivity. *Cell Cycle* 15 (19):2557–60.

Brielmeier, M., J. M. Bechet, M. H. Falk et al. 1998. Improving stable transfection efficiency: Antioxidants dramatically improve the outgrowth of clones under dominant marker selection. *Nucleic Acids Res* 26 (9):2082–5.

Brielmeier, M., J. M. Bechet, S. Suppmann et al. 2001. Cloning of phospholipid hydroperoxide glutathione peroxidase (PHGPx) as an anti-apoptotic and growth promoting gene of Burkitt lymphoma cells. *Biofactors* 14 (1–4):179–90.

Brosnan, J. T., and M. E. Brosnan. 2006. The sulfur-containing amino acids: An overview. *J Nutr* 136 (6 Suppl):1636S–40S.

Brutsch, S. H., C. C. Wang, L. Li et al. 2015. Expression of inactive glutathione peroxidase 4 leads to embryonic lethality, and inactivation of the ALOX15 gene does not rescue such knock-in mice. *Antioxid Redox Signal* 22 (4):281–93.

Conrad, M., J. P. Angeli, P. Vandenabeele, and B. R. Stockwell. 2016. Regulated necrosis: Disease relevance and therapeutic opportunities. *Nat Rev Drug Discov* 15 (5):348–66.

Conrad, M., and H. Sato. 2012. The oxidative stress-inducible cystine/glutamate antiporter, system X (c)(–): Cystine supplier and beyond. *Amino Acids* 42 (1):231–46.

Conrad, M., M. Schneider, A. Seiler, and G. W. Bornkamm. 2007. Physiological role of phospholipid hydroperoxide glutathione peroxidase in mammals. *Biol Chem* 388 (10):1019–25.

Dalton, T. P., M. Z. Dieter, Y. Yang et al. 2000. Knockout of the mouse glutamate cysteine ligase catalytic subunit (GCLC) gene: Embryonic lethal when homozygous, and proposed model for moderate glutathione deficiency when heterozygous. *Biochem Biophys Res Commun* 279 (2):324–9.

Dixon, S. J., K. M. Lemberg, M. R. Lamprecht et al. 2012. Ferroptosis: An iron-dependent form of nonapoptotic cell death. *Cell* 149 (5):1060–72.

Dixon, S. J., G. E. Winter, L. S. Musavi et al. 2015. Human haploid cell genetics reveals roles for lipid metabolism genes in nonapoptotic cell death. *ACS Chem Biol* 10 (7):1604–9.

Doll, S., and M. Conrad. 2017. Iron and ferroptosis: A still ill-defined liaison. *IUBMB Life* 69 (6):423–34.

Doll, S., B. Proneth, Y. Y. Tyurina et al. 2017. ACSL4 dictates ferroptosis sensitivity by shaping cellular lipid composition. *Nat Chem Biol* 13 (1):91–8.

Dolma, S., S. L. Lessnick, W. C. Hahn, and B. R. Stockwell. 2003. Identification of genotype-selective antitumor agents using synthetic lethal chemical screening in engineered human tumor cells. *Cancer Cell* 3 (3):285–96.

Eagle, H. 1955. Nutrition needs of mammalian cells in tissue culture. *Science* 122 (3168):501–14.

Eagle, H., K. A. Piez, and V. I. Oyama. 1961. The biosynthesis of cystine in human cell cultures. *J Biol Chem* 236:1425–8.

Ellis, J. M., J. L. Frahm, L. O. Li, and R. A. Coleman. 2010. Acyl-coenzyme a synthetases in metabolic control. *Curr Opin Lipidol* 21 (3):212–17.

Falk, M. H., L. Hultner, A. Milner et al. 1993. Irradiated fibroblasts protect Burkitt lymphoma cells from apoptosis by a mechanism independent of bcl-2. *Int J Cancer* 55 (3):485–91.

Falk, M. H., T. Meier, R. D. Issels et al. 1998. Apoptosis in Burkitt lymphoma cells is prevented by promotion of cysteine uptake. *Int J Cancer* 75 (4):620–5.

Friedmann Angeli, J. P., B. Proneth, and M. Conrad. 2016. Glutathione peroxidase 4 and ferroptosis. In *Selenium: Its Molecular Biology and Role in Human Health*, edited by D. L. Hatfield, U. Schweizer, P. A. Tsuji and V. N. Gladyshev, 511–21. New York: Springer International Publishing.

Friedmann Angeli, J. P., M. Schneider, B. Proneth et al. 2014. Inactivation of the ferroptosis regulator Gpx4 triggers acute renal failure in mice. *Nat Cell Biol* 16 (12):1180–91.

Gao, M., P. Monian, Q. Pan et al. 2016. Ferroptosis is an autophagic cell death process. *Cell Res* 26 (9):1021–32.

Gao, M., P. Monian, N. Quadri et al. 2015. Glutaminolysis and transferrin regulate ferroptosis. *Mol Cell* 59 (2):298–308.

Gascon, S., E. Murenu, G. Masserdotti et al. 2016. Identification and successful negotiation of a metabolic checkpoint in direct neuronal reprogramming. *Cell Stem Cell* 18 (3):396–409.

Hangauer, M. J., V. S. Viswanathan, M. J. Ryan et al. 2017. Drug-tolerant persister cancer cells are vulnerable to GPx4 inhibition. *Nature* 551 (7679):247–50.

Hou, W., Y. Xie, X. Song et al. 2016. Autophagy promotes ferroptosis by degradation of ferritin. *Autophagy* 12 (8):1425–8.

Imai, H., F. Hirao, T. Sakamoto et al. 2003. Early embryonic lethality caused by targeted disruption of the mouse PHGPx gene. *Biochem Biophys Res Commun* 305 (2):278–86.

Ingold I., C. Berndt, S. Schmitt et al. 2018. Selenium utilization by GPx4 is required to prevent hydroperoxide-induced ferroptosis. *Cell* 172 (3):409–22.e21. doi:10.1016/j.cell.2017.11.048.

Jennis, M., C. P. Kung, S. Basu et al. 2016. An African-specific polymorphism in the TP53 gene impairs p53 tumor suppressor function in a mouse model. *Genes Dev* 30 (8):918–30.

Jhee, K. H., and W. D. Kruger. 2005. The role of cystathionine beta-synthase in homocysteine metabolism. *Antioxid Redox Signal* 7 (5–6):813–22.

Jiang, L., N. Kon, T. Li et al. 2015. Ferroptosis as a p53-mediated activity during tumour suppression. *Nature* 520 (7545):57–62.

Kagan, V. E., G. Mao, F. Qu et al. 2017. Oxidized arachidonic and adrenic PEs navigate cells to ferroptosis. *Nat Chem Biol* 13 (1):81–90.

Kim, S. E., L. Zhang, K. Ma et al. 2016. Ultrasmall nanoparticles induce ferroptosis in nutrient-deprived cancer cells and suppress tumour growth. *Nat Nanotechnol* 11 (11):977–85.

Lee, J., E. S. Kang, S. Kobayashi et al. 2017. The viability of primary hepatocytes is maintained under a low cysteine-glutathione redox state with a marked elevation in ophthalmic acid production. *Exp Cell Res* 361 (1):178–91.

Li, Q., X. Han, X. Lan et al. 2017. Inhibition of neuronal ferroptosis protects hemorrhagic brain. *JCI Insight* 2 (7).

Li, Y., P. Maher, and D. Schubert. 1997a. Requirement for CGMP in nerve cell death caused by glutathione depletion. *J Cell Biol* 139 (5):1317–24.

Li, Y., P. Maher, and D. Schubert. 1997b. A role for 12-lipoxygenase in nerve cell death caused by glutathione depletion. *Neuron* 19 (2):453–63.

Linkermann, A., R. Skouta, N. Himmerkus et al. 2014. Synchronized renal tubular cell death involves ferroptosis. *Proc Natl Acad Sci U S A* 111 (47):16836–41.

Mai, T. T., A. Hamai, A. Hienzsch et al. 2017. Salinomycin kills cancer stem cells by sequestering iron in lysosomes. *Nat Chem* 9 (10):1025–33.

Mandal, P. K., A. Seiler, T. Perisic et al. 2010. System X(c)⁻ and thioredoxin reductase 1 cooperatively rescue glutathione deficiency. *J Biol Chem* 285 (29):22244–53.

Nomura, K., H. Imai, T. Koumura et al. 2000. Mitochondrial phospholipid hydroperoxide glutathione peroxidase inhibits the release of cytochrome c from mitochondria by suppressing the peroxidation of cardiolipin in hypoglycaemia-induced apoptosis. *Biochem J* 351 (Pt 1):183–93.

Ou, W., R. S. Mulik, A. Anwar et al. 2017. Low-density lipoprotein docosahexaenoic acid nanoparticles induce ferroptotic cell death in hepatocellular carcinoma. *Free Radic Biol Med* 112:597–607.

Rennekamp, A. J. 2017. The ferrous awakens. *Cell* 171 (6):1225–7.

Sato, H., A. Shiiya, M. Kimata et al. 2005. Redox imbalance in cystine/glutamate transporter-deficient mice. *J Biol Chem* 280 (45):37423–9.

Sato, H., M. Tamba, T. Ishii, and S. Bannai. 1999. Cloning and expression of a plasma membrane cystine/glutamate exchange transporter composed of two distinct proteins. *J Biol Chem* 274 (17):11455–8.

Schubert, D., H. Kimura, and P. Maher. 1992. Growth factors and vitamin E modify neuronal glutamate toxicity. *Proc Natl Acad Sci U S A* 89 (17):8264–7.

Seiler, A., M. Schneider, H. Forster et al. 2008. Glutathione peroxidase 4 senses and translates oxidative stress into 12/15-lipoxygenase dependent- and AIF-mediated cell death. *Cell Metab* 8 (3):237–48.

Shah, R., K. Margison, and D. A. Pratt. 2017. The potency of diarylamine radical-trapping antioxidants as inhibitors of ferroptosis underscores the role of autoxidation in the mechanism of cell death. *ACS Chem Biol* 12 (10):2538–2545.

Shi, Z. Z., J. Osei-Frimpong, G. Kala et al. 2000. Glutathione synthesis is essential for mouse development but not for cell growth in culture. *Proc Natl Acad Sci U S A* 97 (10):5101–6.

Stockwell, B. R., J. P. Friedmann Angeli, H. Bayir et al. 2017. Ferroptosis: A regulated cell death nexus linking metabolism, redox biology, and disease. *Cell* 171 (2):273–85.

Tan, S., D. Schubert, and P. Maher. 2001. Oxytosis: A novel form of programmed cell death. *Curr Top Med Chem* 1 (6):497–506.

Tuo, Q. Z., P. Lei, K. A. Jackman et al. 2017. Tau-mediated iron export prevents ferroptotic damage after ischemic stroke. *Mol Psychiatry* 22 (11):1520–30.

Ursini, F., M. Maiorino, M. Valente et al. 1982. Purification from pig liver of a protein which protects liposomes and biomembranes from peroxidative degradation and exhibits glutathione peroxidase activity on phosphatidylcholine hydroperoxides. *Biochim Biophys Acta* 710 (2):197–211.

Viswanathan, V. S., M. J. Ryan, H. D. Dhruv et al. 2017. Dependency of a therapy-resistant state of cancer cells on a lipid peroxidase pathway. *Nature* 547 (7664):453–7.

Wenzel, S. E., Y. Y. Tyurina, J. Zhao et al. 2017. PEBP1 wardens ferroptosis by enabling lipoxygenase generation of lipid death signals. *Cell* 171 (3):628–641 e26.

Xie, Y., S. Zhu, X. Song et al. 2017. The tumor suppressor p53 limits ferroptosis by blocking DPP4 activity. *Cell Rep* 20 (7):1692–1704.

Yang, W. S., K. J. Kim, M. M. Gaschler et al. 2016. Peroxidation of polyunsaturated fatty acids by lipoxygenases drives ferroptosis. *Proc Natl Acad Sci U S A* 113 (34):E4966–75.

Yang, W. S., R. Sriramaratnam, M. E. Welsch et al. 2014. Regulation of ferroptotic cancer cell death by GPx4. *Cell* 156 (1–2):317–31.

Yang, W. S., and B. R. Stockwell. 2008. Synthetic lethal screening identifies compounds activating iron-dependent, nonapoptotic cell death in oncogenic-RAS-harboring cancer cells. *Chem Biol* 15 (3):234–45.

Yant, L. J., Q. Ran, L. Rao et al. 2003. The selenoprotein GPx4 is essential for mouse development and protects from radiation and oxidative damage insults. *Free Radic Biol Med* 34 (4):496–502.

Zilka, O., R. Shah, B. Li et al. 2017. On the mechanism of cytoprotection by ferrostatin-1 and liproxstatin-1 and the role of lipid peroxidation in ferroptotic cell death. *ACS Cent Sci* 3 (3):232–43.

9 Peroxiredoxin 6 as Glutathione Peroxidase

Yefim Manevich
Medical University of South Carolina

CONTENTS

9.1 HISTORY OF THE DISCOVERY OF PRDX6

Peroxiredoxins (Rhee et al., 2005), particularly the thioredoxin-specific peroxidases and peroxiredoxin VI (Prdx6, PRDX6, TSA, HORF6, 1-Cys Prx, aiPLA2, AOP2, NS GPx, p29, PRX), were discovered in the NIH laboratory of Earl Stadtman by Sue Goo Rhee (Kang et al., 1998). Interestingly, Prdx6 had earlier been isolated from a rat olfactory epithelium in Russia (Peshenko et al., 1996), but it was not specified as a peroxiredoxin because of its GSH-dependent activity. At the same time, Hitoshi Shichi from the Wayne State University also reported discovery of a novel non-Se glutathione peroxidase (NS GPX) with MW of ~25,064 kDa in bovine eye (Singh and Shichi, 1998). The same protein was also discovered as a novel acidic, Ca-independent phospholipase A_2 (aiPLA$_2$) by Aron B. Fisher in the Institute for Environmental Medicine of the University of Pennsylvania (Wang et al., 1994).

9.2 THE REDUCING AGENT FOR PRDX6 PEROXIDASE ACTIVITY

Upon its discovery, peroxidase activity of Prdx6 required DTT as a reducing agent, which, at that point, made this enzyme questionable as a physiological antioxidant *in vivo* (Kang et al., 1998). Three independent laboratories: one in Russia (Peshenko et al., 1996) and two in the USA [at Wayne State University (Singh and Shichi, 1998) and at the University of Pennsylvania (Wang et al., 1994; Fisher et al., 1999)] have shown peroxidase activity of a partially purified Prdx6 with

DOI: 10.1201/9781351261760-11

143

GSH as a reducing agent. Additionally, the activity of the *Escherichia coli*-expressed bovine eye Prdx6 (with H_2O_2 as a substrate and GSH as a reducing agent) was lower when compared with an *in vitro* translated protein "due to differences in the purity of enzyme protein" (Singh and Shichi, 1998). Later, this activity was no longer detected with the expressed homogeneous protein. PLA_2 activity of Prdx6 was observed only under acidic condition (pH ~ 4.0) and apparently was independent from a reducing agent. Partially purified Prdx6 from bovine lung has shown to have peroxidase activity with H_2O_2 and PLHP as substrates and GSH as a reducing agent, but homogeneous Prdx6 purified from the same bovine lung was inactive as peroxidase (Manevich et al., 2004). *E. coli*-expressed Prdx6, when purified to homogeneity, lacked any peroxidase activity under the same experimental conditions. The nonphysiological nature of the reducing agent (DTT) and the abolishment of a GPx activity of Prdx6 upon its purification (after its heterologous expression or isolation from the natural sources) led to the hypothesis of a specific GSH-binding/transporting agent as a reductant. Following this hypothesis, we have shown that GSH in complex with glutathione-*S*-transferase P1-1 (GSTP1-1) is a specific physiological reductant of Prdx6 (Manevich et al., 2004). A reduction of an oxidized Prdx6 (sulfenic acid) by the GSH/GSTP1-1 complex is relatively well documented and has been confirmed by different laboratories (Manevich et al., 2004; Ralat et al., 2006, 2008; Fisher, 2011; Manevich et al., 2013; Yun et al., 2014; Fisher, 2017; Rhee et al., 2012). The functional interaction of GSTP1-1 and Prdx6, which results in peroxidase activity of the latter has been shown in the intact MCF-7 cells by a Duolink Proximity Ligation Assay (Zhou et al., 2013). Additional evidence of the involvement of the GSTP1-1/ GSH complex in the Prdx6 peroxidase function has been shown by the GSTP1-1 natural allelic variation that affects peroxidase activation (Manevich et al., 2013).

Regardless of the original hypothesis regarding the GSH-involving reductant, Cyclophilin A (CPA) has been reported as a Prdx6-binding and reducing/activating agent. CPA was discovered as a binding partner of Prdx6 in an overlay assay (Lee et al., 2001). The SDS PAGE of expressed and purified Prdx6 was superfused with total cell lysate and the Prdx6-specific band was analyzed by MS and antibodies against Prdx6 and CPA. The results showed that CPA binds to monomeric Prdx6. Since a Prdx6 under PAGE conditions is denatured, it is difficult to evaluate the specificity of its binding to an intact CPA. Additionally, the discovery of CPA as a reducing agent does not address GSH-dependence of Prdx6 peroxidase activity in semipure preparations. Finally, after the original report about CPA's role as a reductant for the Prdx6 (and other peroxiredoxins) was published, there have not been any subsequent publications supporting this discovery.

Additionally, ascorbate was also reported as a nonthiol reductant of Prdx6 (Monteiro et al., 2007). Ascorbate is a natural nonenzymatic antioxidant, but it is not synthetized by the human body and therefore its content depends on an adequate diet. The actual ascorbate concentration determines its anti/prooxidant properties (Manevich et al., 1997). It is unlikely that the activity of a major antioxidant enzyme in astrocytes is dependent on a diet. Conversely, GSH is the most abundant natural redox buffer and its intracellular concentration (5–10 mM) is comparable only with that of the physiological buffer, i.e., phosphate. This fact is indicative of the role of GSH as a major reducing agent for the thiol-based antioxidant protection in the human body. Again, ascorbate as a reductant does not explain the GPx activity of the semipure Prdx6 preparations. Nevertheless, the role of ascorbate as a none-thiol reductant for thiol-based and redox-regulated antioxidant enzymes generally, and for Prdx6 specifically, needs to be further evaluated.

Recently, the resolving role of GSH for the thioredoxin-dependent peroxidase activity of Prx1p (*Saccharomyces cerevisiae* mitochondrial homolog of Prdx6) has been reported. In this case, GSH is substituting the missing resolving cysteine of typical 2-Cys peroxiredoxins, without its consumption in a catalytic cycle. It is an interesting study, which demonstrates a GSH-dependent thioredoxin peroxidase activity of this Prdx6 homolog specific to yeast mitochondria (Pedrajas et al., 2016). The above function of GSH in yeast mitochondria does not follow the GSTP1-1/GSH-mediated heterodimerization/reduction of Prdx6 because of the missing GSTP1-1 or its analog in this particular localization (Sheehan et al., 2001). Obviously, this study conceptually corroborates the general GSH-dependence of the Prdx6 and its homologs in all biological kingdoms.

9.3 STRUCTURE OF PRDX6

The secondary, tertiary and quaternary structure of Prdx6 is crucial for understanding the mechanism of its catalytic reduction of organic peroxides, and especially PLHP. Upon the discovery of Prdx6 the only crystal structure of a homodimeric human Prdx6 with catalytic Cys47 in sulfenic acid state and the Cys91/Ser mutation was published (Choi et al., 1998). The Prdx6 homodimer contains two identical polypeptide chains (224 aa each). The secondary structure of a Prdx6 monomer is 29% α-helical and 22% β-strand, where human and mouse Prdx6 have a conserved catalytic Cys47 and an additional Cys91, while rat Prdx6 has a singular conserved Cys47. Each monomer's polypeptide chain resembles a thioredoxin fold, as a major structural element (Schröder et al., 1998). Recently it has been shown that neither reduction nor oxidation into sulfinic state of the catalytic Cys47 sulfenate affects the structure of Prdx6 peroxidatic or PLA_2 catalytic sites or this protein's homodimerization (Kim et al., 2016). A homodimer of Prdx6, be it reduced or oxidized up to the sulfinic state, consists of two antiparallel monomers with a substantial interface surface (18%–21% of an accessible area of each monomer), involving both hydrophobic and polar interaction. The data is similar to the sulfenic Prdx6 structure that had been reported earlier (Choi et al., 1998) (Figure 9.1). Under the crystallization conditions (~1 mM of homogeneous Prdx6), the preferential state of a reduced and sulfinic protein is that of a homodimer (Kim et al., 2016), although in a complex intracellular environment and at much lower (~μM or lower) concentrations Prdx6 homodimer could presumably dissociate into monomers. In fact, the dynamic light scattering (DLS) analysis has shown a dissociation of the Prdx6, as well as of GSTP1-1, homodimers into their corresponding monomers in solution (Ralat et al., 2008). This dissociation of the homodimers and a subsequent formation of the Prdx6-GSTP1-1 heterodimer were facilitated by GSH (2.5 mM) in the corresponding solutions (Ralat et al., 2006). At the time of an intensive study of the Prdx6 reduction/reactivation mechanisms, it was not clear which state (monomeric or homodimeric) corresponded to the peroxidase-competent enzyme in solution. In respect to the similar function

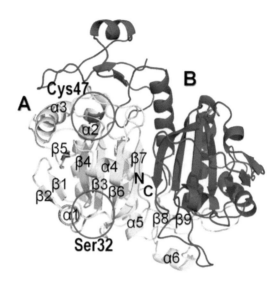

FIGURE 9.1 Tertiary structure of Prdx6 homodimer. Monomers A and B are presented and catalytic sites for their peroxidase (Cys47) and PLA$_2$ (Ser32) activities are presented in red circles. (Reprinted from (Kim et al., 2016) with permission.)

of GPx4, which is a monomeric protein, and the reducing agent GSH being common to GPx4 and Prdx6, the exact oligomeric state of the latter has attracted attention.

Interestingly, over-oxidation of catalytic cysteine to sulfinic acid in classic 2-Cys peroxiredoxins is reversible under reduction/reactivation by sulfiredoxin (Biteau et al., 2003). In contrast, an overoxidation of the peroxidatic Cys47 to sulfinic acid in Prdx6 is irreversible and results in permanent inactivation of Prdx6 as a peroxidase (Woo et al., 2005).

9.4 THE HYDROPEROXIDE SUBSTRATE OF PRDX6 AND ITS BINDING SITES

Since PLHP were shown to be the substrates for the Prdx6 peroxidase as well as PLA$_2$ activities (Fisher et al., 1999; Chen et al., 2000), it deemed important to find out how the structure of this enzyme is designed in order to effectively perform both catalyses. H$_2$O$_2$ and short organic peroxides in solution are obviously reacting with the peroxidatic cysteine nonspecifically, but the specificity of Prdx6 for PLHP requires specific binding of these substrates. The phospholipid nature of the substrates and the PLA$_2$ function were considered reasonable starting points of the studies. Prdx6 binding of a phospholipid molecule in a lipid bilayer most likely follows an interfacial catalysis (IFC) similar to other PLA$_2$ (Jain et al., 1989). The first step of the IFC is a recognition of a PLHP on a membrane surface by the Prdx6 specific site and its binding to the protein with a consequent alignment of sn-2 ester bond to the PLA$_2$ catalytic triad. PLHP resembles the general structure of a phospholipid with a major difference in the stereo-structure and the polarity of its sn-2-peroxidized acyl chain (Hyvönen et al., 1997). A repulsion of sn-2-peroxidised acyl chain outside of the bilayer due to the high polarity of –COOH group has been shown (Greenberg et al., 2008). A specific site on the Prdx6's surface should recognize and to bind to the –COOH group protruding into an aqueous media. For Prdx6, the PLA_2 catalytic site was determined to comprise His26, Ser32 and Asp140. By a site-specific mutagenesis, this triad verified to be responsible for both, PLHP binding (by FRET and the size-specific ultrafiltration) and activities as peroxidase and phospholipase (Manevich et al., 2007). This triad, thus, serves as the recognizing/binding site for the PLHP (in peroxidized micelles and liposomes) and for DPPC (in regular, but negatively charged liposomes) (Manevich et al., 2007), simultaneously combines the PLA_2 and the peroxidase activities of Prdx6, and explains Prdx6' specificity for the reduction of PLHP in membranes. At the time of these investigations, the enzyme in its reduced/active state was still assumed to likely be monomeric. A major reason for this assumption was a newly discovered mechanism of the GSTP1-1-catalyzed reduction/reactivation of oxidized Prdx6 by GSH, which results in the reduced/active monomers.

The crystal structure of the Prdx6 monomer (in sulfenic acid state) shows that the PLA_2 catalytic triad (His26, Ser32 and Asp140) and the peroxidatic Cys47 are located on the opposite sides of a protein globule (Figure 9.1). To explain both catalytic activities, one might think of accommodating the sn-2 peroxidized acyl chain inside of the protein interior to align its –COOH group to Cys47 for a reduction. The assumption, is, however, is unlikely in view of the high polarity of the –COOH group and a high hydrophobicity of an inner protein core. Interestingly, CD analysis of Prdx6 with DPPC (where the sn-2-acyl chain is more hydrophobic compared to the peroxidized one), show a minimal disturbance of the protein secondary structure (Manevich et al., 2007). The CD analyses reveal that the Prdx6 binding to the peroxidized or regular liposomes has specific effects on this protein's secondary structure (Figure 9.2). For peroxidized liposomes, the β-strands of Prdx6 were affected more when compared to the regular negatively-charged liposomes. Conversely, with regular negatively-charged liposomes, the α-helices of Prdx6 were specifically affected. The former effect presumably corresponds to the alignment of the -COOH group of substrate to the peroxidatic Cys47 (situated on and surrounded by β-strands) for catalysis, and the latter effect most likely corresponds to a redistribution of charges on the surface of Prdx6 due to the liposome's negative charge and the accommodation of the hydrophobic sn-2-acyl chain ester bond to Ser32 for hydrolysis. Additionally, the most recent Prdx6 crystal structures demonstrate that a homodimer corresponds to the reduced/active enzyme and that Cys47 from one monomer and Ser32 from another one are on the same plain protein surface (Figure 9.3). A connecting groove was found between the peroxidatic Cys47 and the hydrolytic Ser32 (with distance ~28.0 Å) preferentially populated with positively-charged amino acids (Figure 9.3, panel A). Importantly, a close surrounding of Ser32 at the end of positively-charged groove is negatively charged. Conversely, the close proximity of Cys47 is positively charged. The lengths of the most common polyunsaturated fatty acids: linoleic and linolenic, are ~10.7 and ~7.6 Å, respectively (Fowle, 1921). Approximate length of 1-palmitoyl-2-linoleoyl-sn-glycero-3-phosphoethanolamine (from polar head to the end of its acyl chain) is

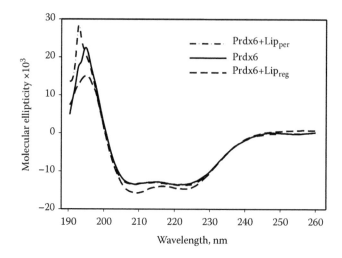

FIGURE 9.2 CD analysis of the Prdx6 secondary structure upon its binding to the peroxidized or regular negatively charged liposomes. The experimental conditions were similar to those presented earlier (Manevich et al., 2007) with a slight modification. Unilamellar liposomes consisting of DPPC/egg PC/cholesterol/PG (50:25:15:10, mol/mol) were prepared by extrusion under pressure. Peroxidized liposomes were prepared by substitution of egg PC and DG with 1-palmytoyl-2-linolenoyl-sn-glycero-3-phosphocholine (PLPC) and their exposure to a Cu^{2+}/ASC mixture (Manevich et al., 1997) for 20 min. The sizes of the liposome were analyzed by DLS (DLS 90 Plus Particle Size Analyzer; Brookhaven Instruments, Holtsville, NY) before and after the Cu^{2+}/Asc treatment and the addition of Prdx6. Our analysis demonstrated a homogeneous population of liposomes with a diameter of ~100–120 nm, which represented ~95% of the total number of particles measured. Liposome peroxidation was detected spectrophotometrically, as a ratio A_{235}/A_{205} before and after treatment with the Cu^{2+}/Asc mixture. The protein samples (10 µM in 20 mM PB, pH 7.4) were maintained at 25°C. CD Spectra were recorded before and after the addition of the indicated liposomes with three repeats, and averaged results after subtraction of an appropriate background are presented here.

~18.9–20.0 Å in the lipid hexagonal phase in bilayer (Thurmond et al., 1993), which is in the range above the size of a groove. It is plausible that the binding of a negatively-charged –COOH group of the phospholipid sn-2-acyl chain to the positively charged vicinity of Cys47, is a first step in Prdx6 homodimer binding to PLHP in a membrane. Thus, Prdx6 becomes anchored by the peroxidized sn-2-acyl chain to the membrane surface and the former goes along the surface groove until the sn-2 ester bond becomes aligned with the PLA₂ catalytic triad (His26, Ser32 and Asp140). That could explain the small disturbance of Prdx6 secondary structure after the PLHP substrate binding (Figure 9.2). It has been shown that the distance between His26 and Asp140 in a Prdx6 homodimer is ~9.7 Å, which is about two times longer than that of a lysosomal PLA₂ (~5.0 Å) (Kim et al., 2016). This could be a reason for Prdx6's relatively slow hydrolysis of a phospholipid with maximum activity at acidic conditions (pH ~ 4.0). Adherence of Prdx6 to the membrane surface, as well as phosphorylation of Prdx6's Thr177, may cause a structural rearrangement, which

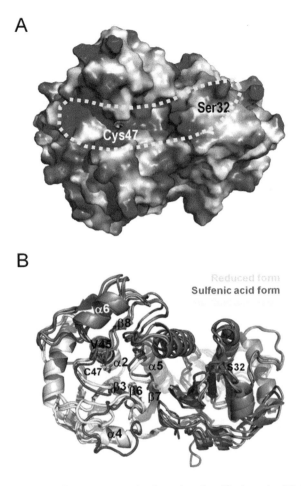

FIGURE 9.3 Prx6 homodimer structure in the reduced, sulfenic and sulfinic acids states. Panel A: Electrostatic charge distribution on the homodimer of Prx6 surface in the sulfenic acid state. The peroxidatic (Cys47) and PLA2 (Ser32) catalytic moieties and their connecting groove are indicated by the yellow dotted line. Panel B: Prx6 monomer structure in the reduced, sulfenic and sulfinic acids states of a homodimer and the connecting loop between α-helix 4 and β-strand 6 (shown in lime, dark red and cyan, respectively). (Reprinted from (Kim et al., 2016) with permission.)

could bring the PLA$_2$ triad to regular dimensions and facilitate hydrolysis with a consequent shift to an optimum activity in the physiological range (pH = 7.4) (Wu et al., 2009). It is interesting that only negatively-charged liposomes (PG/DPPC mixture) were substrates for Prdx6' PLA$_2$ activity (Wu et al., 2009). This result corresponds to the requirement of the positively-charged groove described above, on the Prdx6 dimer surface for both of its catalyses. All these data support the dual functionality of Prdx6 as a result of its unique homodimeric structure through the interdependence of the peroxidatic Cys47 and the hydrolytic His26 and Ser32 for phospholipid substrate binding and for both catalyses. The recently published structure of Prdx6 demonstrates that the oxidation of Cys47 to sulfenic or sulfinic acids does not affect the structure of peroxidase or lipase catalytic sites, and only the Leu114 – Arg132 loop connecting α_4 and β_5 (Figure 9.3, panel B) moves ~10 Å in a sulfenic acid state versus a reduced and a sulfinic states (Kim et al., 2016). Regardless of this loop flexibility, this move may affect the heterodimerization of Prdx6 with GSTP1-1 and so it will be important to find out the structure of the Prdx6/GSTP1-1/±GSH complex in future studies. Another important aspect of both of Prdx6 activities, is the effect of the products of the corresponding reactions on homo/hetero-dimerization. Obviously, a reduction of the –COOH group of an sn-2-acyl chain of a PLHP will eliminate its negative charge, and the acyl chain with the neutral hydroxyl will be pulled back into a bilayer because of its higher hydrophobicity. This will consequently release Prdx6 (oxidized to a sulfenic/sulfinic acid state) from the surface of a membrane. The intact tertiary and quaternary structure of the sulfinic Prdx6 (Kim et al., 2016) results in unchanged specific binding of peroxidized sn-2-acyl chain of PLHP. However, since a reduction of the –COOH group is impossible, Prdx6 will catalyze a hydrolysis of the latter sn-2 ester bond and a homodimeric enzyme with bound fatty acid hydroperoxide (FAHP) will be released from the membrane's surface. It is known, that a direct reduction of the PLHP by the specific peroxidase activity is ~ 10^4 times more effective, when compared to the PLA_2-calyzed hydrolysis (Zhao et al., 2003). All data mentioned above would explain a relatively slow PLA_2 activity of Prdx6 as a supporting mechanism of biological membrane functionality protection under the severe oxidant stress, when an enzyme becomes over-oxidized. In the presence of transition metals (Cu^{2+}, Fe^{2+} etc.), FAHP could serve as a site-specific source of OH-radical - the most damaging ROS. The fate of the FAHP bound to the Prdx6 homodimer surface, the mechanisms of its release, and Prdx6-mediated hydrolysis of regular phospholipids (DPPC) are presently unknown and definitely require additional studies. Finally, it is the PLA_2 activity of Prdx6, which makes the peroxidase function of this unique enzyme specific for the PLHP substrate.

9.5 CATALYTIC CYCLE OF PRDX6 REDUCTION OF ORGANIC HYDROPEROXIDES

The original observation that Prdx6 is active with DTT as a reducing agent (Kang et al., 1998) was not investigated in greater details because of its nonphysiological nature. Assuming that the active form of DTT [$pK_a = 9.2$ and 10.2 (Singh et al., 1995)] at pH > 7.0 is a negatively-charged thiolate anion (TA), a direct reaction of TA with Cys47-SOH would result in a disulfide bond (Cys47-SSR) and an additional DTT (TA) molecule would reduce a disulfide into sulfhydryl (Cys47-SH), reactivating a Prdx6 peroxidase function. A positively-charged surrounding of a Cys47 sulfenate on the surface of a monomeric Prdx6 would attract TA, thus following a catalytic paradigm common to all peroxiredoxins: a disulfide/sulfhydryl redox transformation. There is at least some similarity between the DTT- and the GSTP1-1/GSH complex-mediated Prdx6 reduction/reactivation. The crystal structure of GSTP1-1 in complex with GSH show very specific hydrogen bonding of the latter sulfur atom involving catalytic Tyr7, two water molecules, Arg13, Tyr108, Gly205, Asn204, and

Tyr103. It has been shown that a sulfur atom of GSH becomes protonated due to the lack of appropriate countercharge upon its specific binding to GSTP1-1 (Prade et al., 1997). A deprotonation of GSH leads to negatively-charged thiolate (TR, GS$^-$), which could potentially facilitate stirring of the GSTP1-1/GS$^-$ complex to the positively-charged surrounding of the Cys47 sulfenate on the Prdx6 monomer surface. Structurally Prdx6 sulfenate is different from its reduced and sulfinic states, but only in terms of an extension of its flexible loop, connecting α-helix 4 and β-strand 6 (aa 114–132, Figure 9.3, panel B). The close vicinities of both catalytic sites and their overall structure remain intact. Thus, this loop could be one of the factors in the GSTP1-1/GS$^-$ heterodimerization with the Prdx6 sulfenate monomer (Kim et al., 2016). Although requiring additional studies and verification, it supports the hypothesis that a specific structure of the Prdx6 sulfenate is critical for its heterodimerization with GSTP1-1/GS$^-$. After a heterodimer of Prdx6 and GSTP1-1/GS$^-$ is formed, an alignment of TR to Cys47 sulfenate of Prdx6 occurs. This results in the glutathionylation of the Prdx6 peroxidatic cysteine (Cys47-SSG). There are several models and explanations of the Prdx6 heterodimerization with and a consequent glutathionylation by the GSTP1-1/GS$^-$ complex [for review see (Deponte, 2013)]. When first proposed, a physiological function of the GSTP1-1 of the GSH-binding to overcome the accessibility barrier for the Prdx6 Cys47-SOH glutathionylation was unusual. Indeed, this was a first example showing that Prdx6 underwent glutathionylation through a heterodimerization with the GSTP1-1/GS$^-$ complex. But, there were other examples of GSTP1-1 as well as GSTμ heterodimerization with JNK1 (Adler et al., 1999; Pettigrew et al., 2001; Wang et al., 2001). Also, there have been other examples of the GSTP1-1-mediated glutathionylation of other proteins (Tew, 2007; Tew et al., 2011; Grek et al., 2013; Carvalho et al., 2016). There are also a fast-growing number of examples of the role that the GSTP1-1 plays in the regulation of proteins structure/function through their glutathionylation.

Here is our current view of the GSTP1-1-mediated catalytic glutathionylation and its consequent reduction/reactivation of Prdx6. The complex of GSTP1-1 with glutathionylated Prdx6 (-Cys47-SSG) could be relatively stable due to the high affinity of GSTP1-1 to GSH [$K_D \sim 6.9$ nM, (Waterboer et al., 2005)], which has been observed earlier by the SDS PAGE under none-reducing conditions (Ralat et al., 2006). The site-specific mutation of Tyr7 of the GSTP1-1 or of Cys47 of the Prdx6, leads to the disappearance of Prdx6/GSTP1-1 heterodimer, which also occurs under a reducing (DTT) SDS PAGE conditions (Ralat et al., 2006). It is conceivable that another imbalance of charges in GSTP1-1 within the Prdx6-SSG/GSTP1-1heterodimer will result in a dissociation of a GSTP1-1 monomer from glutathionylated Prdx6 (similar to the release of a glutathionylated electrophile from GSTP1-1 due to its original glutathione *S*-transferase activity), leaving substantial changes in the structure of the Prdx6 glutathionylated monomer. These changes will likely prevent enzyme homodimerization and will facilitate access of free GSH (~5.0–10.0 mM intracellularly) to the close proximity of Cys47-SSG, with consequent deglutathionylation of the latter and the release of GSSG. Deglutathionylation results in Prdx6 reduction/reactivation and in a stabilization of a reduced enzyme homodimeric state. This mechanism of Prdx6 Cys47-SOH reduction follows a general catalytic paradigm of all peroxiredoxins: The disulfide/sulfhydryl redox transformation and the

GSTP1-1/GS⁻ complex mimics a resolving sulfhydryl for the typical 2-Cys perox-iredoxins. The catalytic cycle of Prdx6 requires two molecules of GSH and thus resembles the catalytic cycle of classic glutathione peroxidases. All these data are summarized in the proposed catalytic cycle of Prdx6 (Figure 9.4).

There are substantial differences in peroxidase activity of Prdx6 and GPx4: Prdx6 activity with PLHP (H_2O_2) as a substrate is ~3.2–6.0 μmol/min/mg protein (K_m = 120–150 mM; (Fisher et al., 1999; Manevich et al., 2004; Ralat et al., 2008) and GPx4 activity with H_2O_2 as a substrate is ~30.0–310.0 μmol/min/mg protein [K_m is infinite (Maiorino et al., 1990; Kernstock et al., 2008; Han et al., 2013)]. The structure and oligomerization-driven and cysteine-dependent peroxidase function of Prdx6 and the structure-driven and seleno-cysteine-dependent catalysis of GPx4 underline the differences in activities described above. It is conceivable that a lower peroxidase activity of Prdx6 is compensated by its high specificity to the reduction of the PLHP in supporting the homeostasis of biomembranes. However, the catalytic seleno-cysteine of GPx4 does not undergo over-oxidation (see Chapter 3); it will therefore perform well even under severe oxidant stress. Conversely, Prdx6 could be irreversibly over-oxidized/inactivated to a sulfinic state under such stress. It would be important to study the cooperation/competition of these two enzymes under normal and pathological conditions using an advanced genetic and biochemical/biophysical methods.

Several practical hints how to study Prdx6 peroxidase function may be proposed. It deems important to determine the redox state of the particular enzyme preparation: reduced, oxidized to sulfenate or oxidized to sulfinate/sulfonate. The first two states will require the presence of GSTP1-1 pre-incubated with GSH (GSTP1-1 should be reduced as well). The third state is terminal and could not be reduced/reactivated.

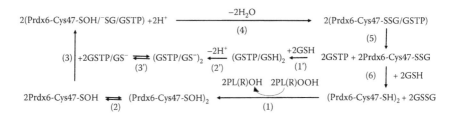

FIGURE 9.4 A catalytic cycle of the Prdx6 reduction of organic peroxides. Step 1: cata-lytic reduction of phospholipid (PLOOH) or other (ROOH) organic peroxides by the Prdx6 reduced homodimer. Step 2: dissociation of the Prdx6 sulfenic homodimer [(Prdx6-Ces47-SOH)₂] into monomers. Step 3: heterodimerization of the Prdx6 sulfenate monomer with the GSTP1-1 monomer complex with the deprotonated glutathione (Prdx6Cys47-SOH/GS⁻/GSTP). Step 4: glutathionylation of the Prdx6 catalytic Cys47 (Cys47-SSG) inside of the Prdx6/GSTP1-1 complex (GSH is still bound to the GST). Step 5: dissociation of the Prdx6 glutathionylated monomers and GSTP1-1 from the homodimer. Step 6: deglutathionyla-tion of the Prdx6 monomer by free GSH; Step 7: homodimerization of the reduced/active Prdx6 monomer. GSTP1-1 complex with GSH generation. Step 1′: generation of the GSTP1-1 homodimer complex with GSH((GSTP/CSH)₂). Step 2′: deprotonation of GSH inside of the GSTP1-1/GSH homodimer [(GSTP/GS⁻)]₂. Step 3′: dissociation of the GSTP1-1 homodimer with deprotonated GSH into monomers (GSTP/GS⁻).

With respect to the biophysical studies of Prdx6 itself, the incubation of sulfenic Prdx6 with excess of DTT and a consequent extensive dialysis to remove DTT, could be recommended. A sulfenic state of Prdx6 could be determined with the use of the specific fluorescent probes (Poole et al., 2007; Paulsen et al., 2013). Sulfinic/sulfonic states of Prdx6 could be detected with the appropriate antibodies.

9.6   BIOLOGICAL AND BIOMEDICAL SIGNIFICANCE OF PRDX6

The biological significance of Prdx6 is defined by its unique antioxidant protection of the structural and functional integrity of cellular membranes. The manifestation of the oxidative damage to the biological membranes is lipid peroxidation (LP). LP is a radical chain reaction which occurs in a phospholipid bilayer under oxidative stress. It affects the unsaturated acyl chains of phospholipids through a production of hydroperoxyl groups (–COOH) (see also Chapter 7). This autocatalytic reaction has three steps: initiation, propagation and termination. Hydroxyl radicals ($^*OH$) and singlet oxygen ($^1O_2$) are the initiators of LP. The sources of $^*OH$ include γ-irradiation, Fenton and Haber-Weiss chemistry, while $^1O_2$ is generated by UV light and superoxide by the xanthine oxidase system (McCord and Fridovich, 1969; Kellogg et al., 1975), by the NADPH oxidase (NOX) of activated leukocytes (Babior et al., 1973) and NOX-type enzymes in general (Sies et al., 2017), by the respiratory chain (Loschen et al., 1974), autoxidation processes (Misra, 1974), and microsomal redox processes (Richter et al., 1977). Moreover, an initiation of lipid peroxidation by lipoxygenases is also discussed (see Chapters 7 and 8). The propagation step occurs inside of a bilayer through the peroxyl radical formation in close proximity to the phospholipid acyl chains and continues until a radical scavenging (termination) stops it. It is known that the hydroperoxyl group (–COOH) of sn-2 acyl chain of a phospholipid floats to the surface of a bilayer due to its high polarity. Thus, the whole peroxidized sn-2 acyl chain gets exposed outside of membrane, forming so-called "whiskers" (Greenberg et al., 2008). This obviously affects the packing order of phospholipids in the glycocalyx and decreases fluidity of the membrane (Sevanian et al., 1983). All these parameters affect membrane functionality directly through their effect on membrane-associating proteins: receptors, channels, pores etc. The membrane-associated proteins are surrounded by phospholipid molecules and the mission of the latter is to keep an intact conformation of the former. It is interesting to note that each membrane-associated protein is surrounded by a monolayer of a specific phospholipid, which provides the adequate surface tension to support the conformation corresponding to the optimal protein function (Molotkovsky et al., 1982; Bergelson et al., 1985). During the ligand–receptor binding, channels opening/closing or membrane-associated enzyme activation/inactivation, the conformation(s) of the corresponding proteins is (are) dynamically changing and, in intact membranes, the phospholipids absorb these changes. These dynamic changes in the phospholipid bilayer resemble the circular waves left on the water surface, after throwing a rock into it. Assuming that the phospholipid bilayer remains intact, these "waves" somehow interfere with each other and produce a stable dynamic pattern of a functionally sound membrane. It is plausible that an existing homeostatic pattern of the phospholipid dynamics in a membrane has specific basic frequencies, and membrane proteins conformation dynamics could be "in" or "out" of the resonance with those frequencies. The resonance corresponds with entropy and results in a substantial modulation of the original bilayer dynamics potentially causing the augmentation/reduction of bio-mechanical signals. This could be induced/inhibited by pathologies resulting from peroxidation of the membrane phospholipids. Thus, an earlier reduction of LP is important, in order to prevent the severity of membrane malfunction, which in the case of plasma membrane, obviously, may cause cells death through necrosis, apoptosis, ferroptosis, autophagy etc.

There are only two known enzymes capable of specifically reducing peroxidized membrane phospholipids: GPx4 (PHGPx) and Prdx6. GPx4 is a monomeric Sec(seleno-Cys)-containing glutathione peroxidase. A translation of the GPx4 results in three enzymes with specific and distinct intracellular localizations: in the nucleous (nGPx4), in the mitochondria (mGPx4) and in cytosol (cGPx4) (Maiorino et al., 2003). It has been shown that cGPx4 expression is vital with respect to the reduction of PLHP and a preservation of the spermatozoon head/tail integrity (Ursini et al., 1999; Foresta et al., 2002). Recently a mechanism of GPx4-catalyzed reduction of cardiolipin [1,1'2,2'-tetra-oleoyl cardiolipin (TOCL)] hydroperoxide in liposomes has been demonstrated, which is based on the electrostatic interaction between negatively charged polar head of the TOCL and large positively charged surface of GPx4 with consequent steering of the –COOH group to the catalytic Sec46 vicinity. Surprisingly, it was stated that TOCL hydroperoxide is the best substrate for GPx4, which hypothetically could suggest that mitochondria are the main sites of the GPx4-supported cellular vitality (Cozza et al., 2017).

In general, the mechanism of GPx4 binding to the phospholipid substrate in a membrane is similar to that of Prdx6 and is determined by the electrostatic interaction between the negative charge of a polar head of TOCLHP (GPx4) or –COOH group of PLHP (Prdx6) and a diffused positive charge (GPx4) or a positively charged groove (Prdx6) on the corresponding protein surface. The next step of catalysis corresponds to the reduction of the -COOH group of substrate into hydroxyl and a simultaneous oxidation of the catalytic Sec46 (GPx4) or Cys47 (Prdx6) into the selenenate or sulfenate, respectively. The subsequent step of catalysis, in principal, is similar and requires the glutathionylation of Sec46-SeOH (Gpx4) by free GSH or of Cys47-SOH (Prdx6) by GSTP1-1/GS⁻ complex. Such glutathionylation causes conformational changes in both enzymes and opens access to an additional molecule of the free GSH, which through the disulfide/thiol exchange reduces catalytic Sec46-SeSG (GPx4) or Cys47-SSG (Prdx6) into corresponding chalcogenyls and regenerates the ground state enzymes in both cases. From a standpoint of chemical kinetics, the –SeOH group of GPx4 is a stronger electrophile compared to the –SOH group of Prdx6 and that facilitates its glutathionylation by free GSH. The kinetic competence of Prdx6 glutathionylation is based on the catalytic activation of GSH by GSTP1-1 into GS⁻ and its consequent reaction with the Cys47–SOH group. It would be interesting to directly compare the speed as well as the rate-limiting steps of the Prdx6 and GPx4 catalytic reduction of PLHP in a membrane under the same conditions. Since the deletion of GPx4 is lethal, it would also be interesting to find out if overexpression of Prdx6 and GSTP1-1 could rescue the defective phenotype.

With respect to the biomedical significance, it is important to understand that in humans GPX4 and Prdx6 are the intracellular enzymes that specifically target intracellular organelles. The GPx4 gene is expressed as cytosolic, mitochondrial and nuclear isoenzymes, while Prdx6 targets the lysosomes and the lamellar bodies. GPx4 is highly expressed in male reproductive organs, and is substantially expressed in the endocrine tissues, lung, GIT, kidney, and bladder tissues [(Uhlén et al., 2015), www.proteinatlas.org/ENSG00000167468-GPX4/tissue]. Prdx6 is highly expressed in the brain, endocrine tissues, bone marrow, the immune system, lung, liver, gall-bladder, pancreas, gastro-intestinal tract, kidney and urinary bladder, male and female organs, adipose and soft tissue, and is sufficiently expressed in the skin and muscles [(Uhlén et al., 2015), www.proteinatlas.org/ENSG00000117592-PRDX6/tissue]. In lung Prdx6 represents ~0.1% of the total proteins. Recently, Prdx6 was detected in the cerebrospinal fluid, which signifies its functional secretion (Manevich et al., 2014). Merely a simple comparison between the expressions of these two anti-oxidant enzymes corresponds to their potential importance for normal homeostasis.

Additionally, it has been found that Prdx6 expression in liver (Reddy et al., 2006), as well as its oxidation in human erythrocytes (O'Neill et al., 2011), follows a circa-dian rhythm. It has been shown that the impairment of "the circadian clock in the brain causes oxidative stress, astrocyte activation, axonal terminal degeneration, and a disrupted resting state functional connectivity". Brain pathology in the circadian clock-disrupted mice is caused by oxidative damage (increase in LP) in susceptible regions. An alteration of Prdx6 expression in the brains of "positive limb of the circa-dian clock"-compromised mice might indicate its involvement in the neuronal redox homeostasis and prevention of neurodegeneration (Musiek et al., 2013). This study could be more conclusive, if it had included an analysis of the cerebrospinal fluid, where Prdx6 oxidation was shown to be critical for the neurological consequences of traumatic brain injury (Manevich et al., 2014).

A very recent study described a phenotype of the diabetes mellitus in Prdx6 KO mice, which indicates a possible crucial role of this enzyme in the pathogenesis of cardio-metabolic diseases. This study also outlines a general involvement of Prdx6 in the pathophysiology of chronic noncommunicable diseases (NCDs). NCDs are "multifactorial diseases, where a combination of genetic, epigenetic and environ-mental factors contributes to pathogenesis". NCDs mainly comprise of diabetes mellitus, cardiovascular diseases, chronic obstructive pulmonary disease (COPD), cancer, and neurodegenerative diseases, which combined account for the morbidity of more than 80% of the population worldwide. Recent findings suggest an increase in oxidative stress and the age-related loss of the antioxidant protection as the com-mon factors for all these pathologies. Through its unique combination of the GSH-peroxidase and PLA$_2$ activities, Prdx6 is thus implicated in pathogenesis of NCDs (Pacifici et al., 2018).

9.7   CONCLUDING REMARKS

Since the evolutionary selection of an oxidative phosphorylation as the most effi-cient way of ATP generation at the time of an adequate concentration of oxygen in the earth's atmosphere, the insufficiencies of oxygen consumption/utilization have become an important issue. The oxidative stress as a misbalance between the generation of the reactive oxygen species (ROS, byproducts of oxidative phosphorylation) and their elimination by antioxidant systems, has become one of the most commonly quoted reasons for innumerable diseases and disorders. Prdx6 is an antioxidant enzyme that, by virtue of its structure, belongs to the peroxiredoxins but, by its peroxidase function with GSH as a reducing agent, may be considered to be a glutathione peroxidase. This enzyme appears to be a specific protector of bio-membranes against oxidative stress and, in this respect, is most similar to GPx4. The biological significance of this enzyme has only begun to show its general involvement in the pathophysiology of the oxidant stress-originated diseases and disorders. Two complimentary activities of this enzyme (GSH-peroxidase and PLA$_2$) are signifying its pivotal role in the antioxidant protection of the phospholipid bilayer in biomembranes. The possible involvement of Prdx6 in the pathophysiology of multiple diseases might result in the development of some novel and efficient therapies. For example, the overexpression of Prdx6 in cancer cells presumably corresponds with their survival under oxidant stress, so the inhibition or compromised expression of Prdx6 will cause these cells to perish. Conversely, the suppression of Prdx6 expression as result of aging is considered to cause neurodegenerative diseases, including Parkinson's and Alzheimer's diseases. The unique properties of Prdx6 described above lead us to reasonably expect a great progress in related practical biomedical applications.

REFERENCES

Adler, V., Z. Yin, S. Y. Fuchs et al. 1999. Regulation of JNK signaling by GSTp. *EMBO J* 18:1321–34.

Babior, B. M., R. S. Kipnes, and J. T. Curnutte. 1973. Biological defense mechanisms. The production by leukocytes of superoxide, a potential bactericidal agent. *J Clin Invest* 52:741–44.

Bergelson, L.D., J. G. Molotkovsky, and Y. Manevich. 1985. Lipid-specific fluorescent probes in studies of biological membranes *Chem Phys Lipids* 37:165–95.

Biteau, B., J. Labarre, and M. B. Toledano. 2003. ATP-dependent reduction of cysteine sulfinic acid by *S. cerevisiae* sulphiredoxin *Nature* 425:980–4.

Carvalho, A.N., C. Marques, R. C. Guedes et al. 2016. *S*-Glutathionylation of Keap1: A new role for glutathione *S*-transferase pi in neuronal protection. *FEBS Lett* 590:1455–66.

Chen, J.W., C. Dodia, S. I. Feinstein, M. K. Jain and A.B. Fisher. 2000. 1-Cys peroxiredoxin, a bifunctional enzyme with glutathione peroxidase and phospholipase A2 activities *J Biol Chem* 275:28421–7.

Choi, H.J., S.W. Kang, C.H.Yang, S.G. Rhee, and S.E. Ryu. 1998. Crystal structure of a novel human peroxidase enzyme at 2.0 Å resolution *Nat Str Biol* 5:400–6.

Cozza, G., M. Rossetto, V. Bosello-Travain et al. 2017. Glutathione peroxidase 4-catalyzed reduction of lipid hydroperoxides in membranes: The polar head of membrane phospholipids binds the enzyme and addresses the fatty acid hydroperoxide group toward the redox center. *Free Rad Biol Med* 112:1–11.

Deponte, M. 2013. Glutathione catalysis and the reaction mechanisms of glutathione-dependent enzymes. *Biochim Biophys Acta* 1830:3217–66.

Fisher, A.B. 2011. Peroxiredoxin 6: A bifunctional enzyme with glutathione peroxidase and phospholipase A$_2$ activities *Antioxid Redox Signal* 15:831–44.

Fisher, A.B. 2017. Peroxiredoxin 6 in the repair of peroxidized cell membranes and cell signaling *Arch Biochem Biophys* 617:68–83.

Fisher, A.B., C. R. Dodia, Y. Manevich, J. W. Chen, and S. I. Feinstein. 1999. Phospholipid hydroperoxides are substrates for non-selenium glutathione peroxidase *J Biol Chem* 274:21326–34.

Fowle, F. E. 1921. Smithsonian miscellaneous collections. *Smithsonian Institution Publ* 71 (1):400.

Greenberg, M.E., X.-M. Li, B. G. Guriu et al. 2008. The lipid whisker model of the structure of oxidized cell membranes. *J Biol Chem* 283:2385–96.

Grek, C.L., J. Zhang, Y. Manevich, D. Townsend, and K. Tew. 2013. Causes and consequences of cysteine *S*-glutathionylation *J Biol Chem* 288:26497–504.

Foresta, C., L. Flohé, A. Garolla et al. 2002. Male fertility is linked to the selenoprotein phospholipid hydroperoxide glutathione peroxidase. *Biol Reprod* 67:967–71.

Han, X., Z. and Y. Yu. Fan. 2013. Expression and characterization of recombinant human phospholipid hydroperoxide glutathione peroxidase. *IUBMB Life* 65:951–6.

Hyvönen, M.T., T. T. Rantala, and M. Ala-Korpela. 1997. Structure and dynamic properties of di-unsaturated 1-palmitoyl-2-linoleoyl-sn-glycero-3-phosphatidylcholine lipid bilayer from molecular dynamics simulation *Biophys J* 73:2907–23.

Jain, M.K., and O. G. Berg. 1989. The kinetics of interfacial catalysis by phospholipase A_2 and regulation of interfacial activation: Hopping versus scooting. *Biochim Biophys Acta –Lipids Lipid Metabolism* 1002:127–56.

Kang, S.W., I.C. Baines, and S. G. Rhee. 1998. Characterization of a mammalian peroxiredoxin that contains one conserved cysteine *J Biol Chem* 273:6303–11.

Kellogg, E. W. III., and I. Fridosvich. 1975. Superoxide, hydrogen peroxide, and singlet oxygen in lipid peroxidation by a xanthine oxidase system *J Biol Chem* 250:8812–17.

Kernstock, R.M., and A. W. Girotti. 2008. New strategies for the isolation and activity determination of naturally occurring type-4 glutathione peroxidase. *Prot Expr Pur* 62:216–22.

Kim, K.H., W. Lee, and E. E. Kim. 2016. Crystal structure of human peroxiredoxin 6 in different oxidation states *Biochem Biophys Res Commun* 477:717–21.

Lee, S.P., Y. S. Hwang, Y. J. Kim et al. 2001. Cyclophilin A binds to peroxiredoxins and activates its peroxidase activity. *J Biol Chem* 276:29826–32.

Loschen, G., A. Azzi, C. Richter, and L. Flohé. 1974. Superoxide radicals as precursors of mitochondrial hydrogen peroxide *FEBS Lett* 42:68–72.

Maiorino, M., C. Gregolin, and F. Ursini. 1990. Phospholipid hydroperoxide glutathione peroxidase *Meth Enzymol* 186:448–57.

Maiorino, M., M. Scapin, F. Ursini et al. 2003. Distinct promoters determine alternative transcription of gpx-4 into phospholipid-hydroperoxide glutathione peroxidase variants. *J Biol Chem* 278:34286–90.

Manevich Y., S. I. Feinstein, and A. B. Fisher. 2004. Activation of the antioxidant enzyme 1-Cys peroxiredoxin requires glutathionylation mediated by heterodimerization with GST. *Proc Natl Acad Sci USA* 101:3780–85.

Manevich, Y., K. D. Held, and J. E. Biaglow. 1997. Coumarin-3-carboxylic acid as a detector for hydroxyl radicals generated chemically and by gamma radiation. *Rad Res* 148:580–91.

Manevich, Y., S. Hutchens, P. V. Halushka et al. 2014. Peroxiredoxin VI oxidation in cerebrospinal fluid correlates with traumatic brain injury outcome. *Free Rad Biol Med* 72:210–21.

Manevich, Y., S. Hutchens, K. D. Tew, and D. M. Townsend. 2013. Allelic variants of glutathione *S*-transferase P1-1 differentially mediate the peroxidase function of peroxiredoxin VI and alter membrane lipid peroxidation *Free Rad Biol Med* 54:62–70.

Manevich, Y., K. S. Reddy., T. Shuvaeva, S. I. Feinstein, and A. B. Fisher. 2007. Structure and phospholipase function of peroxiredoxin 6: Identification of the catalytic triad and its role in phospholipid substrate binding. *J Lip Res* 48:2306–18.

McCord, J. M., and I. Fridovich. 1969. Superoxide dismutase. An enzymic function for erythrocuprein (hemocuprein). *J Biol Chem* 244:6049–55.

Misra, H. P. 1974. Generation of superoxide free radical during the autoxidation of thiols. *J Biol Chem* 249:2151–5.

Molotkovsky, J. G., Y. Manevich, E. N. Gerasimova, and L. D. Bergelson. 1982. Differential study of phosphatidylcholine and sphingomyelin in human high-density lipoproteins with lipid-specific fluorescent probes. *Eur J Biochem* 122:573–9.

Monteiro, G., B. B. Horta, D. C. Pimenta, O. Augusto, and L. E. S. Netto. 2007. Reduction of 1-Cys peroxiredoxins by ascorbate changes the thiol-specific antioxidant paradigm, revealing another function of vitamin C. *Proc Natl Acad Sci USA* 104:4886–91.

Musiek, E.S., M. M. Lim, G. Yang, A. Q. Bauer, and G. A. FitzGerald. 2013. Circadian clock proteins regulate neuronal redox homeostasis and neurodegeneration. *J Clin Invest* 123:5389–400.

O'Neill, J. S., and A. B. Reddy. 2011. Circadian clocks in human red blood cells. *Nature* 469:498–503.

Pacifici, F., D. Della Morte, B. Capuani et al. 2018. Peroxiredoxin 6, a multitask antioxidant enzyme involved in the pathophysiology of chronic non-communicable diseases. *Antioxid Red Signal* (published online 02 Jan 2018). doi:10.1089/ars.2017.7427.

Paulsen, C. E., and K. S. Carroll. 2013. Cysteine-mediated redox signaling: Chemistry, biology, and tools for discovery *Chem Rev* 113:4633–79.

Pedrajas, J. R., B. McDonagh, F. Hernandez-Torres et al. 2016. Glutathione is the resolving thiol for thioredoxin peroxidase activity of 1-Cys peroxiredoxin without being consumed during the catalytic cycle. *Antioxid Redox Signal* 24:115–28.

Peshenko, I. V., V. I. Novoselov, V. A. Evdokimov et al. 1996. Novel 28-kDa secretory protein from the rat olfactory epithelium. *FEBS Lett* 381:12–14.

Pettigrew, N. E., and R. F. Colman. 2001. Heterodimer of glutathione *S*-transferase can form between isoenzyme classes π and μ. *Arch Biochem Biophys* 369:225–30.

Poole, L.B., C. Klomsiri, S. A. Knaggs et al. 2007. Fluorescent and affinity-based tools to detect cysteine sulfenic acid formation in proteins. *Bioconjug Chem* 18:2004–17.

Prade, L., R. Huber, T. H. Manoharan, W. E. Fahl, and W. Reuter. 1997. Structures of class pi glutathione *S*-transferase from human placenta in complex with substrate, transition-state analogue and inhibitor *Struct* 5:1287–95.

Ralat, L.A., Y. Manevich, A. B. Fisher, and R.F. Colman. 2006. Direct evidence for the formation of a complex between 1-cysteine peroxiredoxin and glutathione *S*-transferase π with activity changes in both enzymes *Biochemistry* 45:360–72.

Ralat, L. A., S. A. Misquitta, Y. Manevich, A. B. Fisher, and R. Colman. 2008. Characterization of the complex of glutathione *S*-transferase pi and 1-cysteine peroxiredoxin *Arch Biochem Biophys* 474:109–18.

Reddy, A. B., N. A Karp, E. S. Maywood et al. 2006. Circadian orchestration of the hepatic proteome. *Current Biol* 16:1107–15.

Rhee, S. G., H. Z. Chae, and K. Kim, 2005. Peroxiredoxins: A historical overview and speculative preview of novel mechanisms and emerging concepts in cell signaling. *Free Rad Biol Med* 38:1543–52.

Rhee, S. G., H. A. Woo, I. S. Kil, and S. H. Bae. 2012. Peroxiredoxin functions as a peroxidase and a regulator and sensor of local peroxides *J Biol Chem* 287:4403–10.

Richter, C., A. Azzi, U. Weser, and A. Wendel. 1977. Hepatic microsomal dealkylations. Inhibition by a tyrosine-copper (II) complex provided with superoxide dismutase activity *J Biol Chem* 252:5061–6.

Schröder, E., and C. P. Ponting. 1998. Evidence that peroxiredoxins are novel members of the thioredoxin fold superfamily *Prot Sci* 7:2465–68.

Sevanian, A., S. F. Muakkassah-Kelly, and S. Montestruque. 1983. The influence of phospholipase A$_2$ and glutathione peroxidase on the elimination of membrane lipid peroxides *Arch Biochem Biophys* 223:441–52.

Sheehan, D., G. Meade, V.M. Foley, and C.A. Dowd. 2001. Structure, function and evolution of glutathione transferases: Implications for classification of non-mammalian members of an ancient enzyme superfamily. *Biochem J* 360:1–16.

Sies, H., C. Berndt, and D. P. Jones. 2017. Oxidative stress. *Ann Rev Biochem* 86:715–48.

Singh, A.K., and H. Shichi. 1998. Novel glutathione peroxidase in bovine eye. *J Biol Chem* 273:26171–78.

Singh, R., and G.M. Whitesides. 1995. Reagents for rapid reduction of disulfide bonds in proteins. In *Techniques in Protein Chemistry VI*, 259–266. Academic Press Inc. https://gmwgroup.harvard.edu/pubs/pdf/440.pdf.

Tew, K. 2007. Redox in redux: Emergent roles for glutathione *S*-transferase P (GSTP) in regulation of cell signaling and *S*-glutathionylation. *Biochem Pharm* 73:11257–69.

Tew, K.D., Y. Manevich, C. Grek et al. 2011. The role of glutathione *S*-transferase P in signaling pathways and *S*-glutathionylation in cancer. *Free Rad Biol Med* 51:299–313.

Thurmond, R.L, G. Lindblom, and M. F. Brown. 1993. Curvature, order, and dynamics of lipid hexagonal phases studied by deuterium NMR spectroscopy. *Biochemistry* 32:5394–410.

Uhlén M., L. Fagerberg, B. M. Hallström et al. 2015. Tissue-based map of human proteome. *Science* 347:394–404. www.proteinatlas.org/ENSG00000167468-GPX4/tissue; www.proteinatlas.org/ENSG00000117592-PRDX6/tissue.

Ursini, F., S. Heim, M. Kies et al. 1999. Dual function of the selenoprotein PHGPx during sperm maturation. *Science* 285:1393–6.

Wang, T., P. Arifoglu, Z. Ronai, and K. D. Tew. 2001. Glutathione *S*-transferase P1-1 (GST P1-1) inhibits c-Jun N-terminal kinase (JNK1) signaling through interaction with the C terminus. *J Bio Chem* 276:20999–1003.

Wang, R., C. R. Dodia, M. K. Jain, and A.B. Fisher. 1994. Purification and characterization of a calcium-independent acidic phospholipase A2 from rat lung. *Biochem J* 304:131–7.

Waterboer, T., P. Sehr, K. M. Michael et al. 2005. Multiplex human papillomavirus serology based on in situ-purified glutathione *S*-transferase fusion proteins. *Clin Chem* 5:1845–53.

Woo, H. A., W. Jeong, T.-S. Chang et al. 2005. Reduction of cysteine sulfinic acid by sulfiredoxin is specific to 2-cys peroxiredoxins. *J Biol Chem* 280:3125–8.

Wu, Y., S. I. Feinstein, Y. Manevich et al. 2009. Mitogen-activated protein kinase mediated phosphorylation of peroxiredoxin 6 regulates its phospholipase A$_2$ activity. *Biochem J* 419:669–79.

Yun, H.M., K. R. Park, H. P. Lee et al. 2014. PRDX6 promotes lung tumor progression via its GPx and iPLA2 activities. *Free Rad Biol Med* 69:367–7.

Zhao, L., H. P. Wang, H. J. Zhang et al. 2003. L-PHGPx expression can be suppressed by antisense oligodeoxynucleotides. *Arch Biochem Biophys* 417:212–8.

Zhou, S., Y.-C. Lien, T. Shuvaeva et al. 2013. Functional interaction of glutathione *S*-transferase pi and peroxiredoxin 6 in intact cells. *Int J Biochem Cell Biol* 45:401–7.

10 Glutathione Peroxidases and the Thyroid Gland

Lutz Schomburg

Charité—Universitätsmedizin Berlin

CONTENTS

10.1 INTRODUCTION

The thyroid gland is an exceptional tissue specialized in accumulating the trace element iodide and using it for the biosynthesis of thyroid hormones (THs). To this end, iodide becomes activated via oxidation with hydrogen peroxide. Isozymes of the selenium-dependent glutathione peroxidase (GPx) family are prime candidates for balancing the peroxide concentrations and spatially restricting their oxidizing activity to the colloid lumen of thyroid follicles where TH biosynthesis proceeds. There is little information on a specific role of GPx1 or GPx4 in thyroid biology. GPx2 expression is nearly absent in thyrocytes, whereas GPx3 and the other actively secreted selenoprotein, i.e., selenoprotein P (SELENOP), are abundant. The GPx may protect the thyrocytes during the intracellular transport of the peroxide-generating machinery to the apical membrane and after secretion into the colloidal lumen. This chapter discusses the delicate balance of anabolic enzymes for TH biosynthesis, peroxides, and the detoxifying GPx.

10.2 THE THYROID GLAND—DISTINCTIVE MORPHOLOGICAL AND BIOCHEMICAL FEATURES

The thyroid gland is a central organ of the endocrine system, and responsible for the synthesis of THs. As a typical endocrine organ, it is regulated by stimulating hormones from the superior endocrine gland, i.e., thyrotropin (TSH) from the pituitary, and it is subject to negative feedback control by the hormones it is producing. Hereby, its

DOI: 10.1201/9781351261760-12

activity becomes dynamically adapted to the current needs of the organism (Fliers et al., 2006). THs are essential endocrine messengers, responsible for development, growth, thermogenesis, and metabolic processes (Ortiga-Carvalho et al., 2016). Moreover, they enable, as so-called permissive factors, the regular functioning of other endocrine axes (Mann and Plant, 2010). Diseases of the thyroid gland are common, especially in women. Among adult females, the total prevalence of thyroid dysfunction is >10%, approximately half of which due to autoimmune diseases like Graves' hyperthyroidism or Hashimoto's thyroiditis causing thyroid tissue destruction (Hoogendoorn et al., 2006; Bliddal et al., 2015; Schultheiss et al., 2015).

The thyroid is the only endocrine gland where the major hormone is not produced inside the endocrine cells, but in the extracellular space. The reason for this disease-relevant peculiarity is the biochemistry of TH biosynthesis. THs are the only molecules in the human body that depend on the trace element iodine for their biosynthesis and activity. Iodine belongs to the halogens in the periodic table, group 7, with seven outer electrons, i.e., just one electron short of the noble gases. It is taken up by the regular nutrition mainly in the form of iodide (I^-) and as such constitutes a stable anion and inert electrolyte usually not causing adverse reactions, not even when administered in relatively high amounts during radiocontrast-based diagnostic procedures (Schabelman and Witting, 2010). In the clinics, isotopes of I^- are used for tissue-specific radioiodine therapy in hyperthyroidism or thyroid malignancies with little side effects (Ross, 2011).

Under normal conditions, iodide is not prone to undergoing chemical reactions. It is for this reason, that very drastic reaction conditions are needed in order to oxidize iodide to becoming reactive and usable for TH biosynthesis. To this end, thyrocytes take advantage of the strong oxidant hydrogen peroxide and are producing a steady supply of this semistable reactive oxygen species (ROS) into the colloid lumen (Song et al., 2007). It is likely that the resulting concentrations needed for efficient iodination of thyroglobulin (TG) and the by-products generated during the hydrogen peroxide-mediated iodide oxidation step are too reactive to be compatible with regular cell functioning and long-term survival. A targeted export of the TH synthesis machinery out of the cell and into a constrained extracellular reaction vessel, named the colloid lumen, appears as a meaningful way to solve this problem.

Interestingly, the protein concentration within the colloid is extraordinarily high in the range of 100–400 mg/mL (Hayden et al., 1970). The colloid should therefore be viewed as a highly structured dense and protein-rich milieu. It is characterized by a specific spatial arrangement of the carrier protein TG and the enzymes needed for iodinating its tyrosyl residues to mono- and di-iodinated tyrosyl (mono-iodinated tyrosyl (MIT) and di-iodinated tyrosyl (DIT)) moieties and finally to TG-attached TH. Histochemical preparations show some vague morphological resemblance to the spermatic cord supporting sperm maturation or the growing ovarian follicles before ovulation. All of these structures represent extracellular compartments surrounded by endocrine active cells. However, the physiological reason for generating these protein-rich extracellular compartments is fundamentally different, as the thyroid gland uses this structure for an energy-consuming accumulation of iodide and enabling a drastic and threatening chemical reaction that is otherwise difficult to control inside the cells. In contrast, the spermatic cord or ovulatory follicle rather developed in order to provide a highly nourishing and safe extracellular environment for the protected maturation of the primary cells founding the coming generation of life.

10.3 BIOSYNTHESIS OF THYROID HORMONES

The essential oxidation of iodide in the follicular lumen results from the interplay of a number of specialized thyroid enzymes and their concerted interaction in the generation, safe metabolism, and control of hydrogen peroxide and TG (Figure 10.1). The thyroid gland expresses at least three members of the family of NADPH oxidases (NOX), i.e., NOX4 and the dual oxidases DUOX1 and DUOX2. The correct folding, transport, and activity of the DUOX enzymes depend on the two maturation factors DUOXA1 and DUOXA2. Inherited defects in any of these enzymes have been shown as underlying reasons for a dysfunctional gland and congenital hypothyroidism (Grasberger, 2010).

The detailed characterization of transgenic animals revealed that the DUOX/ DUOXA enzyme complex is not only responsible for enabling TH biosynthesis but also of high relevance, e.g., for the gastric epithelium, where it affects helicobacter infection and resulting inflammation (Grasberger et al., 2013). Professional phagocytes are another well-characterized cell type known for their expression of NOX isozymes and an intentional generation of hydrogen peroxide and superoxide (De Deken et al., 2014). In contrast to thyrocytes, however, these cells of the immune defense are generating hydrogen peroxide and resulting ROS in order to destroy the invading pathogens, extracellularly or within intracellular vesicles after endocytosis. As immune cells are relatively short-lived in comparison to thyrocytes, they are obviously not in need of strictly avoiding self-damage from these potentially harmful molecules. As thyrocytes are highly differentiated and the thyroid gland shows a very limited capacity to self-renewal and proliferation, it may be in need of the aforementioned strategy of a well-structured peroxide generation and usage in a targeted process efficiently controlled by the interacting thyroid proteins. GPx may provide the necessary balance, spatial control, and protection from surplus hydrogen peroxide in order to safeguard thyrocyte integrity and ROS-dependent signaling, both during passage of the peroxide generating enzyme machinery to the apical membrane and within the extracellular thyroid colloid.

The biosynthesis of TH depends on the thyroid-specific large globular protein TG that is secreted into the follicular lumen as the scaffold for TH biosynthesis. There is evidence that TG also acts as a feedback regulatory signal for TH biosynthesis, affecting DUOX expression, hydrogen peroxide tone, and follicle activities. This is reflected in varying sizes, thyrocyte protrusions, and other morphological features of differently active follicles within the same thyroid gland (Yoshihara et al., 2012). Correct folding and trafficking of TG is partly enabled by an intrinsic chaperone domain of TG with a cholinesterase-like structure (Lee et al., 2008). The reuptake of iodinated TG is not fully understood, but proceeds via partial extracellular proteolysis, endocytosis, and subsequent degradation involving members of the cathepsin family of proteases (Brix et al., 1996).

The transport of iodide into the follicular lumen is actively accomplished by the sodium-iodide symporter (NIS) taking advantage of the sodium gradient across the

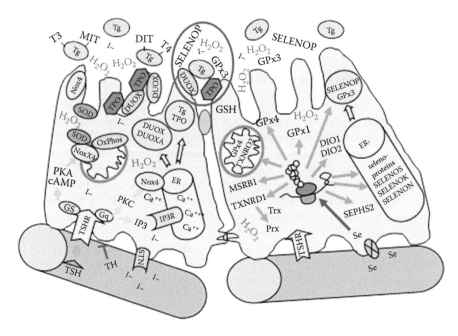

FIGURE 10.1 Schematic view on TH biosynthesis by thyroxisomes. Thyrocytes are mainly controlled by thyrotropin (TSH) receptor (TSHR) activation and TH feedback control, involving among other signaling components, the G-proteins Gs and Gq, along with cyclic AMP (cAMP), protein kinase A (PKA), protein kinase C (PKC), inositol 3-phosphate (IP3), and the IP3 receptor (IP3R). The essential trace elements iodide (I) and selenium (Se) are taken up from the nutrition via the sodium-driven Na$^+$/I$^-$ symporter (NIS) and poorly characterized mechanisms, respectively. Hydrogen peroxide (H$_2$O$_2$) is generated by the thyroid oxidases NOX4 as well as two dual oxidases with associated chaperones (DUOX/DUOXA). Some peroxides are also generated as unavoidable by-products of mitochondrial oxidative phosphorylation (OxPhos). The peroxide-generating activities (thyrocyte to the left) are counter balanced by protecting enzyme systems including the superoxide dismutase (SOD), peroxiredoxins (Prx), thioredoxin (Trx), and a number of selenoproteins (thyrocyte to the right). The selenoenzymes deiodinase type 1 (DIO1) and 2 (DIO2) modify the pattern of TH synthesized, while the selenophosphate synthetase 2 (SEPHS2) contributes to selenoprotein biosynthesis. Colloidal thyroid peroxidase (TPO) together with the scaffold protein thyroglobulin (TG) belongs to the core proteins involved in TH biosynthesis. Together with the substrates I$^-$ and H$_2$O$_2$ and likely in association with the secreted glutathione peroxidases GPx3 and SELENOP, functional units at the border of the apical membrane and the protein-dense colloidal lumen are formed, so-called thyroxisomes (indicated by the red circle). Unfortunately, neither the role of the intracellular GPx1 and GPx4 isozymes nor the other intracellular (methionine sulfoxide reductase B1, MSRB1) or ER-resident selenoproteins (SELENOS, SELENOK, SELENON, and others) are functionally understood with respect to their specific contribution to thyroid diseases.

thyrocyte membrane (Darrouzet et al., 2014). Once within the thyrocytes, iodide follows the concentration and electrical gradient into the follicular lumen through a number of different transport proteins including pendrin, the cystic fibrosis transmembrane conductance regulator, anoctamin (ANO1), and the sodium multivitamin transporter, respectively (De La Vieja and Santisteban, 2018). Once inside the colloid lumen, iodide becomes enzymatically consumed by the thyroid peroxidase (TPO) for coupling to TG (Dunn and Dunn, 2001). As this reaction needs energy, it proceeds close to the apical cell membrane of thyrocytes, linking intracellular energy metabolism to extracellular peroxide generation (Donko et al., 2014). The close interaction of the major peroxide producing oxidases DUOX1 and DUOX2 with their maturation factors DUOXA1 and DUOXA2 ensures a timely and well-controlled peroxide generation outside the cell, preventing an aberrant activity during their intracellular passage. To this end, the complex of DUOX enzymes and DUOXA chaperones are coexpressed and physically associated during passage (Ohye and Sugawara, 2010). Control of DUOX activity also involves Ca binding to the intracellular heiix-loop-helix (EF-hand) motifs. An inadequately high concentration of iodide is capable of inhibiting hydrogen peroxide generation and thereby TH biosynthesis via the so-called Wolff–Chaikoff effect (Burgi, 2010).

The hydrogen peroxide generated is mainly used by TPO. Under resting conditions, most of the TPO is localized in secretory vesicles close to the apical membrane, from where it can be recruited by TSH-stimulated membrane fusion and exposing of its active site into the colloidal lumen. There is evidence that the DUOX and TPO enzymes are building a protein complex in which the reactive intermediates are transferred from one reaction site to the other in a highly efficient interrelated manner avoiding leakiness and potential damage to the immediate environment (Song et al., 2010). The generally high degree of thyroid cancer and autoimmune thyroid diseases as compared to other endocrine organs argues for a sensible and delicate balance between surplus and insufficient generation and consumption of reactive iodine and oxygen intermediates. A second and most important reaction of the hydrogen peroxide-consuming TPO enzymes constitutes in the coupling of two mono-iodo-tyrosyl (MIT) and di-iodo-tyrosyl (DIT) residues, respectively, to form the iodinated thyronyl residues as TG-attached TH precursors. The interrelated reactions together with some experimental evidence for a spatial colocalization of the essential enzymes argue for a functional thyroid-specific unit encompassing at least TPO, the DUOX complex along with their substrates hydrogen peroxide and iodide as a relay structure accepting colloidal TG as substrate. This assembly of interacting components is conceptually known as a thyroxisome (Song et al., 2007; Colin and Gerard, 2010), and it can be assumed that the extracellular GPx form another important component of these functional units (Figure 10.1, red circle).

10.4 EXPRESSION PATTERN AND POTENTIAL ROLE OF THYROID GLUTATHIONE PEROXIDASES

In view of the essential generation of hydrogen peroxide for TH biosynthesis, it is not surprising that the thyroid gland expresses a variety of protecting enzymes, many of which implicated in the antioxidative defense systems like the thioredoxin reductases (TXNRD), peroxiredoxins (Prx), catalase, and the Se-dependent GPx (Schweizer et al., 2008). The isozymes of the selenoprotein families TXNRD and GPx are considered of central importance for thyroid integrity and function, as they are highly expressed and directly connected to the Prx via thioredoxin (Trx) as the common cofactor. This notion concurs with the finding that among all human tissues, the thyroid gland contains probably the highest concentration of selenium (Se), typically reflecting high expression levels of selenoproteins. Accordingly, mRNA analyses have indicated high mRNA concentrations of GPx1, GPx3, and GPX4, but surprisingly little or no GPx2 transcripts (Fagerberg et al., 2014). In an unbiased analysis of human transcripts in different organs, the expression levels of GPx1 and GPx4 appear similarly high as in many other human tissues, whereas the transcripts encoding the secreted isozyme GPx3 are exceptionally enriched, second only to the kidneys where most of circulating GPx3 originates (Olson et al., 2010). This high expression rate of thyroid GPx3 was investigated in some more detail in order to elucidate its potential function.

In thyroid disease, especially in thyroid tumors, GPx3 expression is diminished, indicating some relation to thyroid activity and differentiation grade (Hasegawa et al., 2002; Zhao et al., 2015). This notion is supported by reports on its endocrine regulation, as the hormonal activation of G-protein coupled receptors causing elevated intracellular calcium and IP3 signaling results in increased GPx3 expression and secretion. It is unresolved in how far thyroid GPx3 contributes to the circulating GPx3 activities in blood. However, some GPx3 is secreted into the colloid lumen to the site of TH biosynthesis (Howie et al., 1995). In view of its enzymatic activity as an efficient catalyst of redox reactions readily accepting hydrogen peroxide, it was hypothesized that thyroid GPx3 contributes to the protection of thyroid cells (Schmutzler et al., 2007). If this was the case, some thyroid phenotype should become apparent upon *Gpx3* gene inactivation. However, no overt defects in growth, development, or thyroid axis were reported from the respective mouse model (Olson et al., 2010; Malinouski et al., 2012). Similarly, in human genome wide association studies (GWAS), *GPx3* gene variants have not been identified in relation to thyroid diseases (Simmonds, 2013). Some support for a protective role of GPx isozymes was provided when mice with conditional inactivation of general selenoprotein biosynthesis were characterized. To this end, the gene encoding the selenocysteine-specific tRNA (*Trsp*) was specifically inactivated in thyrocytes by CRE-mediated recombination either driven by *Pax8* or *Tg*-gene expression. Surprisingly, despite an almost complete abrogation of biosynthesis of functional selenoenzymes (Diol, GPx), the thyroids showed an almost normal morphological appearance and circulating TH were not deranged (Chiu-Ugalde et al., 2012). However, increased staining for oxidative stress (4-hydroxynonenal) and oxidative/nitrosative protein modifications (4-nitro-tyrosine) were detected, supporting the relevance of thyroid selenoproteins for gland integrity and avoidance of oxidative damage.

Besides the expected role of protecting the thyroid gland, the targeted secretion of GPx3 into the colloid lumen to the place of TH biosynthesis also raised the interesting question for an active participation in TG storage, iodination, and sequestration. In analogy to the fascinating report on GPx4 serving in male reproduction both as a protective selenoenzyme in spermatids and as a covalently cross-linked multimeric structural component in spermatozoa (Ursini et al., 1999), the idea of GPx3 contributing to the spatial organization of colloidal TG was tested. Colloid was purified from human thyroid and analyzed for GPx3 content and mode of fixation. Thyroid GPx3 was liberated from preparation of thyroid follicles upon incubation with mild detergents, whereas reductive conditions breaking potential mixed selenodisulfide bonds were not required (Schmutzler et al., 2007). These findings verified the presence of thyroid GPx3 within the colloid but failed to support the notion on the enzyme playing some structural role in TG biochemistry. Collectively, secreted GPx3 together with the secreted Se-rich transporter SELENOP may contribute to intracolloidal control of hydrogen peroxide concentrations and activity, as both enzymes are capable of glutathione-dependent reduction of peroxides (Saito et al., 1999). It is thus conceivable that SELENOP also serves as an extracellular GPx within the thyroid. The enzymatic activity relies on some reducing substrate to completing the redox cycle, as shown *in vitro* (Palazzolo and Ely, 2015). However, there are no indications for regulated glutathione concentrations in the colloid that may serve as substrates for GPx or SELENOP or as a direct competitor to iodide oxidation. Moreover, mouse studies indicated no impairment of the thyroid axis in *Selenop* knockout mice (Schomburg et al., 2006), again challenging the suspected importance of selenoproteins for thyroid protection and during TH biosynthesis.

From a more nutritional perspective, strong effects of iodine deficiency were observed on the expression of thyroid selenoproteins including GPx1 and GPx4, indicating some interaction between iodine and Se supply for thyroid activity (Mitchell et al., 1996). This notion is supported by the drastic phenotype observed in children with combined deficiency in these two essential trace elements, giving rise to the devastating syndrome of myxoedematous cretinism (Contempre et al., 1991). Notably, high doses of iodine are inducing thyroid necrosis, which may even proceeds to fibrosis in Se deficiency, supporting the protective role of selenoproteins during iodine overload (Contempre et al., 1996). Similarly, some positive effects of Se supplementation on disease activity are described, especially in patients with Hashimoto's thyroiditis (Schomburg, 2012). However, from such clinical studies, it is not possible to assign the effects of increased Se availability to increased thyroid selenoprotein biosynthesis or elevated thyroid GPx expression. The thyroid gland appears to belong to the organs with the highest priority in times of limited Se supply, i.e., residing in the top league of tissues with respect to the hierarchical Se supply (Beckett et al., 1991; Schomburg and Schweizer, 2009), largely independent of the general Se status (Aaseth et al., 1990). For these reasons, intervention studies in humans with Se aiming to increasing selenoprotein expression for health effects are not providing conclusive insights into thyroid biology.

Similarly, a specific role of thyroid GPx4 for TH biosynthesis or thyrocyte function and survival cannot easily been worked out, as both the thyroid gland as a tissue and also GPx4 as a selenoenzyme are preferentially supplied with Se (Wingler et al., 1999; Schomburg and Schweizer, 2009). In view of its ubiquitous expression and essential function for cell survival (Ingold et al., 2018), it can be assumed that thyroid GPx4 is indispensable for thyroid cells. However, under normal conditions, there are no indications for a thyroid-specific regulation of GPx4 expression, and the large genome-wide association studies failed to indicate a specific role of *GPx4* in thyroid diseases (Simmonds, 2013). Similarly, there was no increased apoptosis rates upon genetic thyroid-specific inactivation of *Trsp* impairing thyrocyte selenoprotein expression in general (Chiu-Ugalde et al., 2012).

In contrast to GPx4, expression of the ubiquitous isozyme GPx1 is more variable and responsive and strongly dependent on the availability of Se for its biosynthesis and stability of its mRNA (Sun et al., 2000). Again, there are no indications for a specific relevance of this enzyme for the thyroid gland from the mouse model with genetic *Gpx1* knockout (Ho et al., 1997). Similarly, different *GPx1* genotypes proved unrelated to thyroid autoimmunity (De Farias et al., 2015), despite some reports on a functional polymorphism in *GPx1* with direct relation to cancer (Lubos et al., 2011) or Kashin-Beck disease risk (Xiong et al., 2010).

Besides the *GPx*, other selenoprotein gene variants have been reported to modifying Hashimoto's thyroiditis risk, e.g., polymorphisms in the *SELENOS* gene (Santos et al., 2014). However, this interaction was not replicated in a recent case–control study (Xiao et al., 2017). Still, reduced *GPx1* expression has been reported in thyroid cancer as compared to intact thyroids, which may be of pathological relevance for malignant transformation (Metere et al., 2018). In view of the tight dependence of *GPx1* expression on Se supply, one may speculate that this enzyme is responsible for translating some of the positive health effects of Se supplementation on the thyroid. Even though the study results are far from conclusive, several intervention studies reported an improved quality-of-life, reduced autoantibody titers and diminished disease symptoms in thyroid patients, i.e., in destructive Hashimoto's thyroiditis as well as in Graves' hyperthyroidism with or without orbitopathy (Schomburg, 2012; Effraimidis and Wiersinga, 2014).

10.5 CONCLUSION

Despite the likely role of *GPx* isozymes and *SELENOP* in thyroid biology, malignant transformation as well as autoimmune disease and course, our current insights into the relative importance of the different selenoproteins is very limited. The reasons for this uncertainty are several fold; a lack of reliable *in vitro* models mimicking the spatial organization of thyroid follicles, the prioritized Se supply of the thyroid gland making clinical supplementation studies difficult to interpret in relation to thyroid selenoproteins, and a lack of relevant interactions in genetic studies linking the different *GPx* genotypes to thyroid disease risks.

REFERENCES

Aaseth, J., H. Frey, E. Glattre et al. 1990. Selenium concentrations in the human thyroid gland. *Biol Trace Elem Res* 24 (2):147–52.

Beckett, G. J., F. E. Peterson, K. Choudhury et al. 1991. Inter-relationships between selenium and thyroid hormone metabolism in the rat and man. *J Trace Elem Electrolytes Health Dis* 5 (4):265–7.

Bliddal, S., M. Boas, L. Hilsted et al. 2015. Thyroid function and autoimmunity in Danish pregnant women after an iodine fortification program and associations with obstetric outcomes. *Eur J Endocrinol* 173 (6):709–18.

Brix, K., P. Lemansky, and V. Herzog. 1996. Evidence for extracellularly acting cathepsins mediating thyroid hormone liberation in thyroid epithelial cells. *Endocrinology* 137 (5):1963–74.

Burgi, H. 2010. Iodine excess. *Best Pract Res Clin Endocrinol Metab* 24 (1):107–15.

Chiu-Ugalde, J., E. K. Wirth, M. O. Klein et al. 2012. Thyroid function is maintained despite increased oxidative stress in mice lacking selenoprotein biosynthesis in thyroid epithelial cells. *Antioxid Redox Signal* 17 (6):902–13.

Colin, I., and A. C. Gerard. 2010. The thyroid angiofollicular units, a biological model of functional and morphological integration. *Bull Mem Acad R Med Belg* 165 (5–6):218–28; discussion 228–30.

Contempre, B., O. Le Moine, J. E. Dumont et al. 1996. Selenium deficiency and thyroid fibrosis. A key role for macrophages and transforming growth factor beta (TGF-beta). *Mol Cell Endocrinol* 124 (1-2):7–15.

Contempre, B., J. Vanderpas, and J. E. Dumont. 1991. Cretinism, thyroid hormones and selenium. *Mol Cell Endocrinol* 81 (1–3):C193–5.

Darrouzet, E., S. Lindenthal, D. Marcellin et al. 2014. The sodium/iodide symporter: State of the art of its molecular characterization. *Biochim Biophys Acta* 1838 (1 Pt B):244–53.

De Deken, X., B. Corvilain, J. E. Dumont, and F. Miot. 2014. Roles of DUOX-mediated hydrogen peroxide in metabolism, host defense, and signaling. *Antioxid Redox Signal* 20 (17):2776–93.

de Farias, C. R., B. R. Cardoso, G. M. B. de Oliveira et al. 2015. A randomized-controlled, double-blind study of the impact of selenium supplementation on thyroid autoimmunity and inflammation with focus on the GPx1 genotypes. *J Endocrinol Invest* 38 (10):1065–74.

De la Vieja, A., and P. Santisteban. 2018. Role of iodide metabolism in physiology and cancer. *Endocr Relat Cancer* 25 (4):R225–45.

Donko, A., S. Morand, A. Korzeniowska et al. 2014. Hypothyroidism-associated missense mutation impairs NADPH oxidase activity and intracellular trafficking of Duox2. *Free Radic Biol Med* 73:190–200.

Dunn, J. T., and A. D. Dunn. 2001. Update on intrathyroidal iodine metabolism. *Thyroid* 11 (5):407–14.

Effraimidis, G., and W. M. Wiersinga. 2014. Mechanisms in endocrinology. Autoimmune thyroid disease: Old and new players. *Eur J Endocrin* 170 (6):R241–52.

Fagerberg, L., B. M. Hallstrom, P. Oksvold et al. 2014. Analysis of the human tissue-specific expression by genome-wide integration of transcriptomics and antibody-based proteomics. *Mol Cell Proteomics* 13 (2):397–406.

Fliers, E., A. Alkemade, W. M. Wiersinga, and D. F. Swaab. 2006. Hypothalamic thyroid hormone feedback in health and disease. *Prog Brain Res* 153:189–207.

Grasberger, H. 2010. Defects of thyroidal hydrogen peroxide generation in congenital hypothyroidism. *Mol Cell Endocrinol* 322 (1–2):99–106.

Grasberger, H., M. El-Zaatari, D. T. Dang, and J. L. Merchant. 2013. Dual oxidases control release of hydrogen peroxide by the gastric epithelium to prevent *Helicobacter felis* infection and inflammation in mice. *Gastroenterology* 145 (5):1045–54.

Hasegawa, Y., T. Takano, A. Miyauchi et al. 2002. Decreased expression of glutathione peroxidase mRNA in thyroid anaplastic carcinoma. *Cancer Lett* 182 (1):69–74.

Hayden, L. J., J. M. Shagrin, and J. A. Young. 1970. Micropuncture investigation of the anion content of colloid from single rat thyroid follicles. A micromethod for the simultaneous determination of iodide and chloride in nanomole quantities. *Pflugers Arch* 321 (2):173–86.

Ho, Y. S., J. L. Magnenat, R. T. Bronson et al. 1997. Mice deficient in cellular glutathione peroxidase develop normally and show no increased sensitivity to hyperoxia. *J Biol Chem* 272 (26):16644–51.

Hoogendoorn, E. H., A. R. Hermus, F. de Vegt et al. 2006. Thyroid function and prevalence of anti-thyroperoxidase antibodies in a population with borderline sufficient iodine intake: Influences of age and sex. *Clin Chem* 52 (1):104–11.

Howie, A. F., S. W. Walker, B. Akesson et al. 1995. Thyroidal extracellular glutathione peroxidase: A potential regulator of thyroid-hormone synthesis. *Biochem J* 308 (Pt 3):713–17.

Ingold, I., C. Berndt, S. Schmitt et al. 2018. Selenium utilization by GPX4 is required to prevent hydroperoxide-induced ferroptosis. *Cell* 172 (3):409–22 e21.

Lee, J., B. Di Jeso, and P. Arvan. 2008. The cholinesterase-like domain of thyroglobulin functions as an intramolecular chaperone. *J Clin Invest* 118 (8):2950–8.

Lubos, E., J. Loscalzo, and D. E. Handy. 2011. Glutathione Peroxidase-1 in health and disease: From molecular mechanisms to therapeutic opportunities. *Antioxid Redox Signal* 15 (7):1957–97.

Malinouski, M., S. Kehr, L. Finney et al. 2012. High-resolution imaging of selenium in kidneys: A localized selenium pool associated with glutathione peroxidase 3. *Antioxid Redox Signal* 16 (3):185–92.

Mann, D. R., and T. M. Plant. 2010. The role and potential sites of action of thyroid hormone in timing the onset of puberty in male primates. *Brain Res* 1364:175–85.

Metere, A., F. Frezzotti, C. E. Graves et al. 2018. A possible role for selenoprotein glutathione peroxidase (GPx1) and thioredoxin reductases (TrxR1) in thyroid cancer: Our experience in thyroid surgery. *Cancer Cell Int* 18:7.

Mitchell, J. H., F. Nicol, G. J. Beckett, and J. R. Arthur. 1996. Selenoenzyme expression in thyroid and liver of second generation selenium- and iodine-deficient rats. *J Mol Endocrinol* 16 (3):259–67.

Ohye, H., and M. Sugawara. 2010. Dual oxidase, hydrogen peroxide and thyroid diseases. *Exp Biol Med (Maywood)* 235 (4):424–33.

Olson, G. E., J. C. Whitin, K. E. Hill et al. 2010. Extracellular glutathione peroxidase (Gpx3) binds specifically to basement membranes of mouse renal cortex tubule cells. *Am J Physiol Renal Physiol* 298 (5):F1244–53.

Ortiga-Carvalho, T. M., M. I. Chiamolera, C. C. Pazos-Moura, and F. E. Wondisford. 2016. Hypothalamus-pituitary-thyroid axis. *Compr Physiol* 6 (3):1387–428.

Palazzolo, D. L., and E. A. Ely. 2015. Arsenic trioxide and reduced glutathione act synergistically to augment inhibition of thyroid peroxidase activity in vitro. *Biol Trace Elem Res* 165 (1):110–17.

Ross, D. S. 2011. Radioiodine therapy for hyperthyroidism. *N Engl J Med* 364 (6):542–50.

Saito, Y., T. Hayashi, A. Tanaka et al. 1999. Selenoprotein P in human plasma as an extracellular phospholipid hydroperoxide glutathione peroxidase. Isolation and enzymatic characterization of human selenoprotein P. *J Biol Chem* 274 (5):2866–71.

Santos, L. R., C. Duraes, A. Mendes et al. 2014. A polymorphism in the promoter region of the selenoprotein S gene (SEPS1) contributes to Hashimoto's thyroiditis susceptibility. *J Clin Endocrinol Metab* 99 (4):E719–23.

Schabelman, E., and M. Witting. 2010. The relationship of radiocontrast, iodine, and seafood allergies: A medical myth exposed. *J Emerg Med* 39 (5):701–7.

Schmutzler, C., B. Mentrup, L. Schomburg et al. 2007. Selenoproteins of the thyroid gland: Expression, localization and possible function of glutathione peroxidase 3. *Biol Chem* 388 (10):1053–9.

Schomburg, L. 2012. Selenium, selenoproteins and the thyroid gland: Interactions in health and disease. *Nat Rev Endocrinol* 8 (3):160–71.

Schomburg, L., C. Riese, M. Michaelis et al. 2006. Synthesis and metabolism of thyroid hormones is preferentially maintained in selenium-deficient transgenic mice. *Endocrinology* 147 (3):1306–13.

Schomburg, L., and U. Schweizer. 2009. Hierarchical regulation of selenoprotein expression and sex-specific effects of selenium. *Biochim Biophys Acta* 1790 (11):1453–62.

Schultheiss, U. T., A. Teumer, M. Medici et al. 2015. A genetic risk score for thyroid peroxidase antibodies associates with clinical thyroid disease in community-based populations. *J Clin Endocrinol Metab* 100 (5):E799–807.

Schweizer, U., J. Chiu, and J. Köhrle. 2008. Peroxides and peroxide-degrading enzymes in the thyroid. *Antioxid Redox Signal* 10 (9):1577–92.

Simmonds, M. J. 2013. GWAS in autoimmune thyroid disease: Redefining our understanding of pathogenesis. *Nat Rev Endocrinol* 9 (5):277–87.

Song, Y., N. Driessens, M. Costa et al. 2007. Roles of hydrogen peroxide in thyroid physiology and disease. *J Clin Endocrin Metab* 92 (10):3764–73.

Song, Y., J. Ruf, P. Lothaire et al. 2010. Association of duoxes with thyroid peroxidase and its regulation in thyrocytes. *J Clin Endocrinol Metab* 95 (1):375–82.

Sun, X., P. M. Moriarty, and L. E. Maquat. 2000. Nonsense-mediated decay of glutathione peroxidase 1 mRNA in the cytoplasm depends on intron position. *Embo J* 19 (17):4734–44.

Ursini, F., S. Heim, M. Kiess et al. 1999. Dual function of the selenoprotein PHGPx during sperm maturation. *Science* 285 (5432):1393–6.

Wingler, K., M. Bocher, L. Flohe et al. 1999. mRNA stability and selenocysteine insertion sequence efficiency rank gastrointestinal glutathione peroxidase high in the hierarchy of selenoproteins. *Eur J Biochem* 259 (1–2):149–57.

Xiao, L., J. H. Yuan, Q. M. Yao et al. 2017. A case-control study of selenoprotein genes polymorphisms and autoimmune thyroid diseases in a Chinese population. *Bmc Medical Genetics* 18 (1):54.

Xiong, Y. M., X. Y. Mo, X. Z. Zou et al. 2010. Association study between polymorphisms in selenoprotein genes and susceptibility to Kashin-Beck disease. *Osteoarthr Cartil* 18 (6):817–24.

Yoshihara, A., T. Hara, A. Kawashima et al. 2012. Regulation of dual oxidase expression and H_2O_2 production by thyroglobulin. *Thyroid* 22 (10):1054–62.

Zhao, H., J. Li, X. Li et al. 2015. Silencing *GPX3* expression promotes tumor metastasis in human thyroid cancer. *Curr Protein Pept Sci* 16 (4):316–21.

Part III

Conjugations and Isomerizations

11 Glutathione Transferases
From the Test Tube to the Cell

Bengt Mannervik and Birgitta Sjödin
Stockholm University

CONTENTS

DOI: 10.1201/9781351261760-14

175

11.1 INTRODUCTION

Enzymology is a cornerstone in the molecular life sciences. Besides providing fundamental understanding of the dynamics of life processes, enzymology has essential applications in drug discovery and biotechnology. Glutathione transferases (GSTs) are ubiquitous and versatile enzymes with crucial physiological functions and are excellently suited to explore many of the above-mentioned aspects. Based on the cutting-edge advances in molecular life sciences, it is now possible to move enzyme research from the test tube into the cell. Genes encoding GSTs in variant forms can be chemically synthesized, and expressed enzyme proteins can be produced and inserted directly into cells or produced intracellularly from transfected DNA. Organisms can be genetically engineered to acquire an altered "GSTome," and precision gene surgery can be performed with novel techniques such as CRISPR/Cas9.

11.2 BACKGROUND

11.2.1 The Versatility of GST Functions

Enzymes not only govern rates and specificities of metabolic reactions but also, via interactions with other biomolecules, are intimately involved in networks of cellular signal transduction. Particularly noteworthy in functional breadth and complexity are the GSTs, which have evolved to not only catalyze diverse chemical reactions but also serve as intracellular transporters of heme and other relatively small molecules, as well as to regulate the activity of protein kinases affecting the life expectancy of a cell (Josephy and Mannervik, 2006). Accordingly, GSTs are involved in different networks of interactions, which to different degrees may be entangled. Clearly, the cellular systems have to master parallel processing of a wide variety of inputs (Singh, 2015). Surprisingly, GST proteins can move in and out of live mammalian cells and participate in intercellular trafficking. As an example from pathology, parasites such as the human liver fluke *Opithorchis viverrini* secrete GSTs, which in the bile duct serve as mitogens that through AKT and ERK signaling can promote cell proliferation and thereby induce cholangio-carcinoma (Daorueang et al., 2012).

11.2.2 GSTs as Model System

The GSTs are promiscuous in their substrate acceptance and collectively they detoxify various xenobiotics or metabolic by-products that otherwise could be harmful. The protective effects of GSTs against xenobiotics are also exemplified by herbicide resistance in plants (Cummins et al., 2011) and insecticide resistance in flies (Low et al., 2007). There are multiple forms of GSTs in higher organisms and both cytosolic and membrane-bound GSTs occur in different numbers. In humans 17 different genes encode cytosolic GSTs, and in poplar (*Populus trichocarpa*) 81 (Lan et al., 2009) and in potato (*Solanum tuberosum*) 90 (Islam et al., 2018) homologous GST genes have been annotated. The intrinsic catalytic promiscuity of GSTs makes evolution to high activity with new substrates facile by mutations of only a few amino acids in the GST protein (Pettersson et al., 2002). Enzyme evolution occurs both in natural systems and in protein engineering. Studies of GSTs are therefore well suited to illustrate and clarify a range of important aspects of enzymology including protein evolution for novel functions, drug discovery, behavior of enzymes in the intracellular milieu, relationships to diseases and cell death, as well as biotechnical applications.

More than 13,000 nonredundant GST gene sequences have been identified in the biosphere (Mashiyama et al., 2014) and at least 30,000 scientific papers on GSTs have been published such that any attempt to cover the "GSTome" (Mannervik, 2012) extensively would be preposterous. We have therefore made a more personal selection of topics that we consider significant for this chapter. Numerous recent reviews cover other aspects of GST research (Singhal et al., 2015; Hollman et al., 2016; Kumar et al., 2017; Mohana et al., 2017; Perperopoulou et al., 2017).

11.3   A NOTE ON NOMENCLATURE

GSTs discovered in hepatic tissues were catalyzing the conjugation of aromatic substrates (Booth et al., 1961; Combes et al., 1961) leading to the name glutathione S-aryltransferase. This designation was subsequently generalized to glutathione S-transferase (Boyland et al., 1969). However, the latter nomenclature was recognized as a misnomer, since sulfur is not transferred but a glutathionyl (GS$^-$) group. In fact, the enzymes catalyzing conjugation and addition reactions can be regarded as "glutathionyl transferases." The currently recommended nomenclature (Mannervik et al., 2005) designates the enzymes as glutathione transferase without the prefix "S−"; the abbreviation GST is still accurate.

In humans and other mammals, the GSTs have been grouped into membrane-bound (microsomal) and soluble (cytosolic) proteins. The former are members of the MAPEG (membrane-associated proteins in eicosanoid and glutathione metabolism) family (Jakobsson et al., 1999) and are composed of three subunits and further described by Morgenstern et al. in Chapter 13. A mitochondrial GST is an outlier with a somewhat different protein fold than the soluble enzymes (Ladner et al., 2004).

The soluble GSTs are generally dimers, although some monomeric plant GSTs are known (Lallement et al., 2014). The dimeric GSTs occur in numerous classes of homologous sequences. Proteins of the same class can form heterodimers as well as homodimers (Mannervik et al., 1982) and can, thus, be identified, for example, as GST A1-1, GST A1-2, and GST A2-2 by their composition of subunits 1 and 2, which are encoded by the *GSTA1* and *GSTA2* genes of the Alpha class.

11.4   INACTIVATION OF TOXIC COMPOUNDS

11.4.1   DETOXIFICATION OF XENOBIOTICS

GSTs catalyze the detoxification of xenobiotics, compounds foreign to the organism, including numerous mutagenic compounds that cause cancer, such as polyaromatic hydrocarbons studied by Boyland and coworkers (Boyland et al., 1969). The abundant forms of the human enzymes, GST A1-1, GST M1-1, and GST P1-1, are all active with the potently carcinogenic diolepoxide of benzo(a)pyrene (Robertson et al., 1986) resulting from oxidation by cytochrome P450 enzymes. The GST substrates also include toxic organic plant components, such as isothiocyanates present in edible vegetables (Zhang et al., 1995). However, isothiocyanates provide a positive feedback, since the compounds also prevent toxicity and carcinogenesis by induction of protective enzymes (including GSTs) via transcriptional gene activation (Dinkova-Kostova, 2013). In addition, GSTs inactivate alkylating anticancer drugs, reactions, which in tumor cells contribute to resistance against chemotherapy (Larsson et al., 2010).

11.4.2   DETOXIFICATION OF ENDOGENOUSLY PRODUCED TOXICANTS

Although GSTs were first recognized as major components in the cellular defense against xenobiotic electrophiles, it became obvious that their protective functions included toxic products arising from oxidative metabolism of lipids, nucleic acids, catecholamines, and other physiologically relevant molecules (Mannervik, 1986; Berhane et al., 1994). Two enzymes, in particular, seem to have evolved to conjugate and inactivate specific endogenous toxicants. GST A4-4 displays prominent activity with 4-hydroxynonenal and other alkenals formed by lipid peroxidation (Hubatsch et al., 1998). GST M2-2 is distinguished by remarkable activity with aminochrome and dopamine orthoquinone (Segura-Aguilar et al., 1997; Dagnino-Subiabre et al., 2000) arising from dopamine, as well as with similar orthoquinones derived from other catecholamines (Baez et al., 1997). The glutathione conjugation of orthoquinones counteracts their propensity to generate reactive oxygen species via extensive redox cycling, which is linked to Parkinson's disease and other degenerative conditions.

Notwithstanding established important roles of the enzymes in detoxification, the elimination of the three loci of the Mu, Pi, and Theta class GST genes, encoding 14 out of the 22 GST enzymes in the mouse has no obvious detrimental effects on normal development, well-being, or fertility of the animals (Xiang et al., 2014). The members of these classes may have their most important functions in the protection against environmental toxins and oxidative stress not imposed under laboratory conditions. The results are also indicative of robust backup functions, in agreement with the finding that the null alleles of human *GSTM1* and *GSTT1* have modest phenotypic consequences (Josephy, 2010). It was originally proposed that the frequent absence of the Mu class enzyme would influence the capacity of different individuals to metabolize and detoxify various carcinogenic polyaromatic hydrocarbons (Warholm et al., 1981), but subsequent genomic studies indicate that susceptibilities are dependent on a variety of gene sequence variations, and are not limited to the null genotype (Moyer et al., 2007).

11.5   ROLES IN INTERMEDIARY METABOLISM

11.5.1   Eicosanoid-Derived Signal Substances

Arachidonic acid and other polyunsaturated fatty acids are not only undergoing oxidative processes to give toxic electrophiles including 4-hydroxyalkenals but also the source of prominent cellular signaling molecules such as prostaglandins, leukotrienes, and thromboxanes, as outlined in Chapter 13. Leukotriene A_4 is an epoxide resulting from the action of 5-lipoxygenase on arachidonic acid, and leukotriene C_4 is the corresponding glutathione adduct. Soluble GSTs can catalyze this synthesis (Söderström et al., 1985), but the physiologically relevant conjugating enzyme is a separate membrane-bound leukotriene C synthase (Söderström et al., 1988). The leukotriene C synthase is established as a distinct enzyme in the MAPEG family.

However, another branch in the metabolism of arachidonic acid leads to the prostaglandins via cyclooxygenase-catalyzed oxidation to the 9,11-endoperoxide prostaglandin H_2, and this reactive product can give rise to a number of other signal molecules including the 9-hydroxy-11-keto derivative prostaglandin D_2 (Smith et al., 2011). Prostaglandin D_2 synthase was discovered as a cytosolic glutathione-dependent isomerase (Christ-Hazelhof et al., 1979) and has subsequently been called hematopoietic prostaglandin D_2 synthase (H-PGDS). A second nonhomologous PGDS catalyzes the same reaction, but has a different tissue distribution and is not dependent on glutathione (Smith et al., 2011). The latter enzyme shows homology to the members of the lipocalin family and is referred to as lipocalin-type PGDS (L-PGDS). By contrast, H-PGDS has prominent functional and structural similarities to the soluble GSTs and is a member of the Sigma class, encompassing members in animal species ranging from nematodes and insects to mammals. Like in the other glutathione-dependent isomerases in the GST, superfamily glutathione is not consumed but serves only as a cofactor.

11.5.2   Isomerization in Aromatic Amino Acid Catabolism

An early report of a glutathione-mediated *cis–trans* isomerization instrumental in the catabolism of aromatic amino acids indicated an unknown biochemical reaction (Edwards et al., 1956). The responsible enzyme was subsequently identified as the novel Zeta class GST Z1-1, ubiquitously present in organisms ranging from plants to mammals (Board et al., 1997). Obviously, GST Z1-1 plays a pivotal role in intermediary metabolism, preventing the accumulation of maleyl-acetoacetate and its by-products maleyl-acetone and succinyl-acetone in the body. Although not lethal, the disruption of the *Gstz1* gene in the mouse caused severe toxicity when the animals were challenged with dietary phenylalanine (Board et al., 2011).

11.5.3   Role in Steroid Hormone Production

Steroidogenesis begins with cholesterol and, via multiple steps, leads to production of steroid hormones such as progesterone and testosterone (Payne et al., 2004). One of the late steps in the synthesis of these hormones is formation of Δ^5-unsaturated

3-ketosteroids in a pyridine-nucleotide-dependent reaction catalyzed by 3β-hydroxysteroid dehydrogenase (Samuels et al., 1951), followed by a double-bond isomerization (Talalay et al., 1955). The double-bond isomerization of Δ^5-pregnene-3,20-dione (Δ^5-PD) to Δ^4-PD is the last step in progesterone biosynthesis. In the synthesis of testosterone, the double-bond isomerization from Δ^5-androstene-3,17-dione (Δ^5-AD) leads to the last precursor of testosterone, Δ^4-AD (Figure 11.1). The 3β-hydroxysteroid dehydrogenase has the prerequisite isomerase function, but the observed activity is modest in comparison with the high catalytic efficiency of the human or equine GST A3-3 (Johansson et al., 2002; Lindström et al., 2018). Indeed, in vitro suppression of the cellular GST steroid isomerase activity by either enzyme inhibitors or RNA interference diminishes progesterone production to a large extent (Raffalli-Mathieu et al., 2008). Furthermore, pharmacological administration of the glucocorticoid dexamethasone to stallions suppressed serum testosterone levels in parallel with both the GSTA3 mRNA concentration and Δ^5-AD isomerase activity in cytosolic testis extracts (Ing et al., 2014).

From a chemical mechanistic perspective, it should be noted that there is a fundamental difference in the double-bond isomerization reactions catalyzed by GST Z1-1 and GST A3-3. The Zeta class enzyme is involved in a *cis–trans* rotational rearrangement enabled by a transient nucleophilic addition of glutathione to the double bond. By contrast, GST A3-3 utilizes the sulfur of glutathione as a base, which allows migration of the double bond in the B ring of the steroid to the A ring. This isomerization is facilitated by a hydrogen bond from the nitrogen of the glycine residue of glutathione to the 3-oxo group of the substrate (Dourado et al., 2014), also revealing a dual function of glutathione in catalysis.

Remarkably, GSTE14, a member of the Epsilon class, a GST class present in *Drosophila* but not in mammals, is involved in the biosynthesis of the main insect steroid

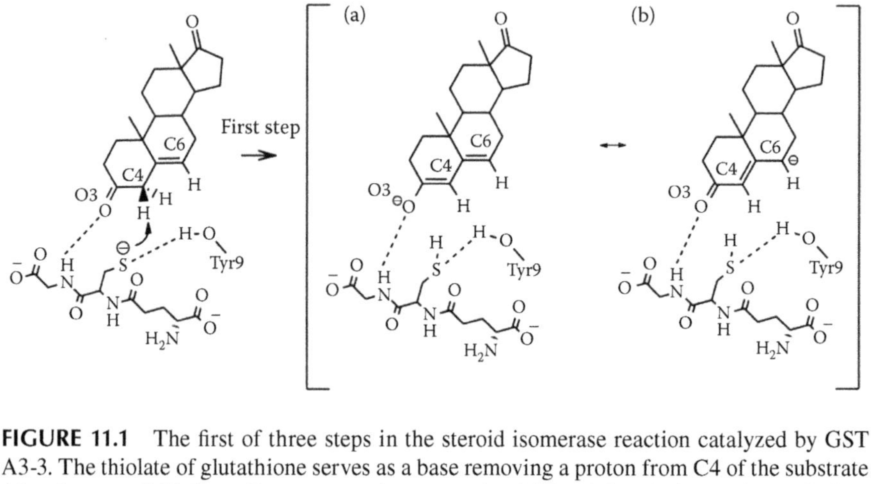

FIGURE 11.1 The first of three steps in the steroid isomerase reaction catalyzed by GST A3-3. The thiolate of glutathione serves as a base removing a proton from C4 of the substrate Δ^5-androstene-3,17-dione. In a concerted manner, the –NH– of the glycine moiety polarizes the O3 of the substrate to promote the dienolate intermediate (a), which facilitates the migration of the double bond between C5 and C6 from the B ring. Subsequently, the active-site Tyr9, following a rearrangement, protonates C6 (Dourado et al., 2014).

hormone ecdysterone (Chanut-Delalande et al., 2014; Enya et al., 2014).The examples aforementioned show that GSTs catalyze reactions relevant to normal physiological processes, thereby expanding the scope of GST enzymology beyond detoxification.

11.6 FUNCTIONAL GROUPS IN BOTH GLUTATHIONE AND PROTEIN EMPOWER GST CATALYSIS

It has been noted that the soluble GSTs can be divided into two main categories distinguishable by the active-site residue interacting with the sulfur of glutathione, one category displaying either serine or cysteine and the other featuring tyrosine (Atkinson et al., 2009). The Alpha class members belong to the second category, and based on site-directed mutagenesis substituting phenylalanine for tyrosine, it was concluded that the hydroxyl group of the latter was not strictly essential but contributed to catalysis, possibly by stabilizing the thiolate of enzyme-bound glutathione via hydrogen bonding (Stenberg et al., 1991).

Surprisingly, the ionization of the glutathione thiol was found not to be accomplished by the active-site tyrosine or any other residue of the protein, but to be due to the α-carboxylate of the γ-glutamyl group of glutathione itself (Widersten et al., 1996; Gustafsson et al., 2001). In fact, computational studies evidenced also the involvement of an active-site water molecule bridging the sulfur of glutathione with the carboxyl group (Dourado et al., 2008).

The GSTs with serine or cysteine in the active site have not been studied to the same extent as those featuring a tyrosine. However, serine is generally considered to stabilize the glutathione thiolate by hydrogen bonding in a similar fashion as tyrosine, whereas cysteine could be redox active and form a covalent bond with a reactant. The two members of the omega class are the only human GSTs presenting cysteine in the active site, and formation of a mixed disulfide with glutathione has been demonstrated as a reaction intermediate in catalysis (Brock et al., 2013).

11.7 REGULATION OF GST GENE EXPRESSION

A key player in the regulation of GST gene expression is the nuclear factor erythroid-2-related factor 2, Nrf2, which is responsible for activation of the transcription of over 500 genes in the human genome, most of which have cytoprotective functions. This nuclear transcription factor binds to the antioxidant response element ARE of the DNA to promote mRNA synthesis (Suzuki et al., 2015). The Kelch-like ECH-associated protein 1 (Keap1) in the cytoplasm functions as a negative regulator by binding of Nrf2 and thereby destining Nrf2 to ubiquitinylation and proteasomal degradation in the absence of chemical stress. Electrophiles and oxidants, including GST substrates, bind to Keap1, thus releasing Nrf2 to escape degradation and enter the nucleus as a result of exposure to the toxicants. The cadre of enzymes induced in general provide protection of the exposed tissue to carcinogenesis and other pathophysiological conditions. However, in neoplastic cells, the Nrf2/Keap1 system may actually promote the carcinogenic process and thereby have undesired consequences (Suzuki et al., 2015; Pandey et al., 2017).

An added level of complexity is the modulation of Nfr2 and Keap1 by a number of microRNAs (Cheng et al., 2013) indicating a mechanism for fine-tuning the expression of the diverse GSTs and other enzymes at the posttranslational level.

In cellular systems there is generally crosstalk between different signaling pathways leading to modulation of the resulting outcomes (Singh, 2015). In this connection and with reference to the role of Alpha class GSTs and steroid hormone biosynthesis, it should be noted that the expression of the enzymes is regulated by the steroidogenic factor 1 (SF-1) in human cells (Matsumura et al., 2013).

11.8 PROTEIN ENGINEERING AND DIRECTED MOLECULAR EVOLUTION OF GSTS

11.8.1 BEYOND THE GENETIC CODE

In the vein of the earliest applications of site-directed mutagenesis applied to enzymes (Winter et al., 1982), the functions of numerous amino acid residues in various GSTs have been explored, as exemplified by the active-site tyrosine in human GST A1-1 (Stenberg et al., 1991). Drawing on the relatively small size of fluorine (van der Waals radius 1.35 Å) versus hydrogen (1.2 Å), steric perturbations resulting from fluorine substitution should be relatively small. This was exploited in the successful site-specific replacement of the active-site tyrosine by four different fluorinated tyrosines in order to influence the pK_a of the hydroxyl group and to provide evidence for its proposed function of hydrogen bonding to the glutathione thiolate in the active site (Thorson et al., 1998). The role of the active-site tyrosine was also suggested by a nonspecific substitution of all 14 tyrosines in rat GST M1-1 by 3-fluorotyrosine (Parsons et al., 1996). In a similar investigation, substituting 5-fluoro-tryptophan for all four tryptophan residues resulting in a 4-fold increased turnover number, apparently due to an enhanced rate of product release (Parsons et al., 1998).

11.8.2 COOPERATIVE PROTEIN SUBUNITS

Most soluble GSTs are dimeric proteins and the question whether an isolated subunit could be catalytically functional has frequently been asked. A set of 10 mutations were therefore introduced in the subunit–subunit interface of human GST P1-1 to prevent dimerization and the monomeric protein was produced (Abdalla et al., 2002). Based on physicochemical parameters, the GSTP1 monomer was properly folded, but the protein showed no enzyme activity. Binding studies suggested that the H-site was functional and could bind hydrophobic substrates, whereas the binding of glutathione was impaired in spite of the fact that none of the G-site residues had been mutated (Abdalla et al., 2002). This lack of activity of the monomer is in accord with other studies suggesting that the two subunits of the functional GST P1-1 cooperate, possibly via networks of water molecules (Hegazy et al., 2004; Hegazy et al., 2006). GSTs are not typical allosteric enzymes, but several studies demonstrate that GSTs can display cooperativity under certain physical conditions (Caccuri et al., 1999) or with select substrates (Lien et al., 2001), as well as in the sequestration of the toxic nitric oxide derivative dinitrosyl-diglutathionyl-iron complex (Bocedi et al., 2016).

11.8.3 Active-Site Mimicry in Engineered GSTs

Based on the premise that the substrate selectivity of GSTs is largely governed by the amino acids in the H-site, several successful attempts have been made to mimic the high activity of a chosen GST by installing corresponding residues in the H-site of a homologous low-activity enzyme of the same GST class. The catalytic efficiency of human GST M1-1 is 2700-fold higher than that of GST M2-2 in the conjugation of *trans*-stilbene oxide, and the mutation Thr210→Ser in GST M2-2 rendered the enzyme selectively more active with the same substrate by 200-fold (Ivarsson et al., 2003). Apart from mimicking the structure of the active site of GST M1-1 and its activity with *trans*-stilbene oxide by this point mutation, investigating all 19 possible residue-210 substitutions in GST M2-2 revealed that a point mutation in the active site can enable or disable alternative catalytic reactions without necessarily altering already established activities with other substrates (Norrgård et al., 2006). This finding demonstrates a significant evolutionary plasticity useful for the emergence of diverse activities of the same protein.

Another example of mimicry is the generation of the GIMFhelix mutant of GST A1-1, comprising the replacement of four amino acids and the C-terminal helix by those present in GST A4-4 (Nilsson et al., 2000). The properties of the mutant were similar to the typical high catalytic activity of GST A4-4 with 4-hydroxyalkenals and the characteristic low pK_a (~7) of the active-site tyrosine. Analyses of crystals demonstrate the structural similarities of GIMFhelix and GST A4-4 (Balogh et al., 2009).

A third case involved the augmentation of the steroid isomerase activity of human GST A2-2, which is lower by three orders of magnitude than that of GST A3-3 (Pettersson et al., 2002). Five residues in GST A2-2 were changed into the corresponding active-site residues of GST A3-3 enabling the steroid substrate to bind in a catalytically favorable orientation (Tars et al., 2010).

11.9 EVOLUTION OF CATALYTIC ACTIVITIES IN NEW DIRECTIONS

11.9.1 Engineering GSTs for Fine-Chemical Synthesis

GSTs are generally promiscuous in their acceptance of alternative substrates, and catalyze alkylation, arylation, thiocarbamoylation, transacylation, reduction, transnitrosylation, isomerization, and various addition reactions with different efficiencies dependent on their structure (Kurtovic et al., 2008). For possible biotechnical applications, it may therefore be possible to enhance a desirable function by protein engineering and in vitro evolution. For example, the catalytic activity of GST M2-2 with indene 1,2-oxide, relevant to the synthesis of the drug Crixivan (indinavir), was enhanced approximately 100-fold by iterative saturation mutagenesis (Norrgård et al., 2011). The highest indene 1,2-oxide activity was obtained with the double mutant Thr210→Gly/Ile10→Cys. However, with five alternative substrates, undergoing mechanistically different reactions, other mutants displayed higher activity. It is noteworthy that the stereoselectivity of epoxide reactions can be manipulated by simple chemical modifications of suitable side chains in the GST structure (Ivarsson et al., 2007).

11.9.2 EVOLUTION OF GSTS FOR PRODRUG ACTIVATION

Human GSTs have been shown to activate different thiopurine prodrugs to release 6-mercaptopurine, which subsequently serves as an antimetabolite interfering with nucleotide metabolism and nucleic acid biosynthesis. Azathioprine (Imuran), 6-[(1-methyl-4-nitro-1H-imidazol-5-yl)thio]-1H-purine, has been in clinical use for more than half a century. It is activated most efficiently by GSTs A1-1, A2-2, and M1-1, whereas 10 other human GSTs were significantly less effective or lacked detectable activity (Eklund et al., 2006). Other prodrugs releasing 6-mercaptopurine are *cis*-6-(2-acetylvinylthio)purine and *trans*-6-(2-acetylvinylthio)-guanine, but with these compounds GSTs A1-1 and A2-2 are not particularly active, whereas GST M1-1 is an efficient catalyst (Eklund et al., 2007). Clinical data show that *GSTM1 positive* patients are more liable to adverse side effects of azathioprine than patients presenting with the *GSTM1 null* genotype (Stocco et al., 2007). This observation can, at least in part, be attributed to higher total GST activity resulting in elevated release of 6-mercaptopurine and more extensive consumption of glutathione in the presence of the GST M1-1 enzyme.

For potential therapeutic purposes (see Section 11.14), it was considered worthwhile to enhance the GST activity with azathioprine. GST A2-2, the most efficient enzyme with this substrate (Eklund et al., 2006), was therefore subjected to various mutational strategies (Modén et al., 2014). The allelic gene variant *GSTA2*E*, encoding the most efficient GST protein, was engineered by a structure-based approach in which three of its H-site residues were mutated. The resulting focused mutant library consisting of 864 possible amino acid combinations was screened with azathioprine, and several highly active triple-point mutants were isolated (Zhang et al., 2012b). The most active variant displayed 70-fold higher catalytic efficiency than the parental GSTA2-2*E enzyme.

In order to map the theoretically available evolutionary trajectories leading, one mutation at a time, from the parent GST A2-2 to the most efficient mutant, all six intermediate mutants were constructed and assayed with eight alternative substrates in addition to azathioprine (Zhang et al., 2012a). Conspicuously, all of the six trajectories showed a monotonically increasing activity with azathioprine, but monotonically decreasing activities, or peaks and valleys, with some of the other substrates in the multidimensional fitness landscape. Furthermore, epistatic effects of the mutations on catalytic activity were noted, which were variable in sign and magnitude depending on the substrate used, showing that epistasis is a multidimensional quality (Zhang et al., 2012a).

11.10 EVOLVING QUASI-SPECIES OF ENZYMES

The examples of mutagenesis aforementioned are based on well-reasoned rational aspects, but in nature, mutations are largely caused by stochastic processes. The rationale of the latter is that mutants are more or less randomly produced and those found useful are retained. Based on the studies of populations of evolving RNA viruses, Eigen and coworkers (Eigen et al., 1988) proposed that survival of the fittest should apply not to an individual but to a population of related mutants called the quasi-species. The concept of quasi-species was adopted for the evolution of GSTs in mutant libraries obtained by DNA shuffling of homologous sequences (Emrén et al., 2006; Runarsdottir et al., 2010). Importantly, this approach incorporates information about activities with several alternative substrates, a multivariate scheme that in many aspects reflects natural evolution. Both the design of iterative mutagenesis for new generations of enzymes and the analysis of evolutionary trajectories in multivariate dimensions are facilitated by regarding the quasi-species, rather than the "best" enzyme variant, as the relevant evolving unit (Mannervik et al., 2009).

11.11   INFOLOGS AS NOVEL INFORMATION-OPTIMIZED MUTANTS FOR ENZYME EVOLUTION

Natural evolution is based on stochastically generated quasi-species. By contrast, a rational primary-structure-guided approach of directed enzyme evolution has been designed (Govindarajan et al., 2015). Suitable amino acid substitutions are selected by phylogenetic analysis and combined, via chemical gene synthesis, into a set of maximally information-rich gene variants called infologs. The relative contribution of each substitution is determined across multiple catalytic dimensions, providing the basis for predictive functional models with broad applicability for bioengineering (Musdal et al., 2017). This novel method for enzyme engineering combines machine learning and synthesis of a modest number of genes and provides multivariate modeling of protein sequence-function in a cost-effective manner.

11.12   EXPRESSION OF GSTS IN PLANTS FOR PHYTOREMEDIATION

A variety of biochemical and biological approaches are drawing on enzymes that catalyze the biotransformation of organic pollutants, such as pesticides, explosives, and other xenobiotic substances occurring in the environment (Abhilash et al., 2009). Appropriate biotransformation render these chemical compounds nontoxic and facilitate their elimination. Genomics studies have demonstrated that GSTs in rice, during various stages of development, can counteract many of the stress challenges indicated earlier (Jain et al., 2010) suggesting that plants could be engineered with suitable GSTs and be used for phytoremediation. For example, overexpression of GSTL2 in rice provided resistance to the herbicides glyphosate and chlorsulfuron (Hu, 2014), and experiments involving transgenic overexpression of rice GSTU4 afforded tolerance to salinity and oxidative stress in *Arabidopsis thaliana* (Sharma et al., 2014). These and other investigations demonstrate the importance of GST enzymes to resistance against various chemical challenges, not only in animals but also in plants.

An example relevant to phytoremediation applications involves the explosive 2,4,6-trinitrotoluene (TNT). The environmental pollutant TNT has for decades

Growth of transgenic *A. thaliana*
expressing GSTE6 from *D. melanogaster*
on agar plates with 15 µM TNT

wt   DmGSTE6

FIGURE 11.2   Comparison of *Arabidopsis thaliana* wild-type plantlets (left) and transgenic plantlets expressing *Drosophila melanogaster* GSTE6 (right) grown in the presence of 1,3,5-trinitrotoluene (TNT) showing the protective effect of the GST transgene. (From collaborative study with the laboratory of Neil Bruce, York University, UK; cf. Tzafestas et al., 2017.)

spread over large areas as a result of large-scale military and industrial activities. TNT can be metabolized by GST-catalyzed glutathione conjugation, and *A. thaliana* plantlets overexpressing one of its GST enzymes deplete the growth medium of TNT and inactivate the compound (Gunning et al., 2014). However, the plant GSTs have modest activity and GSTs optimized by methods of biochemistry and molecular genetics are in demand (Figure 11.2).

The insect enzymes GSTE6 and GSTE7 from *Drosophila melanogaster* were demonstrated to be orders of magnitude more efficient in the detoxification of TNT than the available plant GSTs (Mazari et al., 2016). The gene encoding GSTE6 was therefore expressed in *A. thaliana* to obtain a phytoremediation model system. Plants transgenetically expressing the *Drosophila* GSTE6 were more resistant to TNT than both unmodified plants and the *Arabidopsis* lines overexpressing the plant GSTU24 and GSTU25 (Tzafestas et al., 2017). Also, the uptake of TNT from the growth medium was enhanced in plants expressing the transgene. For actual field applications plants more robust than *Arabidopsis* will obviously be required.

11.13   INTERCELLULAR TRAFFICKING OF GSTS

In a study of protein transduction domains using a *Schistosoma japonicum* GST as intended cargo to be delivered into COS7 cells, the GST protein was unexpectedly taken up even in the absence of an added protein transduction domain (Namiki et al., 2003). Similar cellular GST uptake was demonstrated with the cell lines HeLa, NIH3T3, and PC12. In an investigation using members of different classes within the GST structural superfamily and numerous additional cell lines, the results were extended (Morris et al., 2009). Experiments also indicated that the GSTs enter cells through an energy-dependent process involving endocytosis, and GST protein was found to colocalize with transferrin in the cells implying that the endocytosis process involves clathrin-coated pits. Further, incisive studies of truncated forms of human GST M2-2 showed that the globular C-terminal domain (GST-C) is responsible for the cell translocation. In particular, it was noted that alteration of the conformational stability of GST-C, governed by the α6-helix, can significantly influence cellular uptake efficiency (Morris et al., 2011). GST-C has possible medical applications in the treatment of heart arrhythmia, since it binds selectively to the ryanodine receptor RyR2 (Hewawasam et al., 2010; Samarasinghe et al., 2015) and influences contractility and calcium transients in cardiomyocytes (Hewawasam et al., 2016).

We have verified the remarkable phenomenon of GST uptake in several cell lines including neuroblastoma SH-SYS5 cells (Figure 11.3). The wild-type GST proteins are taken up in a catalytically functional state in the SH-SYS5 cells, as demonstrated with human GST M2-2 and the neurotoxic orthoquinone substrate aminochrome (Cuevas et al., 2015). Uptake of GST M2-2 provided protection against cell death caused by aminochrome, and the protective effect was counteracted by antibodies directed to the enzyme. Remarkably, the protective outcome was obtained not only by administration of the purified recombinant GST M2-2, but also by the enzyme secreted into the culture medium by astrocytoma U373MG cells. GST M2-2 occurs as a constitutive enzyme in the U373MG cells, in which it acts as an endogenous protective agent (Huenchuguala et al., 2014). The conclusion is that the two cell types can communicate via excreted GST M2-2 such that U373MG cells can protect SH-SYS5 cells against the toxic aminochrome (Cuevas et al., 2015). By extrapolation, astrocytes could similarly protect neurons via intercellular GST trafficking in the nervous system.

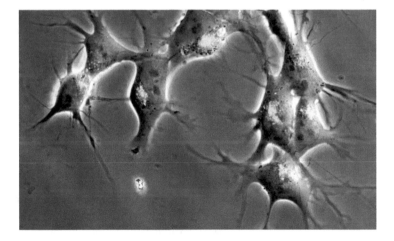

FIGURE 11.3 Uptake of human GST T1-1 in living neuroblastoma SH-SYS5 cells. The cells were incubated for 1.5 h in DMEM with 400 nM hGST T1-1 labeled with Oregon Green 488 (OG) in 37°C and 5% CO_2. Trypan Blue was used to quench any extracellular fluorescence remaining after washing of the cells.

Extracellular vesicles, first described in the form of prostasomes in prostatic fluid (Ronquist et al., 1985), occur in several physiological settings. The vesicles could be single-membrane exosomes (Van Niel et al., 2018) or multivesicular bodies (Von Bartheld et al., 2011) and may be relevant to normal as well as pathological states, including blood coagulation, inflammation, neuronal communication, and tumorigenesis. The GST excretion described earlier may possibly occur via exosomes or other membrane vesicles, and the uptake into cells may be effected via the reverse process. Considering the well-known ligandin function of GSTs (Litwack et al., 1971) intercellular trafficking of bound ligands might thus be mediated by GST proteins.

11.14   BIOMARKER APPLICATIONS OF GSTS

Early studies of carcinogenesis in rat liver demonstrated that foci (Kitahara et al., 1984) and preneoplastic nodules (Jensson et al., 1985) express a protein, now known as GST P1-1, which is not present in normal hepatocytes. Even if the same GST phenotype was not observed in the development of human hepatocarcinoma, the use of anti-GST P1-1 antibodies to detect hepatic lesions was found useful in immunochemical tests of potentially genotoxic agents in the rat model system. However, many human neoplasias other than primary liver cancer do express elevated GST P1-1 concentrations, and diagnostic tests for other tumors have been proposed (Kodate et al., 1986). The finding that different regions of the kidney diverge in their composition of the multiple GSTs present opportunities for differential diagnosis for various lesions (Rozell et al., 1993) using body fluids in additioin to histology (Hao et al., 1994). However, the release of GSTs is quite variable and can depend on various factors, not all of which indicate disease.

A more robust method applicable to diagnosis of prostate cancer is based on measurement of the methylated promoter region of the *GSTP1* gene, which is a signature for the downregulation of the enzyme in this tumor (Wu et al., 2017). Prostate cells are shed in the urine, and their DNA can be analyzed for the hypermethylation characterizing the cancer cells.

11.15   ANTIBODY DIRECTED ENZYME
PRODRUG THERAPY (ADEPT)

A recent promising development in oncology is the use of biologicals, in particular monoclonal antibodies with specificity for epitopes that distinguish tumors from normal tissues (Bhutani et al., 2013). Some therapeutic antibodies themselves afford significant antitumor activity, but new generations of antibodies carry payloads such as a drug, a radionuclide, or a toxin in order to achieve enhanced therapeutic effects (Teicher et al., 2011). Examples of antibodies conjugated with proteins include the use of ADEPT that releases the toxic agent from a prodrug to the targeted tumor (Afshar et al., 2009; Tietze et al., 2009). ADEPT has the potential to significantly improve drug efficacy and reduce adverse side-effects (Sharma et al., 2017).

It should be possible to target cellular tumor receptors by binding proteins fused with a highly active GST followed by administration of the prodrug. The high GST activity will give a focused and concentrated release of the active cytotoxic drug in the tumor tissue. Among the several novel GST-activated drugs (Ruzza et al., 2013), the prodrug TLK286/Telcyta has a particular advantage (Morgan et al., 1998). The activation of Telcyta occurs without involving glutathione as a cosubstrate and is therefore independent of the ambient concentration of glutathione. By contrast, the reactions of GSTs with most other substrates require glutathione in a concentration-dependent manner.

The underlying rationale for the development of the glutathione derivative Telcyta was its selective affinity for GST P1-1 (Lyttle et al., 1994), the GST enzyme that is often overexpressed in cancer cells (Mannervik et al., 1987). Linking the GST to a target-seeking antibody would increase the selectivity of action and obviate the requirement of high expression of the enzyme in the targeted tumor (Figure 11.4). The prodrug is activated by the GST to release an active phosphoramide mustard similar to the alkylating agent released from cyclophosphamide, a drug widely used in cancer chemotherapy. Telcyta has undergone multiple clinical trials in cancer patients and not demonstrated toxicity above that of other alkylating drugs.

It can be assumed that GST bound extracellularly can fulfill its assigned role to kill the target cell, but also that the liberated toxic drug will cause a bystander effect on proximal tumor tissue (Dachs et al., 2009). Unpublished experiments (B. Sjödin and B. Mannervik) show that scFv-anti-CD123 fused with a GST protein is taken up by neuroblastoma cells in culture (cf. Figure 11.3). Specific uptake into the receptor-presenting cells would make the treatment especially powerful.

In order to target diverse tumors, we have invented a generalized tripartite therapeutic toolbox consisting of different combinations of prodrug (substrate)–GST enzyme–binding protein (Figure 11.5). A designated GST could be engineered for high efficiency with a preferred prodrug, as in the example of GST A2-2 and azathioprine (Section 11.9.2). The redesigned GST is then coupled to an antibody or another binding protein with selective affinity for the tumor target.

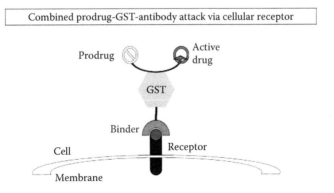

FIGURE 11.4 Scheme of directed combination-treatment using a prodrug activated by a GST linked to a binding protein, which selectively recognizes a receptor or other epitope on the target cell. The active drug will be focused to the cell and the neighboring tissue.

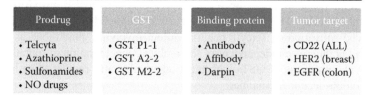

Different prodrugs activated by GSTs can be combined with different GST variants linked to diverse binding proteins targeting different tumor targets in various combinations.

Prodrug	GST	Binding protein	Tumor target
• Telcyta	• GST P1-1	• Antibody	• CD22 (ALL)
• Azathioprine	• GST A2-2	• Affibody	• HER2 (breast)
• Sulfonamides	• GST M2-2	• Darpin	• EGFR (colon)
• NO drugs			

FIGURE 11.5 Molecular tripartite toolbox composed of alternative prodrugs, selective activating GSTs, and alternative binding proteins recognizing different tumors. Using different combinations of the components, the therapeutic applications can be tailored for individual requirements.

11.16 FUTURE DIRECTIONS

11.16.1 STUDIES AT THE MOLECULAR LEVEL

GSTs can conveniently be obtained via gene synthesis and heterologous expression and be subjected to structural and functional investigations in combination with redesign and molecular evolution. Fluorophore-labeled GSTs as well as fluorogenic substrates (Shibata et al., 2013) will enable single-molecule characterization of the interactions of GSTs with alternative substrates as well as inhibitors, both *in vitro* and in living cells (Xie et al., 2008). Of particular importance are the effects induced by the intracellular milieu on catalysis and the interaction with inhibitors (Fu et al., 2014).

11.16.2 STUDIES AT THE CELLULAR LEVEL

Introduction of GST proteins into living cells via endocytosis or via transfection from eukaryote expression vectors perturbs the composition of the proteome by mechanisms that appear unrelated to the catalytic activities of GSTs, which can be verified by treatments with incapacitated mutants. Networks of cellular signaling may be influenced by GSTs and studies of proteomes as well as transcriptomes can explore these newly discovered phenomena in cells. Functionally important complexes between GSTs and c-Jun N-terminal kinase (JNK), apoptosis signal-regulating kinase 1 (ASK1), and other protein kinases have been reported (Singh, 2015). Decreased GST levels can be accomplished by gene silencing or elimination performed by the CRISPR/Cas9 approach (Jinek et al., 2012). By eliminating defined GSTs, the possible compensatory effects of other enzymes can be evaluated.

11.16.3 THE FRUIT FLY AS A MODEL ORGANISM

Drosophila melanogaster has emerged as one of the most effective biological systems for investigations of gene function in eukaryotes, and is increasingly used to model human diseases. Remarkably, a study of *Caenorhabditis elegance, D. melanogaster,* and mouse (*Mus musculus*) demonstrated that a subset of GSTs, but not the entire GSTome in each species was overexpressed in long-lived animals (McElwee et al., 2007). An integrated and comprehensive investigation of the GSTs in flies could clarify key issues related to longevity assurance and concomitantly provide essential information on biochemical processes preventing various degenerative diseases. The various GSTs are differentially expressed and the enzyme composition changes from tissue to tissue and during ontogenesis. Overexpression of GSTE7 via injection of a plasmid carrying the corresponding gene into fly embryos has already been accomplished (Mazari et al., 2014). Intriguingly, the transgenic females overexpressing GSTE7 demonstrated an enhanced egg-laying both in the absence and the presence of the toxic allyl isothiocyanate (Figure 11.6). The effect on the oviposition rate is independent of the presence or absence of toxic allyl isothiocyanate, and surprisingly also obtained with the catalytically incapacitated mutant enzyme GSTE7_S12F. The results demonstrate consequential cellular activities of GSTs other than catalysis of chemical reactions.

Underlying the expression of the proteome is the transcriptional activation of genes. Methods are now available for characterization of the global transcriptome via sequence analysis. It is even possible to study transcripts by noninvasive capturing and sequencing of mRNA from live single cells (Lovatt et al., 2014). Apparently,

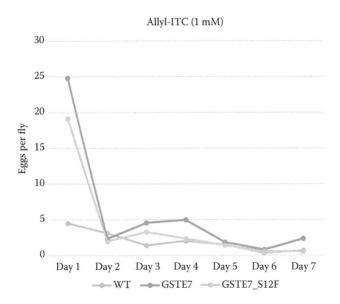

FIGURE 11.6 Significant increase in egg-laying frequency in *Drosophila melanogaster* following transgenesis with the GSTE7 gene. Overexpression of catalytically active GSTE7 as well as the catalytically incapacitated mutant GSTE7_S12F enhanced oviposition. In the experiment shown, the toxic GST substrate allyl isothiocyanate was added to the standard fly food, but the same effect of transgenesis was obtained in the absence of allyl isothiocyanate. The similar effects of the active GSTE7 and the mutant show that catalytic GST activity was irrelevant for enhanced oviposition. (Unpublished data from A.M.A. Mazari, O. Dahlberg, B. Mannervik and M. Mannervik; cf. Mazari et al., 2014.)

the tissue microenvironment shapes the transcriptomic landscape of individual cells. Mosaicism of GST expression in kidney epithelial cells has been observed (Rozell et al., 1993). Incisive examination of the GSTome using transcriptome profiling of single cells resident in their natural microenvironment would help to illuminate the multitude of GST functions.

11.17   SIGNIFICANCE OF GST RESEARCH

Enzymes are key players in all cellular processes in every living organism. In spite of a wealth of knowledge, the entire functional scope of enzymes is still incompletely understood. For example, approximately half of all pharmaceutical drugs are directed against enzymes, but the interactions of targeted enzymes with the full complement of molecules encountered in the cellular context is largely unchartered. The traditional biochemical approach involves isolation of the enzyme of interest and examination of its interaction with its cognate ligands. However, in the intracellular milieu, a protein is surrounded by thousands of different molecules, and we now understand that every enzyme molecule has to cope with both cognate and noncognate partners. Noncognate interactions may give rise to malfunctioning biochemical systems and disease, and in the field of pharmacology, a lack of specificity may cause adverse side reactions that jeopardize the beneficial effects of a drug. Furthermore, numerous application in biotechnology are based on enzymology, and the design of enzymes for new purposes still presents a challenge. Many of these questions are relevant to the GSTs as such. However, the enzymes also lend themselves to research that goes beyond glutathione biochemistry and illustrate principles and phenomena of general significance.

ACKNOWLEDGMENTS

Bengt Mannervik is grateful to Leopold Flohé for the unusual privilege of having been invited to give three lectures at the conference in Tübingen in 1973 and report on our research on glyoxalase-I (Mannervik, 1974a), glutathione reductase (Mannervik, 1974b), and thioltransferase (Mannervik et al., 1974). Our investigations have been generously supported by grants from the Swedish Research Council, the Swedish Cancer Society, and the Swedish Childhood Cancer Foundation.

REFERENCES

Abdalla, A. M., C. M. Bruns, J. A. Tainer et al. 2002. Design of a monomeric human glutathione transferase GSTP1, a structurally stable but catalytically inactive protein. *Protein Eng* 15 (10):827–34.

Abhilash, P. C., S. Jamil, and N. Singh. 2009. Transgenic plants for enhanced biodegradation and phytoremediation of organic xenobiotics. *Biotechnol Adv* 27 (4):474–88.

Afshar, S., T. Olafsen, A. M. Wu, and S. L. Morrison. 2009. Characterization of an engineered human purine nucleoside phosphorylase fused to an anti-her2/neu single chain Fv for use in ADEPT. *J Exp Clin Cancer Res* 28:147. doi:10.1186/1756-9966-28-147.

Atkinson, H. J., and P. C. Babbitt. 2009. Glutathione transferases are structural and functional outliers in the thioredoxin fold. *Biochemistry* 48 (46):11108–16.

Baez, S., J. Segura-Aguilar, M. Widersten et al. 1997. Glutathione transferases catalyse the detoxication of oxidized metabolites (o-quinones) of catecholamines and may serve as an antioxidant system preventing degenerative cellular processes. *Biochem J* 324 (Pt 1):25–8.

Balogh, L. M., I. Le Trong, K. A. Kripps et al. 2009. Structural analysis of a glutathione transferase A1-1 mutant tailored for high catalytic efficiency with toxic alkenals. *Biochemistry* 48 (32):7698–704.

Berhane, K., M. Widersten, A. Engström et al. 1994. Detoxication of base propenals and other alpha, beta-unsaturated aldehyde products of radical reactions and lipid peroxidation by human glutathione transferases. *Proc Natl Acad Sci U S A* 91 (4):1480–4.

Bhutani, D., and U. N. Vaishampayan. 2013. Monoclonal antibodies in oncology therapeutics: Present and future indications. *Expert Opinion on Biological Therapy* 13 (2):269–82.

Board, P. G., and M. W. Anders. 2011. Glutathione transferase zeta: Discovery, polymorphic variants, catalysis, inactivation, and properties of Gstz1-/- mice. *Drug Metab Rev* 43 (2):215–25.

Board, P. G., R. T. Baker, G. Chelvanayagam, and L. S. Jermiin. 1997. Zeta, a novel class of glutathione transferases in a range of species from plants to humans. *Biochem J* 328 (Pt 3):929–35.

Bocedi, A., R. Fabrini, M. Lo Bello et al. 2016. Evolution of negative cooperativity in glutathione transferase enabled preservation of enzyme function. *J Biol Chem* 291 (52):26739–49.

Booth, J., E. Boyland, and P. Sims. 1961. An enzyme from rat liver catalysing conjugations with glutathione. *Biochem J* 79 (3):516–24.

Boyland, E., and L. F. Chasseaud. 1969. The role of glutathione and glutathione *S*-transferases in mercapturic acid biosynthesis. *Adv Enzymol Relat Areas Mol Biol* 32:173–219.

Brock, J., P. G. Board, and P. J. Oakley. 2013. Structural insights into omega-class glutathione transferases: A snapshot of enzyme reduction and identification of a non-catalytic ligandin site. *PLOS ONE* 8 (4):e60324. doi:10.1371/journal.pone.0060324.

Caccuri, A. M., G. Antonini, P. Ascenzi et al. 1999. Temperature adaptation of glutathione *S*-transferase P1-1. A case for homotropic regulation of substrate binding. *J Biol Chem* 274 (27):19276–80.

Chanut-Delalande, H., Y. Hashimoto, A. Pelissier-Monier et al. 2014. Pri peptides are mediators of ecdysone for the temporal control of development. *Nat Cell Biol* 16 (11):1035–44.

Cheng, X., C. H. Ku, and R. C. Siow. 2013. Regulation of the Nrf2 antioxidant pathway by microRNAs: New players in micromanaging redox homeostasis. *Free Radic Biol Med* 64:4–11.

Christ-Hazelhof, E., and D. H. Nugteren. 1979. Purification and characterisation of prostaglandin endoperoxide D-isomerase, a cytoplasmic, glutathione-requiring enzyme. *Biochim Biophys Acta* 572 (1):43–51.

Combes, B., and G. S. Stakelum. 1961. A liver enzyme that conjugates sulfobromophthalein sodium with glutathione. *J Clin Invest* (6):981–8.

Cuevas, C., S. Huenchuguala, P. Munoz et al. 2015. Glutathione transferase-M2-2 secreted from glioblastoma cell protects SH-SY5Y cells from aminochrome neurotoxicity. *Neurotox Res* 27 (3):217–28.

Cummins, I., D. P. Dixon, S. Freitag-Pohl et al. 2011. Multiple roles for plant glutathione transferases in xenobiotic detoxification. *Drug Metab Rev* 43 (2):266–80.

Dachs, G. U., M. A. Hunt, S. Syddall et al. 2009. Bystander or no bystander for gene directed enzyme prodrug therapy. *Molecules* 14 (11):4517–45.

Dagnino-Subiabre, A., B. K. Cassels, S. Baez et al. 2000. Glutathione transferase M2-2 catalyzes conjugation of dopamine and dopa o-quinones. *Biochem Biophys Res Commun* 274 (1):32–6.

Daorueang, D., P. Thuwajit, S. Roitrakul et al. 2012. Secreted *Opisthorchis viverrini* gluta-thione *S*-transferase regulates cell proliferation through AKT and ERK pathways in cholangiocarcinoma. *Parasitol Int* 61 (1):155–61.

Dinkova-Kostova, A. T. 2013. Chemoprotection against cancer by isothiocyanates: A focus on the animal models and the protective mechanisms. *Natural Products in Cancer Prevention and Therapy* 329:179–201.

Dourado, D. F., P. A. Fernandes, B. Mannervik, and M. J. Ramos. 2008. Glutathione transfer-ase: New model for glutathione activation. *Chemistry* 14 (31):9591–8.

Dourado, D. F., P. A. Fernandes, B. Mannervik, and M. J. Ramos. 2014. Isomerization of Delta5-androstene-3,17-dione into Delta4-androstene-3,17-dione catalyzed by human glutathione transferase A3-3: A computational study identifies a dual role for glutathi-one. *J Phys Chem A* 118 (31):5790–800.

Edwards, S. W., and W. E. Knox. 1956. Homogentisate metabolism: The isomerization of maleylacetoacetate by an enzyme which requires glutathione. *J Biol Chem* 220 (1):79–91.

Eigen, M., J. McCaskill, and P. Schuster. 1988. Molecular quasi-species. *J Phys Chem* 92 (24):6881–91.

Eklund, B. I., S. Gunnarsdottir, A. A. Elfarra, and B. Mannervik. 2007. Human glutathione transferases catalyzing the bioactivation of anticancer thiopurine prodrugs. *Biochem Pharmacol* 73 (11):1829–41.

Eklund, B. I., M. Moberg, J. Bergquist, and B. Mannervik. 2006. Divergent activities of human glutathione transferases in the bioactivation of azathioprine. *Mol Pharmacol* 70 (2):747–54.

Emrén, L. O., S. Kurtovic, A. Runarsdottir et al. 2006. Functionally diverging molecular quasi-species evolve by crossing two enzymes. *Proc Natl Acad Sci U S A* 103 (29):10866–70.

Enya, S., T. Ameku, F. Igarashi et al. 2014. A Halloween gene noppera-bo encodes a glutathi-one *S*-transferase essential for ecdysteroid biosynthesis via regulating the behaviour of cholesterol in *Drosophila*. *Sci Rep* 4:6586.

Fu, D., J. Zhou, W. S. Zhu et al. 2014. Imaging the intracellular distribution of tyrosine kinase inhibitors in living cells with quantitative hyperspectral stimulated Raman scattering. *Nat Chem* 6 (7):615–23.

Govindarajan, S., B. Mannervik, J. A. Silverman et al. 2015. Mapping of amino acid substitu-tions conferring herbicide resistance in wheat glutathione transferase. *ACS Synth Biol* 4 (3):221–7.

Gunning, V., K. Tzafestas, H. Sparrow et al. 2014. *Arabidopsis* glutathione transferases U24 and U25 exhibit a range of detoxification activities with the environmental pollutant and explosive, 2,4,6-trinitrotoluene. *Plant Physiol* 165 (2):854–65.

Gustafsson, A., P. L. Pettersson, L. Grehn et al. 2001. Role of the glutamyl alpha-carboxylate of the substrate glutathione in the catalytic mechanism of human glutathione transfer-ase A1-1. *Biochemistry* 40 (51):15835–45.

Hao, X. Y., V. M. Castro, J. Bergh et al. 1994. Isoenzyme-specific quantitative immunoas-says for cytosolic glutathione transferases and measurement of the enzymes in blood-plasma from cancer-patients and in tumor-cell lines. *Biochim Biophys Acta* 1225 (2):223–30.

Hegazy, U. M., U. Hellman, and B. Mannervik. 2006. Replacement surgery with unnatural amino acids in the lock-and-key joint of glutathione transferase subunits. *Chem Biol* 13 (9):929–36.

Hegazy, U. M., B. Mannervik, and G. Stenberg. 2004. Functional role of the lock and key motif at the subunit interface of glutathione transferase P1-1. *J Biol Chem* 279 (10):9586–96.

Hewawasam, R., D. Liu, M. G. Casarotti et al. 2010. The structure of the C-terminal helical bundle in glutathione transferase M2-2 determines its ability to inhibit the cardiacry-anodine receptor. *Biochem Pharmacol* 80 (3):381–8.

Hewawasam, R., D. Liu, M. G. Casarotti et al. 2016. The GSTM2 C-terminal domain depresses contractility and Ca^{2+} transients in neonatal rat ventricular cardiomyocytes. *PLOS One* 11 (9):e0162415. doi:10.1371/journal.pone.0162415.

Hollman, A. L., P. B. Tchounwou, and H. C. Huang. 2016. The association between gene-environment interactions and diseases involving the human GST superfamily with SNP variants. *Int J Environ Res Public Health* 13 (4):379.

Hu, T. Z. 2014. A glutathione S-transferase confers herbicide tolerance in rice. *Crop Breeding and Applied Biotechnology* 14 (2):76–81.

Hubatsch, I., M. Ridderström, and B. Mannervik. 1998. Human glutathione transferase A4-4: An alpha class enzyme with high catalytic efficiency in the conjugation of 4-hydroxynonenal and other genotoxic products of lipid peroxidation. *Biochem J* 330 (Pt 1):175–9.

Huenchuguala, S., P. Munoz, P. Zavala et al. 2014. Glutathione transferase mu 2 protects glioblastoma cells against aminochrome toxicity by preventing autophagy and lysosome dysfunction. *Autophagy* 10 (4):618–30.

Ing, N. H., D. W. Forrest, P. K. Riggs et al. 2014. Dexamethasone acutely down-regulates genes involved in steroidogenesis in stallion testes. *J Steroid Biochem Mol Biol* 143:451–9.

Islam, M. S., M. Choudhury, A. K. Majlish et al. 2018. Comprehensive genome-wide analysis of glutathione S-transferase gene family in potato (*Solanum tuberosum* L.) and their expression profiling in various anatomical tissues and perturbation conditions. *Gene* 639:149–62.

Ivarsson, Y., A. J. Mackey, M. Edalat et al. 2003. Identification of residues in glutathione transferase capable of driving functional diversification in evolution. A novel approach to protein redesign. *J Biol Chem* 278 (10):8733–8.

Ivarsson, Y., M. A. Norrgård, U. Hellman, and B. Mannervik. 2007. Engineering the enantioselectivity of glutathione transferase by combined active-site mutations and chemical modifications. *Biochim Biophys Acta* 1770 (9):1374–81.

Jain, M., C. Ghanashyam, and A. Bhattacharjee. 2010. Comprehensive expression analysis suggests overlapping and specific roles of rice glutathione S-transferase genes during development and stress responses. *BMC Genomics* 11:73.

Jakobsson, P. J., R. Morgenstern, J. Mancini et al. 1999. Common structural features of MAPEG —A widespread superfamily of membrane associated proteins with highly divergent functions in eicosanoid and glutathione metabolism. *Protein Sci* 8 (3):689–92.

Jensson, H., L. C. Eriksson, and B. Mannervik. 1985. Selective expression of glutathione transferase isoenzymes in chemically induced preneoplastic rat hepatocyte nodules. *FEBS Lett* 187 (1):115–20.

Jinek, M., K. Chylinski, I. Fonfara et al. 2012. A programmable dual-RNA-guided DNA endonuclease in adaptive bacterial immunity. *Science* 337 (6096):816–21.

Johansson, A.-S., and B. Mannervik. 2002. Active-site residues governing high steroid isomerase activity in human glutathione transferase A3-3. *J Biol Chem* 277 (19):16648–54.

Josephy, P. D. 2010. Genetic variations in human glutathione transferase enzymes: Significance for pharmacology and toxicology. *Hum Genomics Proteomics* 2010:876940.

Josephy, P. D., and B. Mannervik. 2006. Chapter 10: Glutathione transferases. In *Molecular Toxicology*, 2nd ed., 333–64. New York: Oxford University Press.

Kitahara, A., K. Satoh, K. Nishimura et al. 1984. Changes in molecular forms of rat hepatic glutathione S-transferase during chemical hepatocarcinogenesis. *Cancer Res* 44 (6):2698–703.

Kodate, C., A. Fukushi, T. Narita et al. 1986. Human placental form of glutathione S-transferase (GST-pi) as a new immunohistochemical marker for human colonic carcinoma. *Jpn J Cancer Res* 77 (3):226–9.

Kumar, A., D. K. Dhull, V. Gupta et al. 2017. Role of glutathione-S-transferases in neurological problems. *Expert Opin Ther Pat* 27 (3):299–309.

Kurtovic, S., A. Shokeer, and B. Mannervik. 2008. Emergence of novel enzyme quasi-species depends on the substrate matrix. *J Mol Biol* 382 (1):136–53.

Ladner, J. E., J. F. Parsons, C. L. Rife et al. 2004. Parallel evolutionary pathways for glutathione transferases: Structure and mechanism of the mitochondrial class kappa enzyme rGSTK1-1. *Biochemistry* 43 (2):352–61.

Lallement, P. A., E. Meux, J. M. Gualberto et al. 2014. Structural and enzymatic insights into Lambda glutathione transferases from *Populus trichocarpa*, monomeric enzymes constituting an early divergent class specific to terrestrial plants. *Biochem J* 462:39–52.

Lan, T., Z. L. Yang, X. Yang et al. 2009. Extensive functional diversification of the *Populus* glutathione *S*-transferase supergene family. *Plant Cell* 21 (12):3749–66.

Larsson, A.-K., A. Shokeer, and B. Mannervik. 2010. Molecular evolution of Theta-class glutathione transferase for enhanced activity with the anticancer drug 1,3-bis-(2-chloroethyl)-1-nitrosourea and other alkylating agents. *Arch Biochem Biophys* 497 (1–2):28–34.

Lien, S., A. Gustafsson, A. K. Andersson, and B. Mannervik. 2001. Human glutathione transferase A1-1 demonstrates both half-of-the-sites and all-of-the-sites reactivity. *J Biol Chem* 276 (38):35599–605.

Lindström, H., S. M. Peer, N. H. Ing, and B. Mannervik. 2018. Characterization of equine GST A3-3 as a steroid isomerase. *J Steroid Biochem Mol Biol* 178, 117–26. doi:10.1074/j.jsbmb.2017.11.011.

Litwack, G., B. Ketterer, and I. M. Arias. 1971. Ligandin: A hepatic protein which binds steroids, bilirubin, carcinogens and a number of exogenous organic anions. *Nature* 234 (5330):466–7.

Lovatt, D., B. K. Ruble, J. Lee et al. 2014. Transcriptome in vivo analysis (TIVA) of spatially defined single cells in live tissue. *Nat Methods* 11 (2):190–6.

Low, W. Y., H. L. Ng, C. J. Morton et al. 2007. Molecular evolution of glutathione *S*-transferases in the genus *Drosophila*. *Genetics* 177 (3):1363–75.

Lyttle, M. H., A. Satyam, M. D. Hocker et al. 1994. Glutathione-*S*-transferase activates novel alkylating agents. *J Med Chem* 37 (10):1501–7.

Mannervik, B. 1974a. Glyoxalase I. Kinetic mechanism and molecular properties. In *Glutathione*, edited by H.C. Benöhr, L. Flohé, H. Sies, H.D. Waller and A. Wendel, 78–89. Stuttgart: Georg Thieme Publishers.

Mannervik, B. 1974b. Possible kinetic mechanism of glutathione reductase from yeast. In *Glutathione*, edited by H.C. Benöhr, L. Flohé, H. Sies, H.D. Waller and A. Wendel, 114–20. Stuttgart: Georg Thieme Publishers.

Mannervik, B. 1986. Glutathione and the evolution of enzymes for detoxication of products of oxygen-metabolism. *Chemica Scripta* 26b:281–84.

Mannervik, B. 2012. Five decades with glutathione and the GSTome. *J Biol Chem* 287 (9):6072–83.

Mannervik, B., P. G. Board, J. D. Hayes et al. 2005. Nomenclature for mammalian soluble glutathione transferases. *Methods Enzymol* 401:1–8.

Mannervik, B., V. M. Castro, U. H. Danielson et al. 1987. Expression of class Pi glutathione transferase in human malignant melanoma cells. *Carcinogenesis* 8 (12):1929–32.

Mannervik, B., and S. A. Eriksson. 1974. Enzymatic reduction of mixed disulfides and thiosulfate esters. In *Glutathione*, edited by H.C. Benöhr, L. Flohé, H. Sies, H.D. Waller and A. Wendel, 120–31. Stuttgart: Georg Thieme Publishers.

Mannervik, B., and H. Jensson. 1982. Binary combinations of four protein subunits with different catalytic specificities explain the relationship between six basic glutathione *S*-transferases in rat liver cytosol. *J Biol Chem* 257 (17):9909–12.

Mannervik, B., A. Runarsdottir, and S. Kurtovic. 2009. Multi-substrate-activity space and quasi-species in enzyme evolution: Ohno's dilemma, promiscuity and functional orthogonality. *Biochem Soc Trans* 37 (Pt 4):740–4.

Mashiyama, S. T., M. M. Malabanan, E. Akiva et al. 2014. Large-scale determination of sequence, structure, and function relationships in cytosolic glutathione transferases across the biosphere. *PLoS Biol* 12 (4):e1001843.

Matsumura, T., Y. Imamichi, T. Mizutani et al. 2013. Human glutathione *S*-transferase A (GSTA) family genes are regulated by steroidogenic factor 1 (SF-1) and are involved in steroidogenesis. *FASEB J* 27 (8):3198–208.

Mazari, A. M. A., O. Dahlberg, B. Mannervik, and M. Mannervik. 2014. Overexpression of glutathione transferase E7 in *Drosophila* differentially impacts toxicity of organic isothiocyanates in males and females. *PLOS ONE* 9 (10): e110103.

Mazari, A. M. A., and B. Mannervik. 2016. *Drosophila* GSTs display outstanding catalytic efficiencies with the environmental pollutants 2,4,6-trinitrotoluene and 2,4-dinitrotoluene. *Biochem Biophys Rep* 5:141–5.

McElwee, J. J., E. Schuster, E. Blanc et al. 2007. Evolutionary conservation of regulated longevity assurance mechanisms. *Genome Biol* 8 (7):R132.

Modén, O., and B. Mannervik. 2014. Glutathione transferases in the bioactivation of azathioprine. *Adv Cancer Res* 122:199–244.

Mohana, K., and A. Achary. 2017. Human cytosolic glutathione-*S*-transferases: Quantitative analysis of expression, comparative analysis of structures and inhibition strategies of isozymes involved in drug resistance. *Drug Metab Rev* 49 (3):318–37.

Morgan, A. S., P. E. Sanderson, R. F. Borch et al. 1998. Tumor efficacy and bone marrow-sparing properties of TER286, a cytotoxin activated by glutathione *S*-transferase. *Cancer Res* 58 (12):2568–75.

Morris, M. J., S. J. Craig, T. M. Sutherland et al. 2009. Transport of glutathione transferase-fold structured proteins into living cells. *Biochim Biophys Acta* 1788 (3):676–85.

Morris, M. J., D. Liu, L. M. Weaver et al. 2011. A structural basis for cellular uptake of GST-fold proteins *PLOS ONE* 6 (3) doi:10.1371/journal.pone.0017864.

Moyer, A. M., O. E. Salavaggione, S. J. Hebbring et al. 2007. Glutathione *S*-transferase T1 and M1: Gene sequence variation and functional genomics. *Clin Cancer Res* 13 (23):7207–16.

Musdal, Y., S. Govindarajan, and B. Mannervik. 2017. Exploring sequence-function space of a poplar glutathione transferase using designed information-rich gene variants. *Protein Eng Des & Select* 30 (8):543–9.

Namiki, S., T. Tomida, M. Tanabe et al. 2003. Intracellular delivery of glutathione S-transferase into mammalian cells. *Biochem Biophys Res Commun* 305 (3):592–7.

Nilsson, L. O., A. Gustafsson, and B. Mannervik. 2000. Redesign of substrate-selectivity determining modules of glutathione transferase A1-1 installs high catalytic efficiency with toxic alkenal products of lipid peroxidation. *Proc Natl Acad Sci U S A* 97 (17):9408–12.

Norrgård, M. A., Y. Ivarsson, K. Tars, and B. Mannervik. 2006. Alternative mutations of a positively selected residue elicit gain or loss of functionalities in enzyme evolution. *Proc Natl Acad Sci U S A* 103 (13):4876–81.

Norrgård, M. A., and B. Mannervik. 2011. Engineering GST M2-2 for high activity with indene 1,2-oxide and indication of an H-site residue sustaining catalytic promiscuity. *J Mol Biol* 412 (1):111–20.

Pandey, P., A. K. Singh, M. Singh et al. 2017. The see-saw of Keapl-Nrf2 pathway in cancer. *Crit Rev Oncol Hematol* 116:89–98.

Parsons, J. F., and R. N. Armstrong. 1996. Proton configuration in the ground state and transition state of a glutathione transferase-catalyzed reaction inferred from the properties of tetradeca(3-fluorotyrosyl)glutathione transferase. *J Am Chem Soc* 118 (9):2295–6.

Parsons, J. F., G. Xiao, G. L. Gilliland, and R. N. Armstrong. 1998. Enzymes harboring unnatural amino acids: Mechanistic and structural analysis of the enhanced catalytic activity of a glutathione transferase containing 5-fluorotryptophan. *Biochemistry* 37 (18):6286–94.

Payne, A. H., and D. B. Hales. 2004. Overview of steroidogenic enzymes in the pathway from cholesterol to active steroid hormones. *Endocr Rev* 25 (6):947–70.

Perperopoulou, F., F. Pouliou, and N. E. Labrou. 2017. Recent advances in protein engineering and biotechnological applications of glutathione transferases. *Crit Rev Biotechnol*:1–18.

Pettersson, P. L., A.-S. Johansson, and B. Mannervik. 2002. Transmutation of human glutathione transferase A2-2 with peroxidase activity into an efficient steroid isomerase. *J Biol Chem* 277 (33):30019–22.

Raffalli-Mathieu, F., C. Orre, M. Stridsberg et al. 2008. Targeting human glutathione transferase A3-3 attenuates progesterone production in human steroidogenic cells. *Biochem J* 414:103–9.

Robertson, I. G. C., C. Guthenberg, B. Mannervik, and B. Jernström. 1986. Differences in stereoselectivity and catalytic efficiency of 3 human glutathione transferases in the conjugation of glutathione with 7-beta, 8-alpha-dihydroxy-9-alpha, 10-alpha-oxy-7,8,9,10-tetrahydrobenzo(a)pyrene. *Cancer Res* 46 (5):2220–4.

Ronquist, G., and I. Brody. 1985. The prostasome: Its secretion and function in man. *Biochim Biophys Acta* 822 (2):203–18.

Rozell, B., H. A. Hansson, C. Guthenberg et al. 1993. Glutathione transferases of classes alpha, mu and pi show selective expression in different regions of rat kidney. *Xenobiotica* 23 (8):835–49.

Runarsdottir, A., and B. Mannervik. 2010. A novel quasi-species of glutathione transferase with high activity towards naturally occurring isothiocyanates evolves from promiscuous low-activity variants. *J Mol Biol* 401 (3):451–64.

Ruzza, P., and A. Calderan. 2013. Glutathione transferase (GST)-activated prodrugs. *Pharmaceutics* 5 (2):220–31.

Samarasinghe, K., D. Liu, P. Tummala, et al. 2015. Glutathione transferase M2 variants inhibit ryanodine receptor function in adult mouse cardiomyocytes. *Biochem Pharmacol* 97 (3):269–80. doi:10.1016/j.bcp.2015.08.004.

Samuels, L. T., M. L. Helmreich, M. B. Lasater, and H. Reich. 1951. An enzyme in endocrine tissues which oxidizes Δ^5-3 hydroxy steroids to α, β unsaturated ketones. *Science* 113 (2939):490–1.

Segura-Aguilar, J., S. Baez, M. Widersten et al. 1997. Human class Mu glutathione transferases, in particular isoenzyme M2-2, catalyze detoxication of the dopamine metabolite aminochrome. *J Biol Chem* 272 (9):5727–31.

Sharma, R., A. Sahoo, R. Devendran, and M. Jain. 2014. Over-expression of a rice tau class glutathione *S*-transferase gene improves tolerance to salinity and oxidative stresses in *Arabidopsis*. *PLOS ONE* 9 (3) doi:10.1371/journal.pone.0092900.

Sharma, S. K., and K. D. Bagshawe. 2017. Antibody directed enzyme prodrug therapy (ADEPT): Trials and tribulations. *Adv Drug Deliv Rev* 118:2–7.

Shibata, A., Y. Nakano, M. Ito et al. 2013. Fluorogenic probes using 4-substituted-2-nitrobenzenesulfonyl derivatives as caging groups for the analysis of human glutathione transferase catalyzed reactions. *Analyst* 138 (24):7326–30.

Singh, S. 2015. Cytoprotective and regulatory functions of glutathione *S*-transferases in cancer cell proliferation and cell death. *Cancer Chemother Pharmacol* 75 (1):1–15.

Singhal, S. S., S. P. Singh, P. Singhal et al. 2015. Antioxidant role of glutathione *S*-transferases: 4-hydroxynonenal, a key molecule in stress-mediated signaling. *Toxicol Appl Pharmacol* 289 (3):361–70.

Smith, W. L., Y. Urade, and P. J. Jakobsson. 2011. Enzymes of the cyclooxygenase pathways of prostanoid biosynthesis. *Chem Rev* 111 (10):5821–65.

Söderström, M., S. Hammarström, and B. Mannervik. 1988. Leukotriene C synthase in mouse mastocytoma cells. An enzyme distinct from cytosolic and microsomal glutathione transferases. *Biochem J* 250 (3):713–8.

Söderström, M., B. Mannervik, L. Örning, and S. Hammarström. 1985. Leukotriene C4 formation catalyzed by three distinct forms of human cytosolic glutathione transferase. *Biochem Biophys Res Commun* 128 (1):265–70.

Stenberg, G., P. G. Board, and B. Mannervik. 1991. Mutation of an evolutionarily conserved tyrosine residue in the active-site of a human class alpha-glutathione transferase. *FEBS Lett* 293 (1–2):153–5.

Stocco, G., S. Martelossi, A. Barabino et al. 2007. Glutathione-S-transferase genotypes and the adverse effects of azathioprine in young patients with inflammatory bowel disease. *Inflammatory Bowel Diseases* 13 (1):57–64.

Suzuki, T., and M. Yamamoto. 2015. Molecular basis of the Keap1-Nrf2 system. *Free Radic Biol Med* 88 (Pt B):93–100.

Talalay, P., and V. S. Wang. 1955. Enzymic isomerization of delta5-3-ketosteroids. *Biochim Biophys Acta* 18 (2):300–1.

Tars, K., B. Olin, and B. Mannervik. 2010. Structural basis for featuring of steroid isomerase activity in alpha class glutathione transferases. *J Mol Biol* 397 (1):332–40.

Teicher, B. A., and R. V. J. Chari. 2011. Antibody conjugate therapeutics: Challenges and potential. *Clin Cancer Res* 17 (20):6389–97.

Thorson, J. S., I. Shin, E. Chapman et al. 1998. Analysis of the role of the active site tyrosine in human glutathione transferase A1-1 by unnatural amino acid mutagenesis. *J Amer Chem Soc* 120 (2):451–2.

Tietze, L. F., and B. Krewer. 2009. Antibody-directed enzyme prodrug therapy: A promising approach for a selective treatment of cancer based on prodrugs and monoclonal antibodies. *Chem Biol Drug Des* 74 (3):205–11.

Tzafestas, K., M. M. Razalan, I. Gyulev et al. 2017. Expression of a *Drosophila* glutathione transferase in *Arabidopsis* confers the ability to detoxify the environmental pollutant, and explosive, 2,4,6-trinitrotoluene. *New Phytol* 214 (1):294–303.

van Niel, G., G. D'Angelo, and G. Raposo. 2018. Shedding light on the cell biology of extracellular vesicles *Nat Rev Mol Cell Biol* 19 (4):213–228.

Von Bartheld, C. S., and A. L. Altick. 2011. Multivesicular bodies in neurons: Distribution, protein content, and trafficking functions. *Prog Neurobiol* 93 (3):313–40.

Warholm, M., C. Guthenberg, B. Mannervik, and C. von Bahr. 1981. Purification of a new glutathione *S*-transferase (transferase μ) from human liver having high activity with benzo(alpha)pyrene-4,5-oxide. *Biochem Biophys Res Commun* 98 (2):512–9.

Widersten, M., R. Björnestedt, and B. Mannervik. 1996. Involvement of the carboxyl groups of glutathione in the catalytic mechanism of human glutathione transferase A1-1. *Biochemistry* 35 (24):7731–42.

Winter, G., A. R. Fersht, A. J. Wilkinson et al. 1982. Redesigning enzyme structure by site-directed mutagenesis: Tyrosyl tRNA synthetase and ATP binding. *Nature* 299 (5885):756–8.

Wu, D. J., J. Ni, J. Beretov et al. 2017. Urinary biomarkers in prostate cancer detection and monitoring progression. *Crit Rev Oncol Hematol* 118:15–26.

Xiang, Z., J. N. Snouwaert, M. Kovarova et al. 2014. Mice lacking three loci encoding 14 glutathione transferase genes: A novel tool for assigning function to the GSTP, GSTM, and GSTT families. *Drug Metab Dispos* 42 (6):1074–83.

Xie, X. S., P. J. Choi, G. W. Li et al. 2008. Single-molecule approach to molecular biology in living bacterial cells. *Annu Rev Biophys* 37:417–44.

Zhang, W., D. F. A. R. Dourado, P. A. Fernandes et al. 2012a. Multidimensional epistasis and fitness landscapes in enzyme evolution. *Biochemical J* 445:39–46.

Zhang, W., O. Modén, K. Tars, and B. Mannervik. 2012b. Structure-based redesign of GST A2-2 for enhanced catalytic efficiency with azathioprine. *Chem Biol* 19 (3):414–21.

Zhang, Y., R. H. Kolm, B. Mannervik, and P. Talalay. 1995. Reversible conjugation of isothiocyanates with glutathione catalyzed by human glutathione transferases. *Biochem Biophys Res Commun* 206 (2):748–55.

12 Protein *S*-Glutathionylation and Glutathione *S*-Transferase P

Kenneth D. Tew

Medical University of South Carolina

CONTENTS

12.1 INTRODUCTION

While the principles associated with thiol, disulfide exchange and disulfide bonding have been appreciated for many decades, the principles underlying adding glutathione (GSH) to proteins as a posttranslational modification (PTM) are more recent. A PubMed search shows that although the terminology of *S*-thiolation appeared in a number of articles in the mid-1980s (e.g., Grimm et al., 1985; Sies et al., 1987), the term "glutathionylation" emerged first in a 1992 study describing *S*-thiolation of liver proteins (Nakagawa et al., 1992). *S*-Glutathionylation is a reversible PTM of cysteine residues, resulting in a subsequent increase in molecular mass of ~305.6d (from GSH) and a net increase in negative charge (glutamic acid). This PTM can serve to protect the cysteine against extensive or extended oxidative damage, but can also create distinct structural and functional changes in the modified protein. Some of the earliest literature reports of functional consequences (such as inhibition of enzyme activities) of *S*-glutathionylation include inhibition of Protein Kinase C (Ward et al., 1998; Ward et al., 2000) and DNA-dependent protein kinase (Shen et al., 1999). There are now nearly 1,000 papers published in the

DOI: 10.1201/9781351261760-15

21st century that have various degrees of focus on *S*-glutathionylation. While ongoing efforts continue to identify additional proteins subject to this PTM, indications are that there will be hundreds, not thousands, of proteins that are substrates. This "glutathionome" has been partially described and characterized in earlier reviews (e.g., Xiong et al., 2011; Zhang et al., 2014a) and a complete discussion of this is outside the scope of this chapter.

12.2 *S*-GLUTATHIONYLATION

The pK_a of the receptive cysteine thiol is critical for *S*-glutathionylation and the reaction is favored if the localized three-dimensional environment is basic (low pK_a), as may be provided by vicinal arginines, histidines, or lysines (Grek et al., 2013). The typical cysteine thiol pK_a is ~8.5, but such values can range from 3.5 to >12, depending on the surrounding amino acids, emphasizing the three-dimensional space. Decreased cysteine pK_a values may be attributed to stabilization of the thiolate anion by electron withdrawing groups, or an adjacent positive charge, whereas thiolate pK_a may be higher when adjacent to negatively charged residues, or buried within protein folds (Hatahet and Ruddock, 2009; Bechtel and Weerapana, 2017). *S*-Glutathionylation can proceed either spontaneously or enzymatically (Gallogly and Mieyal, 2007; Mailloux and Willmore, 2014). Nonenzymatically, a number of possible mechanisms could ensue. For example, (i) a thiol-disulfide exchange can occur between the protein thiol (PSH) and glutathione disulfide (GSSG). The kinetics to the formation of *S*-glutathionylated protein (PSSG) can be a product of the

equilibrium constant of the reaction (K_{mix}), expressed as $\dfrac{[PSSG][GSH]}{[PSH][GSSG]}$. For any

protein, the magnitude of protein *S*-glutathionylation $\left(\dfrac{[PSSG]}{[PSH]}\right)$ is contingent upon

the local GSH:GSSG ratio (Gilbert, 1995, 1990). For cysteines with K_{mix} ~1, the intracellular GSH:GSSG ratio would have to decline markedly (e.g., from 100:1 to 1:1) to achieve 50% conversion of PSH to PSSG (Gilbert, 1995). Such immoderate conditions are rare in cells, therefore for the majority of proteins, spontaneous PSSG formation through PSH and GSSG exchange is not customary (Gilbert, 1990). Some proteins such as c-Jun, have a K_{mix} ~13 and thus 50% of c-Jun may be *S*-glutathionylated when the GSH: GSSG ratio approximates 13 (Klatt et al., 1999). (ii) A protein thiol may be oxidized to a thiyl radical (PS$^{\bullet}$), that can react with GSH to form a thiyl radical glutathionyl intermediate (PSSG$^{\bullet-}$) which can then react with O_2 to form PSSG. In the same vein, GS$^{\bullet}$ can react with PSH to form PSSG$^{\bullet-}$. Thiyl radical-mediated protein *S*-glutathionylation has been indicated to occur in cells (Gallogly and Mieyal, 2007), potentially involving glutaredoxin 1 and 2 (Grx1 and Grx2) (Starke et al., 2003; Gallogly et al., 2008). (iii) A protein thiol may be oxidized by reactive oxygen species (ROS) such as H_2O_2 to a sulfenic acid (PSOH) which then reacts with GSH to form PSSG. However, under physiological conditions, intracellular H_2O_2 levels are usually submicromolar (10^{-7} to 10^{-9} M) (Arbault et al., 1997); thus, in cells, spontaneous *S*-glutathionylation through this mechanism would occur only slowly. (iv) Nitric oxide (NO) is itself a weak thiol oxidant. However, either S-nitrosylation (PSNO) or S-glutathionylation (PSSG) can be promoted through secondary generation of reactive nitrogen species (RNS). In addition, PSH may be modified by nitrosoglutathione (GSNO) to form PSNO and/or PSSG. *In vitro*, proteins such as glyceraldehyde-3-phosphate dehydrogenase are susceptible to both S-nitrosylation and S-glutathionylation by GSNO, whereas alcohol dehydrogenase and actin are merely S-nitrosylated (Giustarini et al., 2005). Our own earlier work showed that treatment of cells with PABA/NO [a GST activated diazeniumdiolate prodrug (*O*2-[2,4-dinitro-5-(*N*-methyl-*N*-4-carboxyphenylamino)phenyl]1-*N*, *N*-dimethylamino)diazen-1-ium-1,2-diolate), 9) caused a dose-dependent increase in intracellular NO levels, with subsequent undetectable S-nitrosylation, but high levels of S-glutathionylation of such proteins as β-lactate dehydrogenase, Rho GDP dissociation inhibitor β, ATP synthase, elongation factor 2, protein disulfide isomerase (PDI), nucleophosmin-1, chaperonin, actin, protein tyrosine phosphatase 1B, and glucosidase II (Townsend et al., 2006; Townsend et al., 2009b). Since the S–N bond of GSNO has polarity, with the sulfur more negatively charged than the nitrogen, nucleophilic attack of the protein thiolate anion (PS⁻) on nitrogen to form PSNO should be favored over sulfur to form PSSG (Konorev et al., 2000; Giustarini et al., 2005). Together, such data evince that two distinct pools of S-nitrosylated proteins may exist, one GSH stable and another GSH labile, and subject to rapid conversion to S-glutathionylated products. However, such properties as might favor implementation of PSSG versus PSNO have yet to be established.

Within the glutathionome, clusters of related proteins that possess focused functionalities in cell pathways do seem to be foci for modification and in many cases, these include enzymes with catalytically active cysteines. Examples include those in protein folding and stability, NO regulation, redox homeostasis, energy metabolism, and glycolysis. Other clusters are ion channels, calcium pumps, and calcium binding proteins; protein kinases and phosphatases, which function in a variety of signaling pathways; various transcription factors; ras proteins; cytoskeletal proteins, e.g., actin, particularly during periods of cytoskeletal restructuring associated with either motility or cell growth (Bowers et al., 2012). In some instances, S-glutathionylation can lead to inhibition, or negative control of, enzymatic activity/protein function. Such proteins that fall into this category include GAPDH (Mohr et al., 1999), eNOS (Chen et al., 2010), the heteromeric Kir4.1-Kir5.1 channel (Jin et al., 2012), tyrosine phosphatase (Abdelsaid and El-Remessy, 2012), MAPK phosphatase 1 (Kim et al., 2012), mitochondrial thymidine kinase 2 (Sun et al., 2012), sodium, potassium ATPase (Petrushanko et al., 2012), or protein-disulfide isomerase (Xiong et al., 2012). In each case, three-dimensional location of the modified cysteine is a critical determinant of the impact of the modification. As examples, carbonic anhydrase III can be either inhibited or activated contingent on the cysteine involved (Cabiscol and Levine., 1996). Intriguingly, while canonical phosphorylation and dephosphorylation pathways are cyclical, numerous kinases and phosphatases are secondarily regulated by S-glutathionylation, providing layered control of phosphorus-based signaling by sulfur, following stimuli that cause homeostatic changes in ROS/RNS.

12.3   GLUTATHIONE *S*-TRANSFERASE P

At the genetic level, the glutathione *S*-transferase P (GSTP) gene demonstrates several distinct, but ultimately overlapping mechanisms of pleiotropy (Paaby and Rockman, 2013). For example, the gene product interacts with multiple other proteins (molecular gene pleiotropy), mutations generally impact the resulting phenotype in various ways (developmental pleiotropy), and resultant phenotypes alter various measures of organismal fitness (developmental pleiotropy). Perhaps then, the ligandin nomenclature applied to the early liver alpha and mu family members (Litwack et al., 1971) is relevant to GSTP, since there are, by now, many examples of protein: protein interactions with plausible biological consequences. Indeed, there exists a significant literature detailing the promiscuous interactions of GSTP with target proteins. While on the surface, this may support the idea that GSTP exhibits chaperone activity, a closer examination of the results might advocate a more specific conclusion. Coimmunoprecipation using GSTP antibodies, accompanied by proteomic analyses will frequently be used to identify whichever protein is the primary theme of a particular study's focus. However, attention to the identity of those proteins that are coprecipitated with GSTP can reveal functional commonality. It seems possible that these are not merely chaperone proteins, but that GSH in the G binding site of GSTP may be representing a precursor step to the process of *S*-glutathionylation.

Because there is such an extensive literature on GSTP, there is a tendency to assume that most of its biological functions are already known. Moreover, since the GSTP knockout mouse is able to breed and maintain viability (although the knockout has characteristics quite distinct from the wild type), many scientists eschew the notion that GSTP is a protein indispensable to cell survival. Perhaps functional redundancies provided by the large family of related isozymes does reduce the criticality of removing one, however, numerous features of GSTP infer that it does not behave like the other members of the GST gene family. When considered in context with the other members of the isozyme family, it has a number of characteristics that make it distinct. For example, it is distinguished by a comparatively restricted substrate detoxification potential (Zhang et al., 2014a); it has been extensively linked with cancer drug resistance—even when the selecting resistance drug is not a substrate (Tew and Townsend, 2011); it is overexpressed in a number of different cancers even to the point of being considered as a tumor marker (Tew and Townsend, 2011); it has unusual tissue distribution, even being absent, or poorly expressed, in hepatocytes from the major detoxification organ, the liver (Hayes et al., 1991). Each of these attributes suggests that the innate biological acumen of GSTP might not be restricted to (or perhaps be unrelated to) straightforward detoxification reactions. In this context, collaborative studies from our group have extended those cellular operations (indicating pleiotropic functionalities) that might be ascribed to the protein. There is now an expanded literature on its role as a switch for the signaling elements of c-jun terminal kinase [JNK; (Adler et al., 1999)]. Of most relevance to this chapter, genetic and pharmaceutical analyses have exposed a leading role for GSTP in mediating the forward reaction of *S*-glutathionylation (Townsend et al., 2009a).

12.4   GSTP KNOCKOUT MOUSE

As discussed, while S-glutathionylation can occur spontaneously, the rates and magnitude are greatly enhanced by the catalytic activity of GST, primarily GSTP (Manevich et al., 2004; Ralat et al., 2006; Townsend et al., 2009a; Wetzelberger et al., 2010; De Luca et al., 2011; Klaus et al., 2013; Ye et al., 2017). Perhaps the best evidence supporting the importance of GSTP in mediating S-glutathionylation is provided by results from the GSTP knockout mouse. When either fibroblasts or bone marrow dendritic cells (BMDC) derived from GSTP knockout mice are exposed to oxidative stress, both global levels and specific types of protein S-glutathionylation are diminished, inferring that GSTP facilitates this PTM (Townsend et al., 2009a; Zhang et al., 2014b; Ye et al., 2017). To this end, a number of specific target proteins have been identified and these include peroxiredoxin (Prdx) 6 (Manevich et al., 2004; Ralat et al., 2006), aldose reductase (Wetzelberger et al., 2010), histone H3 (De Luca et al., 2011), AMP-activated protein kinase (AMPK) (Klaus et al., 2013), 78-kDa glucose-regulated protein/immunoglobulin heavy chain binding protein (GRP78/BiP), PDI, calnexin, calreticulin, endoplasmin, and sarco/endoplasmic reticulum Ca^{2+}-ATPase (SERCA) (Ye et al., 2017). GSTP is considered a cytosolic enzyme; however, it is also found in the nucleus (Kamada et al., 2004), mitochondrial (Goto et al., 2009) and endoplasmic reticulum (ER) (Ye et al., 2017) compartments. Catalytically, through abstraction of the proton by Tyr7, GSH bound to GSTP (at the G site) lowers the pK_a of the cysteine residue of GSH [from 9.2 (Tajc et al., 2004) to ~6.3 (Dirr et al., 1994)], producing a thiolate anion (GS⁻) at the active site. While this thiolate can be presented to small molecule substrates, receptive cysteines can also be targeted for transfer. Accordingly, cells engineered to express a Tyr7 mutant of GSTP have greatly diminished levels of S-glutathionylation when exposed to oxidative stress (Townsend et al., 2009a). One such example is provided by reactivation of 1-Cys Prdx6, where oxidation of the catalytic Cys47 of Prdx6 has been associated with loss of peroxidase activity. Heterodimerization of Prdx6 with GSH-saturated GSTP mediates Cys47 S-glutathionylation in Prdx6 and is followed by subsequent spontaneous reduction of the mixed disulfide, restoring peroxidase activity (Manevich et al., 2004; Ralat et al., 2006, see also Chapter 9). As an additional example, even under mild oxidative conditions, equivalent to a normal, unstressed cell environment, either GSTP or GSTM can interact with AMPK, catalyze its S-glutathionylation at Cys299 and Cys304 and cause conformational changes that result in activation of AMPK activity (Klaus et al., 2013). Common ground implies that those characteristics that define GSH saturation of GSTP might be influential in governing the apparently broad range of GSTP interacting proteins (Zhang et al., 2014a) and as a consequence, eventual levels of S-glutathionylation. In other words, if a protein colocalizes with GSTP (as in immunoprecipitation) this might serve as at least one indication that it is a substrate for S-glutathionylation. One could also argue that localized concentrations of oxidative stress may directly regulate the rate and extent of S-glutathionylation. However, teleologically, there might be selective advantage in engineering controlled adjustments of the forward and reverse reactions, maintaining homeostasis as a cell goes through oscillations in endogenous or exogenous ROS/RNS exposure.

By providing the opportunity for selection of adaptive traits, evolution could more readily have established *S*-glutathionylation as a contributory rheostat of response that might be transformed from a primarily primitive defensive response (to external stimuli), into an ROS-based endogenous signaling network (e.g., based on H_2O_2 signaling). To this end, we have suggested the existence of a cycle with regulated forward and reverse reactions.

12.5   *S*-GLUTATHIONYLATION CYCLE

Figure 12.1 is a representation of the *S*-glutathionylation cycle, illustrating those reactions that contribute to both the forward and reverse reactions. The forward reaction has been described earlier, so this section will focus on reversal or deglutathionylation. Initially, it is worth considering the rationale for distinguishing characteristics of the GST isozymes that might distinguish their respective roles in the forward (GSTP) versus the reverse reaction [GSTO (omega)]. Specific to GSTP, there are catalytic cysteines near the surface of the substrate channel and these would seem to be more available as GS⁻ donors, thereby expediting the forward reaction. In GSTO, the catalytic cysteine of each subunit is reasonably buried within the substrate channel and may perhaps be better suited as a GS⁻ acceptor,

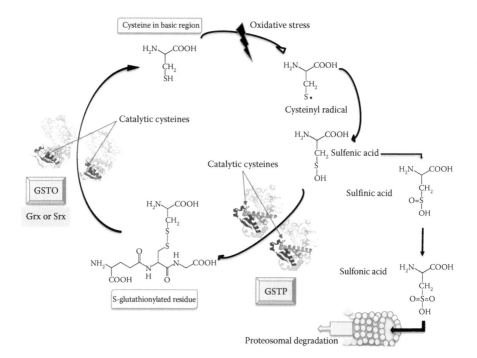

FIGURE 12.1 *S*-glutathionylation cycle. Protein *S*-glutathionylation can occur spontaneously, but the rate and magnitude are enhanced by the catalytic activity of GSTP. Deglutathionylation can be achieved by Grx, Srx, or GSTO. For GSTP and GSTO, dimers are shown.

thereby enabling the reverse reaction. In addition to GSTO (Menon et al., 2013), deglutathionylation can be accomplished by glutaredoxin (Mannervik et al., 1980; Gladyshev et al., 2001; Lundberg et al., 2001) or sulfiredoxin (Srx) (Findlay et al., 2006; Park et al., 2009). GSTO1-1 has structural similarities to Grx, including a thioredoxin-like domain and the glutathionyl stabilization site, where GSH can form a disulfide bond with a conserved active site cysteine (Menon and Board, 2013). GSTO has a relatively open active-side pocket that could potentially accommodate a peptide as a substrate (Board et al., 2000), implying that GSTO might catalyze deglutathionylation in two steps. First, the Cys^{32} on GSTO attacks PSSG to form Cys^{32}-SSG, releasing reduced PSH. Second, Cys^{32}-SSG is recycled with GSH to form GSSG, liberating GSTO. This mechanism has been described fairly recently (Board and Menon, 2016), whereas the influence of Grx on the deglutathionylation step has been portrayed in more detail.

Grx1 is both cytosolic (~1 µM) and in the mitochondrial intermembrane space (~0.1 µM); Grx2 is in the mitochondrial matrix (~1 µM) (Pai et al., 2007; Gallogly et al., 2009), but they share ~34% sequence identity and contain active site CXXC motifs (CSYC Grx2; CPYC Grx1), (Martin, 1995) and glutathionyl stabilization sites (Gladyshev et al., 2001; Lundberg et al., 2001). Grx catalyzes deglutathionylation through two successive thiol-disulfide exchange reactions (Gallogly et al., 2009). The Grx N-terminal active-site cysteine thiolate (Grx-S$^-$) attacks the glutathionyl sulfur of PSSG, forming a Grx–SSG intermediate, releasing reduced PSH. This step is relatively fast and limiting to the overall rate of the catalyzed reaction (Gravina and Mieyal, 1993; Gallogly et al., 2008). The rate enhancement by Grx over noncatalyzed rates has been attributed to the unusually low pK_a of the active-site cysteine (leaving group in this step) of Grx 1 (~3.5) or Grx 2 (~4.6) (Gallogly et al., 2008), compared to the typical pK_a ~ 8.5. For classic thiol-disulfide exchange reactions, each 1 pH unit decrease in the pK_a of the leaving group thiol forecasts a ~4-fold increase in the rate constant, which explains much of the rate enhancement of Grx1-mediated deglutathionylation [$4^{\Delta pKa} = 4^5 = 1024$-fold] (Gallogly et al., 2008). Besides the leaving group effect, an additional enhancement exists for GSH as the second substrate, which suggests an enzyme-induced increase in nucleophilicity of the GS− for the Grx-SSG intermediate (Gallogly et al., 2009). Grx activities are also impacted by local environmental conditions influenced by pH and ROS. Because pH values in cytosol (~7.4), mitochondrial intermembrane space (~7.0) and mitochondrial matrix (~8.0) differ, Grx activities can vary among subcellular compartments. Accordingly, compared to cytosol, Grx activity would be ~66% lower in the mitochondrial intermembrane space and 2.5 times higher in the mitochondrial matrix (Gallogly et al., 2009). Local production of ROS may also favor catalysis of thiyl radical scavenging or protein S-glutathionylation by Grx. Finally, the orientation of the γ-glutamyl moiety of GSH appears essential for Grx selectivity (Peltoniemi et al., 2006). The nonglutathionyl component of the disulfide substrate is sterically unrestricted and thus Grx can operate as a general deglutathionylating enzyme, perhaps distinguishing it from GSTO or sulfiredoxin. However, rate constants can vary more than two orders of magnitude among PSSG substrates and preferable substrates are distinguished by high accessibility of the S-glutathionylated site and low pK_a of the cysteine (Jensen et al., 2014).

12.6 FUNCTIONAL *S*-GLUTATHIONYLATION: MYELOPREOLIFERATION

Although the literature is becoming more replete with examples of physiological processes or human pathologies linked with dysregulated *S*-glutathionylation, our lab has for some time focused on the relationship between redox homeostasis and bone marrow proliferation, differentiation, and function. In this respect, a short discussion of how this PTM regulates hematopoietic events will serve as a primary example of functional significance. The principle that redox conditions govern bone marrow cell proliferation was first advanced over a half-century ago, when cystine and cysteine were shown to be crucial in maintaining balance between hematopoietic stem cell (HSC) quiescence, self-renewal, and lineage commitment (Baldini et al., 1953). Subsequently, it became clear that thiol levels inversely correlated with the capacity of ROS to attenuate HSC self-renewal or differentiation. Bone marrow has two HSC populations defined by low and high intracellular ROS. High ROS cells lack repopulating capacity and are lineage restricted. Treatment with N-acetyl cysteine (NAC) restores functional activity to this high ROS population (Jang and Sharkis, 2007). An operative link with GSTP was first identified when genetic ablation of GSTP in the knockout mouse produced a phenotype characterized, in part, by an increase in myeloid cell differentiation and proliferation, evinced by higher numbers of circulating leukocytes (Gate et al., 2004). Studies with these mice revealed that, when treated with a specific GSTP inhibitor, *Telintra* [γ-glutamyl-*S*-(benzyl)-cysteinyl-*R*-(-) phenyl glycine diethyl ester], peripheral white blood cell numbers were augmented only in the wild-type mice and not in the GSTP-deficient mice that lacked the specific GSTP drug target (Ruscoe et al., 2001). These early observations instilled a direct cause: effect association between GSTP and myeloproliferation, and our own recent data have provided a more explicit link with *S*-glutathionylation and have also served to provide a canonical example of how this PTM alters protein function (Zhang et al., 2018). BMDC isolated from GSTP knockout animals convey the properties of increased cell division rates, differentiation and global decreases in their protein *S*-glutathionylation (Zhang et al., 2014b). These same cells also have increased ROS levels and decreased GSH:GSSG ratios and also, when compared with wild-type cells through a signaling primer array experiment, showed that ablation of GSTP was coupled with significantly enhanced expression of estrogen receptor alpha [ERα; Zhang et al. (2018)]. This observation, in of itself, is of consequence, since ERα signaling is directly invested in pathways of both development and maturation of BMDC and also the general sustainability of an immune response (Kovats, 2012). Moreover, ERα broadly impacts glucose metabolism and metabolic gene expression patterns that stimulate resting T cells to proliferate and differentiate into mature T effector cells (Michalek et al., 2011). While a number of events link ERα with enhanced proliferation and functions of BMDC, the altered expression patterns in GSTP knockout animals predicated a *direct link* between GSTP and ERα activities. In concurrence with this, GSTP forms a complex with ERα, stimulating *S*-glutathionylation of ERα at Cys[221,245,417, and 447] and this produces an altered ERα binding affinity for estradiol and 3-fold reduced overall binding potential (receptor density and affinity). Thus, in the GSTP knockout, there is a diminished level of *S*-glutathionylation and this corresponds with enhanced ERα functionalities. BMDC differentiated by granulocyte-macrophage colony-stimulating factor elevate their ERα levels, and this is particularly exaggerated in cells from GSTP knockout mice. When stimulated with lipopolysaccharide, these same BMDC exhibit: (i) augmented endocytosis, maturation rate, cytokine secretion, and T-cell activation; (ii) heightened glucose uptake and glycolysis; (iii) increased Akt signaling (in the mTOR pathway); and (iv) decreased AMPK-mediated phosphorylation of proteins. Taken together, these findings suggest that quantitative changes in levels of GSTP-mediated *S*-glutathionylation of ERα control BMDC differentiation and alter metabolic function in dendritic cells. This is one example of the growing list of proteins, where functional alterations are interpolated by *S*-glutathionylation of critical cysteine residues. As a prelude to the next frontier linking redox and GST biochemistry with hematopoiesis, there now appear to be significant nexuses between expression of microsomal GST's and control of myeloproliferation, either in zebrafish or mice (Bräutigam et al., 2018). These findings are entirely consistent with the historical context of redox impact on bone marrow as a tissue and provide a platform for considering therapeutic contingencies in hematopoietic diseases. The GSTP inhibitor *Telintra* has advanced into Phase II human clinical trials for the treatment of myelodysplastic syndrome, with initially encouraging results (Raza et al., 2012).

12.7   CONCLUSIONS

The concepts that cellular GSH levels perform primarily as buffers against ROS/RNS conditions or that GSH:GSSG ratios can be interpreted as direct determinants of cell survival likely requires nuanced interpretation. Arguably, the presence or absence of "activated" cysteines may be a limiting or, at least, controlling step in maintenance of cellular redox (thiol) homeostasis. The facile cycling of phosphate groups on serines, threonines, and tyrosines controls a plethora of biological functions. Why then cannot similarly profound affects be associated with a sulfur-based *S*-glutathionylation cycle? This might seem particularly pertinent given that many of these same kinases and phosphatases are themselves subject to secondary regulation by *S*-glutathionylation. Moreover, GSTP has always been an enigmatic member of the GST gene superfamily, with properties not always commensurate with proscribed detoxification functions. As a contributory component of the *S*-glutathionylation cycle, the high levels of expression of this GSTP isozyme in highly proliferating tissues or tumors can be more readily understood. Certainly, the characteristic bone marrow phenotype of the GSTP knockout mouse is consistent with regulatory control through *S*-glutathionylation of certain key proteins. Besides, either genetic (knockout or overexpression) or pharmaceutical (*Telintra*) manipulation of GSTP advance incontrovertible evidence for the involvement of GSTP in *S*-glutathionylation. Independent of exogenous ROS, endogenous oxidative signaling is tightly regulated and the *S*-glutathionylation cycle would appear to provide a permissible, tunable platform affording some degree of biological flexibility. As the field advances, there should develop a database of proteins, the structure and/or function of which are sensitive to this cysteine PTM.

ACKNOWLEDGMENTS

Supported by grants from the National Institutes of Health (CA08660, CA117259, NCRR P20RR024485) and support from the South Carolina Centers of Excellence program.

In addition to recognizing Professor Flohé's 80th birthday, I would also like to acknowledge the retirement of my longtime friend and colleague Dr. Yefim Manevich, with whom, through the years, I shared many insightful discussions.

REFERENCES

Abdelsaid, M. A., and A. B. El-Remessy. 2012. S-glutathionylation of LMW-PTP regulates VEGF-mediated FAK activation and endothelial cell migration. *J Cell Sci* 125 (Pt 20):4751–60.

Adler, V., Z. Yin, S. Y. Fuchs et al. 1999. Regulation of JNK signaling by GSTp. *EMBO J* 18 (5):1321–34.

Arbault, S., P. Pantano, N. Sojic et al. 1997. Activation of the NADPH oxidase in human fibroblasts by mechanical intrusion of a single cell with an ultramicroelectrode. *Carcinogenesis* 18 (3):569–74.

Baldini, M., and C. Sacchetti. 1953. Effect of cystine and cysteine on human bone marrow cultured in medium deficient in amino acids. *Rev Hematol* 8 (1):3–19.

Bechtel, T. J., and E. Weerapana. 2017. From structure to redox: The diverse functional roles of disulfides and implications in disease. Proteomics 17 (6):1–27.

Board, P. G., M. Coggan, G. Chelvanayagam et al. 2000. Identification, characterization, and crystal structure of the Omega class glutathione transferases. *J Biol Chem* 275 (32):24798–806.

Board, P. G., and D. Menon. 2016. Structure, function and disease relevance of Omega-class glutathione transferases. *Arch Toxicol* 90 (5):1049–67.

Bowers, R. R., Y. Manevich, D. M. Townsend, and K. D. Tew. 2012. Sulfiredoxin redox-sensitive interaction with S100A4 and non-muscle myosin IIA regulates cancer cell motility. *Biochemistry* 51 (39):7740–54.

Bräutigam, L., J. Zhang, K. Dreij et al. 2018. A redox biology enzyme essential for development and hematopoetic stem cell differentiation in mice and zebrafish. *Redox Biol* 17:171–9.

Cabiscol, E., and R. L. Levine. 1996. The phosphatase activity of carbonic anhydrase III is reversibly regulated by glutathiolation. *Proc Natl Acad Sci U S A* 93 (9):4170–4.

Chen, C. A., T. Y. Wang, S. Varadharaj et al. 2010. S-glutathionylation uncouples eNOS and regulates its cellular and vascular function. *Nature* 468 (7327):1115–18.

de Luca, A., N. Moroni, A. Serafino et al. 2011. Treatment of doxorubicin-resistant MCF7/Dx cells with nitric oxide causes histone glutathionylation and reversal of drug resistance. *Biochem J* 440 (2):175–83.

Dirr, H., P. Reinemer, and R. Huber. 1994. X-ray crystal structures of cytosolic glutathione S-transferases. Implications for protein architecture, substrate recognition and catalytic function. *Eur J Biochem* 220 (3):645–61.

Findlay, V. J., D. M. Townsend, T. E. Morris et al. 2006. A novel role for human sulfiredoxin in the reversal of glutathionylation. *Cancer Res* 66 (13):6800–6.

Gallogly, M. M., and J. J. Mieyal. 2007. Mechanisms of reversible protein glutathionylation in redox signaling and oxidative stress. *Curr Opin Pharmacol* 7 (4):381–91.

Gallogly, M. M., D. W. Starke, A. K. Leonberg et al. 2008. Kinetic and mechanistic characterization and versatile catalytic properties of mammalian glutaredoxin 2: Implications for intracellular roles. *Biochemistry* 47 (42):11144–57.

Gallogly, M. M., D. W. Starke, and J. J. Mieyal. 2009. Mechanistic and kinetic details of catalysis of thiol-disulfide exchange by glutaredoxins and potential mechanisms of regulation. *Antioxid Redox Signal* 11 (5):1059–81.

Gate, L., R. S. Majumdar, A. Lunk, and K. D. Tew. 2004. Increased myeloproliferation in glutathione S-transferase pi-deficient mice is associated with a deregulation of JNK and Janus kinase/STAT pathways. *J Biol Chem* 279 (10):8608–16.

Gilbert, H. F. 1990. Molecular and cellular aspects of thiol-disulfide exchange. *Adv Enzymol Relat Areas Mol Biol* 63:69–172.

Gilbert, H. F. 1995. Thiol/disulfide exchange equilibria and disulfide bond stability. *Methods Enzymol* 251:8–28.

Giustarini, D., A. Milzani, G. Aldini et al. 2005. S-nitrosation versus S-glutathionylation of protein sulfhydryl groups by S-nitrosoglutathione. *Antioxid Redox Signal* 7 (7–8):930–9.

Gladyshev, V. N., A. Liu, S. V. Novoselov et al. 2001. Identification and characterization of a new mammalian glutaredoxin (thioltransferase), Grx2. *J Biol Chem* 276 (32):30374–80.

Goto, S., M. Kawakatsu, S. Izumi et al. 2009. Glutathione S-transferase pi localizes in mitochondria and protects against oxidative stress. *Free Radic Biol Med* 46 (10):1392–403.

Gravina, S. A., and J. J. Mieyal. 1993. Thioltransferase is a specific glutathionyl mixed disulfide oxidoreductase. *Biochemistry* 32 (13):3368–76.

Grek, C. L., J. Zhang, Y. Manevich et al. 2013. Causes and consequences of cysteine S-glutathionylation. *J Biol Chem* 288 (37):26497–504.

Grimm, L. M., M. W. Collison, R. A. Fisher, and J. A. Thomas. 1985. Protein mixed-disulfides in cardiac cells. S-thiolation of soluble proteins in response to diamide. *Biochim Biophys Acta* 844 (1):50–4.

Hatahet, F., and L. W. Ruddock. 2009. Protein disulfide isomerase: A critical evaluation of its function in disulfide bond formation. *Antioxid Redox Signal* 11 (11):2807–50.

Hayes, P. C., L. May, J. D. Hayes, and D. J. Harrison. 1991. Glutathione S-transferases in human liver cancer. *Gut* 32 (12):1546–9.

Jang, Y. Y., and S. J. Sharkis. 2007. A low level of reactive oxygen species selects for primitive hematopoietic stem cells that may reside in the low-oxygenic niche. *Blood* 110 (8):3056–63.

Jensen, K. S., J. T. Pedersen, J. R. Winther, and K. Teilum. 2014. The pKa value and accessibility of cysteine residues are key determinants for protein substrate discrimination by glutaredoxin. *Biochemistry* 53 (15):2533–40.

Jin, X., L. Yu, Y. Wu et al. 2012. S-Glutathionylation underscores the modulation of the heteromeric Kir4.1-Kir5.1 channel in oxidative stress. *J Physiol* 590 (21):5335–48.

Kamada, K., S. Goto, T. Okunaga et al. 2004. Nuclear glutathione S-transferase pi prevents apoptosis by reducing the oxidative stress-induced formation of exocyclic DNA products. *Free Radic Biol Med* 37 (11):1875–84.

Kim, H. S., S. L. Ullevig, D. Zamora et al. 2012. Redox regulation of MAPK phosphatase 1 controls monocyte migration and macrophage recruitment. *Proc Natl Acad Sci U S A* 109 (41):E2803–12.

Klatt, P., E. P. Molina, M. G. De Lacoba et al. 1999. Redox regulation of c-Jun DNA binding by reversible S-glutathiolation. *FASEB J* 13 (12):1481–90.

Klaus, A., S. Zorman, A. Berthier et al. 2013. Glutathione S-transferases interact with AMP-activated protein kinase: Evidence for S-glutathionylation and activation in vitro. *PLoS One* 8 (5):e62497.

Konorev, E. A., B. Kalyanaraman, and N. Hogg. 2000. Modification of creatine kinase by S-nitrosothiols: S-nitrosation vs. S-thiolation. *Free Radic Biol Med* 28 (11):1671–8.

Kovats, S. 2012. Estrogen receptors regulate an inflammatory pathway of dendritic cell differentiation: Mechanisms and implications for immunity. *Hormones and behavior* 62 (3):254–62.

Litwack, G., B. Ketterer, and I. M. Arias. 1971. Ligandin: A hepatic protein which binds steroids, bilirubin, carcinogens and a number of exogenous organic anions. *Nature* 234 (5330):466–7.

Lundberg, M., C. Johansson, J. Chandra et al. 2001. Cloning and expression of a novel human glutaredoxin (Grx2) with mitochondrial and nuclear isoforms. *J Biol Chem* 276 (28):26269–75.

Mailloux, R. J., and W. G. Willmore. 2014. S-glutathionylation reactions in mitochondrial function and disease. *Front Cell Dev Biol* 2:68.

Manevich, Y., S. I. Feinstein, and A. B. Fisher. 2004. Activation of the antioxidant enzyme 1-CYS peroxiredoxin requires glutathionylation mediated by heterodimerization with pi GST. *Proc Natl Acad Sci U S A* 101 (11):3780–5.

Mannervik, B., and K. Axelsson. 1980. Role of cytoplasmic thioltransferase in cellular regulation by thiol-disulphide interchange. *Biochem J* 190 (1):125–30.

Martin, J. L. 1995. Thioredoxin—A fold for all reasons. *Structure* 3 (3):245–50.

Menon, D., and P. G. Board. 2013. A role for glutathione transferase Omega 1 (GSTO1-1) in the glutathionylation cycle. *J Biol Chem* 288 (36):25769–79.

Michalek, R. D., V. A. Gerriets, A. G. Nichols et al. 2011. Estrogen-related receptor-alpha is a metabolic regulator of effector T-cell activation and differentiation. *Proc Natl Acad Sci U S A* 108 (45):18348–53.

Mohr, S., H. Hallak, A. de Boitte et al. 1999. Nitric oxide-induced S-glutathionylation and inactivation of glyceraldehyde-3-phosphate dehydrogenase. *J Biol Chem* 274 (14):9427–30.

Nakagawa, Y., P. Moldeus, and I. A. Cotgreave. 1992. The S-thiolation of hepatocellular protein thiols during diquat metabolism. *Biochem Pharmacol* 43 (12):2519–25.

Paaby, A. B., and M. V. Rockman. 2013. The many faces of pleiotropy. *Trends Genet* 29 (2):66–73.

Pai, H. V., D. W. Starke, E. J. Lesnefsky et al. 2007. What is the functional significance of the unique location of glutaredoxin 1 (GRx1) in the intermembrane space of mitochondria? *Antioxid Redox Signal* 9 (11):2027–33.

Park, J. W., J. J. Mieyal, S. G. Rhee, and P. B. Chock. 2009. Deglutathionylation of 2-Cys peroxiredoxin is specifically catalyzed by sulfiredoxin. *J Biol Chem* 284 (35):23364–74.

Peltoniemi, M. J., A. R. Karala, J. K. Jurvansuu et al. 2006. Insights into deglutathionylation reactions. Different intermediates in the glutaredoxin and protein disulfide isomerase catalyzed reactions are defined by the gamma-linkage present in glutathione. *J Biol Chem* 281 (44):33107–14.

Petrushanko, I. Y., S. Yakushev, V. A. Mitkevich et al. 2012. S-glutathionylation of the Na, K-ATPase catalytic alpha subunit is a determinant of the enzyme redox sensitivity. *J Biol Chem* 287 (38):32195–205.

Ralat, L. A., Y. Manevich, A. B. Fisher, and R. F. Colman. 2006. Direct evidence for the formation of a complex between 1-cysteine peroxiredoxin and glutathione S-transferase pi with activity changes in both enzymes. *Biochemistry* 45 (2):360–72.

Raza, A., N. Galili, S. E. Smith et al. 2012. A phase 2 randomized multicenter study of 2 extended dosing schedules of oral ezatiostat in low to intermediate-1 risk myelodysplastic syndrome. *Cancer* 118 (8):2138–47.

Ruscoe, J. E., L. A. Rosario, T. Wang et al. 2001. Pharmacologic or genetic manipulation of glutathione S-transferase P1-1 (GSTpi) influences cell proliferation pathways. *J Pharmacol Exp Ther* 298 (1):339–45.

Shen, H., M. P. Schultz, and K. D. Tew. 1999. Glutathione conjugate interactions with DNA-dependent protein kinase. *J Pharmacol Exp Ther* 290 (3):1101–6.

Sies, H., R. Brigelius, and P. Graf. 1987. Hormones, glutathione status and protein S-thiolation. *Adv Enzyme Regul* 26:175–89.

Starke, D. W., P. B. Chock, and J. J. Mieyal. 2003. Glutathione-thiyl radical scavenging and transferase properties of human glutaredoxin (thioltransferase). Potential role in redox signal transduction. *J Biol Chem* 278 (17):14607–13.

Sun, R., S. Eriksson, and L. Wang. 2012. Oxidative stress induced S-glutathionylation and proteolytic degradation of mitochondrial thymidine kinase 2. *J Biol Chem* 287 (29):24304–12.

Tajc, S. G., B. S. Tolbert, R. Basavappa, and B. L. Miller. 2004. Direct determination of thiol pKa by isothermal titration microcalorimetry. *J Am Chem Soc* 126 (34):10508–9.

Tew, K. D., and D. M. Townsend. 2011. Regulatory functions of glutathione S-transferase P1-1 unrelated to detoxification. *Drug Metab Rev* 43 (2):179–93.

Townsend, D. M., V. J. Findlay, F. Fazilev et al. 2006. A glutathione S-transferase pi-activated prodrug causes kinase activation concurrent with S-glutathionylation of proteins. *Mol Pharmacol* 69 (2):501–8.

Townsend, D. M., Y. Manevich, L. He et al. 2009a. Novel role for glutathione S-transferase pi. Regulator of protein S-Glutathionylation following oxidative and nitrosative stress. *J Biol Chem* 284 (1):436–45.

Townsend, D. M., Y. Manevich, L. He et al. 2009b. Nitrosative stress-induced s-glutathionylation of protein disulfide isomerase leads to activation of the unfolded protein response. *Cancer Res* 69 (19):7626–34.

Ward, N. E., D. S. Pierce, S. E. Chung et al. 1998. Irreversible inactivation of protein kinase C by glutathione. *J Biol Chem* 273 (20):12558–66.

Ward, N. E., J. R. Stewart, C. G. Ioannides, and C. A. O'Brian. 2000. Oxidant-induced S-glutathiolation inactivates protein kinase C-alpha (PKC-alpha): A potential mechanism of PKC isozyme regulation. *Biochemistry* 39 (33):10319–29.

Wetzelberger, K., S. P. Baba, M. Thirunavukkarasu et al. 2010. Postischemic deactivation of cardiac aldose reductase: Role of glutathione S-transferase P and glutaredoxin in regeneration of reduced thiols from sulfenic acids. *J Biol Chem* 285 (34):26135–48.

Xiong, Y., Y. Manevich, K. D. Tew, and D. M. Townsend. 2012. S-Glutathionylation of Protein Disulfide Isomerase Regulates Estrogen Receptor alpha Stability and Function. *Int J Cell Biol* 2012:273549.

Xiong, Y., J. D. Uys, K. D. Tew, and D. M. Townsend. 2011. S-glutathionylation: From molecular mechanisms to health outcomes. *Antioxid Redox Signal* 15 (1):233–70.

Ye, Z. W., J. Zhang, T. Ancrum et al. 2017. Glutathione S-transferase P-mediated protein S-glutathionylation of resident endoplasmic reticulum proteins influences sensitivity to drug-induced unfolded protein response. *Antioxid Redox Signal* 26 (6):247–261.

Zhang, J., C. Grek, Z. W. Ye et al. 2014a. Pleiotropic functions of glutathione S-transferase P. *Adv Cancer Res* 122:143–75.

Zhang, J., Z. W. Ye, W. Chen et al. 2018. S-Glutathionylation of estrogen receptor alpha affects dendritic cell function. *J Biol Chem*. 298 (12):4366–80.

Zhang, J., Z. W. Ye, P. Gao et al. 2014b. Glutathione S-transferase P influences redox and migration pathways in bone marrow. *PLOS ONE* 9 (9):e107478.

13 The Role of Glutathione in Biosynthetic Pathways and Regulation of the Eicosanoid Metabolism

Ralf Morgenstern, Jesper Z. Haeggström, and Per-Johan Jakobsson
Karolinska Institutet

Leopold Flohé
Universidad de la República

CONTENTS

13.1 INTRODUCTION

Glutathione (GSH) plays a highly complex role in eicosanoid metabolism being a cofactor of terminal prostaglandin synthases and a substrate for leukotriene synthases, modulating enzyme activities, contributing to transcriptional activation, silencing key enzymes such as cyclooxygenases and lipoxygenases, as well as protecting related enzymes from oxidative inactivation. As to the basics of the biosynthesis of prostaglandins and leukotrienes and its history, we refer to comprehensive reviews and Nobel lectures (Samuelsson, 1983; Vane, 1983; Samuelsson et al., 2007;

DOI: 10.1201/9781351261760-16

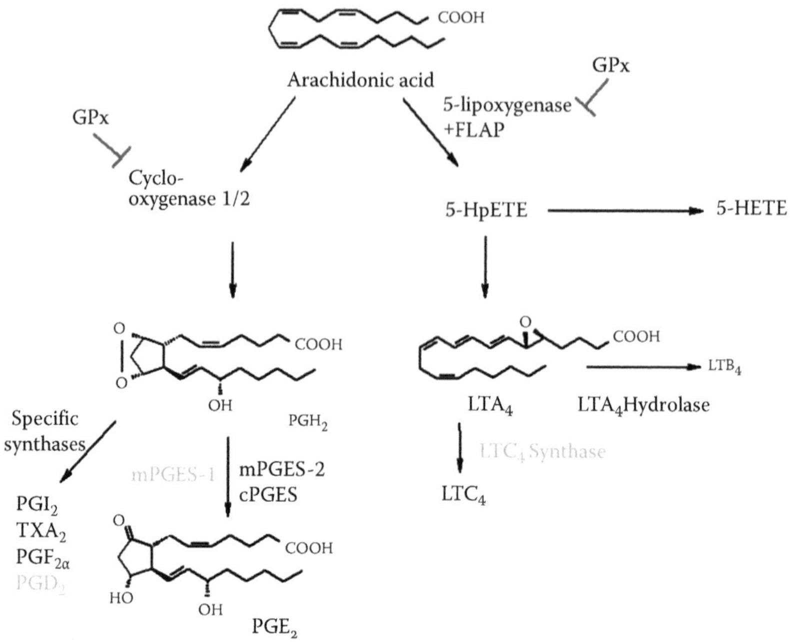

FIGURE 13.1 Eicosanoid biosynthesis with GSH dependent enzyme reactions in blue and GPx "peroxide tone" control of cyclooxygenase and 5-lipoxygenase in red. GPx: GSH peroxidase, LT: leukotriene, PG: prostaglandin, PGES: Prostaglandin E synthase, PGI$_2$: prostacyclin, TXA$_2$: thromboxane, HPETE: hydroperoxy tetraenoic acid, HETE: hydroxytetraenoic acid, FLAP: 5-lipoxygenase-activating protein.

Haeggström and Funk, 2011; Smith et al., 2011; Di Gennaro and Haeggström, 2012). Early findings will only be recalled, if more recent amendments in the field demand a reconsideration of seemingly clear facts. For convenience, the main arachidonic acid-based biosynthetic pathways are depicted in Figure 13.1 with GSH-interactions indicated.

13.2 SILENCING OF LIPOXYGENASES BY GLUTATHIONE PEROXIDASES

Already in the early 1970s Lands and coworkers postulated that the activity of cyclooxygenase was under the control of the extracellular "peroxide tone." They could completely inhibit the enzyme from sheep seminal vesicles by adding GSH and glutathione peroxidase (GPx) and made the same observation with the soy bean lipoxygenase (Lands et al., 1971). The concept was later confirmed for 5-lipoxygenase from rat basophilic leukemia cells (Haurand and Flohé, 1988) and 15-lipoxygenase from rabbit reticulocytes (Schnurr et al., 1996) suggesting that lipoxygenases are in general prone to such inhibition. Mechanistically, a trace of product, likely prostaglandin G$_2$, has to oxidize the porphyrin iron of the cyclooxygenase and thereby activates the enzyme (Hemler and Lands, 1980; Pace-Asciak and Smith, 1983). By analogy, oxidation of the nonheme iron of lipoxygenases is considered to explain the activation by product. Inactivation by product, however, has equally been observed with cyclooxygenases and lipoxygenases (Lands et al., 1971), which indeed point to a critical role of enzymes acting on lipophilic hydroperoxides such as GPxs, glutathione-S-transferases (GSTs), or peroxiredoxins in the biosynthesis of eicosanoids; a complete abrogation of hydroperoxides will silence all lipoxygenases, a sudden rise will kill them (Flohé, 2016).

Which of the hydroperoxide-metabolizing enzymes is most relevant to 5-lipoxygenase inhibition *in vivo* has remained controversial. An *in vivo* role of GPx1 in the inhibition of 5-lipoxygenase in cells, as first shown *in vitro* (Haurand and Flohé, 1988), was later questioned by Weitzel and Wendel for rat basophilic leukemia (RBL) cells. By elegant selenium feeding experiments, which made use of the different position of GPx1 and GPx4 in the hierarchy of selenoproteins, they demonstrated that restoration of GPx4 in RBL cells was sufficient to normalize the overproduction of leukotrienes typically seen in selenium deficiency (Hatzelmann et al., 1989), while a further increase of GPx1 had no effect, and this differential effect of the two GPxs was equally observed in intact rats (Weitzel and Wendel, 1993). Studies on overexpression of GPx4 in RBL cells corroborated the peculiar role of GPx4 (Imai et al., 1998). In human polymorphonuclear leukocytes (PMNs), deprivation of GSH (Hatzelmann and Ullrich, 1987) or the decrease of GPx activity by selenium deprivation (Hatzelmann et al., 1989) dramatically increased 5-lipoxygenase activity. Unfortunately, these approaches do not differentiate between the GPxs affected. In human blood monocytes and Mono Mac 6 cells, it proved to be the GPx1 which inhibited 5-lipoxygenase activity (Straif et al., 2000). In opposite to other leukocytes, for instance neutrophils, the 5-lipoxygenase pathway in B lymphocytes is in dormancy and leukotriene biosynthesis cannot be triggered with calcium ionophore alone. However, by lowering the GSH content by azodicarboxylic acid bis(dimethylamide) or 1-chloro-2,4-dinitrobenzene plus addition of exogenous arachidonic acid, the cells produce high levels of leuokotriene B_4 (LTB_4) (Jakobsson et al., 1991, 1992) However, buthionine sulfoximine did not mimic the other GSH lowering compounds in this respect. In a follow-up study, it was demonstrated that several other thiol reactive compounds could, albeit not all, activate the 5-lipoxygenase pathway in B cells. Furthermore, low micromolar concentrations of hydrogen peroxide could also trigger 5-lipoxygenase activity (Jakobsson et al., 1995). It was reported that B-lymphocytes 5-lipoxygenase was inhibited by selenium and again GPx4 was discussed as the inhibitory cellular factor (Werz and Steinhilber, 1996). Evidently, the inhibition of 5-lipoxygenase can be achieved by different hydroperoxide-metabolizing enzymes, and it will depend on the enzymatic equipment of the cell type under consideration and the nature and intracellular location of the lipoxygenase-activating hydroperoxide, which of the GPxs or other peroxidases come into play.

12-Lipoxygenase activity is also found enhanced by destruction of GPx activity (Hill et al., 1989b). Inhibition of GPx by 1-chloro-2,4 dinitrobenzene enhanced platelet aggregation due to accelerated cyclooxygenase activation with enhanced formation of thromboxane A_2 (Hill et al., 1989a). The type of GPx involved was not addressed in the studies. However, more recently knockout of the extracellular

GPx3, which before had not been found to display any obvious phenotype, was reported to cause a massive platelet-dependent arterial thrombosis (Jin et al., 2011). Although the etiology of this pathology is not entirely understood, it is tempting to speculate that the cyclooxygenase activity in the platelets is also subject to regulation by the extracellular peroxide tone.

The selenium-containing GPxs are considered to generally inhibit oxidant-driven apoptosis and other pathways of programmed cell death (Flohé, 2016; Brigelius-Flohé and Flohé, 2017). Inverse genetic studies suggested that gastrointestinal GPx2 sustains the delicate balance between proliferation and apoptosis along the crypts and villi of the gut epithelium (Florian et al., 2010, see also Chapter 6). A prominent role in counteracting programmed cell death is attributed to GPx4 because of its exceptional ability to reduce hydroperoxyl groups of oxidized arachidonic acid or other polyunsaturated fatty acids even when esterified and firmly integrated in biomembranes. Its substrate specificity spectrum, thus, comprises also the products of 12,15-lipoxygenase. The reduction of 12,15-lipoxygenase products and the inhibition of this lipoxygenase had been implicated in the prevention of apoptosis inducing factor-dependent apoptosis (Seiler et al., 2008), ferroptosis (Friedmann Angeli et al., 2014), and other forms of cell death that are mediated by lipid peroxidation (Brigelius-Flohé and Flohé, 2017). However, this appealing idea is still being discussed controversially (see Chapters 7 and 8).

A general theme of controlling eicosanoid synthesis by regulating the peroxide levels and especially lipid hydroperoxide levels has thus emerged (Flohé, 2016). Selenium levels are paramount, as they control GPx levels and peroxiredoxin activity (Mattmiller et al., 2013; Flohé, 2016; Brigelius-Flohé and Flohé, 2017). The prominent role of GPx4 has been highlighted and its high catalytic efficiency, subcellular distribution, and capacity to reduce phospholipid hydroperoxides directly in the membrane underscored. However, an integral membrane enzyme that shares the capacity to reduce phospholipid hydroperoxides in the membrane is selenium-independent microsomal glutathione transferase 1 (MGST1) (Mosialou et al., 1993a,b). Although the enzyme activity is lower, the enzyme is often present in very high amounts and also widely distributed (Morgenstern et al., 1984). For instance, the levels in rat liver cell membranes are 3% and 5% in the endoplasmic reticulum and outer mitochondrial membrane, respectively. It is known that the enzyme protects cells from oxidative stress (Johansson et al., 2010) and thus is expected to influence the peroxide tone. Whether MGST1 is important in the regulation of eicosanoid metabolism remains to be investigated.

13.3 TRANSCRIPTIONAL REGULATION BY GSH AND GLUTATHIONE PEROXIDASES

Key enzymes of the eicosanoid metabolism, cytosolic phospholipase A2, cyclooxygenase 2 and 5-lipoxygenase (cPLA2, COX-2; 5-LOX), and some of the terminal synthases (inducible PGE synthase; inducible PGD synthase; PGI synthase) are under the control of NF-κB (complete list of target genes with literature under www.bu.edu/nf-kb/gene resources/target-genes). NF-κB in turn is negatively regulated by GPxs plus GSH, first shown by Kretz-Remy et al. (1996) for GPx1, but GPx4 proved to be more efficient in this context (Brigelius-Flohé et al., 1997).

In the gut, GPx2 dampens the expression of COX-2 and microsomal prostaglandin E synthase 1 (mPGES-1) (Banning et al., 2008). Thus, GPx activity negatively controls eicosanoid production both at the transcriptional and enzymatic level [reviewed in Brigelius-Flohé and Kipp (2009) and Mattmiller et al. (2013)]. A picture emerges where redox control inhibits eicosanoid production in normal tissue and is somehow overcome in pathophysiology. Exactly how this control is released when inflammatory signals abound is not known but NADPH oxidase mediated superoxide and hydrogen peroxide formation is certainly a candidate (Pastori et al., 2015).

13.4   GSH AS COFACTOR OR COSUBSTRATE IN EICOSANOID METABOLISM

The sculpting of lipid peroxidation products to specific physiological mediators during evolution involved recruitment of GSH-dependent processes at various stages. One can imagine that in simple early organisms that experienced damage, lipids spontaneously oxidized, as will happen in most cells also today. When the products of peroxidation at some point started to trigger protective processes in neighboring cells, the stage for developing sophisticated signaling cascades was set. Recruitment of more or less chemically labile mediators that are self-destructive could have constituted a first wave, fulfilling the requirement of temporal control with signal termination. The enzymes that protect from reactive lipids [epoxides, endoperoxides, hydroperoxides, and (alpha–beta unsaturated) aldehydes] also terminate such signals. In this scenario, these "terminated signals" [GSH conjugates (Blair, 2006), or reduced/isomerized reactive lipids] replaced the more labile mediators, concomitant with evolution of sensing, and catabolic capacity. It is thus very fitting that the enzyme families protecting from chemical reactivity (glutathione transferases and peroxidases) now also contain highly specialized members that produce and control physiological lipid mediators. The membrane-associated proteins in eicosanoid and glutathione metabolism (MAPEG) family to which MGST1 belongs, contains six members, out of which four have prominent roles in eicosanoid metabolism described below (Jakobsson et al., 1999).

13.4.1   BIOSYNTHESIS OF PROSTAGLANDIN E_2

GSH is required for prostaglandin E_2 (PGE_2) formation from the reactive prostaglandin endoperoxide PGH_2, catalyzed by microsomal prostaglandin E_2 synthase 1 (mPGES-1). This enzyme is highly efficient, inducible and often coregulated with PGH_2 synthase 2, and therefore prominent in pathophysiology such as fever, pain, and inflammation [reviewed in Samuelsson et al. (2007)]. GSH is not consumed during the oxidoreduction/isomerization reaction and functions as a cofactor. Interestingly, mPGES-1 also functions as a GPx and can reduce lipid hydroperoxides, including PGG_2 to PGH_2 (Thoren et al., 2003). Actually mPGES-1 can use either PGH_2 or PGG_2 equally well in the oxidoreduction/isomerization reaction with the product from PGG$_2$ (15-hydroperoxy-PGE$_2$) requiring subsequent reduction to form PGE$_2$. The GPx activity of the enzyme is quite modest and does not appear to impair the activity of cyclooxygenases (by influencing the peroxide tone) as PGH$_2$ is still formed efficiently (Murakami et al., 2000). The K_m for GSH is well within the physiological range (Thoren et al., 2003) and the enzyme binds three GSH molecules per homotrimer (Jegerschold et al., 2008; Sjögren et al., 2013). One-third-of-the-sites-reactivity has been suggested for mPGES-1 (He et al., 2011). However, this needs to be independently confirmed. Based on the high resolution structures and mutagenesis analysis, the active site residues have been identified (Brock et al., 2016; Raouf et al., 2016), and an essential arginine residue is suggested to activate GSH by stabilizing the anionic thiolate form. This role for an arginine appears to be general for MAPEG enzymes [except 5-lipoxygenase activating protein that facilitates LTA$_4$ formation by 5-lipoxygenase in a GSH-independent manner (Evans et al., 2008)].

There are other enzymes reported to catalyze GSH-dependent PGE$_2$ formation, mPGES-2 (Tanikawa et al., 2002), and cytosolic PGE synthase (Tanioka et al., 2000). However, the pathophysiological significance and overall contribution to PGE$_2$ formation appears modest. For mPGES-2 a contribution to PGE$_2$ formation has been questioned based on results with knockout mice, where levels of PGE$_2$ were basically unchanged (Jania et al., 2009). The result can be explained by data showing that the enzyme strongly binds heme, coordinated to its bound GSH (Yamada and Takusagawa, 2007), suggesting that mPGES-2 rather functions as a PGH$_2$ lyase *in vivo* (Takusagawa, 2013). For the cytosolic PGE$_2$ synthase, more specific roles in development and signaling have been demonstrated (Nakatani et al., 2007; Nakatani et al., 2017).

13.4.2 LEUKOTRIENE C$_4$ BIOSYNTHESIS

The role of leukotriene C$_4$ synthase (LTC$_4$S) in GSH-dependent formation of the conjugate LTC$_4$ from the epoxide LTA$_4$ is well established (Hammarström, 1983; Pace-Asciak and Smith, 1983; Piper, 1984; Lam and Austen, 2002; Rinaldo-Matthis and Haeggström, 2010). LTC$_4$ and its downstream metabolites LTD$_4$ and LTE$_4$ are prominent in the pathophysiology of inflammatory conditions such as asthma (Piper, 1983; Di Gennaro and Haeggström, 2012). The enzyme is only found in cells of hematopoietic origin and is also a member of the MAPEG superfamily of enzymes (Jakobsson et al., 1999; Bresell et al., 2005). The chemical mechanism of LTC$_4$S including stoichiometry of GSH binding and reactivity has been well characterized (Rinaldo-Matthis and Haeggström, 2010; Rinaldo-Matthis et al., 2012). Notably and in contrast to other MAPEG members (MGST1 and MGST2), the enzyme does not display one-third-of-the-sites reactivity, as all three active sites in the homotrimer are catalytically competent, stabilizing the GSH thiolate form simultaneously.

Recently, another member of the MAPEG superfamily that produces LTC$_4$, MGST2 (Jakobsson et al., 1996, 1997; Scoggan et al., 1997), was characterized on the enzyme level (Ahmad et al., 2013, 2015). The enzyme also has GPx activity and reduces various lipid hydroperoxides. MGST2 displays a broad distribution but exclusively in cells of nonhematopoietic origin and was demonstrated to mediate endoplasmic reticulum stress and chemotherapy-induced DNA damage (Dvash et al., 2015). Oxidative DNA damage and subsequent cell death is controlled by MGST2-dependent LTC_4 formation, acting on intracellular, nuclear membrane LTC_4 receptors. It was suggested that the selective distribution of MGST2 in nonhematopoietic cells could be utilized therapeutically, by developing inhibitors that prevent cell death in nontarget cells during chemotherapy.

An interesting aspect of LTC_4 is the fact that it acts as a nanomolar inhibitor of MGST1 (Bannenberg et al., 1999). Inhibition of MGST1 might potentiate leukotriene production by increasing the peroxide tone. By this way, MGST1 redox regulation could be involved in inflammation and cell death pathways.

13.4.3 PROSTAGLANDIN D SYNTHASES

Hematopoietic prostaglandin D synthase is a GSH-dependent enzyme belonging to the cytosolic glutathione transferase Sigma family [Thomson et al. (1998) reviewed in Kanaoka and Urade (2003)]. Lipocalin-type Prostaglandin D synthase is not strictly dependent on GSH, accepting various other thiols. Both these enzymes have been reviewed extensively (Smith et al., 2011).

13.4.4 PROSTAGLANDIN $F_{2\alpha}$

The role for GSH in prostaglandin $F_{2\alpha}$ production is indirect, as this mediator is produced by the reduction of either PGE_2 or PGD_2 by aldo-keto reductases (Helliwell et al., 2004).

13.5 GSH AS MODULATOR OF ENZYME ACTIVITY

GSH concentrations in cells can exhibit diurnal variation (Jaeschke and Wendel, 1985) as well as vary across the cell cycle (Li and Oberley, 1998). Whether such variation does have an impact on eicosanoid dynamics is not known. $K_{m(GSH)}$ for the biosynthetic and GPx reactions described earlier are not too far from the physiological levels and an effect is thus possible. As cosubstrate of GPxs, GSH can generally protect redox-sensitive enzymes also including such that are relevant in eicosanoid metabolism, e.g., PGI synthase (Weaver et al., 2001).

13.6 MEDICAL OUTLOOK AND LIST OF UNKNOWNS

GSH-dependent enzymes in eicosanoid production are the focus of intense drug development efforts for targeted removal of PGE_2 (Koeberle and Werz, 2015), PGD (Thurairatnam, 2012), and LTC_4 (Kleinschmidt et al., 2015) with the aim to relieve classical symptoms of inflammation. A role for eicosanoids in tumor development, maintenance and metastasis is explored as a target in tumor treatment or prevention (Radmark and Samuelsson, 2010; Korotkova and Jakobsson, 2014; Ramanan and Doble, 2017). In no case was GSH analogue development pursued, which is reasonable, because of the risk of off target effects. Only one of the developed drugs (against mPGES-1) has yet reached clinical phase 1 trial (Jin et al., 2016). Inhibition of

GPxs, with the exception of GPx4 inhibition for cancer treatment [(Yang et al., 2014; Hangauer et al., 2017; Sakamoto et al., 2017; Viswanathan et al., 2017), see also Chapters 7 and 8], does not appear to be a goal in drug development. However, GPx mimics are investigated for use as anti-inflammatory compounds, potentially correcting redox imbalance (Zhou et al., 2016).

In the future, the precise mechanisms of GSH/GPx-mediated redox control of eicosanoid producing enzymes needs to be unraveled at the transcriptional and enzyme levels, especially how negative control is released. The potential role for MGST1, and physiologically fluctuating GSH levels, in contributing to "peroxide tone" control, needs to be investigated. Also, the potential of new drugs being developed as alternatives to classical non-steroidal anti-inflammatory drugs (NSAIDs) that target biosynthetic enzymes mPGES-1 and PGD synthase as well as the potential of LTC_4S inhibitors, needs to be demonstrated.

REFERENCES

Ahmad, S., D. Niegowski, A. Wetterholm et al. 2013. Catalytic characterization of human microsomal glutathione *S*-transferase 2: Identification of rate-limiting steps. *Biochemistry* 52 (10):1755–64.

Ahmad, S., M. Thulasingam, I. Palombo et al. 2015. Trimeric microsomal glutathione transferase 2 displays one third of the sites reactivity. *Biochim Biophys Acta* 1854 (10 Pt A):1365–71.

Bannenberg, G., S. E. Dahlen, M. Luijerink et al. 1999. Leukotriene C4 is a tight-binding inhibitor of microsomal glutathione transferase-1. Effects of leukotriene pathway modifiers. *J Biol Chem* 274 (4):1994–9.

Banning, A., S. Florian, S. Deubel et al. 2008. GPx2 counteracts PGE2 production by dampening COX-2 and mPGES-1 expression in human colon cancer cells. *Antioxid Redox Signal* 10 (9):1491–500.

Blair, I. A. 2006. Endogenous glutathione adducts. *Curr Drug Metab* 7 (8):853–72.

Bresell, A., R. Weinander, G. Lundqvist et al. 2005. Bioinformatic and enzymatic characterization of the MAPEG superfamily. *FEBS J* 272 (7):1688–703.

Brigelius-Flohé, R., and L. Flohé. 2017. Selenium and redox signaling. *Arch Biochem Biophys* 617:48–59.

Brigelius-Flohé, R., B. Friedrichs, S. Maurer et al. 1997. Interleukin-1-induced nuclear factor kappa B activation is inhibited by overexpression of phospholipid hydroperoxide glutathione peroxidase in a human endothelial cell line. *Biochem J* 328 (Pt 1):199–203.

Brigelius-Flohé, R., and A. Kipp. 2009. Glutathione peroxidases in different stages of carcinogenesis. *Biochim Biophys Acta* 1790 (11):1555–68.

Brock, J. S., M. Hamberg, N. Balagunaseelan et al. 2016. A dynamic Asp-Arg interaction is essential for catalysis in microsomal prostaglandin E2 synthase. *Proc Natl Acad Sci USA* 113 (4):972–7.

Di Gennaro, A., and J. Z. Haeggström. 2012. The leukotrienes: Immune-modulating lipid mediators of disease. *Adv Immunol* 116:51–92.

Dvash, E., M. Har-Tal, S. Barak et al. 2015. Leukotriene C4 is the major trigger of stress-induced oxidative DNA damage. *Nat Commun* 6:10112.

Evans, J. F., A. D. Ferguson, R. T. Mosley, and J. H. Hutchinson. 2008. What's all the FLAP about? 5-Lipoxygenase-activating protein inhibitors for inflammatory diseases. *Trends Pharmacol Sci* 29 (2):72–8.

Flohé, L. 2016. The impact of thiol peroxidases on redox regulation. *Free Radic Res* 50 (2):126–42.

Florian, S., S. Krehl, M. Loewinger et al. 2010. Loss of GPx2 increases apoptosis, mitosis, and GPx1 expression in the intestine of mice. *Free Radic Biol Med* 49 (11):1694–702.

Friedmann Angeli, J. P., M. Schneider, B. Proneth et al. 2014. Inactivation of the ferroptosis regulator GPx4 triggers acute renal failure in mice. *Nat Cell Biol* 16 (12):1180–91.

Haeggström, J. Z., and C. D. Funk. 2011. Lipoxygenase and leukotriene pathways: Biochemistry, biology, and roles in disease. *Chem Rev* 111 (10):5866–98.

Hammarström, S. 1983. Leukotrienes. *Annu Rev Biochem* 52:355–77.

Hangauer, M. J., V. S. Viswanathan, M. J. Ryan et al. 2017. Drug-tolerant persister cancer cells are vulnerable to GPX4 inhibition. *Nature* 551 (7679):247–50.

Hatzelmann, A., M. Schatz, and V. Ullrich. 1989. Involvement of glutathione peroxidase activity in the stimulation of 5-lipoxygenase activity by glutathione-depleting agents in human polymorphonuclear leukocytes. *Eur J Biochem* 180 (3):527–33.

Hatzelmann, A., and V. Ullrich. 1987. Regulation of 5-lipoxygenase activity by the glutathione status in human polymorphonuclear leukocytes. *Eur J Biochem* 169 (1):175–84

Haurand, M., and L. Flohé. 1988. Kinetic studies on arachidonate 5-lipoxygenase from rat basophilic leukemia cells. *Biol Chem Hoppe Seyler* 369 (2):133–42.

He, S., Y. Wu, D. Yu, and L. Lai. 2011. Microsomal prostaglandin E synthase-1 exhibits one-third-of-the-sites reactivity. *Biochem J* 440 (1):13–21.

Helliwell, R. J., L. F. Adams, and M. D. Mitchell. 2004. Prostaglandin synthases: Recent developments and a novel hypothesis. *Prostaglandins Leukot Essent Fatty Acids* 70 (2):101–13.

Hemler, M. E., and W. E. Lands. 1980. Evidence for a peroxide-initiated free radical mechanism of prostaglandin biosynthesis. *J Biol Chem* 255 (13):6253–61.

Hill, T. D., J. G. White, and G. H. Rao. 1989a. Platelet hypersensitivity induced by 1-chloro-2, 4-dinitrobenzene, hydroperoxides and inhibition of lipoxygenase. *Thromb Res* 53 (5):447–55.

Hill, T. D., J. G. White, and G. H. Rao. 1989b. Role of glutathione and glutathione peroxidase in human platelet arachidonic acid metabolism. *Prostaglandins* 38 (1):21–32.

Imai, H., K. Narashima, M. Arai et al. 1998. Suppression of leukotriene formation in RBL-2H3 cells that overexpressed phospholipid hydroperoxide glutathione peroxidase. *J Biol Chem* 273 (4):1990–7.

Jaeschke, H., and A. Wendel. 1985. Diurnal fluctuation and pharmacological alteration of mouse organ glutathione content. *Biochem Pharmacol* 34 (7):1029–33.

Jakobsson, P. J., J. A. Mancini, and A. W. Ford-Hutchinson. 1996. Identification and characterization of a novel human microsomal glutathione *S*-transferase with leukotriene C4 synthase activity and significant sequence identity to 5-lipoxygenase-activating protein and leukotriene C4 synthase. *J Biol Chem* 271 (36):22203–10.

Jakobsson, P. J., R. Morgenstern, J. Mancini et al. 1999. Common structural features of MAPEG—A widespread superfamily of membrane associated proteins with highly divergent functions in eicosanoid and glutathione metabolism. *Protein Sci* 8 (3):689–92.

Jakobsson, P. J., B. Odlander, D. Steinhilber et al. 1991. Human B lymphocytes possess 5-lipoxygenase activity and convert arachidonic acid to leukotriene B4. *Biochem Biophys Res Commun* 178 (1):302–8.

Jakobsson, P. J., K. A. Scoggan, J. Yergey et al. 1997. Characterization of microsomal GST-II by western blot and identification of a novel LTC4 isomer. *J Lipid Mediat Cell Signal* 17 (1):15–19.

Jakobsson, P. J., P. Shaskin, P. Larsson et al. 1995. Studies on the regulation and localization of 5-lipoxygenase in human B-lymphocytes. *Eur J Biochem* 232 (1):37–46.

Jakobsson, P. J., D. Steinhilber, B. Odlander et al. 1992. On the expression and regulation of 5-lipoxygenase in human lymphocytes. *Proc Natl Acad Sci U S A* 89 (8):3521–5.

Jania, L. A., S. Chandrasekharan, M. G. Backlund et al. 2009. Microsomal prostaglandin E synthase-2 is not essential for in vivo prostaglandin E2 biosynthesis. *Prostaglandins Other Lipid Mediat* 88 (3–4):73–81.

Jegerschold, C., S. C. Pawelzik, P. Purhonen et al. 2008. Structural basis for induced formation of the inflammatory mediator prostaglandin E2. *Proc Natl Acad Sci U S A* 105 (32):11110–15.

Jin, R. C., C. E. Mahoney, L. Coleman Anderson et al. 2011. Glutathione peroxidase-3 deficiency promotes platelet-dependent thrombosis in vivo. *Circulation* 123 (18):1963–73.

Jin, Y., C. L. Smith, L. Hu et al. 2016. Pharmacodynamic comparison of LY3023703, a novel microsomal prostaglandin e synthase 1 inhibitor, with celecoxib. *Clin Pharmacol Ther* 99 (3):274–84.

Johansson, K., J. Jarvliden, V. Gogvadze, and R. Morgenstern. 2010. Multiple roles of microsomal glutathione transferase 1 in cellular protection: A mechanistic study. *Free Radic Biol Med* 49 (11):1638–45.

Kanaoka, Y., and Y. Urade. 2003. Hematopoietic prostaglandin D synthase. *Prostaglandins Leukot Essent Fatty Acids* 69 (2–3):163–7.

Kleinschmidt, T. K., M. Haraldsson, D. Basavarajappa et al. 2015. Tandem benzophenone amino pyridines, potent and selective inhibitors of human leukotriene C4 synthase. *J Pharmacol Exp Ther* 355 (1):108–16.

Koeberle, A., and O. Werz. 2015. Perspective of microsomal prostaglandin E2 synthase-1 as drug target in inflammation-related disorders. *Biochem Pharmacol* 98 (1):1–15.

Korotkova, M., and P. J. Jakobsson. 2014. Characterization of microsomal prostaglandin E synthase 1 inhibitors. *Basic Clin Pharmacol Toxicol* 114 (1):64–9.

Kretz-Remy, C., P. Mehlen, M. E. Mirault, and A. P. Arrigo. 1996. Inhibition of I kappa B-alpha phosphorylation and degradation and subsequent NF-kappa B activation by glutathione peroxidase overexpression. *J Cell Biol* 133 (5):1083–93.

Lam, B. K., and K. F. Austen. 2002. Leukotriene C4 synthase: A pivotal enzyme in cellular biosynthesis of the cysteinyl leukotrienes. *Prostaglandins Other Lipid Mediat* 68–69:511–20.

Lands, W. E., R. E. Lee, and W. L. Smith. 1971. Factors regulating the biosynthesis of various prostaglandins. *Ann N Y Acad Sci* 180:107–22.

Li, N., and T. D. Oberley. 1998. Modulation of antioxidant enzymes, reactive oxygen species, and glutathione levels in manganese superoxide dismutase-overexpressing NIH/3T3 fibroblasts during the cell cycle. *J Cell Physiol* 177 (1):148–60.

Mattmiller, S. A., B. A. Carlson, and L. M. Sordillo. 2013. Regulation of inflammation by selenium and selenoproteins: Impact on eicosanoid biosynthesis. *J Nutr Sci* 2:e28.

Morgenstern, R., G. Lundqvist, G. Andersson et al. 1984. The distribution of microsomal glutathione transferase among different organelles, different organs and different organisms. *Biochem Pharmacol.* 33:3609–14.

Mosialou, E., C. Andersson, G. Lundqvist et al. 1993a. Human liver microsomal glutathione transferase. Substrate specificity and important protein sites. *FEBS Lett* 315 (1):77–80.

Mosialou, E., G. Ekström, A. E. Adang, and R. Morgenstern. 1993b. Evidence that rat liver microsomal glutathione transferase is responsible for glutathione-dependent protection against lipid peroxidation. *Biochem Pharmacol* 45 (8):1645–51.

Murakami, M., H. Naraba, T. Tanioka et al. 2000. Regulation of prostaglandin E2 biosynthesis by inducible membrane-associated prostaglandin E2 synthase that acts in concert with cyclooxygenase-2. *J Biol Chem* 275 (42):32783–92.

Nakatani, Y., Y. Hokonohara, S. Kakuta et al. 2007. Knockout mice lacking cPGES/p23, a constitutively expressed PGE2 synthetic enzyme, are peri-natally lethal. *Biochem Biophys Res Commun* 362 (2):387–92.

Nakatani, Y., Y. Miyazaki, and S. Hara. 2017. Cytosolic prostaglandin E synthase is involved in c-Fos expression in rat fibroblastic 3Y1 cells. *Biol Pharm Bull* 40 (11):1963–7.

Pace-Asciak, C. R., and W. L. Smith. 1983. Enzymes in the biosynthesis of the eicosanids: Prostaglandins, thromboxanes, leukotrienes and hydroxy fatty acids. In *The Enzymes*, edited by P. D. Boyer, 543–603. New York and London: Academic Press.

Pastori, D., P. Pignatelli, R. Carnevale, and F. Violi. 2015. Nox-2 up-regulation and platelet activation: Novel insights. *Prostaglandins Other Lipid Mediat* 120:50–5.

Piper, P. J. 1983. Leukotrienes: Possible mediators in bronchial asthma. *Eur J Respir Dis Suppl* 129:45–64.

Piper, P. J. 1984. Formation and actions of leukotrienes. *Physiol Rev* 64 (2):744–61.

Radmark, O., and B. Samuelsson. 2010. Microsomal prostaglandin E synthase-1 and 5-lipoxygenase: Potential drug targets in cancer. *J Intern Med* 268 (1):5–14.

Ramanan, M., and M. Doble. 2017. Transcriptional regulation of mPGES1 in cancer: An alternative approach to drug discovery? *Curr Drug Targets* 18 (1):119–31.

Raouf, J., N. Rafique, M. C. Goodman et al. 2016. Arg126 and Asp49 are essential for the catalytic function of microsomal prostaglandin E2 synthase 1 and Ser127 is not. *PLOS ONE* 11 (9):e0163600.

Rinaldo-Matthis, A., S. Ahmad, A. Wetterholm et al. 2012. Pre-steady state kinetic characterization of thiolate anion formation in human leukotriene C4 synthase. *Biochemistry* 51 (4):848–56.

Rinaldo-Matthis, A., and J. Z. Haeggström. 2010. Structures and mechanisms of enzymes in the leukotriene cascade. *Biochimie* 92 (6):676–81.

Rinaldo-Matthis, A., A. Wetterholm, D. M. Molina et al. 2010. Arginine 104 is a key catalytic residue in leukotriene C-4 synthase. *J Biol Chem* 285 (52):40771–6

Sakamoto, K., S. Sogabe, Y. Kamada et al. 2017. Discovery of GPX4 inhibitory peptides from random peptide T7 phage display and subsequent structural analysis. *Biochem Biophys Res Commun* 482 (2):195–201.

Samuelsson, B. 1983. From studies of biochemical mechanism to novel biological mediators: Prostaglandin endoperoxides, thromboxanes, and leukotrienes. Nobel Lecture, 8 December 1982. *Biosci Rep* 3 (9):791–813.

Samuelsson, B., R. Morgenstern, and P. J. Jakobsson. 2007. Membrane prostaglandin E synthase-1: A novel therapeutic target. *Pharmacol Rev* 59 (3):207–24.

Schnurr, K., J. Belkner, F. Ursini et al. 1996. The selenoenzyme phospholipid hydroperoxide glutathione peroxidase controls the activity of the 15-lipoxygenase with complex substrates and preserves the specificity of the oxygenation products. *J Biol Chem* 271 (9):4653–8.

Scoggan, K. A., P. J. Jakobsson, and A. W. Ford-Hutchinson. 1997. Production of leukotriene C4 in different human tissues is attributable to distinct membrane bound biosynthetic enzymes. *J Biol Chem* 272 (15):10182–7.

Seiler, A., M. Schneider, H. Forster et al. 2008. Glutathione peroxidase 4 senses and translates oxidative stress into 12/15-lipoxygenase dependent- and AIF-mediated cell death. *Cell Metab* 8 (3):237–48.

Sjögren, T., J. Nord, M. Ek et al. 2013. Crystal structure of microsomal prostaglandin E2 synthase provides insight into diversity in the MAPEG superfamily. *Proc Natl Acad Sci U S A* 110 (10):3806–11.

Smith, W. L., Y. Urade, and P. J. Jakobsson. 2011. Enzymes of the cyclooxygenase pathways of prostanoid biosynthesis. *Chem Rev* 111 (10):5821–65.

Straif, D., O. Werz, R. Kellner et al. 2000. Glutathione peroxidase-1 but not -4 is involved in the regulation of cellular 5-lipoxygenase activity in monocytic cells. *Biochem J* 349 (Pt 2):455–61.

Takusagawa, F. 2013. Microsomal prostaglandin E synthase type 2 (mPGES2) is a glutathione-dependent heme protein, and dithiothreitol dissociates the bound heme to produce active prostaglandin E2 synthase in vitro. *J Biol Chem* 288 (14):10166–75.

Tanikawa, N., Y. Ohmiya, H. Ohkubo et al. 2002. Identification and characterization of a novel type of membrane-associated prostaglandin E synthase. *Biochem Biophys Res Commun* 291:884–9.

Tanioka, T., Y. Nakatani, N. Semmyo et al. 2000. Molecular identification of cytosolic prostaglandin E2 synthase that is functionally coupled with cyclooxygenase-1 in immediate prostaglandin E2 biosynthesis. *J Biol Chem* 275 (42):32775–82.

Thomson, A. M., D. J. Meyer, and J. D. Hayes. 1998. Sequence, catalytic properties and expression of chicken glutathione-dependent prostaglandin D2 synthase, a novel class Sigma glutathione *S*-transferase. *Biochem J* 333 (Pt 2):317–25.

Thoren, S., R. Weinander, S. Saha et al. 2003. Human microsomal prostaglandin E synthase-1: Purification, functional characterization, and projection structure determination. *J Biol Chem* 278 (25):22199–209.

Thurairatnam, S. 2012. Hematopoietic prostaglandin D synthase inhibitors. *Prog Med Chem* 51:97–133.

Vane, J. R. 1983. Adventures and excursions in bioassay: The stepping stones to prostacyclin. Nobel Lecture, 8 December 1982. *Biosci Rep* 3 (8):683–711.

Viswanathan, V. S., M. J. Ryan, H. D. Dhruv et al. 2017. Dependency of a therapy-resistant state of cancer cells on a lipid peroxidase pathway. *Nature* 547 (7664):453–7.

Weaver, J. A., J. F. Maddox, Y. Z. Cao et al. 2001. Increased 15-HPETE production decreases prostacyclin synthase activity during oxidant stress in aortic endothelial cells. *Free Radic Biol Med* 30 (3):299–308.

Weitzel, F., and A. Wendel. 1993. Selenoenzymes regulate the activity of leukocyte 5-lipoxygenase via the peroxide tone. *J Biol Chem* 268 (9):6288–92.

Werz, O., and D. Steinhilber. 1996. Selenium-dependent peroxidases suppress 5-lipoxygenase activity in B-lymphocytes and immature myeloid cells. The presence of peroxidase-insensitive 5-lipoxygenase activity in differentiated myeloid cells. *Eur J Biochem* 242 (1):90–7.

Yamada, T., and F. Takusagawa. 2007. PGH2 degradation pathway catalyzed by GSH-heme complex bound microsomal prostaglandin E2 synthase type 2: The first example of a dual-function enzyme. *Biochemistry* 46 (28):8414–24.

Yang, W. S., R. SriRamaratnam, M. E. Welsch et al. 2014. Regulation of ferroptotic cancer cell death by GPX4. *Cell* 156 (1–2):317–31.

Zhou, J., H. Xu, and K. Huang. 2016. Organoselenium small molecules and chromium(III) complexes for intervention in chronic low-grade inflammation and type 2 diabetes. *Curr Top Med Chem* 16 (8):823–34.

14 Nitric Oxide and S-Nitrosoglutathione

Iain L. O. Buxton and Scott D. Barnett
University of Nevada

CONTENTS

14.1 INTRODUCTION

In this chapter, we consider that role of S-nitrosoglutathione (GSNO) in the critical regulation of the actions of nitric oxide in a mammalian system. We have chosen to address smooth muscle as it was in this tissue that the early work on the origins of reactive nitrogen species (RNS) unfolded. Beginning with the work of Robert Furchgott nearly 40 years ago (Furchgott, 1993) the role of nitric oxide in the regulation of biological functions has been clearly established. The classic experiments that first identified an endothelial factor that relaxed vascular smooth muscle (Furchgott and Zawadzki, 1980) together with its identification as nitric oxide and the transduction of its biological effect through cyclic guanosine monophosphate formation established the signaling dogma (Furchgott et al., 1998) and the awarding of the Nobel Prize for Physiology or Medicine in 1998.

The original notion that nitric oxide, a reactive nitrogen oxide with an unpaired electron (•NO), was formed in the endothelial cell and entered the adjacent smooth muscle cell to activate guanylate cyclase leading to the formation of cGMP assumed that this reactive molecule survived time and space to target the smooth muscle guanylyl cyclase selectively. This is an intellectually displeasing notion.

Glutathione is the major thiol in mammalian cells occurring in concentrations up to 10 mM (Meister and Anderson, 1983). In the presence of •NO, glutathione is nitrosated on the free thiol of its cysteine moiety to form GSNO (Figure 14.1).

DOI: 10.1201/9781351261760-17

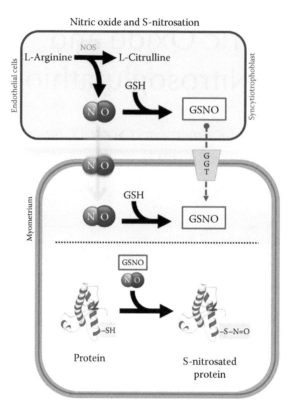

FIGURE 14.1 *S*-Nitrosoglutathione and protein *S*-nitrosation: Nitric oxide, an important mediator of smooth muscle relaxation, is generated by the NOS family of enzymes. Nitric oxide is known to originate in endothelial cells. NO either moves free through the membrane or may be transported as nitrosoglutathione (GSNO) by proteins such as γ-glutamyltransferase (GGT). NO is often stored in the cell as stable *S*-nitrosoglutathione. GSNO then reacts with the thiol group of cysteines. This posttranslational modification is called *S*-nitrosation.

GSNO, first discovered in airway by Gaston et al. (Gaston et al., 1993), is the likely form in which NO crosses the membrane of most cell types to *trans*-nitrosate (*S*-NO) proteins. While the formation of GSNO is likely to be nonenzymatic, its transport and thus its availability to signal locally may be regulated (Zhang and Hogg, 2004; Cowles et al., 2016). Indeed, the entire process of •NO action in cells from its formation by the nitric oxide synthase, its likely passage into target cells as GSNO, to the formation of *S*-nitrosoproteins and their function is highly regulated rather than obligatory as first imagined.

Our interest in GSNO stems from the discovery of a signaling exception to the dogma describing the actions of •NO in smooth muscle (Kuenzli et al., 1996; Buxton, 2004) in the 1990s. Because the proximate cause of preterm birth (PTB) is the contraction of uterine muscle prior to 37 weeks of gestation, regardless of the origin(s) of this early labor, we became interested in the ability of •NO to relax uterine smooth muscle (myometrium). In human myometrium, •NO acts independently of cGMP

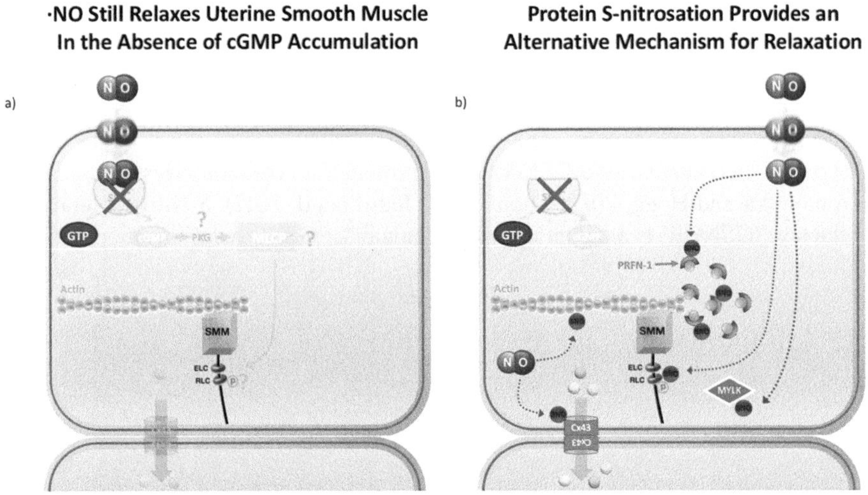

FIGURE 14.2 *S*-Nitrosation as an alternative pathway to myometrial quiescence: (a) Nitric oxide relaxes myometrial tissue independently of cGMP accumulation. (b) We hypothesize the *S*-nitrosation of important contractile-associated proteins (CAPS) by nitric oxide may contribute to the cGMP-independent relaxation of the tissue.

accumulation (Bradley et al., 1998) (Figure 14.2). As we describe in the following sections, this signaling exception has led to an intriguing hypothesis that pregnancy state specific changes in protein *S*-nitrosations by GSNO, rather than cyclic nucleotide accumulation, are responsible for •NO signaling to relax uterine smooth muscle. Addressing this hypothesis offers a therapeutic approach to an otherwise intractable medical dilemma.

14.2   *S*-NITROSOGLUTATHIONE AND *S*-NITROSATION

The study of protein *S*-nitrosations and their influence on normal cell-signaling and in disease states has significantly impacted research in medicine for over 25 years (Stamler et al., 1992; Foster et al., 2009; Broniowska and Hogg, 2012). The detection and quantitation of protein *S*-nitrosations in biological systems is inherently challenging. The biotin switch technique (Jaffrey and Snyder, 2001), in which *S*-nitrosated cysteines are reduced and biotinylated, provides a simple and elegant method for the qualitative detection of *S*-nitrosated proteins. An analysis of a wide variety of so-called *S*-nitrosothiol (RSNO) measurement techniques, including the biotin switch, has established that artifacts are common when measuring RSNOs and it is not always possible to identify which thiols have been *S*-nitrosated (Giustarini et al., 2003). Newer techniques have become available in recent years (Chen et al., 2013; Devarie-Baez et al., 2013), such as tandem mass spectrometry (MS/MS) of *S*-nitrosated protein thiols (Murray et al., 2012; Ulrich et al., 2013), that are highly quantitative. Beyond the problem of quantitation, it has been proposed that other thiol modifications such as dithiol/disulfide exchange, *S*-glutathionylation, and oxidation may affect signaling more readily than do RSNOs (Lancaster, 2008) and should be investigated along with *S*-nitrosation.

As with phosphorylation, *S*-nitrosation regulates cellular mechanisms and affects protein–protein interactions. The intracellular availability of nitric oxide and its functional derivatives, like GSNO, affect protein *S*-nitrosation (Hess et al., 2005; Broniowska and Hogg, 2012; Thomas and Jourd'heuil, 2012). *S*-Nitrosoglutathione reductase (GSNOR) is a potent negative regulator of GSNO levels in smooth muscle (Que et al., 2009). GSNOR utilizes the coenzyme NADH to carry out a 2e$^-$ reduction of GSNO to generate glutathione sulfinamide (Staab et al., 2008a) before it is further reduced back to glutathione by glutathione reductase. The aberrant expression of this enzyme is associated with disease (Jelski and Szmitkowski, 2008; Jelski et al., 2009; Laniewska-Dunaj et al., 2013). In fact, the deletion of the GSNO reductase gene increases both the levels of GSNO and total protein *S*-nitrosation *in vivo* (Liu et al., 2001). As we describe in these pages, the metabolism of GSNO plays an important role in conditions such as preterm labor (PTL). Others have described the role of *S*-nitrosation in disease including Type 2 diabetes (Carvalho-Filho et al., 2005), sickle cell anemia (Bonaventura et al., 1999; Bonaventura et al., 2002), ventricular arrhythmia in individuals with Duchenne muscular dystrophy (Fauconnier et al., 2010), cell death and survival pathways (Iyera et al., 2011), postinfarct cardioprotection (Methner et al., 2014), and many others.

•NO is a RNS that is critical to the normal function of most cells type (Moncada et al., 1991; Radi et al., 1991; Moncada and Higgs, 1993; Beckman and Koppenol, 1996). It is a powerful smooth muscle relaxing agent (Bradley et al., 1998; Buxton et al., 2001; Tomita et al., 2002; Ricciardolo et al., 2004), neuroeffector (Bredt and Snyder, 1992; Corti et al., 2014), and immune system modulator (Macmicking et al., 1997). •NO is likely carried as GSNO from endothelium, and other sources, and acts as a stable NO reserve (Smith and Marletta, 2012; Broniowska et al., 2013). GSNO can transfer its NO moiety to a cysteine thiol, resulting in the posttranslational modification of *S*-nitrosation (Stamler et al., 1992) of a target thiol.

14.3   THE CLINICAL PROBLEM

PTL results in PTB of an underdeveloped fetus in 50% of cases. PTL is defined as labor between 20 and 37 weeks of gestation, with extreme PTB between 22 and 28 weeks (Weismiller, 2000). Spontaneous PTL (sPTL) differs from PTL in that it is idiopathic in nature, meaning no root cause for the labor has been established. PTB can adversely affect fetal development; in particular, the heart, lungs, and brain (Saigal and Doyle, 2008). The earliest case of a severely premature infant to survive of which we are aware, is that of a child born in the United States at 21 weeks, 6 days. This is an extraordinary success of modern medicine considering a full-term pregnancy in humans is 40 weeks. While the medical community continues to enhance its ability to decrease morbidity and mortality in extreme preterm infants (Glass et al., 2015), 60% of all neonatal deaths still occur in infants born prior to 34 weeks of gestation (Glass et al., 2015) and the neonatal isolette remains the most expensive in any hospital. Identifying effective methods to prevent PTB altogether would better serve the infant and the global community.

The cause(s) of PTL are unclear (Table 14.1). Strong associations have been made between circumstances such as maternal stress, being single, and not receiving prenatal care. While such factors may hint at underlying cause, they do not offer obvious mechanistic explanations. While delivery of a preterm infant can suggest the risk of a future PTB, this association is not strong and there does not appear to be the result of fetal abnormality. Because the incidence of PTB has risen in recent decades, it has been suggested that some prematurity may be the result of female "premies" surviving to deliver premature infants themselves (Bhattacharya et al., 2010). Such a possibility suggests that there is a genetic basis of at least some cases of prematurity although genetic analysis has yet to reveal a likely cause (Zhang et al., 2017).

Whatever the underlying mechanism(s) of sPTL leading to PTB, millions of lives are impacted each year and the incidence of PTL is a worldwide problem. PTB places a massive financial burden on society estimated at $26.2 billion annually in the United States alone (Behrman and Butler, 2007). Adjusted for inflation, this figure has risen to 30 billion. Estimates of the cost of PTB cannot accurately account for many of the costs of ongoing medical care for premature infants with unwanted outcomes such as chronic lung disease, cardiovascular disease, or cerebral palsy, nor are these outcomes or their impact tracked to adulthood.

PTB remains the primary cause of neonatal morbidity and hospitalization during pregnancy (Miniño et al., 2006; D'Onofrio et al., 2013; Rundell and Panchal, 2017) and accounts for a majority of pediatric care worldwide (Howson et al., 2013). In the United States alone, greater than 12% of infants are born prematurely, resulting in 20,000 deaths annually (Martin et al., 2011). About half of those premature births are attributed to spontaneous PTL, that is initiation of labour without any known precipitating cause (Goldenberg et al., 2008). Worldwide, the statistics are even more disconcerting. Approximately 13 million infants across the globe are affected by premature birth each year (Behrman and Butler, 2007). Sub-Saharan Africa is of particular concern, where as many as 336,000 of the 1.2 million (28%) births each year result in newborn death (Kinney et al., 2010). In fact, women of African descent are 50% more likely to deliver preterm than women of European descent (Cdc, 2015). While some cohorts are disproportionately affected by sPTL, the emotional and physical consequences of early labor and delivery span all races, nationalities, ages, and socioeconomic groups. It is because of this that solving this complex problem is of paramount importance and we shall see, GSNO has a critical role in a way forward.

Unfortunately, drugs employed to prevent early contractions of the uterus, so-called tocolytics (Figure 14.3), are ineffective. No drugs reliably halt labor in patients who enter labor spontaneously preterm allowing their pregnancies to go to term. In the United States, the use of tocolytics to manage PTL is employed without Food and Drug Administration approval or clear evidence of benefit for acute or maintenance tocolysis (Dodd et al., 2006; Whitworth and Quenby, 2008), and on average are said to delay labor for only 48h (Elvira et al., 2014), a window for antenatal steroid to promote the premature lung to accept air, (Roberts et al., 2017), but hardly a solution to the problem. Microbial infection might initiate PTL

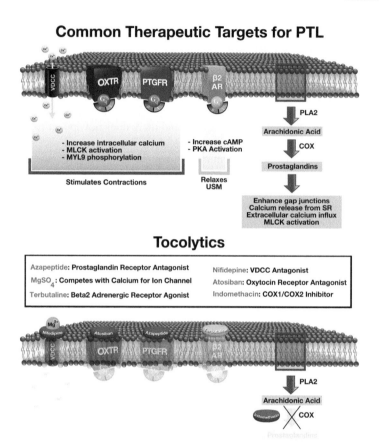

FIGURE 14.3 Common tocolytics used to treat PTL: Tocolytics are drugs used to prevent or to halt labor. Tocolytics generally target essential pathways to contraction, such as the oxytocin receptor (OXTR), prostaglandin receptors (PTGFR) and prostaglandin synthesis (COX), adrenergic receptors (B2AR), and voltage-dependent calcium channels (VDCC). Currently, available tocolytics are not able to delay PTB beyond 48-h.

in some cases, but antibiotic treatment, prophylactic or otherwise, does not prevent PTB (Prince et al., 2014; Vinturache et al., 2016). If we are to advance our understanding of PTL in order to delay or prevent PTB, understanding the biochemical mechanisms underlying relaxation of the uterus is paramount. This point is bolstered when we consider that employing tools such as Terbutaline, which is used clinically to relax airway smooth muscle, or Nifedipine, used to relax vascular smooth muscle, in an effort to prevent PTL are borrowed pharmacology. Even Atosiban, a selective oxytocin–vasopressin receptor antagonist designed specifically to mitigate contractions, is not approved for use in the United States and does not reduce the risk of PTB or improve neonatal outcome (Papatsonis et al., 2005). It is not unreasonable, therefore, to conclude that myometrial relaxation signaling is unique, and a detailed understanding of this signaling is urgently needed. Understanding the failure of drugs currently employed to prevent early labor requires an understanding of smooth muscle.

14.4 SMOOTH MUSCLE CONTRACTION

The stimulation of a smooth muscle contraction varies depending on the type of the muscle group. For example, enteric neurons drive smooth muscle contraction in the digestive tract (Kunze and Furness, 1999). Vascular and airway smooth muscle respond strongly to adrenergic stimulation (Barnes and Liu, 1995), and the muscle of the uterus relies heavily on hormones and prostaglandins (Gimpl and Fahrenholz, 2001). While there is some overlap in the mechanisms that drive contraction in these different smooth muscle types, such as with common G protein-coupled receptor (GPCR) signaling pathways and depolarization-driven Ca^{2+} entry, phenotypic variations dictate unique responses to stimuli and drugs. This means therapeutics intended for one type of smooth muscle may not be effective in other types; an observation that has become exceedingly clear when examining the failure of tocolytics (Giles and Bisits, 2007).

The myometrium is a powerful smooth muscle that acts as the principal expulsive driving force during labor. It is phasic in nature, as opposed to tonic, meaning that contractions are balanced by relaxation, a short time after stimulation (Szal et al., 1994). One such endogenous myometrial stimulator is oxytocin (OT), a peptide hormone primarily generated in the hypothalamus and secreted into the bloodstream via the posterior pituitary gland (Kimura et al., 1992). OT binds to oxytocin receptors (OXTR) in the myometrium (Gimpl and Fahrenholz, 2001). When stimulated, these GPCRs (Gαq/11) activate phospholipase C, in turn generating inositol triphosphate (IP3) and diacylglycerol (DAG) from membrane stores of phosphatidylinositol 4,5-bisphosphate (PIP2). This activates protein kinase C (PKC), as well as other downstream effectors of smooth muscle contraction (Wray, 1993) (Figure 14.4).

GPCR stimulation activates voltage-gated Ca^{2+} channels, which in turn triggers calcium-induced calcium release (CICR) within the cell from the sarcoplasmic reticulum, further depolarizing the membrane. Ca^{2+} binds to calmodulin (CaM), which activates myosin light chain kinase (MLCK) (Bursztyn et al., 2007). MLCK phosphorylates serines S18/S19 of the regulatory light chain (MYL9) of smooth muscle myosin (SMM) (Frearson et al., 1976; Hong et al., 2011).

It is widely accepted that the phosphorylation of MYL9 is the "master switch" that drives smooth muscle contraction, and that the kinase activity of MLCK is balanced by the dephosphorylation of MYL9 by myosin light chain phosphatase (MLCP). Class II SMM contains two interacting heavy chains, each with their own actin-binding head domain along with an essential light chain and a regulatory light chain (MYL9). In its unphosphorylated state, one head of the SMM complex binds to the actin filament, while the complementary actin-binding site on the second head is "blocked" due to interactions with the first head (Baumann et al., 2012). As a result, cross-bridge cycling is prevented, and a contraction cannot occur. Phosphorylation of MYL9 causes a conformation change to the myosin head formerly bound to its "sister" head, which in turn permits actin binding, and cross-bridge cycling can occur.

(a)   Canonical contraction/relaxation in myometrium

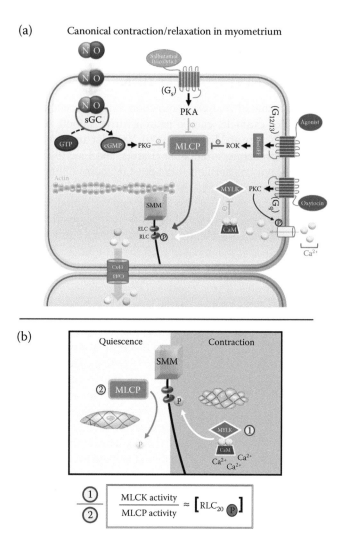

FIGURE 14.4 Canonical contraction and relaxation pathways in smooth muscle: (a) The canonical contraction and relaxation of smooth muscle is mediated by transmembrane receptors, ion channels, small molecules, hormones, and thousands of other molecules. (b) Despite the complex intertwining of pathways that control smooth muscle contraction and relaxation, it is the phosphorylation of MYL9, the 20 kD regulatory light chain of smooth muscle myosin that acts as the "master switch" to contraction, and it is the ratio of kinase (MYLK) to phosphatase (MLCP) that dictates contraction in the cell.

14.5 CANONICAL SMOOTH MUSCLE RELAXATION

In smooth muscle, high levels of cytosolic calcium are not required to maintain contraction. Smooth muscles maintain contractile force until MYL9 is dephosphorylated (Bursztyn et al., 2007). Dephosphorylation of MYL9 is primarily driven by the myosin phosphatase targeting subunit, MYPT1, of MLCP, the predominant phosphatase in smooth muscle. MLCP contains a PP1cδ catalytic subunit, a MYPT1 targeting subunit, as well as an M20 subunit with unknown function (Hudson et al., 2012). MYPT1 itself is phosphorylated at serine and threonine residues S695, T696, S852, and T853, and is constitutively active when not phosphorylated, albeit with low activity, so some contractile stimulation is needed to maintain contraction (Ito et al., 2004). This modulation of MLCP activity is a major component of Ca^{2+} sensitization/desensitization (Somlyo and Somlyo, 2003), and it allows for complex regulation of smooth muscle contraction and relaxation.

MYPT1 is phosphorylated by cAMP-dependent PKA and cGMP-dependent PKG, which activate the enzyme, as well as Rho-Kinase (ROCK or ROK), which inhibits MLCP (Puetz et al., 2009). The relationship among these kinases and their activation pathways is complex. In smooth muscle, one common route to PKA activation is through β2 adrenergic (Gs)-driven cAMP production. PKG is also activated through cyclic nucleotide production, in this case cGMP, which is created as the result of •NO-stimulated soluble guanylate cyclase (sGC). ROK, an inhibitor of MLCP, is activated by the RhoGEF complex, as a result of GPCR12/13 stimulation. While all three of these kinases act upon MYPT1, it should be noted that each of them has a number of phosphorylation targets that either enhance or inhibit contraction (Puetz et al., 2009). PKG action halts when either of its upstream activating factors, •NO or cGMP, are exhausted. •NO itself is generally depleted by either endogenous metabolizers of •NO, such as GSNOR, or the thioredoxin system, while cGMP is metabolized by phosphodiesterases (PDEs). As such, tocolytics that target these systems have been used with varying claims of success. Confounding MLCP function in smooth muscle is the finding that disparate functional isoforms of MLCP exist; specifically, it has been discovered that a leucine-zipper variant of MLCP affects PKG activity and cGMP-mediated relaxation in the cell (Pan, 2010; Yuen et al., 2011). Thus, neither the mechanism whereby PKG activates MLCP, nor the nature of the interaction between these two proteins are completely understood.

Another critical mediator of pMYL9 is Telokin, the 17kD terminal fragment of MLCK that binds to SMM. Telokin is not only a functional domain of MLCK, but also serves as an autonomous, independently translated protein that is transcribed by a separate promoter from *MLCK* (Smith et al., 1998). Telokin serves two known functions. It has an inhibitory effect on MYL9 phosphorylation via competitive binding with MLCK to SMM (Khromov et al., 2006). Second, it aids in MLCP activation (Komatsu et al., 2002). Telokin is thought to be differentially *S*-nitrosated depending of the state of pregnancy, which may affect its function.

14.6   NITRIC OXIDE

Canonically, •NO stimulates the soluble sGC in smooth muscle, catalyzing the formation cGMP, a second messenger analogous to cAMP. Cyclic GMP activates its cognate kinase, PKG, which phosphorylates serine S695 on the catalytic subunit of MLCP, MYPT1 (Puetz et al., 2009). Phosphorylation, *per se*, does not appear to be an obligate requirement for MLCP activation, and thus, some unknown action may serve to activate MLCP, orchestrated by PKG activation. Interestingly, •NO also activates the inducible form of cyclooxygenase-2 (COX2), increasing the synthesis of prostaglandin PGE2 (Salvemini et al., 1993; Kim, 2011) which promotes quiescence via GPCR (Gs) stimulation. PGE2 is also an important mediator of cervical ripening (Keelan et al., 2003), indicating that •NO's role in pregnancy and parturition is multilayered. Another important distinction of uterine smooth muscle is that cGMP action can be compartmentalized (Buxton et al., 2010), and relaxation of the tissue in response to NO is independent of global cGMP accumulation (Buxton, 2004). The reason behind this phenomenon has not been fully elucidated, but our lab has made substantial progress on this front. This signaling exception may have far reaching consequences, and affect how we approach the treatment of sPTL in the future.

As noted, nitric oxide is a powerful endogenous smooth muscle relaxing agent. In pregnant women, the administration of nitric oxide or NO-donors, such as with nitroglycerine transdermal patches, show little (Smith et al., 2007) to no (Nankali et al., 2014) clinical efficacy; and in fact fails to quiesce sPTL myometrium. These findings do not dismiss the role of •NO in the myometrium. •NO functions as an important endogenous mediator of relaxation in myometrium, and there is evidence that the failure of •NO to serve as an effective tocolytic may be related to its metabolism which is dysregulated in women who enter labor spontaneously preterm.

•NO is a power smooth muscle relaxing agent, (Bradley et al., 1998; Buxton et al., 2001; Tomita et al., 2002; Ricciardolo et al., 2004), but also a cardiopulmonary regulator (Liu et al., 2004; Tamargo et al., 2010), neuroeffector (Bredt and Snyder, 1992; Corti et al., 2014), and immune system modulator (Macmicking et al., 1997). •NO is likely carried as GSNO from endothelium, and other sources, and acts as a stable •NO reserve (Smith and Marletta, 2012; Broniowska et al., 2013). GSNO can transfer its NO moiety to a cysteine thiol, resulting in the posttranslational modification (PTM) *S*-nitrosation/*S*-nitrosylation (Stamler et al., 1992) of a target thiol. Protein *S*-nitrosations are also referred to in the literature in a fashion that takes into account protein and nonprotein nitrosations (e.g., RSNO). We employ the term RSNO as it appears in the literature.

•NO availability in the myometrium is regulated by enzymes such as thioredoxin (Trx) and its cognate reductase (TrxR) (Sahlin et al., 2000), carbonyl reductase (Bateman et al., 2008), and the class-III alcohol dehydrogenase, GSNOR or ADH5. GSNOR utilizes the coenzyme NADH to carry out a 2e⁻ reduction of GSNO to generate glutathione sulfinamide (Staab et al., 2008b) before it is further reduced back to glutathione by glutathione reductase (Figure 14.5). A drug that could effectively

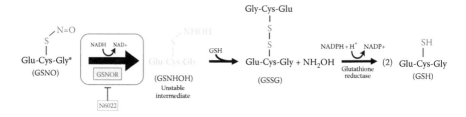

FIGURE 14.5 *S*-Nitrosoglutathione metabolism. *S*-Nitrosoglutathione metabolism by the reductase completes a cycle regenerating glutathione.

lower •NO metabolism, while minimizing adverse or off-target effects—a common problem with currently employed drugs—could function well as a tocolytic.

When considering tocolytic options for the treatment of sPTL, recent studies offer unique insight. First is the observation that, while •NO relaxes full-term human myometrium, •NO's actions are severely blunted in sPTL myometrial tissue (Buxton et al. unpublished observation). This finding reveals a unique, and previously unreported, phenotype of sPTL myometrium. Because •NO fails to relax sPTL tissue, its use as a tocolytic is not appropriate, beyond the known health risks to mother and child (Duckitt et al., 2014). As a result, it is possible that aberrant •NO metabolism in sPTL myometrium may explain sPTL's blunted response to •NO by lowering available NO action(s) and thus favoring contraction? Interestingly, we find that GSNOR expression and activity are increased in sPTL myometrium. The finding that GSNOR dysregulation corresponds to the pathophysiology that is sPTL, suggests a closer look at GSNOR as an important metabolizing agent of GSNO, the stable endogenous form of •NO in the cell. GSNOR dysregulation has long been known to be involved in many disease states (Barnett and Buxton, 2017). Importantly, this enzyme is also dysregulated in some patient cohorts with asthma, which leads to enhanced airway smooth muscle constriction, and resistance to bronchodilators (Henderson and Gaston, 2005; Wu et al., 2007; Choudhry et al., 2010). This, combined with observations that women with asthma exhibit a higher relative risk of PTL and delivery (Doucette and Bracken, 1993), favor GSNOR as protein of interest in understanding GSNO and myometrial function.

If GSNOR expression is dysregulated and elevated in women in labor preterm as we have seen, then GSNO levels will be reduced and the decreased availability of NO action may explain the dysfunction. Beyond investigating GSONR expression as a function of gestational length, it is possible that GSNOR expression in the myometrium varies among race. This question is particularly germane as sPTL is known to disproportionately impact women of African descent (Culhane and Goldenberg, 2011; Hamilton et al., 2015). A study in African American children found that single nucleotide polymorphisms (SNPs) in the GSNOR gene in airway smooth muscle correlates to severe asthma and resistance to treatment (Moore et al., 2009). GSNOR dysregulation may also be more prevalent in this population and impact pregnancy. Future work is needed to explore GSNOR expression in sPTL samples from women of African descent, as well as whether the aforementioned SNPs in the GSNOR gene exists in sPTL samples, as several SNPs in the promoter and 3′ UTR of the gene can result in the aberrant expression of GSNOR (Choudhry et al., 2010).

Perhaps the most intriguing and important outcome of GSNOR dysregulation in myometrium is that GSNOR now presents as a therapeutic target on which to test novel tocolytics that inhibit this enzyme. N6022, a potent and selective inhibitor of GSNOR, is well tolerated in humans and has already been tested in clinical trials as an airway smooth muscle relaxing agent in asthmatics (clinicaltrials.gov— NCT01316315), and for individuals with cystic fibrosis (clinicaltrials.gov—N6022: NCT01746784; N91115: NCT02724527) (Nivalis Therapeutics, 2014). The efficacy of GSNOR inhibition in the human sPTL myometrium is yet to be evaluated, although experiments in guinea pigs identified N6022 as an effective tocolytic, nearly abolishing all contractile force (unpublished observation). Numerous derivatives of N6022 have been identified (Sun et al., 2011) as well as unrelated compounds targeting GSNOR (Ferrini et al., 2013), which may prove useful as tocolytic agents.

Ultimately, the underlying cause(s) of sPTL remain unknown. While the complex mechanisms that drive sPTL have not been fully elucidated, the discovery that GSNO metabolism by GSNOR is dysregulated in the myometrium of women who experience sPTL provides novel insight into this disease state suggesting GSNOR inhibitors as a new class of tocolytics. GSNOR inhibitors not only function to increase endogenous levels of •NO elevating GSNO levels, but by extension, may also increase total SNOs in the cell.

14.7   AN NO-CGMP SIGNALING EXCEPTION

While relaxation of the human uterine muscle following the addition of •NO or NO donors is readily demonstrated in the tissue bath, blockade of the sGC fails to prevent the relaxation (Buxton, 2004; Buxton et al., 2010). Thus, while cGMP is formed and phosphorylation by its cognate kinase can be demonstrated (Tichenor et al., 2003), this is neither necessary nor sufficient for relaxation of the myometrium (Figure 14.6). This was not the failure of cGMP and PKG to activate myosin phosphatase since activation of the particulate guanylate cyclase by uroguanylin does relax the muscle (Buxton et al., 2010). The inevitable conclusion is that the action of global cGMP accumulation in the myometrium is compartmented and that relaxation to NO donors is not the result of the classical notions surrounding smooth muscle contraction–relaxation mechanisms. Compartmentation of the action of cyclic nucleotide in muscle at the cellular level is known, as first demonstrated in 1983 (Buxton and Brunton, 1983). If not cGMP accumulation and subsequent activation of the smooth muscle myosin phosphatase, then what mechanism underlies the ability of NO donors to relax the myometrium?

14.8   PROTEIN S-NITROSATION

If global cGMP accumulation fails to relax human myometrium, the obvious possibility that follows is that nitrosation of critical proteins involved in contraction–relaxation signaling might explain the effect. GSNO can transfer its •NO moiety to a cysteine thiol, resulting S-nitrosation/S-nitrosylation (Stamler et al., 1992). S-Nitrosation describes a thiol (e.g., cysteine) converted to a S-nitrosothiol (RSNO) by a one-electron oxidation from the •NO radical (Smith and Marletta, 2012). The term nitrosylation describes addition of an •NO group to a metal centered protein such as guanylate cyclase (Martínez-Ruiz and Lamas, 2004). Researchers have used both terms to describe •NO addition to a protein thiol. We employ S-nitrosation to refer to protein modifications on cysteine residues.

The posttranslational modification of proteins has long been recognized as a key regulator of cellular function. In recent years, it has been shown that S-nitrosation acts as an important mediator of disease states (Anand and Stamler, 2012). As with phosphorylation, S-nitrosation regulates cellular mechanisms and affects protein–protein interactions, and the emerging field of S-nitrosation, and its effects on protein function represents an exciting new branch of research. S-Nitrosation cannot occur

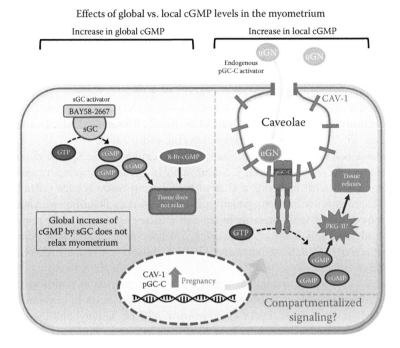

Effects of global vs. local cGMP levels in the myometrium

FIGURE 14.6   Global cGMP elevation does not relax myometrium while localized elevation does: Compartmented actions of cGMP in the myometrium. In the myometrium, the
cyclic nucleotide cGMP can be generated by either soluble guanylate cyclase (sGC), or by
particulate guanylate cyclase type C (pGC-C). A global increase in cGMP by sGC stimulation with BAY58–2667 does not relax the tissue. On the other hand, pGC-C activated by
uroguanylin (uGN) does relax pregnant myometrium. pGC-C is known to form complexes
in cholesterol-rich caveolae. Both pGC-c and CAV-1, and important structural component of
caveolae, are upregulated during pregnancy. It may be this localization with other mediators
of relaxation, such as PKG-II that promotes relaxation of the tissue.

without an available source of •NO (Martínez-Ruiz and Lamas, 2004). Reactive
NO is produced enzymatically in many cell types and will readily form GSNO.
S-Nitrosation occurs when an •NO moiety is covalently added to the thiol side chain
of cysteine residues within proteins and peptides (Hess et al., 2005).

   An analysis of the myometrial S-nitrosoproteome (the host of S-nitrosated proteins)
has revealed that several smooth muscle contractile-associated proteins (CAPS) are
differentially S-nitrosated based upon the state of labor in women (Ulrich et al., 2012).
Included in these S-nitrosated CAP proteins is the regulatory light chain of SMM,
called myosin regulatory light polypeptide 9 (LC20, RLC, or MYL9), as well as the
telokin domain of MLCK, an important domain that affects MYL9 phosphorylation
(pMYL9) through its binding to SMM (Silver et al., 1997). The phosphorylation
of MYL9 by MLCK is at the center of canonical contraction/relaxation pathways
in smooth muscle. The finding that these proteins are differentially S-nitrosated
in sPTL myometrium when exposed to GSNO, an endogenous •NO donor, makes

them an interesting target to study. Unfortunately, there is a dearth of published data describing the functional significance of protein *S*-nitrosation, specifically in regards to its effects on smooth muscle.

S-Nitrosation in the cell is highly dependent on the availability of intracellular GSNO (Smith and Marletta, 2012). As GSNO concentrations increase, so do levels of total protein *S*-nitrosation. Because *GSNOR$^{-/-}$* mice show increased cellular levels of GSNO and SNOs (Liu et al., 2004), it is likely that GSNO, and *S*-nitrosated proteins, are in equilibrium governed by Cys-to-Cys *trans*-nitrosation (Dalle-Donne et al., 2000), and GSNOR mediated denitrosation (Liu et al., 2001).

We have previously shown that many proteins critical to the uterine contraction/ relaxation cycle are differentially *S*-nitrosated in preterm laboring myometrium, as compared to term laboring, and term nonlaboring tissue lysates (Ulrich et al., 2013). There is a conspicuous paucity of functional data detailing whether or not the *S*-nitrosation of CAP proteins in smooth muscles are functionally relevant. To further advance this body of knowledge, we have investigated how actomyosin ATPase activity, MLCK activity, and TREK-1 channel activity are altered in the presence of GSNO.

Muscles are unable to contract without active actomyosin cross-bridge cycling. The actin motility assay measures the translocation of F-actin by myosin (Sellers, 2001), and serves as an indicator of actomyosin ATPase activity, and by extension, cross-bridge cycling. •NO donors have long been known to impact skeletal muscle systems with a general trend towards decreasing contractile force, sliding velocity, and ATPase activity (Bansbach and Guilford, 2016). In fact, *S*-nitrosation of the heavy chain of skeletal and cardiac myosin causes a decrease in actin velocity (Evangelista et al., 2010), and *S*-nitrosation of skeletal myosin affects the catalytic cycle, but does not alter actomyosin affinity (Nogueira et al., 2009). While the structure and function of smooth muscle myosin differs in several critical ways to skeletal and cardiac myosin, and thus prevent direct comparison of GSNO actions in the two systems, the actin monomers that interact with the myosin head are functionally identical between among muscle classes (Harris and Warshaw, 1993). To this point, relevant studies of skeletal SNO-actin have been shown to affect the rate of F-actin formation (Dalle-Donne et al., 2000), and in both skeletal and smooth muscle, actin can be *S*-nitrosated at two sites per monomer and this results in a decrease in sliding velocities by ~24% when using skeletal myosin (Bansbach and Guilford, 2016).

The actin motility assay reports ATPase activity, which is an important metric detailing the physical mechanism of force production (Le Clainche and Carlier, 2004). Upstream of the process, however, is the critical phosphorylation of MYL9. Phosphorylation of MYL9 by MLCK is at the crux of all contractile activity in smooth muscle. Beyond MYL9's phosphorylation, which initiates ATPase activity and cross-bridge cycling (Word, 1995), MYL9 is also required, from a structural sense, for folding into the 10S conformation (Katoh and Morita, 1996), and pMYL9 interaction with the essential light chain enhances lever action of the myosin head to allow for engagement with actin (Ni et al., 2012). There is little doubt that MYL9 is a fundamentally critical component to SMM structure and function. The obligate phosphorylation of MYL9 for cross-bridge cycling is at S19 (Colburn et al., 1988), but less is known about the relevance of PTMs more distal from the

N-terminus where that phosphorylation occurs. Interestingly, through the use of photo-crosslinking, it has been shown that C108, the sole cysteine in MYL9, is unshielded when in the un-phosphorylated state (Mazhari et al., 2004), intimating the possibility that C108 may be susceptible to *S*-nitrosation when the cell is relaxed. Work from our lab has shown that MYL9 is not *S*-nitrosated when the myometrium is quiescent prior to labor (TNL) and that it achieves the highest state of *S*-nitrosation during spontaneous PTL (Ulrich et al., 2012). At first glance, it may appear counter-intuitive that GSNO, an agent known to promote relaxation, increases the rate of MYL9 phosphorylation. However, when we consider that sPTL myometrium is functionally a "disease" state and that MYL9 is highly *S*-nitrosated during sPTL is over term laboring and term nonlaboring myometrium, it is easy to entertain the possibility that *S*-nitrosation of MYL9 is promoting contraction through increased kinase activity. What remains unclear is why MYL9 in sPTL fundamentally more permissive to *S*-nitrosation.

Phosphorylation of MYL9 cannot occur unless MLCK is activated. MLCK activation requires stimulation by calmodulin, which in turn requires high cytosolic Ca^{2+}. That influx of Ca^{2+}, spurred by calcium-induced calcium release, does not occur until the membrane is depolarized. One of the most obvious distinguishing characteristics of the myometrium, when compared to other types of smooth muscle, is that it must remain largely quiescent over the 40 weeks of gestation. The myometrium employs many mechanisms to achieve this end, from oxytocin receptor modulation (Takemura et al., 1996) a shift in progesterone receptor ratios to decreased CAP protein production (Tan et al., 2012).

One such mechanism is the increase in the expression of a stretch-activated potassium channel, TREK-1 (Tichenor et al., 2005), which helps to maintain a negative cell membrane potential by pumping potassium out of the cell (Figure 14.7). This action prevents an influx of Ca^{2+}. Our laboratory has previously determined that splice variants of TREK-1 affect trafficking of full-length functional TREK-1 in sPTL (Cowles et al., 2015), and computational analysis of TREK-1 using the program GPS-SNO (Xue et al., 2010), suggests a high likelihood that TREK-1 may be *S*-nitrosated at C414 on the C-terminus, in the same region as other key PTMs that alter TREK-1 function by enhancing membrane hyperpolarization. Future experiments such as those that test the effect of mutation of C414 to determine if GSNO still activates the channel can help to determine the contribution of TREK-1 to myometrial quiescence.

14.9 SUMMARY

We have added to the inevitable conclusion highlighted in this volume that glutathione is a pleiotropic molecule that has a remarkable spectrum of control of mammalian biology. In its role as GSNO, it offers a mechanistic understanding of disease, and its metabolism provides targets for drug development. While the underlying causes of sPTL remain largely unknown, we here present at least a novel mechanism that sheds light on the relaxation of the myometrium through protein *S*-nitrosation. Our data have revealed that total protein *S*-nitrosations in sPTL myometrium are decreased relative to term tissue, which serves as an indicator of

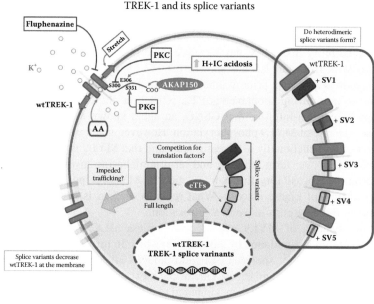

FIGURE 14.7 TREK-1 splice variants alter its trafficking and function: Native TREK-1 assembles as a homodimer and acts as an outward rectifying potassium channel, which in turn hyperpolarizes the membrane, maintaining quiescence during pregnancy. There are five known splice variants of TREK-1 that decrease full-length TREK-1's trafficking to the membrane, resulting in decreased TREK-1 currents.

•NO availability, and implicates sPTL as a "disease state" (Barnett et al., 2018). In an effort to elucidate the functional implications of protein S-nitrosation, we determined that actomyosin ATPase activity is retarded in the presence of GSNO, MLCK activity is enhanced, and that the outward rectifying K+ channel, TREK-1, exhibits an increase in current. Our finding that GSNOR and by extension protein S-nitrosation are dysregulated in the myometrium of women undergoing sPTL affords an opportunity to investigate a new class of drugs that increase the availability of endogenous •NO in the cell.

REFERENCES

Anand, P., and J. S. Stamler. 2012. Enzymatic mechanisms regulating protein S-nitrosylation: Implications in health and disease. *J Mol Med (Berl)* 90:233–44.

Bansbach, H. M., and W. H. Guilford. 2016. Actin nitrosylation and its effect on myosin driven motility. *AIMS Mol Sci* 3:426–38.

Barnes, P., and S. Liu. 1995. Regulation of pulmonary vascular tone. *Pharmacol Rev* 47:87–131.

Barnett, S. D., and I. L. O. Buxton. 2017. The role of S-nitrosoglutathione reductase (GSNOR) in human disease and therapy. *Crit Rev Biochem Mol Biol* 52:340–54.

Barnett, S. D., C. R. Smith, C. C. Ulrich et al. 2018. S-Nitrosoglutathione reductase underlies the dysfunctional relaxation to nitric oxide in preterm labor. *Nat-Sci Rep*, in press.

Bateman, R. L., D. Rauh, B. Tavshanjian, and K. M. Shokat. 2008. Human carbonyl reductase 1 is an *S*-nitrosoglutathione reductase. *J Biol Chem* 283:35756–62.

Baumann, B. A. J., D. W. Taylor, Z. Huang et al. 2012. Phosphorylated smooth muscle heavy meromyosin shows an open conformation linked to activation. *J Mol Biol* 415:274–87.

Beckman, J. S., and W. H. Koppenol. 1996. Nitric oxide, superoxide, and peroxynitrite: The good, the bad, and ugly. *Am J Physiol* 271:C1424–37.

Behrman, R. E., and A. S. Butler. 2007. *Preterm Birth: Causes, Consequences, and Prevention*, 772. Washington, DC: National Academies Press.

Bhattacharya, S., E. Amalraj Raja, E. Ruiz Mirazo et al. 2010. Inherited predisposition to spontaneous preterm delivery. *Obstet Gynecol* 115:1125–33.

Bonaventura, C., G. Ferruzzi, R. D. Stevens, and S. Tesh. 1999. Effects of *S*-nitrosation on oxygen binding by normal and sickle cell hemoglobin. *J Biol Chem* 274:24742–8.

Bonaventura, C., G. Godette, G. Ferruzzi et al. 2002. Responses of normal and sickle cell hemoglobin to *S*-nitroscysteine: Implications for therapeutic applications of NO in treatment of sickle cell disease. *Biophys Chem* 98:165–81.

Bradley, K. K., I. L. Buxton, J. E. Barber et al. 1998. Nitric oxide relaxes human myometrium by a cGMP-independent mechanism. *Am J Physiol* 275:C1668–73.

Bredt, D. S., and S. H. Snyder. 1992. Nitric oxide: A novel neuronal messenger. *Neuron* 8:3–11.

Broniowska, K. A., A. R. Diers, and N. Hogg. 2013. *S*-Nitrosoglutathione. *Biochim Biophys Acta* 1830:3173–81.

Broniowska, K. A., and N. Hogg. 2012. The chemical biology of *S*-nitrosothiols. *Antioxid Redox Signal* 17:969–80.

Bursztyn, L., O. Eytan, A. J. Jaffa, and D. Elad. 2007. Modeling myometrial smooth muscle contraction. *Ann NY Acad Sci* 1101:110–38.

Buxton, I. L. O. 2004. Regulation of uterine function: A biochemical conundrum in the regulation of smooth muscle relaxation. *Mol Pharmacol* 65:1051–9.

Buxton, I. L. O., and L. L. Brunton. 1983. Compartments of cyclic AMP and protein kinase in mammalian cardiomyocytes. *J Biol Chem* 258:10233–9.

Buxton, I. L. O., R. A. Kaiser, N. A. Malmquist, and S. Tichenor. 2001. NO-induced relaxation of labouring and non-labouring human myometrium is not mediated by cyclic GMP. *Br J Pharmacol* 134:206–14.

Buxton, I. L. O., D. Milton, S. D. Barnett, and S. D. Tichenor. 2010. Agonist-specific compartmentation of cGMP action in myometrium. *JPET* 335:256–63.

Carvalho-Filho, M. A., M. Ueno, S. M. Hirabara et al. 2005. *S*-Nitrosation of the insulin receptor, insulin receptor substrate 1, and protein kinase B/Akt: A novel mechanism of insulin resistance. *Diabetes* 54:959–67.

CDC. 2015. 2015 Period linked birth/infant death. Paper read at CDC.

Chen, Y.-J., W.-C. Ching, Y.-P. Lin, and Y.-J. Chen. 2013. Methods for detection and characterization of protein *S*-nitrosylation. *Methods* 62:138–50.

Choudhry, S., L. G. Que, Z. Yang et al. 2010. GSNO reductase and β2-adrenergic receptor gene–gene interaction: Bronchodilator responsiveness to albuterol. *Pharmacogenet Genom* 20:351–58.

Colburn, J. C., C. H. Michnoff, L. C. Hsu et al. 1988. Sites phosphorylated in myosin light chain in contracting smooth muscle. *J Biol Chem* 263:19166–73.

Corti, A., M. Franzini, I. Scataglini, and A. Pompella. 2014. Mechanisms and targets of the modulatory action of *S*-nitrosoglutathione (GSNO) on inflammatory cytokines expression. *Arch Biochem Biophys* 562:80–91.

Cowles, C. L., H. R. Burkin, C. Ulrich et al. 2016. Transport of nitric oxide in human uterin myocytes. *Reprod Sci* 23:216a.

Cowles, C. L., Y.-Y. Wu, S. D. Barnett et al. 2015. Alternatively spliced human TREK-1 variants alter TREK-1 channel function and localization. *Biol Reprod* 93:122.

Culhane, J. F., and R. L. Goldenberg. 2011. Racial disparities in preterm birth. *Semin Perinatol* 35:234–39.

D'Onofrio, B. M., Q. A. Class, M. E. Rickert et al. 2013. Preterm birth and mortality and morbidity: A population-based quasi-experimental study. *JAMA Psychiatry* 70:1231–40.

Dalle-Donne, I., A. Milzani, D. Giustarini et al. 2000. S-NO-actin: S-nitrosylation kinetics and the effect on isolated vascular smooth muscle. *J Muscle Res Cell Motil* 21:171–81.

Devarie-Baez, N. O., D. Zhang, S. Li et al. 2013. Direct methods for detection of protein S-nitrosylation. *Methods* 62:171–6.

Dodd, J. M., C. a. Crowther, M. R. Dare, and P. Middleton. 2006. Oral betamimetics for maintenance therapy after threatened preterm labour. *Cochrane Database Syst Rev* (Online):CD003927.

Doucette, J. T., and M. B. Bracken. 1993. Possible role of asthma in the risk of preterm labor and delivery. *Epidemiology* 4:143–50.

Duckitt, K., S. Thornton, O. P. O'Donovan, and T. Dowswell. 2014. Nitric oxide donors for treating preterm labour. *Cochrane Database Syst Rev* 5:CD002860.

Elvira, E. O. G. van Vliet, E. M. Boormans et al. 2014. Preterm labor: Current pharmacotherapy options for tocolysis. *Expert Opin Pharmacother* 15:787–97.

Evangelista, A. M., V. S. Rao, A. R. Filo et al. 2010. Direct regulation of striated muscle myosins by nitric oxide and endogenous nitrosothiols. *PLOS ONE* 5:e11209.

Fauconnier, J., J. Thireau, S. Reiken et al. 2010. Leaky RyR2 trigger ventricular arrhythmias in Duchenne muscular dystrophy. *Proc Natl Acad Sci U S A* 107:1559–64.

Ferrini, M. E., B. J. Simons, D. J. P. Bassett et al. 2013. S-Nitrosoglutathione reductase inhibition regulates allergen-induced lung inflammation and airway hyperreactivity. *PLOS ONE* 8:e70351.

Foster, M. W., D. T. Hess, and J. S. Stamler. 2009. Protein S-nitrosylation in health and disease: A current perspective. *Trends Mol Med* 15:391–404.

Frearson, N., B. W. W. Focant, and S. V. Perry. 1976. Phosphorylation of a light chain component of myosin from smooth muscle. *FEBS Lett* 63:27–32.

Furchgott, R., L. Ignarro, and F. Murad. 1998. The 1998 Nobel Prize in Physiology or Medicine. www.nobel.se/medicine/laureates/1998/index.html

Furchgott, R. F. 1993. The discovery of endothelium-dependent relaxation. *Circulation* 87 Suppl.:V3–V8.

Furchgott, R. F., and J. V. Zawadzki. 1980. The obligatory role of endothelial cells in the relaxation of arterial smooth muscle by acetylcholine. *Nature* 288:373–6.

Gaston, B., J. Reilly, J. M. Drazen et al. 1993. Endogenous nitrogen oxides and bronchodilator S-nitrosothiols in human airways. *Proc Natl Acad Sci U S A* 90:10957–61.

Giles, W., and A. Bisits. 2007. The present and future of tocolysis. *Best Pract Res Clin Obstet Gynaecol* 21:857–68.

Gimpl, G., and F. Fahrenholz. 2001. The oxytocin receptor system: Structure, function, and regulation. *Physiol Rev* 81:629–83.

Giustarini, D., A. Milzani, R. Colombo et al. 2003. Nitric oxide and S-nitrosothiols in human blood. *Clin Chim Acta* 330:85–98.

Glass, H. C., A. T. Costarino, S. A. Stayer et al. 2015. Outcomes for extremely premature infants. *Anesth Analg* 120:1337–51.

Goldenberg, R. L., J. F. Culhane, J. D. Iams, and R. Romero. 2008. Epidemiology and causes of preterm birth. *Lancet* 371:75–84.

Hamilton, B. E., J. A. Martin, M. J. K. Osterman et al. 2015. National vital statistics reports births: Final data for 2014. *Natl Vital Stat Rep* 64:1–104.

Harris, D. E., and D. M. Warshaw. 1993. Smooth and skeletal muscle actin are mechanically indistinguishable in the in vitro motility assay. *Circ Res* 72:219–24.

Henderson, E. M., and B. Gaston. 2005. SNOR and wheeze: The asthma enzyme? *Trends Mol Med* 11:481–4.

Hess, D. T., A. Matsumoto, S.-O. Kim et al. 2005. Protein S-nitrosylation: Purview and parameters. *Nat Rev Mol Cell Biol* 6:150–66.

Hong, F., B. D. Haldeman, D. Jackson et al. 2011. Biochemistry of smooth muscle myosin light chain kinase. *Arch Biochem Biophys* 510:135–46.

Howson, C. P., M. V. Kinney, L. McDougall, and J. E. Lawn. 2013. Born too soon: Preterm birth matters. *Reprod Health*.

Hudson, C. A., K. J. Heesom, and A. Lopez Bernal. 2012. Phasic contractions of isolated human myometrium are associated with Rho-kinase (ROCK)-dependent phosphorylation of myosin phosphatase-targeting subunit (MYPT1). *Mol Hum Reprod* 18:265–79.

Ito, M., T. Nakano, F. Erdodi, and D. J. Hartshorne. 2004. Myosin phosphatase: Structure, regulation and function. *Mol Cell Biochem* 259:197–209.

Iyera, A. K. V., Y. Rojansakulb, and N. Azada. 2011. Nitrosothiol signaling and protein nitrosation in cell death. *Biol Bull* 221:18–34.

Jaffrey, S. R., and S. H. Snyder. 2001. The biotin switch method for the detection of *S*-nitrosylated proteins. *Sci STKE*. 2001:L1.

Jelski, W., K. Orywal, B. Panek et al. 2009. The activity of class I, II, III and IV of alcohol dehydrogenase (ADH) isoenzymes and aldehyde dehydrogenase (ALDH) in the wall of abdominal aortic aneurysms. *Exp Mol Pathol* 87:59–62.

Jelski, W., and M. Szmitkowski. 2008. Alcohol dehydrogenase (ADH) and aldehyde dehydrogenase (ALDH) in the cancer diseases. *Clin Chim Acta* 395:1–5.

Katoh, T., and F. Morita. 1996. Roles of light chains in the activity and conformation of smooth muscle myosin. *J Biol Chem* 271:9992–6.

Keelan, J. A., M. Blumenstein, R. J. A. Helliwell et al. 2003. Cytokines, prostaglandins and parturition—A review. *Placenta Trophoblast Res* 24:33–46.

Khromov, A. S. A. S., H. H. Wang, N. N. Choudhury et al. 2006. Smooth muscle of telokin-deficient mice exhibits increased sensitivity to Ca^{2+} and decreased cGMP-induced relaxation. *Proc Natl Acad Sci U S A* 103:2440–5.

Kim, S. F. 2011. The role of nitric oxide in prostaglandin biology; update. *Nitric Oxide* 25:255–64.

Kimura, T., O. Tanizawa, K. Mori et al. 1992. Structure and expression of a human oxytocin receptor. *Nature* 356:526–9.

Kinney, M. V., K. J. Kerber, R. E. Black et al. 2010. Sub-Saharan Africa's mothers, newborns, and children: Where and why do they die? *PLoS Med* 7:e1000294.

Komatsu, S., K. Miyazaki, R. A. Tuft, and M. Ikebe. 2002. Translocation of telokin by cGMP signaling in smooth muscle cells. *Am J Physiol Cell Physiol* 283:C752–61.

Kuenzli, K. A., M. E. Bradley, and I. L. Buxton. 1996. Cyclic GMP-independent effects of nitric oxide on guinea-pig uterine contractility. *Br J Pharmacol* 119:737–43.

Kunze, W. A. A., and J. B. Furness. 1999. The enteric nervous system and regulation of intestinal motility. *Ann Rev Physiol* 61:117–42.

Lancaster, J. R. 2008. Protein cysteine thiol nitrosation: Maker or marker of reactive nitrogen species-induced nonerythroid cellular signaling? *Nitric Oxide—Biol Chem* 19:68–72.

Laniewska-Dunaj, M., W. Jelski, K. Orywal et al. 2013. The activity of class I, II, III and IV of alcohol dehydrogenase (ADH) isoenzymes and aldehyde dehydrogenase (ALDH) in brain cancer. *Neurochemical Res* 38:1517–21.

Le Clainche, C., and M. F. Carlier. 2004. Actin-based motility assay. *Curr Protoc Stem Cell Biol* 24 (1):12.7.1–20.

Liu, L., A. Hausladen, M. Zeng et al. 2001. A metabolic enzyme for S-nitrosothiol conserved from bacteria to humans. *Nature* 410:490–4.

Liu, L., Y. Yan, M. Zeng et al. 2004. Essential roles of S-nitrosothiols in vascular homeostasis and endotoxic shock. *Cell* 116:617–28.

MacMicking, J., Q. Xie, and C. Nathan. 1997. Nitric oxide and macrophage function. *Ann Rev Immunol* 15:323–50.

Martin, J. A., B. E. Hamilton, S. J. Ventura et al. 2011. Births: Final data for 2011. *Natl Vital Stat Rep* 62:1–69.

Martínez-Ruiz, A., and S. Lamas. 2004. *S*-Nitrosylation: A potential new paradigm in signal transduction. *Cardiovascular Res* 62:43–52.

Mazhari, S. M., C. T. Selser, and C. R. Cremo. 2004. Novel sensors of the regulatory switch on the regulatory light chain of smooth muscle myosin. *J Biol Chem* 279:39905–14.

Meister, A., and M. E. Anderson. 1983. Glutathione. *Ann Rev Biochem* 52 (1):711–60.

Methner, C., E. T. Chouchani, G. Buonincontri et al. 2014. Mitochondria selective *S*-nitrosation by mitochondria-targeted S-nitrosothiol protects against post-infarct heart failure in mouse hearts. *Eur J Heart Fail* 16:712–7.

Miniño, A. M., M. P. Heron, and B. L. Smith. 2006. Deaths: Preliminary data for 2006. *National Vital Statistics Reports* 54:1–49.

Moncada, S., and A. Higgs 1993. The L-arginine-nitric oxide pathway. *New Engl J Med* 329 (27):2002–12.

Moncada, S., R. M. Palmer, and E. A. Higgs. 1991. Nitric oxide: Physiology, pathophysiology, and pharmacology. *Pharmacol Rev* 43:109–42.

Moore, P. E., K. K. Ryckman, S. M. Williams et al. 2009. Genetic variants of GSNOR and ADRB2 influence response to albuterol in African-American children with severe asthma. *Pediatr Pulmonol* 44:649–54.

Murray, C. I., H. Uhrigshardt, R. N. O'Meally et al. 2012. Identification and quantification of *S*-nitrosylation by cysteine reactive tandem mass tag switch assay. *Mol Cell Proteomics* 11:M111 013441.

Nankali, A., P. K. Jamshidi, and M. Rezaei. 2014. The effects of glyceryl trinitrate patch on the treatment of preterm labor: A single-blind randomized clinical trial. *J Reprod Infertil* 15:71–7.

Ni, S., F. Hong, B. D. Haldeman et al. 2012. Modification of interface between regulatory and essential light chains hampers phosphorylation-dependent activation of smooth muscle myosin. *J Biol Chem* 287:22068–79.

Nivalis Therapeutics, I. 2014. Safety and Pharmacokinetic Study of N6022 in Subjects With Cystic Fibrosis Homozygous for the F508del-CFTR Mutation. Available from https://clinicaltrials.gov/ct2/show/NCT01746784.

Nogueira, L., C. Figueiredo Freitas, G. Casimiro Lopes et al. 2009. Myosin is reversibly inhibited by S-nitrosylation. *Biochem J* 424:221–31.

Pan, D. 2010. The hippo signaling pathway in development and cancer. *Developmental Cell* 19:491–505.

Papatsonis, D., V. Flenady, and S. Cole. 2005. Oxytocin receptor antagonists for inhibiting preterm labour. *Cochrane Database Syst Rev* 2005(3):CD004452.

Prince, A. L., K. M. Antony, D. M. Chu, and K. M. Aagaard. 2014. The microbiome, parturition, and timing of birth: More questions than answers. *J Rep Immunol* 104–105:12–9.

Puetz, S., L. T. Lubomirov, and G. Pfitzer. 2009. Regulation of smooth muscle contraction by small GTPases. *Physiology* 24:342–56.

Que, L. G., Z. Yang, J. S. Stamler et al. 2009. *S*-Nitrosoglutathione reductase. *American J Resp Critical Care Med* 180:226–31.

Radi, R., J. S. Beckman, K. M. Bush, and B. A. Freeman. 1991. Peroxynitrite oxidation of sulfhydryls. The cytotoxic potential of superoxide and nitric oxide. *J Biol Chem* 266:4244–50.

Ricciardolo, F. L. M., P. J. Sterk, B. Gaston, and G. Folkerts. 2004. Nitric oxide in health and disease of the respiratory system. *Physiol Rev* 84:731–65.

Roberts, D., J. Brown, N. Medley, and S. Dalziel. 2017. Antenatal corticosteroids for accelerating fetal lung maturation for women at risk of preterm birth. *Cochrane Database Syst Rev* 6 (1):1–84.

Rundell, K., and B. Panchal. 2017. Preterm labor: Prevention and management. *Am Fam Physician* 95 (6):366–72.

Saigal, S., and L. W. Doyle. 2008. An overview of mortality and sequelae of preterm birth from infancy to adulthood. *Lancet* 371:261–9.

Sahlin, L., H. Wang, B. Lindblom, H. Eriksson, A. Holmgren, A. Blanck. 2000. Thioredoxin expression in human myometrium and fibroids. *Mol Hum Reprod* 6 (1):60–7.

Salvemini, D., T. P. Misko, J. L. Masferrer et al. 1993. Nitric oxide activates cyclooxygenase enzymes. *Proc Natl Acad Sci U S A* 90:7240–4.

Sellers, J. R. 2001. In vitro motility assay to study translocation of actin by myosin. *Curr Protoc Cell Biol* Chapter 13: Unit 13.2.

Silver, D. L., A. V. Vorotnikov, D. M. Watterson et al. 1997. Sites of interaction between kinase-related protein and smooth muscle myosin. *J Biol Chem* 272:25353–9.

Smith, A. F., R. M. Bigsby, R. A. Word, and B. P. Herring. 1998. A 310-bp minimal promoter mediates smooth muscle cell-specific expression of telokin. *Am J Physiol* 274:C1188–95.

Smith, B. C., and M. A. Marletta. 2012. Mechanisms of *S*-nitrosothiol formation and selectivity in nitric oxide signaling. *Curr Opin Chem Biol* 16:498–506.

Smith, G. N., M. C. Walker, A. Ohlsson et al. 2007. Randomized double-blind placebo-controlled trial of transdermal nitroglycerin for preterm labor. *Am J Obstet Gynecol* 196:37.e1–e8.

Somlyo, A. P., and A. V. Somlyo. 2003. Ca^{2+} sensitivity of smooth muscle and nonmuscle myosin II: Modulated by G proteins, kinases, and myosin phosphatase. *Physiol Rev* 83:1325–58.

Staab, C. A., J. Ålander, M. Brandt et al. 2008a. Reduction of *S*-nitrosoglutathione by alcohol dehydrogenase 3 is facilitated by substrate alcohols via direct cofactor recycling and leads to GSH-controlled formation of glutathione transferase inhibitors. *Biochem J* 413:493–504.

Staab, C. A., M. Hellgren, and J. O. Höög. 2008b. Dual functions of alcohol dehydrogenase 3: Implications with focus on formaldehyde dehydrogenase and *S*-nitrosoglutathione reductase activities. *Cell Mol Life Sci* 65:3950–60.

Stamler, J. S., D. I. Simon, J. A. Osborne et al. 1992. *S*-Nitrosylation of proteins with nitric oxide: Synthesis and characterization of biologically active compounds. *Proc Natl Acad Sci U S A* 89:444–8.

Sun, X., J. Qiu, S. a. Strong et al. 2011. Discovery of potent and novel *S*-Nitrosoglutathione reductase inhibitors devoid of cytochrome P450 activities. *Bioorg Med Chem Lett* 21:5849–53.

Szal, S. E., J. T. Repke, E. W. Seely et al. 1994. $[Ca^{2+}]i$ signaling in pregnant human myometrium. *Am J Physiol Endocrinol Metabol* 267:E77–87.

Takemura, M., S. Nomura, T. Nobunaga et al. 1996. Expression of oxytocin receptor in human pregnant myometrium. *Endocrinology* 137:780–5.

Tamargo, J., R. Caballero, R. Gomez, and E. Delpon. 2010. Cardiac electrophysiological effects of nitric oxide. *Cardiovasc Res* 87:593–600.

Tan, H., L. Yi, N. S. Rote et al. 2012. Progesterone receptor-A and -B have opposite effects on proinflammatory gene expression in human myometrial cells: Implications for progesterone actions in human pregnancy and parturition. *J Clin Endocrinol Metabol* 97:E719–30.

Thomas, D. D., and D. Jourd'heuil. 2012. S-Nitrosation: Current concepts and new developments. *Antioxid Redox Signal* 17:934–6.

Tichenor, J. N., E. T. Hansen, and I. L. Buxton. 2005. Expression of stretch-activated potassium channels in human myometrium. *Proc West Pharmacol Soc* 48:44–8.

Tichenor, S. D., N. A. Malmquist, and I. L. O. Buxton. 2003. Dissociation of cGMP accumulation and relaxation in myometrial smooth muscle: Effects of *S*-nitroso-*N*-acetylpenicillamine and 3-morpholinosyndonimine. *Cell Signal* 15:763–72.

Tomita, R., S. Fujisaki, T. Ikeda, and M. Fukuzawa. 2002. Role of nitric oxide in the colon of patients with slow-transit constipation. *Dis Colon Rectum* 45:593–600.

Ulrich, C., D. R. Quillici, K. Schegg et al. 2012. Uterine smooth muscle *S*-nitrosylproteome in pregnancy. *Mol Pharmacol* 81:143–53.

Ulrich, C., D. R. Quilici, K. A. Schlauch, and I. L. O. Buxton. 2013. The human uterine smooth muscle *S*-nitrosoproteome fingerprint in pregnancy, labor, and preterm labor. *Am J Physiol Cell Physiol* 305:C803–16.

Vinturache, A. E., C. Gyamfi-Bannerman, J. Hwang et al. 2016. Maternal microbiome—A pathway to preterm birth. *Semin Fetal Neonatal Med* 21:94–9.

Weismiller, D. G. 2000. Preterm labor. *Am Fam Physician* 59:593–602.

Whitworth, M., and S. Quenby. 2008. Prophylactic oral betamimetics for preventing preterm labour in singleton pregnancies. *Cochrane Database Syst Rev*, Art. No. CD006395.

Word, R. A. 1995. Myosin phosphorylation and the control of myometrial contraction/relaxation. *Semin Perinatol* 19 (1):3–14.

Wray, S. S. 1993. Uterine contraction and physiological mechanisms of modulation. *Am J Physiol* 264:C1–18.

Wu, H., I. Romieu, B. E. Rio-Navarro et al. 2007. Genetic variation in *S*-nitrosoglutathione reductase (GSNOR) and childhood asthma. *J Allergy Clin Immunol* 120:322–8.

Xue, Y., Z. Liu, X. Gao et al. 2010. GPS-SNO: Computational prediction of protein *S*-nitrosylation sites with a modified GPS algorithm. *PLoS One* 5:e11290.

Yuen, S., O. Ogut, and F. V. Brozovich. 2011. MYPT1 protein isoforms are differentially phosphorylated by protein kinase G. *J Biol Chem* 286:37274–9.

Zhang, G., B. Feenstra, J. Bacelis et al. 2017. Genetic associations with gestational duration and spontaneous preterm birth. *New Engl J Med* 377:1156–67. doi:10.1056/NEJMoa1612665.

Zhang, Y., and N. Hogg. 2004. The mechanism of transmembrane *S*-nitrosothiol transport. *Proc Natl Acad Sci U S A* 101:7891–6.

Part IV

The Glutaredoxins

15 The Catalytic Mechanism of Glutaredoxins

Linda Liedgens and Marcel Deponte
Technische Universität Kaiserslautern

CONTENTS

15.1 INTRODUCTION

Glutaredoxins (Grx) exert crucial functions in pro- and eukaryotes. These functions range from redox catalysis and iron-sensing to the biosynthesis of iron–sulfur clusters (Herrero et al., 2007; Mieyal et al., 2008; Outten et al., 2013; Couturier et al., 2015; Berndt et al., 2017; Deponte, 2017). Structurally, Grx are members of the thioredoxin (Trx) superfamily with a reactive cysteine residue at the N-terminus of helix $\alpha2$ within their Trx-fold. In contrast to Trx, Grx share a preference for reduced glutathione (GSH), which can either serve as a reducing agent or as a ligand for iron–sulfur clusters (Deponte, 2013; Couturier et al., 2015; Berndt et al., 2017). The chemistry of the Grx active site cysteine residue is extremely versatile and depends on the protein isoform. For example, in addition to its reduced thiolate state, Grx-S$^-$, the cysteine residue might form an intramolecular disulfide bond, Grx(S$_2$), or undergo glutathionylation, Grx-SSG (i.e., form a mixed disulfide bond with glutathione). Furthermore, the cysteine residue might react with another protein or low-molecular weight (LMW) compound resulting in the formation of a nonglutathione mixed disulfide, Grx-SSR (Lillig et al., 2008; Deponte, 2013; Begas et al., 2015; Begas et al., 2017). In this chapter, we will first review the classification of Grx and introduce the diversity of Grx substrates and enzyme assays followed by a discussion of the enzymatic mechanism of Grx with a focus on recent discoveries.

DOI: 10.1201/9781351261760-19

15.2 CLASSIFICATION OF ENZYMATICALLY ACTIVE AND INACTIVE GRX

As outlined in Chapter 16, multiple different Grx-isoforms are usually found in a bacterial or eukaryotic organism. One way to classify the extremely diverse Grx family is to group Grx-isoforms according to their sequence similarity and their enzymatic activity in standard *in vitro* assays (described in Section 15.3). Based on this classification, most isoforms can be grouped into two major subfamilies: (i) The first Grx subfamily comprises the so-called dithiol Grx, which predominantly share a CPYC-motif at the active site and are sometimes referred to as class I Grx (Lillig et al., 2008; Couturier et al., 2015). This subfamily comprises the canonical Grx-isoforms that were originally characterized from mammals, *Saccharomyces cerevisiae*, and *Escherichia coli* and that are enzymatically active in standard *in vitro* assays as reviewed previously (Deponte, 2013). However, not all homologues of the class I subfamily have a second cysteine residue. For example, the enzymatically active homologues *Sc*Grx6 and *Sc*Grx7 from baker's yeast have a CSYS- and CPYS-motif, respectively (Mesecke et al., 2008a; Izquierdo et al., 2008; Mesecke et al., 2008b). (ii) The second Grx subfamily comprises the so-called monothiol Grx, which predominantly share a CGFS-motif and are sometimes referred to as Grx-like proteins or class II Grx. These proteins bind iron–sulfur clusters, play a role in iron metabolism and can be fused to other Grx- or Trx-domains (Herrero et al., 2007; Lillig et al., 2008; Outten et al., 2013; Couturier et al., 2015). In contrast to class I homologues, class II Grx are enzymatically inactive in standard *in vitro* assays (Tamarit et al., 2003; Deponte et al., 2005; Fernandes et al., 2005; Filser et al., 2008; Zaffagnini et al., 2012).

Some Grx belong to neither the class I nor the class II subfamily. For example, land plants harbor one or several Grx-isoforms with a CC-motif (Gutsche et al., 2015). To the best of our knowledge, these proteins remain to be tested in standard enzyme kinetic assays. Furthermore, there are Grx-like proteins that cannot be easily classified as Grx or Trx homologues, because the boundaries between these and other related protein families are sometimes blurred. For example, *Sc*Grx8 from yeast has a Grx-fold but an unusual Trp14-like WCPDC-motif resulting in a hybrid protein that is only poorly active in standard *in vitro* assays (Eckers et al., 2009; Tang et al., 2014). In summary, Grx can be grouped into two major subfamilies as well as smaller subfamilies, some of which include hybrid proteins. An enzymatic activity in standard *in vitro* assays has so far only been detected for homologues of the dithiol Grx-isoforms.

15.3 GLUTAREDOXIN SUBSTRATES AND ENZYMATIC ASSAYS

There are numerous *in vitro* assays for the analysis of Grx catalysis. These assays are predominantly based on the turnover of model LMW disulfide substrates as well as physiological protein disulfides. The two most commonly used thiol:disulfide oxidoreductase assays measure the GSH-dependent reduction of the glutathionylated model substrate L-cysteine-glutathione disulfide (GSSCys) or the nonglutathione model substrate bis(2-hydroxyethyl)disulfide (HEDS) (Nagai et al., 1968; Mieyal et al.,

1991; Gravina et al., 1993; Holmgren et al., 1995; Mesecke et al., 2008a; Eckers et al., 2009; Begas et al., 2015). However, Grx can also reduce selected oxidized proteins, including glutathionylated model proteins such as bovine serum albumin (Gravina et al., 1993; Gallogly et al., 2008) as well as physiological substrates such as purified ribonucleotide reductase (Holmgren, 1979; Holmgren et al., 1995), phosphoadenylyl-sulfate reductase (Lillig et al., 1999; Lillig et al., 2008; Berndt et al., 2015), and some thiol-dependent peroxidases (Rouhier et al., 2001; Pauwels et al., 2003; Hanschmann et al., 2010; Djuika et al., 2013; Staudacher et al., 2015; Peskin et al., 2016). As reviewed previously, the preferences and reactivities of Grx/substrate couples and thiol–disulfide exchange reactions are determined by the geometric and electrostatic complementarity of the interaction partners (Deponte, 2013; Deponte et al., 2015), as demonstrated for ribonucleotide reductase (Berardi et al., 1999) and phosphoadeny-lylsulfate reductase (Berndt et al., 2015). The complementarity ensures an efficient substrate recruitment, which can be also enhanced by protein tethering or fusion, as shown for a peroxiredoxin from *Haemophilus influenza* (Kim et al., 2003) and redox-sensitive fluorescent proteins (Bjornberg et al., 2006; Gutscher et al., 2008; Morgan et al., 2013).

All of the mentioned redox reactions can be analyzed in GSH-dependent coupled enzymatic assays yielding glutathione disulfide (GSSG) as a product, which is sub-sequently converted by a detection system that consists of glutathione reductase and NADPH. Regardless of the assay, care has to be taken to ensure that the detection system does not become rate-limiting and that the background consumption of NADPH is subtracted. Furthermore, it is advisable to include negative controls by systematically omitting one of the assay components or to perform a mock assay with a Grx mutant that lacks the active site cysteine residue. It may also be revealing to analyze the effects when the assay is started either by the addition of enzyme or by the addition of one of the substrates (Nagai et al., 1968; Mieyal et al., 1991; Holmgren et al., 1995; Mesecke et al., 2008a; Eckers et al., 2009; Djuika et al., 2013; Begas et al., 2015).

Less commonly used Grx assays rely on radiolabeling or high-performance liquid chromatography (HPLC) measurements in order to directly monitor the consumption of substrates or the formation of products at defined time points (Gravina et al., 1993; Srinivasan et al., 1997; Gallogly et al., 2008; Lillig et al., 2008). The advantage of such direct assays is that they do not require an enzymatic detection system with its inherent drawbacks. Furthermore, the reverse reactions, i.e., the Grx-catalyzed oxidation of reduced protein substrates by GSSG, can be also monitored (Starke et al., 2003; Gallogly et al., 2008), which is impossible in the presence of gluta-thione reductase and NADPH. However, in contrast to coupled enzymatic assays, radiolabel- or HPLC-dependent assays are discontinuous and far more labor-intensive and time-consuming.

Another option for monitoring the oxidation- or reduction kinetics of Grx is to genetically fuse a Grx of interest to a redox-sensitive fluorescent protein. Such recombinant constructs have an intrinsic fluorescent reporter that allows for real-time monitoring of the redox state of the attached Grx, for example, after mixing the protein with oxidants or reductants (Bjornberg et al., 2006; Gutscher et al., 2008; Begas et al., 2017). The drawback of the fusion constructs is that their fluorescence cannot be used in steady-state kinetic assays. Noteworthy, fusion constructs between Grx and redox-sensitive fluorescent proteins were originally developed for monitoring the intracellular ratio between GSH and GSSG (Ostergaard et al., 2004; Gutscher et al., 2008; Morgan et al., 2013). Furthermore, a recent pilot experiment suggests that redox-sensitive fluorescent proteins can be used to gain mechanistic insights about peroxiredoxin catalysis *in vivo* (Staudacher et al., 2018). Whether redox-sensitive fluorescent proteins can also be used for the noninvasive mechanistic assessment of Grx catalysis *in vivo* remains to be shown.

In summary, Grx catalysis can be monitored in a variety of enzymatically coupled as well as direct assays *in vitro*. Most of these assays use GSH as an electron donor and either physiological protein disulfide substrates or artificial LMW disulfide substrates such as GSSCys or HEDS as electron acceptors.

15.4   THE DITHIOL MECHANISM OF GLUTAREDOXINS

The catalytic mechanism of Grx depends on the type of substrate and Grx-isoform. Some protein disulfide substrates are reduced by dithiol Grx using a so-called dithiol mechanism (Lillig et al., 2008; Deponte, 2013). This mechanism consists of two major reaction sequences. During the first reaction sequence, the reduced enzyme, Grx_{SH}^{S-}, undergoes a dithiol–disulfide exchange reaction with the protein disulfide substrate. The product of this oxidative reaction sequence, $Grx(S_2)$, is reduced by two molecules of GSH during the second reaction sequence, yielding Grx_{SH}^{S-} and GSSG (Figure 15.1). According to the Cleland nomenclature (Cleland, 1963), this is a uni-uni-bi-uni ping-pong reaction. Exchanged protons are neglected in this nomenclature. However, in reality, the protons and acid–base catalysis play an important role (Srinivasan et al., 1997; Gallogly et al., 2008), because thiolates are the actual nucleophiles and leaving groups of the S_N2 elementary reactions, which are expected to have a negatively charged trigonal bipyramidal transition state with three aligned sulfur atoms (Eckers et al., 2009; Deponte, 2013; Deponte et al., 2015; Begas et al., 2017). Two different mixed disulfides are formed as intermediates. The first mixed disulfide, Grx-SSR, is formed after the nucleophilic attack of the N-terminal active site cysteine thiolate on the protein disulfide bond. The second mixed disulfide, Grx-SSG, is formed when GS^- attacks the N-terminal active site cysteine of $Grx(S_2)$ (Lillig et al., 2008; Deponte, 2013; Deponte et al., 2015). All of the reaction steps in Figure 15.1 are in principle reversible. However, the removal of GSSG by glutathione reductase prevents the backward reaction *in vitro* as well as in the cytosol, nucleus, and other subcellular compartments *in vivo* (Ostergaard et al., 2004; Gutscher et al., 2008; Morgan et al., 2013; Deponte, 2017).

The dithiol mechanism for specific Grx/disulfide substrate couples is predominantly deduced from absent or drastically reduced activities of mutant dithiol Grx with CPYS-motifs. For example, neither *E. coli* ribonucleotide reductase nor phosphoadenylylsulfate reductase are efficiently reduced by such monothiol mutants, whereas wild-type dithiol Grx are enzymatically active in accordance with the dithiol mechanism (Bushweller et al., 1992; Lillig et al., 1999). However, in contrast to *E. coli* ribonucleotide reductase, mouse ribonucleotide reductase is reduced by a mutant dithiol Grx with a CxxS-motif, suggesting a monothiol mechanism for this enzyme/substrate couple

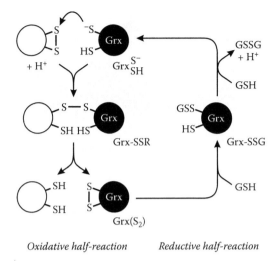

Oxidative half-reaction   *Reductive half-reaction*

FIGURE 15.1   Reaction scheme of the Grx dithiol mechanism.

(Zahedi Avval et al., 2009). This example demonstrates the necessity to experimentally test the mechanism for each Grx/substrate couple. Another noteworthy aspect is that an obligate disulfide mechanism is not restricted to bulky protein disulfide substrates. For example, in contrast to wild-type dithiol *Sc*Grx8, its monothiol mutant *Sc*Grx8[C28S] cannot catalyze the reduction of the LMW disulfide substrate HEDS (Eckers et al., 2009). Hypotheses on the underlying structure–function relationships for dithiol mechanisms are outlined in the following section.

15.5   THE MONOTHIOL MECHANISM OF GLUTAREDOXINS

Grx employ a so-called monothiol mechanism for the reduction of most substrates studied to date. This includes LMW- and protein-disulfide substrates as well as glutathione- and nonglutathione disulfide substrates. The second cysteine residue in the CxxC-motif of dithiol Grx is not only dispensable, as shown for the reduction of mammalian ribonucleotide reductase (Zahedi Avval et al., 2009), but is sometimes even detrimental for catalysis, as has been demonstrated for the reduction of HEDS, GSSCys, or *S*-sulfocysteine by mammalian dithiol Grx and the dithiol *Pf*Grx from *Plasmodium falciparum* (Yang et al., 1991; Yang et al., 1998; Gallogly et al., 2008; Djuika et al., 2013) as well as for the reduction of redox-sensitive yellow fluorescent protein by yeast *Sc*Grx1 (Bjornberg et al., 2006). Furthermore, wild-type *Sc*Grx6 and *Sc*Grx7 both possess only one cysteine residue and efficiently reduce HEDS and GSSCys (Mesecke et al., 2008a; Eckers et al., 2009; Luo et al., 2010; Begas et al., 2015; Begas et al., 2017). Hence, the formation of Grx(S$_2$) is often considered as a side reaction of the monothiol mechanism.

The monothiol mechanism for glutathionylated substrates consists of an oxidative and a reductive half-reaction with two different glutathione moieties (Mieyal et al., 2008; Lillig et al., 2008; Eckers et al., 2009; Deponte, 2013; Begas et al., 2017).

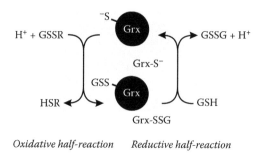

Oxidative half-reaction   Reductive half-reaction

FIGURE 15.2 Reaction scheme of the Grx monothiol mechanism.

During the oxidative half-reaction, the reduced enzyme, Grx-S⁻, attacks the sulfur atom of the glutathione moiety of the disulfide substrate, yielding a thiolate and Grx-SSG. This intermediate is reduced by GS⁻ during the rate-limiting reductive half-reaction, yielding GSSG and Grx-S⁻ (Figure 15.2). According to the Cleland nomenclature, which again neglects crucial proton exchanges and acid–base catalysis during both thiol–disulfide exchange reactions, the monothiol mechanism is a ping-pong bi-bi reaction. The ping-pong mechanism is supported by numerous kinetic studies with a variety of Grx-isoforms and glutathionylated substrates as reviewed previously (Deponte, 2013). However, which structure–function relationships underlie the established substrate preferences of class I Grx for GSSR and GSH (Gravina et al., 1993; Rabenstein et al., 1995; Srinivasan et al., 1997; Peltoniemi et al., 2006; Gallogly et al., 2008; Elgan et al., 2008)? Furthermore, which structure–function relationships result in the absent activities of class II Grx in standard *in vitro* assays?

Based on the comparison of different Grx-isoforms, we previously hypothesized that trigonal bipyramidal transition states of both half-reactions should necessitate different glutathione interaction sites for GSSR and GSH (Figure 15.3), unless the enzyme undergoes drastic conformational changes during catalysis (Eckers et al., 2009; Deponte, 2013). Recently, we were able to demonstrate that enzymatically active *Sc*Grx7 and *Pf*Grx indeed have two glutathione interaction sites, a glutathione scaffold site, which interacts with GSSR, and a glutathione activator site, which interacts with GSH (Begas et al., 2017). The glutathione scaffold site represents the previously reviewed interaction site from numerous crystal and NMR structures (Deponte, 2013). It consists of a G[C/x][D/E/x][D/E/x]-motif at the N-terminus of helix α4, a conserved glutamine residue and further candidate residues within helix α3 as well as loop residues that precede strand β3 (Begas et al., 2017). The glutathione activator site remains to be studied in more detail. So far, it consists of a lysine residue (which also activates the catalytic cysteine residue) at the end of strand β2 and candidate residues within helix α3 that stick out on the protein surface. It is to be noted that modifications of the glutathione scaffold and activator sites could explain the absent activity of class II Grx in standard enzyme kinetic assays (Begas et al., 2017). For example, class II Grx have a significantly altered helix α3 with a bulky WP-motif that might prevent the interaction with GSH (Figure 15.3). The enzymatic inactivity of these

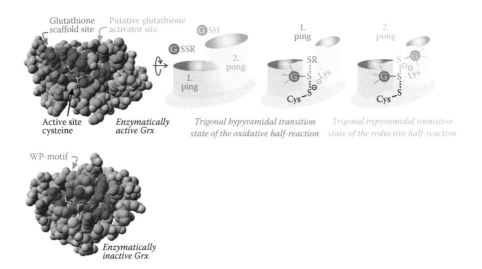

FIGURE 15.3 Model of the two different glutathione interaction sites and transition states of enzymatically active Grx.

proteins might reflect a kinetic uncoupling mechanism that allows iron sensing or the synthesis of iron–sulfur clusters in the presence of millimolar GSH (Deponte, 2017).

Two different glutathione interactions sites could also explain the unusual sequential kinetic patterns that are obtained for the GSH-dependent reduction of the nonglutathione model substrate HEDS (Mieyal et al., 1991; Mesecke et al., 2008a). We recently re-evaluated the reaction kinetics of the HEDS assay, suggesting that Grx-S⁻ can directly react with HEDS, yielding 2-mercaptoethanol and a mixed Grx disulfide with 2-mercaptoethanol, Grx-SSEtOH (Begas et al., 2015). Based on our two-site model and the hypothesis that Grx-SSEtOH is subsequently reduced by GS⁻, the product GSSEtOH would become an inhibitor, because its glutathione moiety occupies the activator instead of the scaffold site. In order to react with Grx-S⁻, GSSEtOH has first to dissociate and re-associate in the correct orientation. The described mechanism could explain the detected kinetics, because a noncompetitive inhibition can resemble a sequential kinetic pattern (Begas et al., 2017).

15.6 TRAPPED REACTION INTERMEDIATES AND FUTURE PERSPECTIVES

A plausible explanation for absent activities between certain substrates and mutant dithiol Grx or wild-type monothiol Grx with CxxS-motifs might be that the Grx-SSR or Grx-SSG species are kinetically trapped (Eckers et al., 2009; Begas et al., 2017; Deponte, 2017). For example, the structure of bulky protein–protein disulfide species might prevent a geometrically productive interaction with GSH as a reducing agent, as has been demonstrated for the protein–peptide disulfide between *Ec*Grx1 and a fragment of *E. coli* ribonucleotide reductase (Berardi et al., 1999).

Under these conditions, the second cysteine residue of dithiol Grx is essential for catalysis in order to liberate the product. Please note that such a resolving model might also apply to additional cysteine residues (Eckers et al., 2009). For example, several class I and class II Grx have a relatively conserved cysteine residue at the N-terminus of helix α4 (Yang et al., 1998; Tamarit et al., 2003; Deponte et al., 2005; Deponte, 2013). This residue was shown to affect the overall enzyme kinetics and to form a disulfide bond with the active site cysteine residue of *Pf*Grx *in vitro* (Djuika et al., 2013). One might also speculate that trapped disulfide-bonded Grx species are involved in redox signaling, in analogy to peroxidase signaling cascades (Brigelius-Flohé and Flohé, 2011). An absent reactivity of Grx-SSR or Grx-SSG might then even be a prerequisite to kinetically uncouple such species from the GSH pool in order to ensure a persistent signal transduction (Eckers et al., 2009; Deponte, 2013, 2017). Nevertheless, it remains to be studied in much more detail whether a steric hindrance due to bulky substrates is the major driving force for the abundance of dithiol Grx or whether the inactivity of a Grx might contribute to redox signaling. Noteworthy, the inability of ScGrx8^{C28S} to catalyze the conversion of the LMW disulfide substrate HEDS suggests that other factors than the size of the substrate might prevent a productive interaction between Grx-SSR or Grx-SSG and GSH (Eckers et al., 2009). Indeed, several data sets indicate that the active site of Grx might have to undergo subtle conformational changes in order to react with GSH (Jao et al., 2006; Yu et al., 2008; Eckers et al., 2009; Deponte, 2013; Begas et al., 2017). Future studies will have to decipher the exact roles of the glutathione interacting residues, the real structures and conformations of the transitions states, and the physiological relevance of the presence or absence of additional cysteine residues and modified interaction sites for catalysis, redox sensing, and iron metabolism.

REFERENCES

Begas, P., L. Liedgens, A. Moseler et al. 2017. Glutaredoxin catalysis requires two distinct glutathione interaction sites. *Nat Commun* 8:14835.

Begas, P., V. Staudacher, and M. Deponte. 2015. Systematic re-evaluation of the bis(2-hydroxyethyl)disulfide (HEDS) assay reveals an alternative mechanism and activity of glutaredoxins. *Chem Sci* 6 (7):3788–96.

Berardi, M. J., and J. H. Bushweller. 1999. Binding specificity and mechanistic insight into glutaredoxin-catalyzed protein disulfide reduction. *J Mol Biol* 292 (1):151–61.

Berndt, C., and C. H. Lillig. 2017. Glutathione, glutaredoxins, and iron. *Antioxid Redox Signal* 27 (15):1235–51.

Berndt, C., J.-D. Schwenn, and C. H. Lillig. 2015. The specificity of thioredoxins and glutaredoxins is determined by electrostatic and geometric complementarity. *Chem Sci* 6 (12):7049–58.

Bjornberg, O., H. Ostergaard, and J. R. Winther. 2006. Mechanistic insight provided by glutaredoxin within a fusion to redox-sensitive yellow fluorescent protein. *Biochemistry* 45 (7):2362–71.

Brigelius-Flohé, R., and L. Flohé. 2011. Basic principles and emerging concepts in the redox control of transcription factors. *Antioxid Redox Signal* 15 (8):2335–81.

Bushweller, J. H., F. Aslund, K. Wuthrich, and A. Holmgren. 1992. Structural and functional characterization of the mutant *Escherichia coli* glutaredoxin (C14----S) and its mixed disulfide with glutathione. *Biochemistry* 31 (38):9288–93.

Cleland, W. W. 1963. The kinetics of enzyme-catalyzed reactions with two or more substrates or products. I. Nomenclature and rate equations. *Biochim Biophys Acta* 67:104–37.

Couturier, J., J. Przybyla-Toscano, T. Roret et al. 2015. The roles of glutaredoxins ligating Fe-S clusters: Sensing, transfer or repair functions? *Biochim Biophys Acta* 1853 (6):1513–27.

Deponte, M. 2013. Glutathione catalysis and the reaction mechanisms of glutathione-dependent enzymes. *Biochim Biophys Acta* 1830 (5):3217–66.

Deponte, M. 2017. The incomplete glutathione puzzle: Just guessing at numbers and figures? *Antioxid Redox Signal* 27 (15):1130–61.

Deponte, M., K. Becker, and S. Rahlfs. 2005. *Plasmodium falciparum* glutaredoxin-like proteins. *Biol Chem* 386 (1):33–40.

Deponte, M., and C. H. Lillig. 2015. Enzymatic control of cysteinyl thiol switches in proteins. *Biol Chem* 396 (5):401–13.

Djuika, C. F., S. Fiedler, M. Schnolzer et al. 2013. *Plasmodium falciparum* antioxidant protein as a model enzyme for a special class of glutaredoxin/glutathione-dependent peroxiredoxins. *Biochim Biophys Acta* 1830 (8):4073–90.

Eckers, E., M. Bien, V. Stroobant et al. 2009. Biochemical characterization of dithiol glutaredoxin 8 from *Saccharomyces cerevisiae:* The catalytic redox mechanism redux. *Biochemistry* 48 (6):1410–23.

Elgan, T. H., and K. D. Berndt. 2008. Quantifying *Escherichia coli* glutaredoxin-3 substrate specificity using ligand-induced stability. *J Biol Chem* 283 (47):32839–47.

Fernandes, A. P., M. Fladvad, C. Berndt et al. 2005. A novel monothiol glutaredoxin (Grx4) from *Escherichia coli* can serve as a substrate for thioredoxin reductase. *J Biol Chem* 280 (26):24544–52.

Filser, M., M. A. Comini, M. M. Molina-Navarro et al. 2008. Cloning, functional analysis, and mitochondrial localization of *Trypanosoma brucei* monothiol glutaredoxin-1. *Biol Chem* 389 (1):21–32.

Gallogly, M. M., D. W. Starke, A. K. Leonberg et al. 2008. Kinetic and mechanistic characterization and versatile catalytic properties of mammalian glutaredoxin 2: Implications for intracellular roles. *Biochemistry* 47 (42):11144–57.

Gravina, S. A., and J. J. Mieyal. 1993. Thioltransferase is a specific glutathionyl mixed disulfide oxidoreductase. *Biochemistry* 32 (13):3368–76.

Gutsche, N., C. Thurow, S. Zachgo, and C. Gatz. 2015. Plant-specific CC-type glutaredoxins: Functions in developmental processes and stress responses. *Biol Chem* 396 (5):495–509.

Gutscher, M., A. L. Pauleau, L. Marty et al. 2008. Real-time imaging of the intracellular glutathione redox potential. *Nat Methods* 5 (6):553–9.

Hanschmann, E. M., M. E. Lonn, L. D. Schutte et al. 2010. Both thioredoxin 2 and glutaredoxin 2 contribute to the reduction of the mitochondrial 2-Cys peroxiredoxin Prx3. *J Biol Chem* 285 (52):40699–705.

Herrero, E., and M. A. de la Torre-Ruiz. 2007. Monothiol glutaredoxins: A common domain for multiple functions. *Cell Mol Life Sci* 64 (12):1518–30.

Holmgren, A. 1979. Glutathione-dependent synthesis of deoxyribonucleotides. Purification and characterization of glutaredoxin from *Escherichia coli*. *J Biol Chem* 254 (9):3664–71.

Holmgren, A., and F. Aslund. 1995. Glutaredoxin. *Methods Enzymol* 252:283–92.

Izquierdo, A., C. Casas, U. Muhlenhoff et al. 2008. *Saccharomyces cerevisiae* Grx6 and Grx7 are monothiol glutaredoxins associated with the early secretory pathway. *Eukaryot Cell* 7 (8):1415–26.

Jao, S. C., S. M. English Ospina, A. J. Berdis et al. 2006. Computational and mutational analysis of human glutaredoxin (thioltransferase): Probing the molecular basis of the low pKa of cysteine 22 and its role in catalysis. *Biochemistry* 45 (15):4785–96.

Kim, S. J., J. R. Woo, Y. S. Hwang et al. 2003. The tetrameric structure of *Haemophilus influenza* hybrid Prx5 reveals interactions between electron donor and acceptor proteins. *J Biol Chem* 278 (12):10790–8.

Lillig, C. H., C. Berndt, and A. Holmgren. 2008. Glutaredoxin systems. *Biochim Biophys Acta* 1780 (11):1304–17.

Lillig, C. H., A. Prior, J. D. Schwenn et al. 1999. New thioredoxins and glutaredoxins as electron donors of 3′-phosphoadenylylsulfate reductase. *J Biol Chem* 274 (12):7695–8.

Luo, M., Y. L. Jiang, X. X. Ma et al. 2010. Structural and biochemical characterization of yeast monothiol glutaredoxin Grx6. *J Mol Biol* 398 (4):614–22.

Mesecke, N., S. Mittler, E. Eckers et al. 2008a. Two novel monothiol glutaredoxins from Saccharomyces cerevisiae provide further insight into iron-sulfur cluster binding, oligomerization, and enzymatic activity of glutaredoxins. *Biochemistry* 47 (5):1452–63.

Mesecke, N., A. Spang, M. Deponte, and J. M. Herrmann. 2008b. A novel group of glutaredoxins in the cis-Golgi critical for oxidative stress resistance. *Mol Biol Cell* 19 (6):2673–80.

Mieyal, J. J., M. M. Gallogly, S. Qanungo et al. 2008. Molecular mechanisms and clinical implications of reversible protein S-glutathionylation. *Antioxid Redox Signal* 10 (11):1941–88.

Mieyal, J. J., D. W. Starke, S. A. Gravina, and B. A. Hocevar. 1991. Thioltransferase in human red blood cells: Kinetics and equilibrium. *Biochemistry* 30 (36):8883–91.

Morgan, B., D. Ezerina, T. N. Amoako et al. 2013. Multiple glutathione disulfide removal pathways mediate cytosolic redox homeostasis. *Nat Chem Biol* 9 (2):119–25.

Nagai, S., and S. Black. 1968. A thiol-disulfide transhydrogenase from yeast. *J Biol Chem* 243 (8):1942–7.

Ostergaard, H., C. Tachibana, and J. R. Winther. 2004. Monitoring disulfide bond formation in the eukaryotic cytosol. *J Cell Biol* 166 (3):337–45.

Outten, C. E., and A. N. Albetel. 2013. Iron sensing and regulation in *Saccharomyces cerevisiae*: Ironing out the mechanistic details. *Curr Opin Microbiol* 16 (6):662–8.

Pauwels, F., B. Vergauwen, F. Vanrobaeys et al. 2003. Purification and characterization of a chimeric enzyme from *Haemophilus influenzae* Rd that exhibits glutathione-dependent peroxidase activity. *J Biol Chem* 278 (19):16658–66.

Peltoniemi, M. J., A. R. Karala, J. K. Jurvansuu et al. 2006. Insights into deglutathionylation reactions. Different intermediates in the glutaredoxin and protein disulfide isomerase catalyzed reactions are defined by the gamma-linkage present in glutathione. *J Biol Chem* 281 (44):33107–14.

Peskin, A. V., P. E. Pace, J. B. Behring et al. 2016. Glutathionylation of the active site cysteines of peroxiredoxin 2 and recycling by glutaredoxin. *J Biol Chem* 291 (6):3053–62.

Rabenstein, D. L., and K. K. Millis. 1995. Nuclear magnetic resonance study of the thioltransferase-catalyzed glutathione/glutathione disulfide interchange reaction. *Biochim Biophys Acta* 1249 (1):29–36.

Rouhier, N., E. Gelhaye, P. E. Sautiere et al. 2001. Isolation and characterization of a new peroxiredoxin from poplar sieve tubes that uses either glutaredoxin or thioredoxin as a proton donor. *Plant Physiol* 127 (3):1299–309.

Srinivasan, U., P. A. Mieyal, and J. J. Mieyal. 1997. pH profiles indicative of rate-limiting nucleophilic displacement in thioltransferase catalysis. *Biochemistry* 36 (11):3199–206.

Starke, D. W., P. B. Chock, and J. J. Mieyal. 2003. Glutathione-thiyl radical scavenging and transferase properties of human glutaredoxin (thioltransferase). Potential role in redox signal transduction. *J Biol Chem* 278 (17):14607–13.

Staudacher, V., C. F. Djuika, J. Koduka et al. 2015. *Plasmodium falciparum* antioxidant protein reveals a novel mechanism for balancing turnover and inactivation of peroxiredoxins. *Free Radic Biol Med* 85:228–36.

Staudacher, V., M. Trujillo, T. Diederichs et al. 2018. Redox-sensitive GFP fusions for monitoring the catalytic mechanism and inactivation of peroxiredoxins in living cells. *Redox Biol* 14:549–56.

Tamarit, J., G. Belli, E. Cabiscol et al. 2003. Biochemical characterization of yeast mitochondrial Grx5 monothiol glutaredoxin. *J Biol Chem* 278 (28):25745–51.

Tang, Y., J. Zhang, J. Yu et al. 2014. Structure-guided activity enhancement and catalytic mechanism of yeast grx8. *Biochemistry* 53 (13):2185–96.

Yang, Y., S. Jao, S. Nanduri et al. 1998. Reactivity of the human thioltransferase (glutaredoxin) C7S, C25S, C78S, C82S mutant and NMR solution structure of its glutathionyl mixed disulfide intermediate reflect catalytic specificity. *Biochemistry* 37 (49):17145–56.

Yang, Y. F., and W. W. Wells. 1991. Identification and characterization of the functional amino acids at the active center of pig liver thioltransferase by site-directed mutagenesis. *J Biol Chem* 266 (19):12759–65.

Yu, J., N. N. Zhang, P. D. Yin et al. 2008. Glutathionylation-triggered conformational changes of glutaredoxin Grx1 from the yeast *Saccharomyces cerevisiae*. *Proteins* 72 (3):1077–83.

Zaffagnini, M., M. Bedhomme, C. H. Marchand et al. 2012. Glutaredoxin s12: Unique properties for redox signaling. *Antioxid Redox Signal* 16 (1):17–32.

Zahedi Avval, F., and A. Holmgren. 2009. Molecular mechanisms of thioredoxin and glutaredoxin as hydrogen donors for mammalian S-phase ribonucleotide reductase. *J Biol Chem* 284 (13):8233–40.

16 The Role of Glutaredoxins in the Brain

Carsten Berndt, Anna Dorothee Engelke, and Klaudia Lepka
Heinrich-Heine-Universität Düsseldorf

Lars Bräutigam
Karolinska Institutet

CONTENTS

16.1 INTRODUCTION

The impact of redox regulation and oxidative damage on development and diseases of the brain is undisputed. However, molecular mechanisms underlying developmental and damaging processes are rare. Therefore, this chapter provides a summary of the current knowledge on the interplay between glutaredoxins (Grx), glutathione (GSH), and iron in the context of the central nervous system.

16.1.1 GLUTAREDOXINS

Grxs are small oxidoreductases originally identified as GSH-dependent electron donors of ribonucleotide reductase (Holmgren, 1976; Luthman et al., 1979). Grxs belong to the thioredoxin (Trx) family, which is characterized by a common structural fold named the Trx-fold consisting of a central 4–5-stranded β-sheet surrounded by 3–4 α-helices (Martin, 1995). The active site of oxidoreductases of the Trx family is located on a loop connecting the first β-sheet and the first α-helix (Lillig and Berndt, 2013). The active site motif distinguishes the two main subclasses of Grxs: dithiol Grxs possess a Cys-X-X-Cys active site, whereas monothiol Grxs possess a Cys-X-X-Ser active site (Lillig et al., 2008). In contrast to dithiol Grxs, the vast majority of monothiol Grxs lack oxidoreductase activity, but are important in different aspects of iron metabolism, e.g., iron–sulfur cluster biosynthesis or iron trafficking (Lillig et al., 2008; Berndt and Lillig, 2017). Most of the monothiol Grxs, but also some dithiol Grxs, were characterized as FeS proteins (Figures 16.1 and 16.2A) (Lillig et al., 2005; Berndt and Lillig, 2017). Dithiol Grxs catalyze two types of reactions: the reduction of protein disulfides via the dithiol mechanism and the reduction of mixed disulfides of proteins and low-molecular weight thiols—in this chapter we will focus on GSH—via the monothiol mechanism (Lillig and Berndt, 2013). The enzymatic activity of Grxs is described in detail in Chapter 15.

In brief, both dithiol and monothiol mechanism reactions (Figure 16.1) are initiated by a nucleophilic attack of the N-terminal active site cysteinyl residue forming a mixed disulfide between the N-terminal cysteine and a cysteine of the target protein or the GSH molecule, respectively. In the second step of the reaction, the disulfide

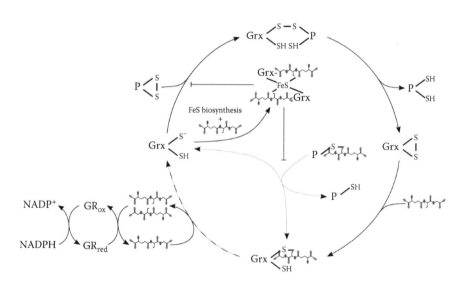

FIGURE 16.1 The Glutaredoxin (Grx) system. Grxs are able to reduce protein disulfides via the dithiol mechanism (black arrows) and glutathionylated proteins via the monothiol mechanism (gray arrows). The oxidized Grx is reduced by glutathione (GSH). Oxidized GSH (GSSG) is reduced by glutathione reductase (GR) and NADPH as final electron donor. Coordination of an FeS cluster inhibits both, dithiol, and monothiol mechanisms.

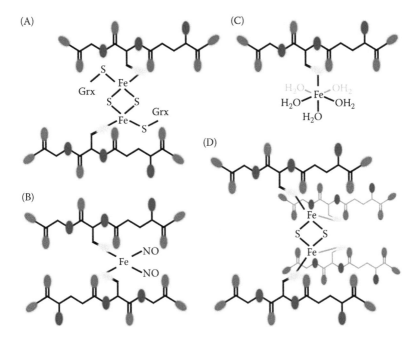

FIGURE 16.2 Glutathione binds iron. (A) FeS cluster coordination in glutaredoxins (Grx), (B) diglutathionyl-dinitrosyl-iron complex, (C) iron(II) glutathione, (D) [[Fe$_2$S$_2$$^{2+}$] (GS$^-$)$_4$]$^{2-}$ cluster.

between Grx and the substrate is reduced by the C-terminal active site cysteine. The resulting disulfide in the active site (disulfide mechanism) and the glutathionylated Grx (monothiol mechanism) is reduced by GSH. Oxidized GSH (GSSG) is reduced by glutathione reductase with NADPH (Figure 16.1).

16.1.2 GLUTATHIONE

Besides other proteins, e.g., GSH peroxidases or peroxiredoxins, Grxs link GSH to physiological functions (Deponte, 2013). So far, not a single physiological function that depends solely on the GSH:GSSG redox couple has been described. Therefore, the functions of GSH need the catalytic activity of specific enzymes (Berndt et al., 2014).

Grxs bind GSH in three different modes: (1) in a transient, intermediary mixed disulfide, (2) in interaction with a second GSH molecule that leads to the reduction of this mixed disulfide, and (3) noncovalently as ligand for the FeS cluster forming the holo-Grx complex (Figure 16.2A). All of these interactions rely on overlapping GSH binding sites on the different Grxs (Begas et al., 2017; Berndt and Lillig, 2017).

Holo-Grx usually consists of two Grx monomers which are linked via the FeS cluster that is ligated by the N-terminal active site cysteines and the cysteines of two

GSH molecules (Berndt et al., 2007; Johansson et al., 2007; Rouhier et al., 2007). Using GSH as nonprotein ligands holo-Grxs are unique FeS proteins. Moreover, GSH is not only necessary for the coordination of the FeS cluster, its concentration is highly important for the stability of holo-Grxs (Berndt et al., 2007; Berndt and Lillig, 2017).

Next to the ligation of iron in holo-Grxs, GSH is able to bind iron in several other complexes. (1) GSH stabilizes dinitrosyl-iron complexes in the form of diglutathionyl-dinitrosyl-iron complexes (Figure 16.2B; Vanin, 2009). (2) GSH binds very effectively iron(II) forming iron(II) glutathione (Figure 16.2C; Hider and Kong, 2011). (3) GSH might be able to bind FeS clusters without Grxs in the form of $[[Fe_2S_2^{2+}](GS^-)_4]^{2-}$ (Figure 16.2D). Such clusters were postulated already in 1972 (Sugiura and Tanaka, 1972) and described 40 years later (Qi et al., 2012).

Via binding of iron, GSH is linked to several iron-dependent physiological pathways and functions. The coordination of FeS clusters in monothiol Grxs links GSH to iron trafficking (Haunhorst et al., 2013), iron sensing (Mühlenhoff et al., 2010), FeS cluster biosynthesis (Rodríguez-Manzaneque et al., 2002; Mühlenhoff et al., 2003; Moseler et al., 2015), and transfer of FeS clusters (Iwema et al., 2009; Bandyopadhyay et al., 2008; Banci et al., 2015). The FeS clusters bound to dithiol Grxs were proposed to act as redox sensor (Lillig et al., 2005). Binding to iron(II) and coordination of $[[Fe_2S_2^{2+}](GS^-)_4]^{2-}$ clusters might be important for iron trafficking as well as iron and FeS cluster storage (Hider and Kong, 2011; Li et al., 2015). Iron(II) glutathione complexes were suggested to be the dominant form of iron in the labile iron pool (Hider and Kong, 2011). Formation of dinitrosyl-iron complexes using GSH molecules as ligands plays a role in iron export (Lok et al., 2012) and protects against cell death (Kim et al., 2000). In addition, diglutathionyl-dinitrosyl-iron complexes promote regenerative processes such as skin wound healing (Shekhter et al., 2015).

The group of Toledano even suggested that GSH is rather important for iron homeostasis and FeS cluster biosynthesis than for redox regulation (Kumar et al., 2011). This suggestion is based on the observation that depletion of GSH in *Saccharomyces cerevisiae* leads to defects in iron metabolism, especially biosynthesis of cytosolic FeS proteins, but not in redox maintenance. The phenotype is rescued by addition of ferric iron, whereas addition of dithiothreitol shows no effect.

A tight regulation of iron homeostasis is needed since iron is a double-edged sword, very toxic, e.g., by catalyzing formation of hydroxyl radicals via Fenton chemistry, and at the same time essential for several cellular processes, e.g., via iron-dependent protein activities. Therefore, disturbances in iron homeostasis are associated with a variety of (patho)physiological problems, including development and diseases of the central nervous system.

16.1.3 THE BRAIN

In 1992, Gregor Eichele raised the question if the human brain will ever be powerful enough to solve the problem of its own creation (Eichele, 1992). Now, 25 years later, it is still an enigma, how our most complex organ develops, matures, and ages. The brain consists mainly of neurons (functioning in signal transduction via axons and synapses, both forming connections between neurons, but also between neurons and muscles), oligodendrocytes (forming the myelin sheath protection around axons and guaranteeing rapid signal transduction), astrocytes (assuring trophic support of neurons and forming part of the blood–brain barrier), microglia (resident immune cells of the central nervous system), and ependymal cells (producing cerebrospinal fluid).

The fundamental process of brain development appears similar in all vertebrates. The neural plate, a streak-like structure on the surface of the developing embryo, thickens at the edges and folds upwards during early embryonic development. Those neural folds migrate towards the midline of the embryo and converge to form a hollow tube underneath the ectoderm. Soon thereafter, this neural tube bulges into the three primary brain vesicles along the anterior–posterior axes: the forebrain, the midbrain, and the hindbrain vesicle. Those balloons develop into five secondary vesicles which eventually give rise to all brain derivatives in the adult (Fitzgerald et al., 2012). The wall of the neural tube contains rapidly dividing neuronal stem cells, which gradually stop multiplying and continue to differentiate into neurons or supportive cells. These subsequently migrate to different parts of the brain and start self-organizing into local circuits in an activity-independent, i.e., genetically hardwired, process. Once these local circuits have formed, neuronal activity and sensory experience will constantly refine the nascent neuronal circuits and eventually lead to the formation of an organ that thinks, perceives, remembers, writes books, and makes us the person we are (Gilbert, 2014).

The importance of reactive oxygen and nitrogen species in neural cell fate decision and neural stem cell maintenance was confirmed by numerous publications (summarized in Prozorovski et al., 2015; Wilson et al., 2017). Several major processes in brain development like axonal outgrowth and myelin production were shown to be redox-dependent (Jana and Pahan, 2005; Gauron et al., 2016).

Even though our brain remains plastic during most of our lives, brain plasticity and brain functionality decreases with age. As in all other tissues, DNA damage and oxidative damage accumulate in the brain, which leads to anatomical changes like thinning of the cortex (Sowell et al., 2003) and biochemical changes including decreased synthesis and reception of neurotransmitters (Hof and Mobbs, 2009). All these factors will eventually lead to a decreased cognitive function and neuropsychological changes which are characteristics for an aging individual (Hof and Mobbs, 2009).

Neurodegenerative diseases mimic accelerated aging. In general, the central nervous system can be damaged by different, although overlapping, events: traumatic injury, neuroinflammation, and neurodegeneration. Independent of the damage paradigm, regeneration and repair are extremely rare in mammals, especially in higher primates, whereas lower vertebrates show a reasonable capacity to regenerate the damaged central nervous system. The limited regeneration in mammals is connected to the formation of a glial scar after trauma, the ineffective removal of damaged myelin after neuroinflammation, and the fact that mammals lack a significant adult neurogenesis to replace lost neurons upon neurodegeneration. In contrast, zebrafish regenerate very well after brain injury (Kizil et al., 2012). Interestingly, zebrafish seems to avoid forming a glial scar (März et al.,

2011; Lenkowski et al., 2013) and the adult fish is still able to build new neurons to repair damaged brain areas (Alunni and Bally-Cuif, 2016).

Redox processes are of high importance during brain diseases. Oxidative stress for instance is a common feature of all neurodegenerative diseases, e.g., familiar amyotrophic lateral sclerosis is often based on a mutation in the gene encoding superoxide dismutase 1 (Kaur et al., 2016). Another example is the secreted nitric oxide and the subsequently formed peroxynitrite within oligodendrocytes, the primary factor causing demyelination (Smith et al., 1999).

In the following sections, we summarize the current knowledge regarding the role of iron, GSH, and Grxs during brain development and brain diseases.

16.2 BRAIN DEVELOPMENT

Iron is an essential element for physiological brain function (Youdim and Green, 1978; Connor and Menzies, 1996). Alterations in iron homeostasis due to genetic mutation in iron-related proteins, iron deficiency, or iron overload represent significant risks for/during brain development with short- as well as long-term consequences. Maternal iron restriction to the fetus affects memory function due to structural changes of the hippocampus. In line, infants suffering from neonatal iron deficiency show delayed cognitive function, e.g., problems with auditory recognition (Deregnier et al., 2000). Moreover, infants born with low ferritin concentrations exhibit poorer early school performance (Tamura et al., 2002). Next to memory and cognition, the temperament is affected by iron starvation (Wachs et al., 2005; Shafir et al., 2006). Although iron deficiency leads to decreased oxygen delivery, the primary effect on brain tissue by iron deficiency was demonstrated to be independent of anemia in the 1970s (Dallman, 1986). Instead, loss of cytochrome c (Deungria et al., 2000), changes in ATP generation, inefficient electron transport (Rao et al., 2003; Ward et al., 2007), and altered fatty acid concentration (Connor and Menzies, 1996; Beard et al., 2003) were reported. These alterations lead to hypomyelination and subsequent problems with signal transduction (Deregnier et al., 2000; Badaracco et al., 2008) as well as to changes in levels of iron-dependent neurotransmitters like dopamine and serotonin (Youdim et al., 1986).

16.2.1 GLUTATHIONE

GSH levels and the ratio between reduced and oxidized GSH is highly regulated during embryonic development (Timme-Laragy et al., 2013; Hansen and Harris, 2015). The GSH concentration, localization, and redox state is related to cell proliferation and differentiation (Firth and Greydanus, 1987; Messina and Lawrence, 1989; Pallardó et al., 2009). In zebrafish embryos, the total GSH concentration and the GSSG:GSH ratio increase slightly during organogenesis and at the beginning of circulation [18–24 hours past fertilization (hpf) (Timme-Laragy et al., 2013)]. Between 24 and 36 hpf, the total GSH level increases around two times and stays at a constant level of about 5 mM (Timme-Laragy et al., 2013).

GSH and its interaction with enzymes regulate several developmental processes. Vascular development, for instance, depends on the (de-)glutathionylation of sirtuin 1, a protein de-acetylase (Bräutigam et al., 2013).

During brain development, GSH seems to be important for the establishment of proper neurotransmission, e.g., formation of myelin sheaths and synapses. Measurements in rats show that GSH levels in the brain increase within hours after birth (Glöckner and Kretzchmar, 1991) with high concentrations in cortex and hippocampus 3 days after birth (Rice and Russo-Menna, 1998) and a peak around 7 days after birth (Nanda et al., 1996). These time points correlate with onset of myelination and intense synaptogenesis. In line, GSH concentration is slightly higher in glia cells (oligodendrocytes are the myelin forming cells and astrocytes are important for the structure of synapses) than in neurons after birth (Rice and Russo-Menna, 1998). Afterwards, the GSH content in the brain is constantly decreasing (Rao et al., 2014).

The inhibition of GSH synthesis by buthionine sulphoximine (BSO) during rat development (postnatal days 5–16) induces impaired spatial memory (Cabungcal et al., 2007). The described cognitive deficits are similar to symptoms found in human schizophrenia patients (Castagné et al., 2004; Cabungcal et al., 2007) (see Section 16.3).

16.2.2 Glutaredoxins

Both functions of Grxs, regulation of redox-dependent pathways as well as iron homeostasis, are of high importance during embryonic development. A regulated transcription of Grxs during mouse development was suggested by Jurado et al. (2003). Knockout of either Grx3 or Grx5 in mice is embryonically lethal (Wingert et al., 2005; Cheng et al., 2011). Whereas Grx3 is needed for iron trafficking (Haunhorst et al., 2013), Grx5 is essential for FeS cluster biosynthesis (Wingert et al., 2005). Due to these functions, lack of both proteins inhibits the maturation and activity of iron-dependent enzymes, such as hemoglobin, in zebrafish, mice, and patients (Wingert et al., 2005; Camaschella et al., 2007; Haunhorst et al., 2013). As mentioned earlier, iron is an essential factor for brain development and disturbed iron homeostasis leads to severe brain dysfunctions.

So far, only one Grx-regulated pathway essential for brain development, the semaphorin 3A signaling pathway, has been described. Semaphorins regulate the direction of cell migration and axonal outgrowth as repulsive guidance molecules. They are highly essential during brain development and convey their extracellular signal via collapsin response mediator protein 2 (CRMP2) (Schmidt and Strittmatter, 2007), a phosphoprotein that interacts with tubulins and microtubules supporting their stabilization and polymerization (Lin et al., 2011; Niwa et al., 2017). Oxidized CRMP2 is reduced by Grx2c, the cytosolic Grx2 isoform (Bräutigam et al., 2011; Gellert et al., 2013; Möller et al., 2017). Since a regulation between oxidized and reduced CRMP2 is needed for axonal outgrowth, cytosolic Grx2 is essential for the formation of a functioning neuronal network (Figures 16.3 and 16.4A) (Bräutigam et al., 2011).

The identified redox switch in CRMP2 is a thiol–disulfide switch of cysteine 504 and affects the protein function via profound structural and conformational changes

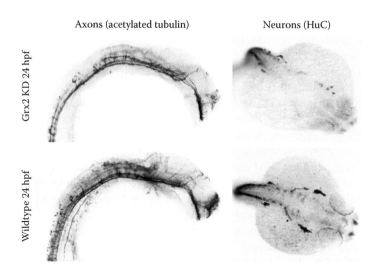

FIGURE 16.3 Glutaredoxin 2 is essential for the formation of a neuronal network. In zebrafish embryos 24 h post fertilization (hpf) without (wildtype) and with glutaredoxin 2 knock-down (Grx2 KD). Axons (antiacetylated antibodies) and neurons (anti-HuC antibodies) were visualized by immunohistochemistry.

including the C-terminal domain containing the active site (Möller et al., 2017). CRMP2 forms a tetramer which is essential for its enzymatic function and which is stabilized by intermolecular disulfides between cysteines 504 (Figure 16.4A) (Gellert et al., 2013). Reduction of these disulfides allows phosphorylation of serine 518 (Figure 16.4A) (Möller et al., 2017).

Next to brain development, this mechanism is also important during brain diseases (see Section 16.3).

16.3 BRAIN DISEASES

Many diseases of the central nervous system are accompanied by iron accumulation (Rouault, 2013). Iron accumulates during normal aging in deposits in human brains (Hallgren and Sourander, 1958); however, in diseased brains, this accumulation is accelerated and more prominent (Craelius et al., 1982).

Many publications describe iron accumulation in neurological disorders, but the knowledge of causes and consequences, e.g., the contribution to disease onset, progression, or symptom manifestation, is very limited. Increased cellular iron levels can (1) induce formation of hydroxyl radicals and thereby damage of biomolecules, (2) lead to lipid peroxidation, and (3) promote aggregation of α-synuclein and prion protein, for instance, hallmarks of Parkinson's and prion disease, respectively (Angelova and Brown, 2015). Furthermore, iron deposition was demonstrated to promote blood–brain barrier leakage and thereby increases the inflammatory response under neurodegenerative pathologies like Alzheimer's and Parkinson's diseases (Liu et al., 2006).

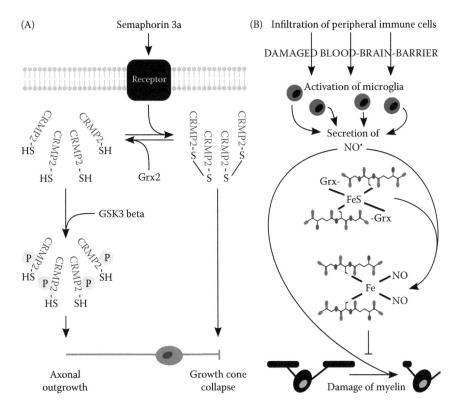

FIGURE 16.4 Glutaredoxin 2 is important during development and disease of the brain. (A) In concert with the semaphorin 3A signaling pathway, glutaredoxin 2 (Grx2) regulates the redox state of collapsin response mediator protein 2 (CRMP2). Whereas the CRMP2-tetramer containing disulfides between cysteines 504 leads to growth cone collapse and thereby changes in direction, reduced and subsequently phosphorylated CRMP2 promotes axonal outgrowth. (B) Activation of microglia upon neuroinflammation leads to secretion of nitric oxide damaging the myelin structure. Presence of FeS-Grx detoxifies nitric oxide by formation of a diglutathionyl-dinitrosyl-iron complex and thereby inhibits demyelination.

Not surprisingly, iron chelation and protection against ferroptosis, an iron-dependent cell death mechanism [Yang et al. (2014); for details, see Chapters 7 and 8], provides decreased disease severity in the animal model of multiple sclerosis (Mitchell et al., 2007) and survival of neurons (Ingold et al., 2018), respectively.

16.3.1 Glutathione

GSH depletion is a normal age-related phenomenon that already has been described for insects, rodents, and humans (Sohal and Weindruch, 1996). Accelerated GSH depletion or a shift towards GSSG is associated with a variety of diseases and especially brain disorders including Alzheimer's disease (Venkateshappa et al.,

2012; Saharan and Mandal, 2014), Parkinson's disease (Pearce et al., 1997), multiple sclerosis (Calabrese et al., 2002; Carvalho et al., 2014), schizophrenia (Fournier et al., 2017), bipolar disorder (Fournier et al., 2017), and depression (Saharan and Mandal, 2014). Not surprisingly, cell death of dopaminergic neurons was induced *in vitro* by depletion of GSH via BSO treatment (Ibi et al., 1999) and *in vivo* via a virus-based shRNA approach leading to downregulation of the glutamate-cysteine ligase (see Chapter 1) in adult rat brains (Garrido et al., 2011).

Brain GSH levels can be measured *in vivo* and noninvasively with ^1H magnetic resonance spectroscopy, which makes it principally a useful biomarker (Mandal et al., 2015), although it cannot be considered as disease-specific one, since many brain diseases show a comparable downregulation of GSH.

Not much is known about the function of GSH during diseases of the central nervous system. One potential function is its role as cosubstrate of GPx4, the main player in ferroptosis. Ferroptosis is induced by two mechanisms: (1) directly via inhibition of GPx4 and (2) indirectly via inhibition of the glutamate/cysteine antiporter x_c^- resulting in reduced levels of GSH (Yang and Stockwell, 2016). GPx4 inhibition subsequently leads to increased formation of iron-mediated lipid peroxidation. Ferroptosis appears to be related to several brain diseases, e.g., Alzheimer's, Parkinson's, and Huntington's diseases, stroke, intracerebral hemorrhage, and traumatic brain injury (Stockwell et al., 2017).

Following traumatic brain injury, inhibition of ferroptosis increases the regeneration by improved glia-to-neuron reprogramming (Gascón et al., 2016). Knockout of GPx4 in forebrain neurons of mice leads to hippocampal neurodegeneration accompanied by deficits in learning and memory as observed during Alzheimer's disease (Hambright et al., 2017). These defects were ameliorated upon treatment with a ferroptosis inhibitor in GPx4 knockout mice (Hambright et al., 2017).

For the interplay between GSH and iron, ferroptosis is a very prominent example in the context of brain diseases. Although iron is the most crucial or at least the best investigated metal connected to neurotoxicity mediated by GSH depletion, also other metals, such as extracellular copper, might be of importance (White and Cappai, 2003).

Although most of the described brain diseases are characterized by decreased GSH levels, boosting GSH synthesis by overexpression of glutamate-cysteine ligase leads to increased amounts of glutathionylated proteins and cell death of dopaminergic neurons (Garrido et al., 2011). Nevertheless, GSH donation is suggested to be a feasible therapeutic strategy. N-acetyl-L-cysteine, a precursor of GSH that passes the blood–brain barrier, improves the cognition in a mouse model of Alzheimer's disease (Farr et al., 2003). In addition, intravenous injection of GSH leads to an improvement of motoric functions in Parkinson's disease patients (Sechi et al., 1996).

16.3.2 Glutaredoxins

Grxs regulate deglutathionylation and thiol–disulfide switches. Dysregulation of these posttranslational modifications is associated with numerous aspects of neurodegeneration like apoptosis and plaque formation (Sabens Liedhegner et al., 2012).

Pathways that were previously proven to be regulated in a Grx-dependent manner were found to be altered upon inflammation-related brain disorders. For example, in Alzheimer's disease, the hypophosphorylated form of CRMP2 is aggregated in disease-specific protein deposits (Uchida et al., 2005; Hensley and Kursula, 2016). Since the phosphorylation of CRMP2 is dependent on Grx2-mediated reduction (see above, Figure 16.4A), increased amounts of hypophosphorylated CRMP2 indicate a decrease in Grx2 levels or activity in Alzheimer patients. Likewise, decreased levels of Grx1 lead to dephosphorylation of the serine-threonine protein kinase Akt1, which inhibits its kinase activity, promotes its degradation, and subsequently induces cell death. The inhibition of this mechanism is important for neuronal survival in neurodegenerative disorders like Parkinson's disease (Ahmad et al., 2014).

In Parkinson's disease, Grx2 was described to be upregulated in the acute phase of the MTPT-induced mouse model (Karunakaran et al., 2007). Changes in the levels of Grxs are also characteristics of other brain disorders like focal ischemia, Alzheimer's disease and multiple sclerosis. After induction of focal ischemia in rat brains, Grx1 level decreased in parallel to the rate of neuronal damage (Takagi et al., 1999). In human patients suffering from Alzheimer's disease, Grx1 is upregulated in healthy neurons but downregulated in degenerating neurons (Akterin et al., 2006). During progression of experimental autoimmune encephalomyelitis, an animal model of multiple sclerosis, Grx2 is upregulated in acute disease phases and downregulated in chronic phases, whereas Grx1 shows the opposite expression pattern (Lepka et al., 2017).

The difficult balance between beneficial and detrimental effects of redox-dependent events is demonstrated by the high number of publications that describe conflicting results, although these were often obtained from different animal models. For example, Grx1 deficiency exacerbates neurodegeneration in a Parkinson's disease model in *Caenorhabditis elegans* (Johnson and Johnson, 2015) and the MTPT model in mice (Kenchappa and Ravindranath, 2003), whereas Miller et al. described a positive correlation between the expression of Grx1 and the onset as well as the severity of Parkinson's disease (Miller et al., 2016). Therefore, the authors suggested Grx1 as potential biomarker for Parkinson's disease.

Besides the above mentioned mechanisms, Grx1 regulates the activities of N-methyl-D-aspartate receptor (Baxter et al., 2015), endothelial nitric oxide synthase (Chen et al., 2013; Shang et al., 2017), and mitochondrial complex 1 (Diwakar et al., 2007). Therefore, lack of Grx1 could lead to Huntington's, Alzheimer's, and Parkinson's diseases and ameliorates neurolathyrism, a motor neuron disease affecting the pyramidal system (Diwakar et al., 2007). The latter is another example of the enzyme dependency of GSH function, since only downregulated Grx1 induced mitochondrial dysfunction in neurolathyrism, but not downregulated GSH (Diwakar et al., 2007).

In the end, we will focus on two pathological conditions illustrating clearly the complex interplay between Grxs, GSH, and iron.

An example of the beneficial interplay between Grxs, GSH, and iron is the attenuation of neuroinflammatory damage of oligodendrocytes upon nitric oxide release by activated microglia (Figure 16.4B). Here, the unique coordination of the FeS cluster using GSH as ligand is important for the protective effect, not the oxidoreductase activity of Grx2 (Lepka et al., 2017). Instead of forming peroxynitrite, nitric oxide disassembles holo-Grx2 and most likely also other holo-Grxs. Thereby, nitric oxide is converted into a diglutathionyl-dinitrosyl-iron complex that shows antiapoptotic effects and positive results in both clinical trials and animal models associated with neuroinflammatory disorders (Chazov et al., 2012).

Data showing that GSH depletion potentiates nitric oxide toxicity of dopaminergic neurons (Ibi et al., 1999) suggest that the aforementioned mechanism is not restricted to oligodendrocytes.

The second example is Parkinson's disease (Figure 16.5). As described for other neurodegenerative diseases, GSH concentration and the GSH:GSSG ratio is decreased in Parkinson's disease, whereas the overall iron concentration is increased. Postmortem human *substantia nigra* specimens of Parkinson patients showed reduced GSH levels (Pearce et al., 1997), while other brain regions did not show altered GSH levels (Dexter et al., 1994; Sian et al., 1994). Interestingly, there is a direct link between GSH and iron (see also Section 16.1): In an iron overload mouse model, increased brain iron concentrations lead to a decrease of the GSH levels (Lan and Jiang, 1997). Moreover, iron chelation blocks neuronal cell death induced by GSH depletion (Chouraqui et al., 2013). Surprisingly, only one publication proved ferroptotic cell death of dopaminergic neurons during Parkinson's disease (Do Van et al., 2016), although many publications link this disease to almost all of the characteristic features of ferroptosis, e.g., iron accumulation, GSH depletion, or oxidative damage (see above). The lack of GSH affects both dithiol and monothiol Grxs, and thereby the regulation of the redox state of protein thiols and iron homeostasis, respectively. A direct link between functions of Grxs 3 and 5 and iron accumulation during Parkinson's disease can be assumed, but has not yet been described. However, GSH-depleted dopaminergic cells have a disturbed iron homeostasis due to inhibition of the mitochondrial isoform of Grx2, Grx2a (Lee et al., 2009). In general, Grxs, especially Grx1, provide protection against dopaminergic neurodegeneration as seen in *C. elegans* Parkinson models (Arodin et al., 2014; Johnson et al., 2015). In Parkinson patients, a reduced Grx1 expression especially in dopaminergic neurons was found (Johnson et al., 2015). Surprisingly, the same research group described also an upregulation of Grx1 activity in patients leading to increased microglia activation and neuroinflammation as well as an earlier disease onset (Miller et al., 2016). Loss of Grx1 activity was connected to the function of the protein deglycase DJ-1 (Johnson et al., 2016). The gene encoding DJ-1 is often mutated in Parkinson patients and leads to the early onset of the disease in animal models (Kalinderi et al., 2016). Based on data obtained in yeast (*S. cerevisiae*) (Silva et al., 2008), it can be assumed that Grx activity is needed for proteasomal function, which is needed to remove accumulating α-synuclein aggregates. These characteristic aggregates (Spillantini et al., 1997) inhibit proteasomal function by themselves and are formed in an iron-dependent manner (Angelova and Brown, 2015). In addition, increased iron levels lead to the formation of hydroxyl radicals via Fenton chemistry and oxidation of dopamin to 6-OHDA-quinone, which can be detoxified by Grxs (Arodin et al., 2014), and subsequently to oxidative damage and mitochondrial dysfunction (Figure 16.5). In line, iron chelators are considered as therapeutic tools against Parkinson's disease (Devos et al., 2014).

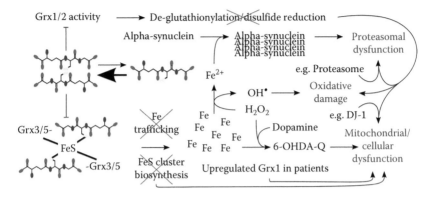

FIGURE 16.5 Several pathogenic processes during Parkinson's disease are regulated by glutathione and glutaredoxins. The interplay between the GSH:GSSG ratio, activities of glutaredoxins (Grx) 1 and 2 as oxidoreductases as well as Grxs 3 and 5 in iron homeostasis influence accumulation of alpha-synuclein aggregates or oxidation of dopamine, for instance, leading subsequently to dysfunction and damage of dopaminergic cells.

16.4 CONCLUSIONS

In summary, Grxs are important/essential proteins during development and diseases of the brain. They can attenuate or exacerbate disease progression. Both subfamilies, mono- and dithiol Grxs, and their specific functions, iron homeostasis and redox regulation, contribute to the described role in the brain. This role of Grxs is closely connected to, and dependent on, GSH and iron. Although some mechanisms affecting development and diseases of the central nervous system have been elucidated in detail, e.g., posttranslational modifications of CRMP2, many mechanisms underlying Grx-dependent phenotypes are unknown.

REFERENCES

Ahmad, F., P. Nidadavolu, L. Durgadoss, and V. Ravindranath. 2014. Critical cysteines in Akt1 regulate its activity and proteasomal degradation: Implications for neurodegenerative diseases. *Free Rad Biol Med* 74:118–28.

Akterin, S., R. F. Cowburn, A. Miranda-Vizuete et al. 2006. Involvement of glutaredoxin-1 and thioredoxin-1 in beta-amyloid toxicity and Alzheimer's disease. *Cell Death Differ* 13:1454–65.

Alunni, A., and L. Bally-Cuif. 2016. A comparative view of regenerative neurogenesis in vertebrates. *Development* 143:741–53.

Angelova, D., and D. Brown. 2015. Iron, aging, and neurodegeneration. *Metals* 5:2070–92.

Arodin, L., A. Miranda-Vizuete, P. Swoboda, and A. P. Fernandes. 2014. Protective effects of the thioredoxin and glutaredoxin systems in dopamine-induced cell death. *Free Rad Biol Med* 73:328–36.

Badaracco, M. E., E. H. Ortiz, E. F. Soto, J. Connor, and J. M. Pasquini. 2008. Effect of transferrin on hypomyelination induced by iron deficiency. *J Neurosci Res* 86:2663–73.

Banci, L., S. Ciofi-Baffoni, K. Gajda, R. Muzzioli, R. Peruzzini, and J. Winkelmann. 2015. N-terminal domains mediate [2Fe-2S] cluster transfer from glutaredoxin-3 to anamorsin. *Nature Chem Biol* 11:772–8.

Bandyopadhyay, S., F. Gama, M. M. Molina-Navarro et al. 2008. Chloroplast monothiol glutaredoxins as scaffold proteins for the assembly and delivery of [2Fe-2S] clusters. *EMBO J* 27:1122–33.

Baxter, P. S., K. F. S. Bell, P. Hasel et al. 2015. Synaptic NMDA receptor activity is coupled to the transcriptional control of the glutathione system. *Nature Commun* 6:6761.

Beard, J. L., J. A. Wiesinger, and J. R. Connor. 2003. Pre- and postweaning iron deficiency alters myelination in Sprague-Dawley rats. *Dev Neurosci* 25:308–15.

Begas, P., L. Liedgens, A. Moseler, A. J. Meyer, and M. Deponte. 2017. Glutaredoxin catalysis requires two distinct glutathione interaction sites. *Nature Commun* 8:14835.

Berndt, C., C. Hudemann, E.-M. Hanschmann et al. 2007. How does iron-sulfur cluster coordination regulate the activity of human glutaredoxin 2? *Antioxid Redox Signal* 9:151–7.

Berndt, C., and C. H. Lillig. 2017. Glutathione, glutaredoxins, and iron. *Antioxid Redox Signal* 27:1235–51.

Berndt, C., C. H. Lillig, and L. Flohé. 2014. Redox regulation by glutathione needs enzymes. *Frontiers Pharmacol* 5:168.

Bräutigam, L., L. D. E. Jensen, G. Poschmann et al. 2013. Glutaredoxin regulates vascular development by reversible glutathionylation of sirtuin 1. *Proc Natl Acad Sci U S A* 110:20057–62.

Bräutigam, L., L. D. Schütte, J. R. Godoy et al. 2011. Vertebrate-specific glutaredoxin is essential for brain development. *Proc Natl Acad Sci U S A* 108:20532–7.

Cabungcal, J.-H., D. Preissmann, C. Delseth et al. 2007. Transitory glutathione deficit during brain development induces cognitive impairment in juvenile and adult rats: Relevance to schizophrenia. *Neurobiol Dis* 26:634–45.

Calabrese, V., G. Scapagnini, A. Ravagna et al. 2002. Nitric oxide synthase is present in the cerebrospinal fluid of patients with active multiple sclerosis and is associated with increases in cerebrospinal fluid protein nitrotyrosine and *S*-nitrosothiols and with changes in glutathione levels. *J Neurosci Res* 70:580–7.

Camaschella, C., A. Campanella, L. de Falco et al. 2007. The human counterpart of zebrafish shiraz shows sideroblastic-like microcytic anemia and iron overload. *Blood* 110:1353–8.

Carvalho, A. N., J. L. Lim, P. G. Nijland, M. E. Witte, and J. van Horssen. 2014. Glutathione in multiple sclerosis: More than just an antioxidant? *Multiple Sclerosis* 20:1425–31.

Castagné, V., M. Rougemont, M. Cuenod, and K. Q. Do. 2004. Low brain glutathione and ascorbic acid associated with dopamine uptake inhibition during rat's development induce long-term cognitive deficit: Relevance to schizophrenia. *Neurobiol Dis* 15:93–105.

Chazov, E. I., O. V. Rodnenkov, A. V. Zorin et al. 2012. Hypotensive effect of Oxacom® containing a dinitrosyl iron complex with glutathione: Animal studies and clinical trials on healthy volunteers. *Nitric Oxide Biol Chem* 26:148–56.

Chen, C.-A., F. de Pascali, A. Basye, C. Hemann, and J. L. Zweier. 2013. Redox modulation of endothelial nitric oxide synthase by glutaredoxin-1 through reversible oxidative post-translational modification. *Biochemistry* 52:6712–23.

Cheng, N.-H., W. Zhang, W.-Q. Chen et al. 2011. A mammalian monothiol glutaredoxin, Grx3, is critical for cell cycle progression during embryogenesis. *FEBS J* 278:2525–39.

Chouraqui, E., A. Leon, Y. Repesse et al. 2013. Deferoxamine blocks death induced by glutathione depletion in PC 12 cells. *Neurotoxicol* 37:221–30.

Connor, J. R., and S. L. Menzies. 1996. Relationship of iron to oligodendrocytes and myelination. *Glia* 17:83–93.

Craelius, W., M. W. Migdal, C. P. Luessenhop, A. Sugar, and I. Mihalakis. 1982. Iron deposits surrounding multiple sclerosis plaques. *Arch Path Lab Med* 106:397–9.

Dallman, P. R. 1986. Biochemical basis for the manifestations of iron deficiency. *Annu Rev Nutr* 6:13–40.

Deponte, M. 2013. Glutathione catalysis and the reaction mechanisms of glutathione-dependent enzymes. *Biochim Biophys Acta* 1830:3217–66.

Deregnier, R. A., C. A. Nelson, K. M. Thomas, S. Wewerka, and M. K. Georgieff. 2000. Neurophysiologic evaluation of auditory recognition memory in healthy newborn infants and infants of diabetic mothers. *J Pediatr* 137:777–84.

de Deungria, M., R. Rao, J. D. Wobken et al. 2000. Perinatal iron deficiency decreases cytochrome c oxidase (CytOx) activity in selected regions of neonatal rat brain. *Pediatric Res* 48:169–76.

Devos, D., C. Moreau, J. C. Devedjian et al. 2014. Targeting chelatable iron as a therapeutic modality in Parkinson's disease. *Antioxid Redox Signal* 21:195–210.

Dexter, D. T., J. Sian, S. Rose et al. 1994. Indices of oxidative stress and mitochondrial function in individuals with incidental Lewy body disease. *Ann Neurol* 35:38–44.

Diwakar, L., R. S. Kenchappa, J. Annepu, and V. Ravindranath. 2007. Downregulation of glutaredoxin but not glutathione loss leads to mitochondrial dysfunction in female mice CNS: Implications in excitotoxicity. *Neurochem Int* 51:37–46.

Do Van, B., F. Gouel, A. Jonneaux et al. 2016. Ferroptosis, a newly characterized form of cell death in Parkinson's disease that is regulated by PKC. *Neurobiol Dis* 94:169–78.

Eichele, G. 1992. Budding thoughts. *Sciences* 32:30–6.

Farr, S. A., H. F. Poon, D. Dogrukol-Ak et al. 2003. The antioxidants alpha-lipoic acid and *N*-acetylcysteine reverse memory impairment and brain oxidative stress in aged SAMP8 mice. *J Neurochem* 84:1173–83.

Firth, E. C., and Y. Greydanus. 1987. Cartilage thickness measurement in foals. *Res Vet Sci* 42:35–46.

Fitzgerald, M. J. T, G. Gruener, and E. Mtui. 2012. *Clinical Neuroanatomy and Neuroscience*. Philadelphia: Saunders/Elsevier.

Fournier, M., A. Monin, C. Ferrari et al. 2017. Implication of the glutamate-cystine antiporter xCT in schizophrenia cases linked to impaired GSH synthesis. *NPJ Schizophrenia* 3:31.

Garrido, M., Y. Tereshchenko, Z. Zhevtsova et al. 2011. Glutathione depletion and overproduction both initiate degeneration of nigral dopaminergic neurons. *Acta Neuropathol* 121:475–85.

Gascón, S., E. Murenu, G. Masserdotti et al. 2016. Identification and successful negotiation of a metabolic checkpoint in direct neuronal reprogramming. *Cell Stem Cell* 18:396–409.

Gauron, C., F. Meda, E. Dupont et al. 2016. Hydrogen peroxide (H_2O_2) controls axon pathfinding during zebrafish development. *Dev Biol* 414:133–41.

Gellert, M., S. Venz, J. Mitlöhner et al. 2013. Identification of a dithiol-disulfide switch in collapsin response mediator protein 2 (CRMP2) that is toggled in a model of neuronal differentiation. *J Biol Chem* 288:35117–25.

Gilbert, S. F. 2014. *Developmental Biology*. Sunderland, MA: Sinauer.

Glöckner, R., and M. Kretzschmar. 1991. Perinatal glutathione levels in liver and brain of rats from large and small litters. *Biol Neonate* 59:287–93.

Hallgren, B., and P. Sourander. 1958. The effect of age on the non-haemin iron in the human brain. *J Neurochem* 3:41–51.

Hambright, W. S., R. S. Fonseca, L. Chen, R. Na, and Q. Ran. 2017. Ablation of ferroptosis regulator glutathione peroxidase 4 in forebrain neurons promotes cognitive impairment and neurodegeneration. *Redox Biol* 12:8–17.

Hansen, J. M., and C. Harris. 2015. Glutathione during embryonic development. *Biochim Biophys Acta* 1850:1527–42.

Haunhorst, P., E.-M. Hanschmann, L. Bräutigam et al. 2013. Crucial function of vertebrate glutaredoxin 3 (PICOT) in iron homeostasis and hemoglobin maturation. *Mol Biol Cell* 24:1895–903.

Hensley, K., and P. Kursula. 2016. Collapsin response mediator protein-2 (CRMP2) is a plausible etiological factor and potential therapeutic target in Alzheimer's disease: Comparison and contrast with microtubule-associated protein tau. *J Alzheimer's Dis* 53:1–14.

Hider, R. C., and X. L. Kong. 2011. Glutathione: A key component of the cytoplasmic labile iron pool. *Biometals* 24:1179–87.

Hof, P. R., and C. V. Mobbs. 2009. *Handbook of the Neuroscience of Aging*. Amsterdam: Elsevier.

Holmgren, A. 1976. Hydrogen donor system for *Escherichia coli* ribonucleoside-diphosphate reductase dependent upon glutathione. *Proc Natl Acad Sci USA* 73:2275–9.

Ibi, M., H. Sawada, T. Kume et al. 1999. Depletion of intracellular glutathione increases susceptibility to nitric oxide in mesencephalic dopaminergic neurons. *J Neurochem* 73:1696–703.

Ingold, I., C. Berndt, S. Schmitt et al. 2018. Selenium utilization by GPX4 Is required to prevent hydroperoxide-induced ferroptosis. *Cell* 172:409–22.

Iwema, T., A. Picciocchi, D. A. Traore et al. 2009. Structural basis for delivery of the intact [Fe2S2] cluster by monothiol glutaredoxin. *Biochemistry* 48:6041–3.

Jana, M., and K. Pahan. 2005. Redox regulation of cytokine-mediated inhibition of myelin gene expression in human primary oligodendrocytes. *Free Radic Biol Med* 39:823–31.

Johansson, C., K. L. Kavanagh, O. Gileadi, and U. Oppermann. 2007. Reversible sequestration of active site cysteines in a 2Fe-2S-bridged dimer provides a mechanism for glutaredoxin 2 regulation in human mitochondria. *J Biol Chem* 282:3077–82.

Johnson, W. M., M. Golczak, K. Choe et al. 2016. Regulation of DJ-1 by glutaredoxin 1 in vivo: Implications for Parkinson's disease. *Biochemistry* 55:4519–32.

Johnson, D. A., and J. A. Johnson. 2015. Nrf2—A therapeutic target for the treatment of neurodegenerative diseases. *Free Radic Biol Med* 88:253–67.

Johnson, W. M., C. Yao, S. L. Siedlak et al. 2015. Glutaredoxin deficiency exacerbates neurodegeneration in *C. elegans* models of Parkinson's disease. *Hum Mol Gen* 24:1322–35.

Jurado, J., M.-J. Prieto-Alamo, J. Madrid-Rísquez, and C. Pueyo. 2003. Absolute gene expression patterns of thioredoxin and glutaredoxin redox systems in mouse. *J Biol Chem* 278:45546–54.

Kalinderi, K., S. Bostantjopoulou, and L. Fidani. 2016. The genetic background of Parkinson's disease: Current progress and future prospects. *Acta Neurol Scand* 134:314–26.

Karunakaran, S., U. Saeed, S. Ramakrishnan, R. C. Koumar, and V. Ravindranath. 2007. Constitutive expression and functional characterization of mitochondrial glutaredoxin (Grx2) in mouse and human brain. *Brain Res* 1185:8–17.

Kaur, S. J., S. R. McKeown, and S. Rashid. 2016. Mutant SOD1 mediated pathogenesis of amyotrophic lateral sclerosis. *Gene* 577:109–18.

Kenchappa, R. S., and V. Ravindranath. 2003. Glutaredoxin is essential for maintenance of brain mitochondrial complex I: Studies with MPTP. *FASEB J* 17:717–9.

Kim, Y. M., H. T. Chung, R. L. Simmons, and T. R. Billiar. 2000. Cellular non-heme iron content is a determinant of nitric oxide-mediated apoptosis, necrosis, and caspase inhibition. *J Biol Chem* 275:10954–61.

Kizil, C., J. Kaslin, V. Kroehne, and M. Brand. 2012. Adult neurogenesis and brain regeneration in zebrafish. *Dev Neurobiol* 72:429–61.

Kumar, C., A. Igbaria, B. D'Autreaux et al. 2011. Glutathione revisited: A vital function in iron metabolism and ancillary role in thiol-redox control. *EMBO J* 30:2044–56.

Lan, J., and D. H. Jiang. 1997. Desferrioxamine and vitamin E protect against iron and MPTP-induced neurodegeneration in mice. *J Neural Transm* 104:469–81.

Lee, D. W., D. Kaur, S. J. Chinta, S. Rajagopalan, and J. K. Andersen. 2009. A disruption in iron-sulfur center biogenesis via inhibition of mitochondrial dithiol glutaredoxin 2 may contribute to mitochondrial and cellular iron dysregulation in mammalian glutathione-depleted dopaminergic cells: Implications for Parkinson's disease. *Antioxid Redox Signal* 1:2083–94.

Lenkowski, J. R., Z. Qin, C. J. Sifuentes et al. 2013. Retinal regeneration in adult zebrafish requires regulation of TGFβ signaling. *Glia* 61:1687–97.

Lepka, K., K. Volbracht, E. Bill et al. 2017. Iron-sulfur glutaredoxin 2 protects oligodendrocytes against damage induced by nitric oxide release from activated microglia. *Glia* 65:1521–34.

Li, J., S. A. Pearson, K. D. Fenk, and J. A. Cowan. 2015. Glutathione-coordinated [2Fe-2S] cluster is stabilized by intramolecular salt bridges. *J Biol Inorganic Chem* 20:1221–27.

Lillig, C. H., and C. Berndt. 2013. Glutaredoxins in thiol/disulfide exchange. *Antioxid Redox Signal* 18:1654–65.

Lillig, C. H., C. Berndt, and A. Holmgren. 2008. Glutaredoxin systems. *Biochim Biophys Acta* 1780:1304–17.

Lillig, C. H., C. Berndt, O. Vergnolle et al. 2005. Characterization of human glutaredoxin 2 as iron-sulfur protein: A possible role as redox sensor. *Proc Natl Acad Sci U S A* 102:8168–73.

Lin, P.-C., P. M. Chan, C. Hall, and E. Manser. 2011. Collapsin response mediator proteins (CRMPs) are a new class of microtubule-associated protein (MAP) that selectively interacts with assembled microtubules via a taxol-sensitive binding interaction. *J Biol Chem* 286:41466–78.

Liu, G., P. Men, P. L. Harris et al. 2006. Nanoparticle iron chelators: A new therapeutic approach in Alzheimer disease and other neurologic disorders associated with trace metal imbalance. *Neurosci Lett* 406:189–93.

Lok, H. C., Y. Suryo Rahmanto, C. L. Hawkins et al. 2012. Nitric oxide storage and transport in cells are mediated by glutathione *S*-transferase P1-1 and multidrug resistance protein 1 via dinitrosyl iron complexes. *J Biol Chem* 287:607–18.

Luthman, M., S. Eriksson, A. Holmgren, and L. Thelander. 1979. Glutathione-dependent hydrogen donor system for calf thymus ribonucleoside-diphosphate reductase. *Proc Natl Acad Sci U S A* 76:2158–62.

Mandal, P. K., S. Saharan, M. Tripathi, and G. Murari. 2015. Brain glutathione levels—A novel biomarker for mild cognitive impairment and Alzheimer's disease. *Biol Psychiatry* 78:702–10.

Martin, J. L. 1995. Thioredoxin—A fold for all reasons. *Structure* 3:245–50.

März, M., R. Schmidt, S. Rastegar, and U. Strähle. 2011. Regenerative response following stab injury in the adult zebrafish telencephalon. *Dev Dyn* 240:2221–31.

Messina, J. P., and D. A. Lawrence. 1989. Cell cycle progression of glutathione-depleted human peripheral blood mononuclear cells is inhibited at S phase. *J Immunol* 143:1974–81.

Miller, G. O., J. B. Behring, S. L. Siedlak et al. 2016. Upregulation of glutaredoxin-1 activates microglia and promotes neurodegeneration: Implications for parkinson's disease. *Antioxid Redox Signa* 25:967–82.

Mitchell, K. M., A. L. Dotson, K. M. Cool et al. 2007. Deferiprone, an orally deliverable iron chelator, ameliorates experimental autoimmune encephalomyelitis. *Multiple Sclerosis* 13:1118–26.

Möller, D., M. Gellert, W. Langel, and C. H. Lillig. 2017. Molecular dynamics simulations and *in vitro* analysis of the CRMP2 thiol switch. *Mol BioSyst* 13:1744–53.

Moseler, A., I. Aller, S. Wagner et al. 2015. The mitochondrial monothiol glutaredoxin S15 is essential for iron-sulfur protein maturation in *Arabidopsis thaliana*. *Proc Natl Acad Sci U S A* 112:13735–40.

Mühlenhoff, U., J. Gerber, N. Richhardt, and R. Lill. 2003. Components involved in assembly and dislocation of iron-sulfur clusters on the scaffold protein Isu1p. *EMBO J* 22:4815–25.

Mühlenhoff, U., S. Molik, J. R. Godoy et al. 2010. Cytosolic monothiol glutaredoxins function in intracellular iron sensing and trafficking via their bound iron-sulfur cluster. *Cell Metabol* 2:373–85.

Nanda, D., J. Tolputt, and K. J. Collard. 1996. Changes in brain glutathione levels during postnatal development in the rat. *Brain Res Dev Brain Res* 94:238–41.

Niwa, S., F. Nakamura, Y. Tomabechi et al. 2017. Structural basis for CRMP2-induced axonal microtubule formation. *Sci Rep* 7:10681.

Pallardó, F. V., J. Markovic, J. L. García, and J. Viña. 2009. Role of nuclear glutathione as a key regulator of cell proliferation. *Mol Aspects Med* 30:77–85.

Pearce, R. K., A. Owen, S. Daniel, P. Jenner, and C. D. Marsden. 1997. Alterations in the distribution of glutathione in the *substantia nigra* in Parkinson's disease. *J Neural Transm* 104:661–77.

Prozorovski, T., R. Schneider, C. Berndt et al. 2015. Redox-regulated fate of neural stem progenitor cells. *Biochim Biophys Acta* 1850:1543–54.

Qi, W., J. Li, C. Y. Chain, G. A. Pasquevich, A. F. Pasquevich, and J. A. Cowan. 2012. Glutathione complexed Fe-S centers. *J Am Chem Soc* 134:10745–8.

Rao, A. R., H. Quach, E. Smith, G. T. Vatassery, and R. Rao. 2014. Changes in ascorbate, glutathione and α-tocopherol concentrations in the brain regions during normal development and moderate hypoglycemia in rats. *Neurosci Lett* 568:67–71.

Rao, R., I. Tkac, E. L. Townsend, R. Gruetter, and M. K. Georgieff. 2003. Perinatal iron deficiency alters the neurochemical profile of the developing rat hippocampus. *J Nutr* 133:3215–21.

Rice, M. E., and I. Russo-Menna. 1998. Differential compartmentalization of brain ascorbateand glutathione between neurons and glia. *Neuroscience* 82:1213–23.

Rodríguez-Manzaneque, M. T., J. Tamarit, G. Bellí, J. Ros, and E. Herrero. 2002. Grx5 is a mitochondrial glutaredoxin required for the activity of iron/sulfur enzymes. *Mol Biol Cell* 13:1109–21.

Rouault, T. A. 2013. Iron metabolism in the CNS: Implications for neurodegenerative diseases. Nature reviews. *Neuroscience* 14:551–64.

Rouhier, N., H. Unno, S. Bandyopadhyay et al. 2007. Functional, structural, and spectroscopic characterization of a glutathione-ligated [2Fe-2S] cluster in poplar glutaredoxin C1. *Proc Natl Acad Sci U S A* 104:7379–84.

Sabens Liedhegner, E. A., X.-H. Gao, and J. J. Mieyal. 2012. Mechanisms of altered redox regulation in neurodegenerative diseases—Focus on *S*-glutathionylation. *Antioxid Redox Signal* 16:543–66.

Saharan, S., and P. K. Mandal. 2014. The emerging role of glutathione in Alzheimer's disease. *J Alzheimer's Dis* 40:519–29.

Schmidt, E. F., and S. M. Strittmatter. 2007. The CRMP family of proteins and their role in Sema3A signaling. *Adv Exp Med Biol* 600:1–11.

Sechi, G., M. G. Deledda, G. Bua et al. 1996. Reduced intravenous glutathione in the treatment of early Parkinson's disease. *Progr Neuro-psychopharm Biol Psychiatry* 20:1159–70.

Shafir, T., R. Angulo-Barroso, A. Calatroni, E. Jimenez, and B. Lozoff. 2006. Effects of iron deficiency in infancy on patterns of motor development over time. *Hum Mov Sci* 25:821–38.

Shang, Q., L. Bao, H. Guo et al. 2017. Contribution of glutaredoxin-1 to *S*-glutathionylation of endothelial nitric oxide synthase for mesenteric nitric oxide generation in experimental necrotizing enterocolitis. *Translat Res J Lab Clin Med* 188:92–105.

Shekhter, A. B., T. G. Rudenko, L. P. Istranov et al. 2015. Dinitrosyl iron complexes with glutathione incorporated into a collagen matrix as a base for the design of drugs accelerating skin wound healing. *Eur J Pharml Sci* 78:8–18.

Sian, J., D. T. Dexter, A. J. Lees et al. 1994. Alterations in glutathione levels in Parkinson's disease and other neurodegenerative disorders affecting basal ganglia. *Ann Neurol* 36:348–55.

Silva, G. M., L. E. Netto, K. F. Discola et al. 2008. Role of glutaredoxin 2 and cytosolic thioredoxins in cysteinyl-based redox modification of the 20S proteasome. *FEBS J* 275:2942–55.

Smith, K. J., R. Kapoor, and P. A. Felts. 1999. Demyelination: The role of reactive oxygen and nitrogen species. *Brain Pathol* 9:69–92.

Sohal, R. S., and R. Weindruch. 1996. Oxidative stress, caloric restriction, and aging. *Science* 273:59–63.

Sowell, E. R., B. S. Peterson, P. M. Thompson et al. 2003. Mapping cortical change across the human life span. *Nature Neurosci* 6:309–15.

Spillantini, M. G., M. L. Schmidt, V. M. Lee et al. 1997. Alpha-synuclein in Lewy bodies. *Nature* 388:839–40.

Stockwell, B. R., J. P. Friedmann Angeli, H. Bayir et al. 2017. Ferroptosis: A regulated cell death nexus linking metabolism, redox biology, and disease. *Cell* 171:273–85.

Sugiura, Y., and H. Tanaka. 1972. Iron-sulfide chelates of some sulfur-containing peptides as model complex of non-heme iron proteins. *Biochem Biophys Res Commun* 46:335–40.

Takagi, Y., A. Mitsui, A. Nishiyama et al. 1999. Overexpression of thioredoxin in transgenic mice attenuates focal ischemic brain damage. *Proc Natl Acad Sci U S A* 96:4131–6.

Tamura, T., R. L. Goldenberg, J. Hou et al. 2002. Cord serum ferritin concentrations and mental and psychomotor development of children at five years of age. *J Pediatr* 140:165–70.

Timme-Laragy, A. R., J. V. Goldstone, B. R. Imhoff et al. 2013. Glutathione redox dynamics and expression of glutathione-related genes in the developing embryo. *Free Rad Biol Med* 65:89–101.

Uchida, Y., T. Ohshima, Y. Sasaki et al. 2005. Semaphorin3A signalling is mediated via sequential Cdk5 and GSK3beta phosphorylation of CRMP2: Implication of common phosphorylating mechanism underlying axon guidance and Alzheimer's disease. *Genes Cells* 10:165–79.

Vanin, A. F. 2009. Dinitrosyl iron complexes with thiolate ligands: Physico-chemistry, biochemistry and physiology. *Nitric Oxide Biol Chem* 21:1–13.

Venkateshappa, C., G. Harish, A. Mahadevan, M. M. Srinivas Bharath, and S. K. Shankar. 2012. Elevated oxidative stress and decreased antioxidant function in the human hippocampus and frontal cortex with increasing age: Implications for neurodegeneration in Alzheimer's disease. *Neurochem Res* 37:1601–14.

Wachs, T. D., E. Pollitt, S. Cueto, E. Jacoby, and H. Creed-Kanashiro. 2005. Relation of neonatal iron status to individual variability in neonatal temperament. *Dev Psychobiol* 46:141–53.

Ward, K. L., I. Tkac, Y. Jing et al. 2007. Gestational and lactational iron deficiency alters the developing striatal metabolome and associated behaviors in young rats. *J Nutr* 137:1043–9.

White, A. R., and R. Cappai. 2003. Neurotoxicity from glutathione depletion is dependent on extracellular trace copper. *J Neurosci Res* 71:889–97.

Wilson, C., E. Muñoz-Palma, and C. González-Billault. 2018. From birth to death: A role for reactive oxygen species in neuronal development. *Semin Cell Dev Biol* 80:43–9.

Wingert, R. A., J. L. Galloway, B. Barut et al. 2005. Deficiency of glutaredoxin 5 reveals Fe-S clusters are required for vertebrate haem synthesis. *Nature* 436:1035–9.

Yang, W. S., R. SriRamaratnam, M. E. Welsch et al. 2014. Regulation of ferroptotic cancer cell death by GPX4. *Cell* 156:317–31.

Yang, W. S., and B. R. Stockwell. 2016. Ferroptosis: Death by lipid peroxidation. *Trends Cell Biol* 26:165–76.

Youdim, M. B., and A. R. Green. 1978. Iron deficiency and neurotransmitter synthesis and function. *Proc Nutr Soc* 37:173–79.

Youdim, M. B., M. A. Sills, W. E. Heydorn, G. J. Creed, and D. M. Jacobowitz. 1986. Iron deficiency alters discrete proteins in rat caudate nucleus and nucleus accumbens. *J Neurochem* 47:794–9.

Part V

Glutathione Derivatives and
Substitutes in Pathogenic
Microorganisms

17 Biosynthesis of Polyamine–Glutathione Derivatives in *Enterobacteria* and *Kinetoplastida*

Marcelo A. Comini

Institut Pasteur de Montevideo

CONTENTS

17.1 DISCOVERY OF GSH–POLYAMINE CONJUGATES AND SIGNIFICANCE

Polyamines, in particular spermidine (Sp), and the tripeptide glutathione (GSH) are predominant metabolites in almost all organisms. The polycationic and aliphatic nature of polyamines favor their interaction with several macromolecules (i.e., nucleic acids, phospholipids, and proteins) and, hence, their involvement in a plethora of fundamental cellular events such as transcription, translation, regulation of enzyme activity, differentiation, and proliferation. In pathogens, the action of polyamines have been linked to microbial carcinogenesis, biofilm formation, escape from phagolysosomes, bacteriocin production, toxin activity, and protection from oxidative and acid stress (Shah and Swiatlo, 2008; Willert and Phillips, 2008). Apart from their regulatory or stabilizing role when forming complexes with different macromolecules, the polyamines also serve as substrates for the synthesis of different alkaloids in plants (Shoji and Hashimoto, 2015) and of deoxyhypusine in eukaryotic

DOI: 10.1201/9781351261760-22

cells (Park, 2006). For instance, deoxyhypusine is incorporated to the eukaryotic translation initiation factor 5A (eIF5A), which transports newly transcribed mRNAs from the nucleus to the cytoplasm.

The tripeptide GSH functions as redox cosubstrate of enzymes that prevent oxidative damage and regulate important cellular processes (see Chapters 2–16). In addition, GSH has been shown to be ligand for iron–sulfur clusters that play structural or catalytic roles in proteins (see Chapters 15, 16, and 18).

In the majority of the living organisms, polyamine and glutathione metabolism do not converge. Pioneer metabolic studies conducted in *Escherichia coli* led to the discovery of glutathione–spermidine conjugates (Dubin, 1959). Several decades later and motivated by the unusual redox and hydroperoxide metabolism of trypanosomatids, Fairlamb and Cerami (Fairlamb and Cerami, 1985) reported the identification of a sulfur-containing, low-molecular weight cofactor, in dialyzed cell-free extracts of African trypanosomes. The structure of these conjugates was determined to be predominantly N^1-mono(glutathionyl)spermidine (Gsp) and N^1, N^8-bis(glutathionyl)spermidine [trypanothione or $T(SH)_2$] (Fairlamb and Cerami, 1985), where the terminal amine(s) of Sp is (are) bound to the carbon atom from the carboxylate moiety of glycine (Figure 17.1).

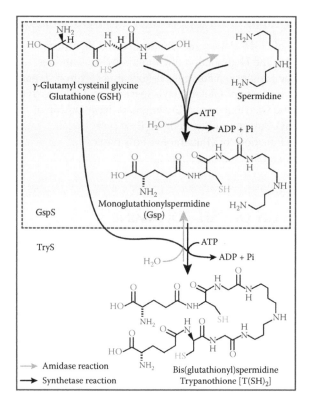

FIGURE 17.1 Structure, biosynthesis, and hydrolysis of glutathionylspermidine conjugates. GspS, monoglutathionylspermidine synthetase/amidase; TryS, trypanothione synthetase/amidase. The reactions catalyzed by GspS are highlighted with a gray box.

In *E. coli*, the mono- but not the bis-glutathionyl derivative was detected (Tabor and Tabor, 1970), whereas in kinetoplastids, in addition to Gsp, significant amounts of T(SH)$_2$ were detected (Fairlamb et al., 1985). Despite the initial high expectations for an important biological role of Gsp in *E. coli*, subsequent experiments proved its dispensability for bacterial growth *in vitro* (Chiang et al., 2010; Chattopadhyay et al., 2013). Nonetheless, these and more recent studies provided novel insights in a potential role of Gsp in bacterial gene expression, adaptation to cold and redox signaling. In contrast, the discovery of the glutathionylspermidine conjugates in kinetoplastids, together with the confirmation of its essentiality for pathogenic species (Comini et al., 2004; Sousa et al., 2014), was of paramount relevance for the parasitological field, because it disclosed the missing link of the thiol-dependent redox system of these organisms and, given its uniqueness, provided several excellent candidates for drug development (Comini and Flohé, 2013).

17.2   BIOSYNTHESIS AND DEGRADATION OF GLUTATHIONYL-POLYAMINES

The covalent joining and the hydrolysis of glutathionyl-polyamines is catalyzed by paralog enzymes denominated monoglutathionyl spermidine synthetase/amidase (GspS) and trypanothione synthetase/amidase (TryS). GspS and TryS were isolated, purified, and characterized from *E. coli* (Bollinger et al. 1995; Kwon et al. 1997) and different kinetoplastids (Smith et al., 1992; Oza et al., 2002a,b; Comini et al., 2003; Oza et al., 2003; Comini et al., 2005; Oza et al., 2005; Leroux et al., 2013; Benitez et al., 2016), These enzymes are bifunctional. They consist of a unique fusion between a peptidase domain, i.e., the C-terminal domain of the NlpC/p60 protein superfamily (Anantharaman and Aravind, 2003) that hydrolyzes mono- and bis(glutathionyl)spermidine back to the original substrates GSH and Sp, and a C-terminal ligase domain with an ATP-grasp fold (Fawaz et al., 2011) that is responsible for the ATP-dependent synthesis of the glutathionyl-polyamines. Addition of a single GSH to Sp yields N^1- and N^8-regioisomers of Gsp and is catalyzed by either GspS or TryS (Figure 17.2). In contrast, conjugation of a second GSH to the free amine of Gsp is performed only by TryS and produces T(SH)$_2$ (Figure 17.1). The structural elements responsible for such substrate selectivity are described in Section 17.3 and Figure 17.3. In each reaction cycle, ATP-hydrolysis is used to activate the glycyl carboxylate of GSH (Koenig et al., 1997). In the second step, the resulting acylphosphate intermediate is attacked by the free amine from the polyamines to produce a new C–N bond [e.g., Gsp or T(SH)$_2$] with the concomitant elimination of ADP and phosphate (Figures 17.1 and 17.2).

Kinetic data available for recombinant GspS from *E. coli* (*Ec*GspS; Bollinger et al., 1995) and *Crithidia fasciculata* (*Cf*GspS; Oza et al., 2002a) show values of Michaelis constant (K_M) for the substrates of the synthetase activity that are within the same order of magnitude (Table 17.1). Interestingly, both enzymes present a comparatively low affinity for the thiol substrate (K_{MGSH} = 240–800 μM) than for the polyamine (K_{MSp} = 60–142 μM) and the nucleotide (K_{MATP} = 100–230 μM). In comparison, the K_M values for each substrate of TryS show large variations between species that, nevertheless, follow the order: $K_{MGsp} <$ or $\approx K_{MATP} < K_{MGSH} < K_{MSp}$

(Table 17.1). The exception to this rule is the enzyme from *C. fasciculata* with a K_M for Gsp (480 μM), which is higher than that for ATP (52–222 μM) (Comini et al., 2005). Noticeably, the crithidial TryS has the larger K_M values for polyamine substrates (>5-fold for Sp and Gsp) when compared to its homologue from pathogenic Kinetoplastids. This may be explained by the higher intracellular abundance of free polyamines (>2 mM) and Gsp in *Crithidia* (0.9–2.3 mM) (Shim and Fairlamb, 1988; Ariyanayagam and Fairlamb, 2001) than in trypanosomatids (≤2 mM for Sp and 0.01–1.2 mM for Gsp) [reviewed in Coombs and Sanderson (1985), Oza et al., (2002b), Ariyanayagam et al.(2003), Krauth-Siegel and Comini (2008), Taylor et al. (2008)]. With Sp as substrate, *Cf*TryS has a nearly three orders of magnitude lower catalytic efficiency ($k_{cat}/K_M = 7.6 \times 10^2$ M^{-1}s^{-1}) than *Cf*GspS ($k_{cat}/K_M = 2.6 \times 10^5$ M^{-1}s^{-1}; Tables 17.1 and 17.2), which agrees with the need of this organism for an independent source of Gsp to support an efficient production of T(SH)$_2$. At variance with *Crithidia*, most trypanosomatids are GspS deficient (e.g., *Trypanosoma brucei*, *Leishmania infantum*, and *Trypanosoma cruzi*) [reviewed in Manta et al. (2018), see also Section 17.4]. In line with its functional role, all TryS show an affinity for Sp that is 1.7- to 120-fold lower than that of GspS, and a comparatively higher affinity for Gsp than for Sp, as noted by the 9- to 75-fold lower K_M values for the first substrate (Table 17.1). The substrate selectivity evolved by TryS is also exemplified by their 2- to 3-order of magnitude higher catalytic efficiency with Gsp ($k_{cat}/K_M = 1.8 \times 10^4$ M^{-1}s^{-1}, 6.6×10^4 M^{-1}s^{-1}, 4.2×10^4 M^{-1}s^{-1} and 5.0×10^4 M^{-1}s^{-1} for *Cf*TryS, *Tb*TryS, *Tc*TryS, and *Lm*TryS, respectively) than with Sp ($k_{cat}/K_M = 5.6 \times 10^2$ M^{-1}s^{-1}, 4.1×10^3 M^{-1}s^{-1}, 5.4×10^3 M^{-1}s^{-1} and 2.1×10^3 M^{-1}s^{-1} for *Cf*TryS, *Tb*TryS, *Tc*TryS and *Lm*TryS, respectively). Also, TryS from pathogenic trypanosomatids are 10-fold more efficient than the crithidial homologue in using Sp as substrate, suggesting that the former organisms may accomplish T(SH)$_2$ biosynthesis without the need of an additional GspS. Interestingly, for *Cf*TryS the turnover of N^8-Gsp ($k_{cat}/K_M = 3.47 \times 10^6$ M^{-1}s^{-1}) was higher than with N^1-Gsp [$k_{cat}/K_M = 5.0 \times 10^5$ M^{-1}s^{-1} (Henderson et al., 1990)]. Docking studies confirmed that the N^8 regioisomer adopts a more stable and productive binding to the Gsp-pocket of *Lm*TryS than the N^1 isomer (Koch et al., 2013). On the other hand, *Ec*GspS has been shown to have a remarkable preference for catalyzing the formation of N^1-Gsp (Bollinger et al., 1995). The preference of TryS for catalyzing the modification of a particular amino-terminal group of Sp is unknown and cannot be inferred from the structural models, as the polyamine binding site is prone to undergo conformational changes (see Section 17.3).

The comparison of the kinetic parameters of TryS from different species is somehow hampered, because assay conditions (buffer, pH, and temperature) have been reported to affect the kinetic behavior of the enzymes (Leroux et al., 2013). However, kinetic data collected under identical assay conditions highlighted important species-specific differences among trypanosomatid TryS in regard to their affinity for substrates and inhibitors (Benitez et al., 2016). *T. brucei* TryS displayed K_M values for ATP and Sp that are >2-fold lower than those of the *T. cruzi* and *L. infantum* enzyme (Table 17.1). Such differences are likely due to the peculiar metabolic and biological features that distinguish the infective stage of these species. *T. brucei* is an extracellular pathogen with a high replication rate

TABLE 17.1

Kinetic Parameters for Synthetase Activity of GspS and TryS

Enzyme	Organism/strain	K_M (μM)					K_i (μM)		
		ATP	GSH	Sp	Gsp	k_{cat} (s^{-1})	GSH	T(SH)$_2$	ADP
GspS	E. coli[a]	100±20	800±150	60±10	NA	7	NR	NR	NR
	C. fasciculata	114±8[b]	242±14[b]	59±12[b]	NA	15.5[b]	NR	480[d]	80[d]
		230±28[c]	694±74[c]	142±14[c]					
TryS	C. fasciculata[e]	52[f]	1175[f]	7424	480	4.2[f]	≥1000[g]	ND	ND
		222[g]	407[g]			8.7[g]			
	L. major[h]	63±2	89±7[f]	940±140	40±4	2	1000±80[f]	ND	ND
	L. infantum[i]	170±23	170±23[f]	1434±135	19±0.2	NR	699±75[f]	ND	90±8[f]
	T. cruzi/Silvio X10[j]	53±3	570±70[f]	625±39	66±3	3.4[f]	1200±100[f]	ND	ND
			190±30[g]			2.8[g]	600±100[g]		
	T. cruzi/Tulahuen0[i]	41±6	140±13[f]	702±70	ND	NR	1690±172[f]	ND	60±6[f]
	T. cruzi/Ninoa[k]	70±40	760±210[f]	860±95	ND	NR	NR	>1000	ND
	T. brucei/427 (MITa1.4)	7.1±0.4[l]	56±11[f,l]	38±5[l]	2.4±0.2[l]	2.9[f,l]	37±7[f,l]	346 (377)[m]	ND
		12 (18)[m]	34 (69)[m]	687±90[m]	32±5[m]	2.8 (2.1)[m]	849 (1085)[m]		
	T. brucei/427[i]	19±5	142±20[f]	238±25	ND	NR	254±45[f]	ND	40±5[f]

Details on the experimental conditions are provided in the table's footnotes and in the pertinent references.

[a] Measured at 37°C in 50 mM Na-PIPES, pH 6.8 (Bollinger et al., 1995).

[b] Measured at 25°C in 100 mM K$^+$-HEPES, pH 7.3 or

[c] In 50 mM Bis-Tris propane/50 mM Tris buffer, pH 7.5 (Oza et al., 2002a).

[d] Measured at 25°C in 50 mM Bis-Tris propane/50 mM Tris buffer, pH 7.5 (Koenig et al., 1997).

[e] Measured at 25°C in 100 mM K$^+$-HEPES, pH 7.2 (Comini et al., 2005).

[f] With Sp as cosubstrate.

[g] With Gsp as cosubstrate.

[h] Measured at 25°C in 50 mM K$^+$-Hepps, pH 7.7 (Oza et al., 2005).

[i] Measured at 20°C–25°C in 100 mM K$^+$-HEPES, pH 7.2 (Benitez et al., 2016).

[j] Measured in 100 mM K$^+$-Hepps, pH 8.0 (Oza et al., 2002b).

[k] Measured at 37°C in 40 mM Na$^+$-HEPES, pH 7.4 (Olin-Sandoval et al., 2012).

[l] Measured at 25°C in 100 mM K$^+$-Hepps, pH 8.0 (Oza et al., 2003).

[m] Measured at 37°C in 10 mM K$^+$-phosphate buffer, pH 7.0, with the concentration referring to values obtained with Sp or, in brackets, with Gsp as cosubstrate (Leroux et al., 2013).

NA, not active. ND, not determined. NR, not reported.

(in vitro doubling time ≤5 h) and, hence, a high demand of reducing power for biosynthetic processes, which exclusively relies on glycolysis for ATP production (Haanstra et al., 2012), and on de novo synthesis of spermidine (Taylor et al., 2008). In contrast, L. infantum and T. cruzi are intracellular pathogens with a significantly lower replication rate [in vitro doubling time ≈14 h for T. cruzi amastigotes (De Rycker et al., 2016)]. They do not strictly depend on glycolysis and can obtain polyamines from the host cell (Carrillo et al., 2006; Hasne and Ullman,

TABLE 17.2

Kinetic Parameters for the Amidase Activity of GspS and TryS

Enzyme	Organism[a]	Gsp		Specific Activity (nmol min^{-1} s^{-1})				
		K_M (μM)	k_{cat}(s-1)	Gsp	T(SH)$_2$	hT(SH)$_2$	GSH-OEt	GSH-OPr
GspS	E. coli[b]	900	2.1	NR	NR	ND	ND	ND
	C. fasciculata	500±175[c]	0.38[c]	1390[d]	74[d]	126[d]	2315[c]	13.1[c]
TryS	C. fasciculata[e]	ND	ND	>400	240	ND	ND	ND
	T. brucei	5600[f]	5.1[f]	2.1[g]	6.4[g]	ND	43[g]	2.9[g]
	T. cruzi[g]	ND	ND	5.4	3.1	3.7	253	6.1

[a] Information on the strain can be obtained from Table 17.1.
[b] Bollinger et al. (1995).
[c] Oza et al. (2002a).
[d] Oza et al. (2003).
[e] Comini et al. (2005).
[f] Leroux et al. (2013).
[g] Oza et al. (2002b).
[h] (TSH)$_2$, N1, N9 bis-glutathionylaminopropylcadaverine or homotrypanothione.
NR, not reported. ND, not determined.

2011). Thus, the higher affinity of TryS for the nucleotide and polyamine substrate warrants the maintenance of the trypanothione pool in African trypanosomes. Another interesting feature reported for trypanosomal TryS is their broad substrate specificity for polyamines. *T. cruzi* TryS has a nearly identical affinity for the long aminopropyl cadaverine ($K_M = 662$ μM) as for spermidine [$K_M = 625$ μM (Oza et al., 2002b)], and *T. brucei* TryS is able to synthesize the mono- and bis-glutathionyl derivatives of spermine (Oza et al., 2003). Homotrypanothione, N^1, N^9 bis-glutathionylaminopropylcadaverine, has been detected in cell extracts of *T. cruzi* (Oza et al., 2002b), which suggests that such substrate promiscuity may be advantageous in a biological context.

Despite the coexistence of opposite hydrolytic and synthetic activities in a single polypeptide, the amidase activity of GspS and TryS is comparatively low to the synthetase (Table 17.2). The amidase activity of GspS has 3.5- to 40-fold lower k_{cat} values than the synthetase domain (Bollinger et al., 1995; Oza et al., 2002a). Such difference translates into a ≥10-fold lower catalytic efficiency of the amidase domain for Gsp than the synthetase domain for its cognate substrates. Also the hydrolytic activity of *Cf*GspS exhibits a marked substrate promiscuity that is not paralleled by the trypanosomal TryS (Table 17.2) and is suggestive of structural differences at the amidase active site of these proteins. In this respect, the amidase activity of dimeric *Ec*GspS (Bollinger et al., 1995), but not of monomeric *Tb*TryS (Leroux et al., 2013), has been shown to be subject of allosteric activation by the synthetase domain, as GSH and ATP increased 15-fold the hydrolytic activity of the former. The amidase activity of *Ec*GspS is also target of redox regulation, as shown by oxidation of a catalytic residue (Cys59), which renders the protein inactive (Chiang et al. 2010).

With Gsp as substrate, the amidase activity of TbTryS was reported to be significantly less catalytically efficient (k_{cat}/K_M for Gsp = 9.1×10^2 $M^{-1}s^{-1}$) than the synthetase activity [k_{cat}/K_M for Gsp = 6.6×10^4 $M^{-1}s^{-1}$ (Leroux et al., 2013). But, at variance with GspS, this is the consequence of a significantly lower affinity of the TbTryS amidase domain for Gsp (K_M Gsp for amidase vs. synthetase domain is 5600 vs. 32 μM) and not due to a lower turnover constant of the amidase reaction (k_{cat} for amidase $vs.$ synthetase = 5.1 vs. 2.1 s^{-1}). Another striking difference is that the amidase from TryS of pathogenic kinetoplastids display a 10- to 1000-fold lower specific activity with Gsp and T(SH)$_2$ than that reported for the TryS or GspS from the nonpathogenic $C.$ $fasciculata$ (Table 17.2). Accordingly, $Crithidia$ shows a more balanced content of Gsp (0.9–2.28 mM) and T(SH)$_2$ (0.37–1.51 mM), whereas the infective stages of most trypanosomatids show a distribution of glutathionyl-polyamines that is largely shifted towards T(SH)$_2$ [reviewed in Krauth-Siegel and Comini (2008)]. The metabolic consequences of the specificity evolved by GspS and TryS are discussed in Section 17.5.

As a hallmark for a house-keeping protein, the biosynthetic activity of GspS and TryS is also tightly regulated (Table 17.1). An increase in temperature (from 25°C to 37°C) increased almost 3-fold the k_{cat} and about ≥50% the substrates' K_M values of TbTryS (Leroux et al., 2013). This regulation may have physiological consequences taking into account that many trypanosomatids, which shuttle from a poikilothermic invertebrate vector to a mammal, and free-living bacteria suffer drastic temperature changes during their life cycle. The synthetase activity of CfGspS is efficiently inhibited by ADP (K_i = 80 μM) and, to a lesser extent, by T(SH)$_2$ [K_i = 480 μM (Koenig et al., 1997)]. ADP proved also to be a potent inhibitor of TryS from different trypanosomatids with K_i values (40–90 μM) close to the corresponding K_M values for ATP [19–170 μM (Benitez et al., 2016)]. However, the contribution of ADP to TryS inhibition is probably irrelevant considering the physiologically high ATP:ADP ratio (3–10:1) in trypanosomatids (Martins et al., 2009; Haanstra et al., 2012). In contrast, while inhibition of TryS by the thiol substrate ($K_{i\,GSH} \geq K_{M\,GSH}$) or product ($K_{i\,T(SH)2}$ = 377 μM and >1 mM for TbTryS and TcTryS, respectively; Table 17.1) is achieved at comparatively higher concentrations of both ligands, their K_i values are close to their physiological concentrations [reviewed in Krauth-Siegel and Comini (2008)]. Thus, T(SH)$_2$ and, to a lesser extent, GSH, are likely to negatively regulate the synthetase activity of TryS in $vivo$. In this context, the physiological relevance of TryS in T(SH)$_2$ homeostasis has been revealed by a comparative metabolic study performed in $T.$ $cruzi$ subjected or not to oxidative stress (Olin-Sandoval et al., 2012). In agreement with the data presented earlier, TryS ranked second, after γ-glutamylcysteine synthetase and before spermidine transport, among the major factors controlling the metabolic flux of T(SH)$_2$.

The mechanistic aspect of the amidase/synthetase catalysis and inhibition are discussed in the next paragraph.

17.3   MECHANISTIC INSIGHTS IN GLUTATHIONYL-POLYAMINE HYDROLYSIS AND SYNTHESIS

So far, there is no evidence for the occurrence of phosphorylated-enzyme intermediates during the catalytic cycle of the synthetases. Instead, GspS and TryS facilitate direct substrate phosphorylation by ATP. Based on kinetic analysis of *Cf*GspS (Koenig et al., 1997) and *Tb*TryS (Leroux et al., 2013), the first biosynthetic reaction appears to obey a ter-reactant mechanism, which implies that catalysis starts once all three substrates are bound to the enzyme. Nevertheless, *Cf*GspS and *Tb*TryS appear to differ in how the enzyme–substrate complex is formed. While the first did not show a marked preference in the order the substrates bind to the active site (rapid equilibrium random mechanism), *Tb*TryS displayed certain preference for binding Sp only after ATP has been bound. The deduced catalytic mechanism was further corroborated for *Cf*GspS by ^{31}P NMR experiments that ruled out the formation glutahionylphosphate, acylphosphate, or any ATP hydrolysis unless all substrates were present (Koenig et al., 1997) (Figure 17.2).

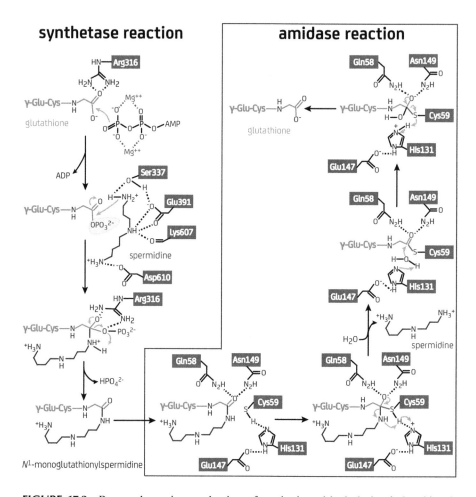

FIGURE 17.2 Proposed reaction mechanism of synthesis and hydrolysis of glutathionyl-spermidine derivatives. The reaction scheme and residues participating in the synthesis and hydrolysis (box) of glutathionyl-polyamine is based on data reported for *Ec*GpS (Bollinger et al., 1995; Lin et al., 1997; Pai et al., 2006; Pai et al., 2011).

For TryS, the second catalytic cycle involves the dissociation of the intermediate product Gsp from the enzyme and its rebinding in an orientation suitable for addition of the second GSH molecule (Comini et al., 2005; Koch et al., 2013; Leroux et al., 2013). The best models extrapolated from kinetic data of *Tb*TryS fit to a rapid equilibrium ter-reactant mechanism, identical to that described for the first reaction (Leroux et al., 2013). The authors also noted that Gsp binds better when GSH has been bound before, suggesting again a certain preferred binding order. This mechanism conflicts with the data obtained for the crithidial enzyme under narrow substrate concentrations [i.e., [GSH] < 1 mM, to avoid substrate inhibition (Comini et al., 2005)]. *Cf*TryS follows a concerted substitution mechanism where a ternary complex is formed by the sequential binding of ATP-Mg^{++}and then GSH, which is followed by the binding and reaction via a ping-pong mechanism of Gsp with the phosphorylated GSH to produce T(SH)$_2$. Additional evidence supporting the decay of the initial complex is based on the ATPase activity of *Cf*TryS that doubles upon GSH addition [ATPase activity is 2.6% of the synthetase activity (Comini et al., 2005)]. In contrast, *Tb*TryS was reported to have a 5-fold lower ATPase activity [0.5% of the synthetase activity (Leroux et al., 2013)]. Although such mechanistic contradictions may certainly arise from the use of distinct experimental approaches, a true mechanistic divergence between the crithidial and trypanosomatid TryS cannot be ruled out considering the remarkably different kinetic parameters of these enzymes (Tables 17.1 and 17. 2) and related structural singularities.

GSH and T(SH)$_2$ act as uncompetitive inhibitors of the synthetase activity by binding with high affinity to the activated enzyme complex at a not yet identified site that becomes only accessible after a structural rearrangement upon phosphorylation of GSH. T(SH)$_2$ is also capable to inhibit the free-enzyme by remaining bound to the product site (Leroux et al., 2013).

The kinetic mechanism operating during the hydrolytic reaction was investigated in detail for *Ec*GspS using chromogenic substrate analogues (Lin et al., 1997). The amidase reaction involves the nucleophilic attack of the enzyme on the isopeptide bond between GSH and Sp with formation of a covalent transition-state intermediate that is followed by release of the amide product and formation of a glutathionyl-enzyme intermediate, which finally breaks down to yield free enzyme and GSH (Figure 17.2). The assays showed that the hydrolysis, and not the formation, of the acyl-enzyme intermediate was the rate-limiting step of the reaction.

17.4 STRUCTURAL FEATURES OF GLUTATHIONYL-SPERMIDINE AMIDASE/SYNTHETASE

Crystallographic data are available for *Ec*GspS (Pai et al., 2006; Pai et al., 2011) and *Leishmania major* TryS [*Lm*TryS (Fyfe et al., 2008)]. Overall GspS and TryS present globular structures, where the N-terminal amidase domain (~190–210 residues) with a α/β fold is connected to the equilateral triangle-shaped C-terminal synthetase domain (~420–450 residues) via a linker region that spans 11 (*Ec*GspS) to 25 (*Lm*TryS) amino acids. *Ec*GspS has been shown to be homodimeric where the inter-subunit interactions involve hydrophobic, salt-bridge, and hydrogen bonds between

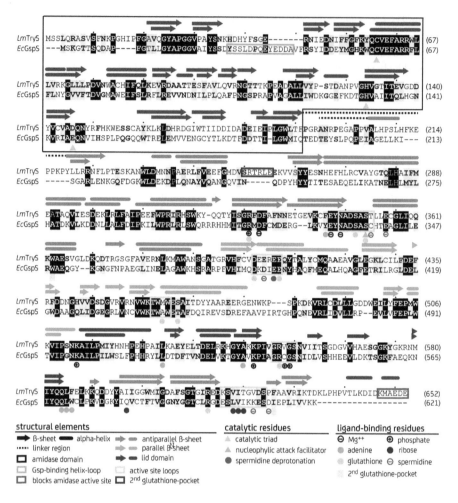

FIGURE 17.3 Sequence alignment and structural features of glutathionyl-polyamine amidase/synthetase. Sequence alignment of *Leishmania major* TryS (*Lm*TryS) and *Escherichia coli* GspS (*Ec*GpS). The different protein domains, structural elements, regions and residues relevant for substrate binding and catalysis are labelled with different symbols and colors.

amino acids from the amidase domain and the synthetase domain of different subunits (Bollinger et al., 1995; Pai et al., 2006). However, most of these residues are not conserved in GspS and TryS from kinetoplastids (Figure 17.3), which agrees with their monomeric behavior (Oza et al., 2002a,b; Comini et al., 2003;Oza et al., 2005; Fyfe et al., 2008).

The amidase domains of GspS and TryS share a papain-like fold (Anantharaman and Aravind, 2003) composed of a core of two α-helices and a β-barrel of seven chains (Pai et al., 2006; Fyfe et al., 2008). At the atomic level, the binding pocket for the polyamine–GSH derivatives is well conserved between the two enzymes and involves several residues from the N-terminal region of this domain. However, a major difference between the amidase active site of *Ec*GspS and *Lm*TryS is their accessibility to the solvent, which is determined by different orientations of their amidase and synthetase domains. In *Ec*GspS, the active site cleft is located on the surface and easily accessible from the solvent, whereas for *Lm*TryS, the C-terminus protrudes into and blocks the catalytic site (discussed later). The position of the residues conforming the catalytic triad in *Ec*GspS (Cys59, His131, and Glu147) and in *Lm*TryS (Cys59, His130, Asp146) is conserved in both proteins. In *Ec*GspS, these residues show an orientation compatible with the catalytic mechanism, while in *Lm*TryS the side chain of His130 is not hydrogen bonded to Asp146 and protrudes towards the substrate pocket hence interfering with ligand binding. Supported by hydrogen bonds between Glu147 and His131, the imidazolium group of histidine helps to deprotonate and enhance the nucleophilicity of Cys59. Two additional conserved residues, Asn149 and Gln58 in *Ec*GspS (Asn148 and Gln58 in *Lm*TryS), are catalytically important, because they increase the electrophilicity of the amide and stabilize the negative charge of the carbonyl oxygen atom at the scissile bond [(Pai et al., 2011) Figure 17.2]. X-ray crystallography allowed to identify a sulfenic acid in Cys59 of *Ec*GspS exposed to H_2O_2, which is responsible for the (reversible) inactivation of the enzyme under oxidative conditions (Chiang et al., 2010).

Based on structural information, other regulatory mechanisms have been proposed for GspS and TryS. *Ec*GspS has a helix-loop region (residues 30–42) near the binding pocket that modulates catalysis and substrate binding. In the substrate-free form, a catalytically incompetent conformation of the enzyme is supported by hydrogen bonds between Tyr30 and Tyr38 and two residues that stabilize different reaction intermediates, namely Asn149 and Gln58 (Pai et al., 2011). Upon substrate binding, these interactions are lost and several acidic residues of this segment go on to interact with the polyamine moiety of the substrate. Furthermore, in the *Ec*GspS dimer the synthetase domain (the side opposite to its active site) contacts and faces the amidase domain and active site, providing additional modulation on the accessibility of substrates to the amidase active site. Such allosteric regulation of *Ec*GspS amidase has been suggested to account for the 15- and 70-fold increase in hydrolytic activity when the enzyme was incubated with substrates or expressed separated from the synthetase domain, respectively (Lin et al., 1997).

A different mechanism should regulate the amidase activity of TryS, as these proteins lack the helix-loop region of *Ec*GspS at the amidase active site and are monomeric. Indeed, for *Lm*TryS a C-terminal extension of the synthetase domain has been shown to block the access to the active site due to the multiple interactions that three acidic residues (Glu650, Asp651, and Glu652) establish with amino acids that shape the amidase binding pocket [His270, Arg375, Arg383, His39, and C59 (Fyfe et al., 2008)]. The high sequence conservation at the C-terminal region of leishmanial and crithidial TryS suggests that these proteins share a common regulatory mechanism. In contrast, trypanosomal TryS completely lack this C-terminal extension (e.g., *Tb*TryS) or its sequence is significantly different from the leishmanial enzyme (*Tc*TryS). Thus, TryS from trypanosomes may control the GSH–polyamine amidase activity by a different mechanism.

Despite the marginal sequence identity of GspS and TryS with the ATP-grasp superfamily of enzymes, their C-terminal synthetase domain presents a folding and

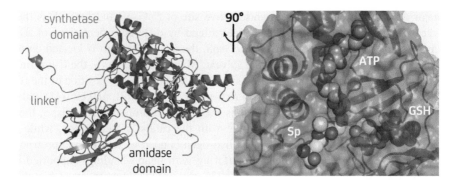

FIGURE 17.4 Structure of glutathionylpolyamine synthetase/amidase enzymes. Left panel: cartoon representation of the backbone structure of *Lm*TryS (PDB 2VOB, Fyfe et al., 2008). Right panel: synthetase active site of *Lm*TryS depicted as surface with docked ADP (callots; backbone C-atoms shown as orange balls), GSH with Gly pointing towards the reaction center (green) and a phosphinate inhibitor (backbone carbons shown as yellow balls) (Pai et al., 2006). The binding sites for the different substrates are indicated.

subdomain arrangement similar to other members of this superfamily (Fawaz et al., 2011). As a matter of fact, *Ec*GspS and *Lm*TryS have a high degree of structural similarity with human glutathione synthetase (PDBcode: 2HGS). The synthetase domain of GspS and TryS comprises three conserved structural units, the parallel β-sheet, antiparallel β-sheet, and a lid domain (Figure 17.3). The active site is located at the central antiparallel β-sheet and presents a triangular shaped cavity with each vertices of the triangle hosting one of the three substrates (Figure 17.4).

In addition, each subdomain, and regions outside, provides several loops of variable length that shape the active site and contribute to substrate binding. A clue to the role played by these elements have been provided by limited-proteolysis assays and kinetic studies, which concluded that the active and substrate-binding sites of the synthetase domain of GspS and TryS undergo large conformational changes upon interaction with ligands (Bollinger et al., 1995; Comini et al., 2005; Leroux et al., 2013). In fact, these disordered regions shift from an open to a closed conformation upon ligand binding in *Ec*GspS (Pai et al., 2006). The particular arrangement of the crystals of *Lm*TryS precluded substrate binding by soaking or cocrystallization approaches. However, the crystallographic data for *Ec*GspS in the presence of substrates, product, or inhibitor shed light on the ligand binding mode at the synthetase domain of both proteins. The antiparallel β-sheet and the lid domain, and loops thereof, form the ATP binding pocket. The adenine ring is anchored at a hydrophobic patch, the ribose moiety interacts with the most C-terminal loop of the antiparallel β-sheet and the phosphates are oriented towards the catalytic site by electrostatic interactions with several positively charged residues, whereas the Mg^{2+} ions are coordinated by acidic amino acids. Several residues from the parallel β-sheet and, to a lesser extent, from the antiparallel β-sheet, contribute with hydrogen bonds and van der Waals interactions to anchor GSH. The polyamine-binding site, which is formed by secondary structure elements from the parallel and antiparallel β-sheet, as well as the C-terminal loop of the latter, extents in a direction opposite to that of the GSH pocket. Noticeably, *Lm*TryS but not *Ec*GspS has an additional cavity next to the polyamine pocket, in the extremity distant from the catalytic site. According to molecular docking and dynamics, this cavity is suited to host the glutathionyl moiety of Gsp (Fyfe et al., 2008; Koch et al., 2013). The element responsible for generating this pocket is an insertion (3–5 residues) within a loop at the N-terminal region of the antiparallel β-sheet domain, which is strictly conserved in all TryS (Manta et al., 2018). In *Ec*GspS, the shorter loop occludes this pocket and prevents Gsp binding.

About seven strictly conserved residues in the ATP-binding site of *Ec*GspS and *Lm*TryS (Figure 17.3) are critical for binding and orienting the triphosphate component of ATP and to activate the γ-phosphate component that will be transferred to the glycyl carboxylate of GSH (Fyfe et al., 2008). For *Lm*TryS (and the equivalent residues in *Ec*GspS), Arg613, located at the interface of the polyamine- and the ATP-binding pocket, together with Arg328 and Ser351, both at the bottom of the catalytic site, are engaged in aligning Sp or Gsp with the activated GSH as well as in generating and stabilizing the transition state intermediate (Fyfe et al., 2008) (Figure 17.2). The Mg^{++} ions serve as Lewis acids to assist the phosphate transfer and compensate the negative charges during catalysis by coordinating the phosphorylated GSH carboxylate (Pai et al. 2011). The binding and activation of Sp for catalysis is better described by the crystal structure of *Ec*GspS with bound inhibitor. In this protein, Sp is anchored by hydrogen bonds with Glu391 (Glu407 in *Lm*TryS), Asp610, and Lys607, as well as four negatively charged residues (Asp387, Asp389, Glu391, and Glu392) of the active site loop located at the N-terminus of the parallel β-sheet domain [Figure 17.3; (Pai et al., 2011)]. Interactions of the terminal amine of Sp with the hydroxyl group of Ser337 and the carboxylate of Glu391 (both conserved in *Lm*TryS, as well as Ser351 and Glu407) contribute to deprotonate the polyamine and, thus, to facilitate the nucleophilic attack on the phosphorylated GSH.

17.5 PHYLOGENETIC DISTRIBUTION AND EVOLUTIONARY RELATIONSHIP OF GLUTATHIONYL-POLYAMINE BIOSYNTHETIC ENZYMES

The occurrence of glutathionyl-polyamine derivatives is reliable documented for a few organisms such as *γ-Proteobacteria, Euglena gracilis*, and *Kinetoplastida* (Tabor and Tabor, 1970; Fairlamb and Cerami, 1985; O'neill et al., 2015). The recent boom of genome sequencing projects contributed to the identification of sequences encoding glutahionyl-polyamine synthetases in phylogenetically close and distant organisms [(Chattopadhyay et al., 2013; Manta et al., 2018) Figure 17.5].

The existence of sequences encoding for separate glutathionyl-polyamine amidase or synthetase domains has not yet been reported. A more exhaustive data mining is required in order to demonstrate whether the domains of these bifunctional enzymes evolved independently. GspS-encoding sequences were detected in several *Eubacteria* (except for the group Chlamydiae) but not in *Archaea* and *Spirochetes*, which suggests that the first group of prokaryotes gave origin to this protein. Within the *Eukaryotes*, the distribution of GspS sequences is restricted to certain species

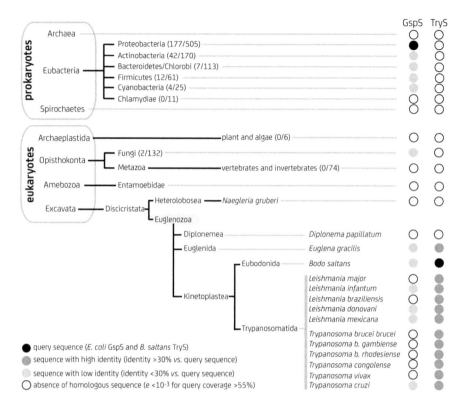

FIGURE 17.5 Phylogenetic distribution of GspS and TryS. Phylogenetic tree constructed based on information published by Chattopadhyay et al. (2013) and Manta et al. (2018).

from the *Euglenozoa* (i.e., *Euglenida* and *Kinetoplastida*) and *Oomycota* class [i.e., the pseudo-fungi *Aphanomyces* and *Saprolegnia* (Gaulin et al., 2008)]. In contrast, TryS sequences are exclusively present in organisms from the *Euglenozoa* phylum: in all *Kinetoplastida*, for which sequence data are available (Manta et al., 2018), and in *Euglena gracilis* (O'neill et al., 2015). Confirming its functional relevance for try-panosomatids (Section 17.6), the TryS gene is placed in a highly conserved genomic context (syntheny), whereas GspS sequences are often fully absent or transcription-ally/translationally silent [e.g., pseudogene or nonexpression (Oza et al., 2005; Sousa et al., 2014)].

The amino acid identity between bacterial and *Euglenida* or *Kinetoplastida* GspS is low (<25%; Chattopadhyay et al., 2013), which contrasts with the higher degree of sequence conservation among *Euglenida* and *Kinetoplastida* GspS (35%–48%). The sequence identity for TryS from different kinetoplastid species is relatively high (62%–72%) and, interestingly, slightly more similar to GspS from *Proteobacteria* (average sequence identity 29%) than from Euglenozoan [average sequence identity 24% (Manta et al., 2018)].

Based on the information above, the following evolutionary scenario has been sug-gested: (i) A gene fusion took place in a *Proteobacteria* ancestor linking a peptidase domain to a C–N ligase with an ATP-grasp fold. (ii) This new gene underwent mutations that led to the acquisition of specificity towards glutathione and polyamine and conjugates thereof. (iii) The GspS gene was horizontally transferred from a γ-proteobacterium to an *Euglenozoa* ancestor, probably during one of the many endosymbiotic events that occurred in *Euglenozoa* (Teixeira et al., 2011; Votýpka et al., 2014). (iv) The inherited sequence was duplicated early in the *Euglenozoa* and, after subsequent rounds of mutations, one of them gained capacity to synthesize bis-glutathionyl polyamines (i.e., TryS). (v) With a thiol-redox metabolism fully sustained by trypanothione, there was no selective pressure left to express a GspS gene, which was consequently lost or silenced in some *Kinetoplastida*.

17.6   BIOLOGICAL FUNCTIONS OF GLUTATHIONYL-POLYAMINES

Although glutathionylspermidine derivatives play both common and different biological roles in *Enterobacteria* and kinetoplastid organisms, a distinctive feature is that they are not indispensable for the prokaryotes but for many trypanosomatids. Applying a simplistic criterion, some functions can be linked to the use of the glutathionyl-polyamines as a store of their components (GSH and Sp) or to employ the thiol group(s) as ligand or redox cofactor.

Regarding the first of these roles, it is important to stress that conjugation of GSH to polyamines will render these metabolites not suitable for degradation by catabolic enzymes, since, except for the amidase activity of GspS/TryS, there are no reports on enzymatic activities being capable to hydrolyze these conjugates. In *E. coli* and *C. fasciculata*, the level of intracellular polyamine-thiols has been reported to change significantly during the different growth phases and culture conditions. For *E. coli*, transition from the exponential to the stationary phase was associated with a ~4-fold accumulation of Gsp that accounted for 95% and 11% of the intracellular spermidine (Dubin, 1959; Tabor and Tabor, 1970; Tabor and Tabor, 1975) and GSH (Smith et al., 1995), respectively. More surprisingly, Gsp rapidly (within 4 h) became the principal low-molecular-weight thiol (80% of the total GSH) when cells were transferred from aerobic to anaerobic growth conditions (Smith et al., 1995). Similarly, a redistribution of Gsp was reported for *C. fasciculata*, where the metabolite accumulated in the stationary phase and was rapidly converted into $T(SH)_2$ upon medium exchange (Shim and Fairlamb, 1988). For both organisms, the change in the pool of free and GSH-bound spermidine was not associated with *de novo* synthesis of polyamines, thus, the amidase activity of GspS and TryS were likely responsible for this re-equilibration. In line with this interpretation, the amidase activity of *Ec*GspS displays optimum activity at a pH higher (pH > 7.5) than the synthetase activity [pH 6.8 (Bollinger et al., 1995)]. Given that polyamines are biogenic compounds with important roles in processes linked to cell proliferation, it is tempting to speculate that GspS and TryS control the levels of the precursor metabolites, Sp and GSH, in response to certain environmental or metabolic stresses (Tabor and Tabor, 1975; Suzuki and Kurihara, 2015). In view of the multiple functions of Sp and GSH in cells, perturbing the homeostasis of these metabolites should have substantial consequences. Accordingly, a comparative transcriptomic study conducted on wildtype and GspS-KO *E. coli* has disclosed a major role for this gene on bacterial physiology and gene expression. In fact, the loss of GspS entailed the upregulation of 76 genes (the major metabolic categories included: sulfur assimilation, glutamine, succinate, polyamine, and arginine metabolism and purine/pyrimidine metabolism) and the downregulation of 35 genes (Chattopadhyay et al., 2013). Adding value to this conclusion, a recent study highlighted the phenotypic relevance of the polyamine and GSH metabolism for the fitness of two strains of an Antarctic *Pseudoalteromonas* species (*γ-Proteobacteria*) isolated from different ecological niches (Mocali et al., 2017). Interestingly, the bacterial strain showing better cryo-tolerance had a more efficient polyamine catabolism and antioxidant capacity and expressed a GspS gene.

As discussed earlier, TryS from trypanosomatids are largely committed to trypanothione biosynthesis due to the higher substrate turnover of the synthetase versus the amidase activity. Although this may argue against a regulatory function of TryS in polyamine/GSH metabolism, a study has suggested that Gsp and $T(SH)_2$ catabolism contributes to maintain the spermidine pool in African trypanosomes with impaired spermidine synthase activity (Willert and Phillips, 2012).

Other nonredox function that glutathionyl-polyamines may fulfill is to bind metals and small molecules using the thiol group of the GSH moiety as ligand. Nitric oxide (NO) is an important signaling molecule that plays a key role in the immune response against various pathogens. NO produces peroxynitrite ($^{\bullet}NO + O_2^{\bullet-} \rightarrow ONOO^-$), which is a more potent oxidant involved in the killing of foreign organisms. The formation of the paramagnetic dinitrosyl-trypanothionyl iron complex was observed in *T. brucei* and *L. infantum* exposed to NO (Bocedi et al., 2010). Compared to the dinitrosyl-diglutathionyl-iron complex, which proved to be a potent irreversible inhibitor of glutathione reductase ($IC_{50} = 4\,\mu M$), the complex formed with $T(SH)_2$ was significantly more stable and, even at millimolar concentrations, did not inactivate trypanothione reductase (Bocedi et al., 2010). Thus, $T(SH)_2$ may contribute to lower the concentration of free NO and, hence, its potential cytotoxicity by complexing it in a harmless form.

In vitro, $T(SH)_2$, and also Gsp, was capable to bind Fe-S clusters with high affinity and to efficiently substitute GSH in the formation of iron–sulfur complexes with trypanosomal glutaredoxins from two different classes and distinct biological functions (reviewed in Comini et al., 2013). One of these partners correspond to a class I cytosolic glutaredoxin that in the apo form has a classical oxidoreductase activity that is silenced upon binding the Fe/S cluster (Ceylan et al., 2010). The protein is fully dispensable for parasite survival in a mammalian host and does not contribute to redox or iron homeostasis (Musunda et al., 2015). This contrast with data obtained for the mammalian homologue that has been shown to donate its Fe/S cluster to complex NO and, hence, to protect oligodendrocytes against peroxynitrite-mediated damage (Lepka et al., 2017). Considering the high stability of the Fe/S clusters formed with $T(SH)_2$, it is thus tempting to speculate that, in trypanosomatids, the dithiol alone may take over the coordination of dinitrosyl-iron complex without the assistance of a glutaredoxin.

The second glutaredoxin with capacity to bind Fe/S clusters using Gsp and, more efficiently, $T(SH)_2$ as ligand, belongs to the class II, is located in the parasite mitochondrion and has a key role in the biogenesis of iron-sulfur cluster proteins (Manta et al., 2013). The participation of glutathionyl-polyamines as a constitutive member of a Fe/S complex with class II glutaredoxins has not been demonstrated *in vivo*, but the *in vitro* data opens the possibility for such a new role in *Proteobacteria* and *Kinetoplastida*.

Because the role of T(SH)$_2$ as redox cofactor is addressed in detail in the next chapter (Chapter 18), only a few important considerations are discussed here for *Enterobacteria*. An important biological difference between trypanosomatids and *E. coli* is that bis-glutathionylspermidine is absolutely essential for the survival (*in vitro* and *in vivo*) of the former (Comini et al., 2004; Ariyanayagam et al., 2005; Sousa et al., 2014), whereas the *Enterobacteria* can fully dispense with Gsp, and even its precursors, as mutants defective in GSH, Sp, and Gsp synthesis growth normally on minimal medium and under aerobic conditions (Greenberg and Demple, 1986; Chattopadhyay et al., 2009; Chattopadhyay et al., 2013). Although GSH and polyamines are typically required for protection against oxidative stress (Chattopadhyay et al., 2003; Masip et al., 2006), Gsp does not confer protection against ROS, UV radiation, or metal toxicity in *E. coli* (Chattopadhyay et al., 2013). This could be expected, because *E. coli* lacks a glutathionylspermidine-related redox system (Smith et al., 1995), and instead, works with the more common redox systems (glutathione reductase/GSH/glutaredoxin and thioredoxin reductase/thioredoxin). Nonetheless, the recent discovery of mixed-disulfides between Gsp and different *E. coli* proteins suggests that Gsp may act at the posttranscriptional level regulating protein activity (Chiang et al., 2010; Chiang et al., 2012). Interestingly, the amidase activity of *Ec*GspS appears to play an important role in reverting this oxidative modification by cleaving off Sp from GSH on proteins previously modified with Gsp, leaving a glutathionylated target, which is then a substrate of Grx that recovers the reduced form of the protein with high efficiency (Chiang et al., 2010). Probably, some of the roles proposed for GspS in the regulation of bacterial gene expression are mediated by redox mechanisms. A recent study reporting the occurrence of mixed disulfides between Gsp/T(SH)$_2$ and proteins in African trypanosomes proposes that the amidase activity of TryS may play a similar role in the hydrolysis of the protein-bound thiol–polyamine complexes (Ulrich et al., 2017). Such alternative role for the amidase activity may be the reason why the *in vivo* growth retardation of a *T. brucei* line devoid of TryS-amidase activity could not be restored in infected animals by a daily bolus of Sp (Wyllie et al., 2009) at a dosage that effectively protected the parasites against chemical inhibition of Sp biosynthesis (Nathan et al., 1981).

REFERENCES

Anantharaman, V., and L. Aravind. 2003. Evolutionary history, structural features and biochemical diversity of the NlpC/P60 superfamily of enzymes. *Genome Biol* 4 (2):R11.

Ariyanayagam, M. R., and A. H. Fairlamb. 2001. Ovothiol and trypanothione as antioxidants in trypanosomatids. *Mol Biochem Parasitol* 115 (2):189–98.

Ariyanayagam, M. R., S. L. Oza, M. L. Guther, and A. H. Fairlamb. 2005. Phenotypic analysis of trypanothione synthetase knockdown in the African trypanosome. *Biochem J* 391 (Pt 2):425–32.

Ariyanayagam, M. R., S. L. Oza, A. Mehlert, and A. H. Fairlamb. 2003. Bis(glutathionyl) spermine and other novel trypanothione analogues in *Trypanosoma cruzi. J Biol Chem* 278 (30):27612–9.

Benitez, D., A. Medeiros, L. Fiestas et al. 2016. Identification of novel chemical scaffolds inhibiting trypanothione synthetase from pathogenic trypanosomatids. *PLoS Negl Trop Dis* 10 (4):e0004617.

Bocedi, A., K. F. Dawood, R. Fabrini et al. 2010. Trypanothione efficiently intercepts nitric oxide as a harmless iron complex in trypanosomatid parasites. *FASEB J* 24 (4):1035–42.

Bollinger, J. M., Jr., D. S. Kwon, G. W. Huisman et al. 1995. Glutathionylspermidine metabolism in *Escherichia coli*. Purification, cloning, overproduction, and characterization of a bifunctional glutathionylspermidine synthetase/amidase. *J Biol Chem* 270 (23):14031–41.

Carrillo, C., G. E. Canepa, I. D. Algranati, and C. A. Pereira. 2006. Molecular and functional characterization of a spermidine transporter (TcPAT12) from *Trypanosoma cruzi*. *Biochem Biophys Res Commun* 344 (3):936–40.

Ceylan, S., V. Seidel, N. Ziebart et al. 2010. The dithiol glutaredoxins of African trypanosomes have distinct roles and are closely linked to the unique trypanothione metabolism. *J Biol Chem* 285 (45):35224–37.

Chattopadhyay, M. K., W. Chen, and H. Tabor. 2013. *Escherichia coli* glutathionylspermidine synthetase/amidase: Phylogeny and effect on regulation of gene expression. *FEMS Microbiol Lett* 338 (2):132–40.

Chattopadhyay, M. K., C. W. Tabor, and H. Tabor. 2003. Polyamines protect Escherichia coli cells from the toxic effect of oxygen. *Proc Natl Acad Sci U S A* 100 (5):2261–5.

Chattopadhyay, M. K., C. W. Tabor, and H. Tabor. 2009. Polyamines are not required for aerobic growth of *Escherichia coli*: Preparation of a strain with deletions in all of the genes for polyamine biosynthesis. *J Bacteriol* 191 (17):5549–52.

Chiang, B. Y., T. C. Chen, C. H. Pai et al. 2010. Protein *S*-thiolation by glutathionylspermidine (Gsp): The role of *Escherichia coli* Gsp synthetase/amidase in redox regulation. *J Biol Chem* 285 (33):25345–53.

Chiang, B. Y., C. C. Chou, F. T. Hsieh et al. 2012. *In vivo* tagging and characterization of *S*-glutathionylated proteins by a chemoenzymatic method. *Angew Chem Int Ed Engl* 51 (24):5871–5.

Comini, M. A., and L. Flohé. 2013. The trypanothione-based redox metabolism of trypanosomatids. In Trypanosomatids *Diseases, Molecular Routes* to *Drug Discovery*, edited by O. Koch, T. Jäger, and L. Flohé, 167–99. Weinheim: Wiley-Blackwell.

Comini, M. A., S. A. Guerrero, S. Haile et al. 2004. Validation of *Trypanosoma brucei* trypanothione synthetase as drug target. *Free Radic Biol Med* 36 (10):1289–302.

Comini, M., U. Menge, and L. Flohé. 2003. Biosynthesis of trypanothione in *Trypanosoma brucei*. *Biol Chem* 384 (4):653–6.

Comini, M., U. Menge, J. Wissing, and L. Flohé. 2005. Trypanothione synthesis in *Crithidia* revisited. *J Biol Chem* 280 (8):6850–60.

Coombs, G. H., and B. E. Sanderson. 1985. Amine production by *Leishmania mexicana*. *Ann Trop Med Parasitol* 79 (4):409–15.

De Rycker, M., J. Thomas, J. Riley et al. 2016. Identification of trypanocidal activity for known clinical compounds using a new *Trypanosoma cruzi* hit-discovery screening cascade. *PLoS Negl Trop Dis* 10 (4):e0004584.

Dubin, D. T. 1959. Evidence for a conjugate between polyamines and glutathione in *E. coli*. *Biochem Biophys Res Commun* 1 (5):262–65.

Fairlamb, A. H., P. Blackburn, P. Ulrich et al. 1985. Trypanothione: A novel bis(glutathionyl)spermidine cofactor for glutathione reductase in trypanosomatids. *Science* 227 (4693):1485–7.

Fairlamb, A. H., and A. Cerami. 1985. Identification of a novel, thiol-containing co-factor essential for glutathione reductase enzyme activity in trypanosomatids. *Mol Biochem Parasitol* 14 (2):187–98.

Fawaz, M. V., M. E. Topper, and S. M. Firestine. 2011. The ATP-grasp enzymes. *Bioorg Chem* 39 (5–6):185–91.

Fyfe, P. K., S. L. Oza, A. H. Fairlamb, and W. N. Hunter. 2008. *Leishmania* trypanothione synthetase-amidase structure reveals a basis for regulation of conflicting synthetic and hydrolytic activities. *J Biol Chem* 283 (25):17672–80.

Greenberg, J. T., and B. Demple. 1986. Glutathione in Escherichia coli is dispensable for resistance to H2O2 and gamma radiation. *J Bacteriol* 168 (2):1026–9.

Haanstra, J. R., A. van Tuijl, J. van Dam et al. 2012. Proliferating bloodstream-form *Trypanosoma brucei* use a negligible part of consumed glucose for anabolic processes. *Int J Parasitol* 42 (7):667–73.

Hasne, M. P., and B. Ullman. 2011. Genetic and biochemical analysis of protozoal polyamine transporters. *Methods Mol Biol* 720:309–26.

Henderson, G. B., M. Yamaguchi, L. Novoa et al. 1990. Biosynthesis of the trypanosomatid metabolite trypanothione: Purification and characterization of trypanothione synthetase from *Crithidia fasciculata*. *Biochemistry* 29 (16):3924–9.

Koch, O., D. Cappel, M. Nocker et al. 2013. Molecular dynamics reveal binding mode of glutathionylspermidine by trypanothione synthetase. *PLOS One* 8 (2):e56788.

Koenig, K., U. Menge, M. Kiess et al. 1997. Convenient isolation and kinetic mechanism of glutathionylspermidine synthetase from *Crithidia fasciculata*. J Biol Chem 272 (18):11908–15. Erratum in: M. Comini, U. Menge and L. Flohé. 2005. *J Biol Chem* 280:7407.

Krauth-Siegel, R. L., and M. A. Comini. 2008. Redox control in trypanosomatids, parasitic protozoa with trypanothione-based thiol metabolism. *Biochim Biophys Acta* 1780 (11):1236–48.

Kwon, D.S., C.H. Lin, S. Chen et al. 1997. Dissection of glutathionylspermidine synthetase/amidase from *Escherichia coli* into autonomously folding and functional synthetase and amidase domains. *J Biol Chem* 272 (4):2429–36.

Lepka, K., K. Volbracht, E. Bill et al. 2017. Iron-sulfur glutaredoxin 2 protects oligodendrocytes against damage induced by nitric oxide release from activated microglia. *Glia* 65 (9):1521–1534.

Leroux, A. E., J. R. Haanstra, B. M. Bakker, and R. L. Krauth-Siegel. 2013. Dissecting the catalytic mechanism of *Trypanosoma brucei* trypanothione synthetase by kinetic analysis and computational modeling. *J Biol Chem* 288 (33):23751–64.

Lin, C. H., D. S. Kwon, J. M. Bollinger, Jr., and C. T. Walsh. 1997. Evidence for a glutathionyl-enzyme intermediate in the amidase activity of the bifunctional glutathionylspermidine synthetase/amidase from *Escherichia coli*. *Biochemistry* 36 (48):14930–8.

Manta, B., M. Bonilla, L. Fiestas et al. 2018. Polyamine-based thiols in trypanosomatids: Evolution, protein structural adaptations, and biological functions. *Antioxid Redox Signal* 28 (6):463–86.

Manta, B., C. Pavan, M. Sturlese et al. 2013. Iron-sulfur cluster binding by mitochondrial monothiol glutaredoxin-1 of *Trypanosoma brucei*: Molecular basis of iron-sulfur cluster coordination and relevance for parasite infectivity. *Antioxid Redox Signal* 19 (7):665–82.

Martins, R. M., C. Covarrubias, R. G. Rojas et al. 2009. Use of L-proline and ATP production by *Trypanosoma cruzi* metacyclic forms as requirements for host cell invasion. *Infect Immun* 77 (7):3023–32.

Masip, L., K. Veeravalli, and G. Georgiou. 2006. The many faces of glutathione in bacteria. *Antioxid Redox Signal* 8 (5–6):753–62.

Mocali, S., C. Chiellini, A. Fabiani et al. 2017. Ecology of cold environments: New insights of bacterial metabolic adaptation through an integrated genomic-phenomic approach. *Sci Rep* 7 (1):839.

Musunda, B., D. Benitez, N. Dirdjaja et al. 2015. Glutaredoxin-deficiency confers blood-stream *Trypanosoma brucei* with improved thermotolerance. *Mol Biochem Parasitol* 204 (2):93–105.

Nathan, H. C., C. J. Bacchi, S. H. Hutner et al. 1981. Antagonism by polyamines of the curative effects of alpha-difluoromethylornithine in *Trypanosoma brucei* infections. *Biochem Pharmacol* 30 (21):3010–3.

O'Neill, E. C., M. Trick, L. Hill et al. 2015. The transcriptome of *Euglena gracilis* reveals unexpected metabolic capabilities for carbohydrate and natural product biochemistry. *Mol Biosyst* 11 (10):2808–20.

Olin-Sandoval, V., Z. Gonzalez-Chavez, M. Berzunza-Cruz et al. 2012. Drug target valida-tion of the trypanothione pathway enzymes through metabolic modelling. *FEBS J* 279 (10):1811–33.

Oza, S. L., M. R. Ariyanayagam, N. Aitcheson, and A. H. Fairlamb. 2003. Properties of trypanothione synthetase from *Trypanosoma brucei*. *Mol Biochem Parasitol* 131 (1):25–33.

Oza, S. L., M. R. Ariyanayagam, and A. H. Fairlamb. 2002a. Characterization of recombi-nant glutathionylspermidine synthetase/amidase from *Crithidia fasciculata*. *Biochem J* 364 (Pt 3):679–86.

Oza, S. L., M. P. Shaw, S. Wyllie, and A. H. Fairlamb. 2005. Trypanothione biosynthesis in *Leishmania major*. *Mol Biochem Parasitol* 139 (1):107–16.

Oza, S. L., E. Tetaud, M. R. Ariyanayagam et al. 2002b. A single enzyme catalyses formation of Trypanothione from glutathione and spermidine in *Trypanosoma cruzi*. *J Biol Chem* 277 (39):35853–61.

Pai, C. H., B. Y. Chiang, T. P. Ko et al. 2006. Dual binding sites for translocation catalysis by *Escherichia coli* glutathionylspermidine synthetase. *EMBO J* 25 (24):5970–82.

Pai, C. H., H. J. Wu, C. H. Lin, and A. H. Wang. 2011. Structure and mechanism of *Escherichia coli* glutathionylspermidine amidase belonging to the family of cysteine; histidine-dependent amidohydrolases/peptidases. *Protein Sci* 20 (3):557–66.

Park, M. H. 2006. The post-translational synthesis of a polyamine-derived amino acid, hypusine, in the eukaryotic translation initiation factor 5A (eIF5A). *J Biochem* 139 (2):161–9.

Shah, P., and E. Swiatlo. 2008. A multifaceted role for polyamines in bacterial pathogens. *Mol Microbiol* 68 (1):4–16.

Shim, H., and A. H. Fairlamb. 1988. Levels of polyamines, glutathione and glutathione-spermidine conjugates during growth of the insect trypanosomatid *Crithidia fascicu-lata*. *J Gen Microbiol* 134 (3):807–17.

Shoji, T., and T. Hashimoto. 2015. Polyamine-derived alkaloids in plants: Molecular elu-cidation of biosynthesis. In *Polyamines*, edited by T. Kusano and H. Suzuki. Tokyo: Springer.

Smith, K., A. Borges, M. R. Ariyanayagam, and A. H. Fairlamb. 1995. Glutathionylspermidine metabolism in *Escherichia coli*. *Biochem J* 312 (Pt 2):465–9.

Smith, K., K. Nadeau, M. Bradley et al. 1992. Purification of glutathionylspermidine and trypanothione synthetases from *Crithidia fasciculata*. *Protein Sci* 1 (7):874–83.

Sousa, A. F., A. G. Gomes-Alves, D. Benitez et al. 2014. Genetic and chemical analyses reveal that trypanothione synthetase but not glutathionylspermidine synthetase is essential for *Leishmania infantum*. *Free Radic Biol Med* 73:229–38.

Suzuki, H., and S. Kurihara. 2015 Polyamine catabolism in prokaryotes. In *Polyamines*, edited by T. Kusano and H. Suzuki. Tokyo: Springer.

Tabor, C. W., and H. Tabor. 1970. The complete conversion of spermidine to a peptide derivative in *Escherichia coli*. *Biochem Biophys Res Commun* 41 (1):232–8.

Tabor, H., and C. W. Tabor. 1975. Isolation, characterization, and turnover of glutathionyl-spermidine from *Escherichia coli*. *J Biol Chem* 250 (7):2648–54.

Taylor, M. C., H. Kaur, B. Blessington et al. 2008. Validation of spermidine synthase as a drug target in African trypanosomes. *Biochem J* 409 (2):563–9.

Teixeira, M. M., T. C. Borghesan, R. C. Ferreira et al. 2011. Phylogenetic validation of the genera *Angomonas* and *Strigomonas* of trypanosomatids harboring bacterial endosymbionts with the description of new species of trypanosomatids and of proteobacterial symbionts. *Protist* 162 (3):503–24.

Ulrich, K., C. Finkenzeller, S. Merker et al. 2017. Stress-induced protein *S*-glutathionylation and *S*-trypanothionylation in African trypanosomes—A quantitative redox proteome and thiol analysis. *Antioxid Redox Signal* 27 (9):517–33.

Votýpka, J., A. Y. Kostygov, N. Kraeva et al. 2014. Kentomonas gen. n., a new genus of endosymbiont-containing trypanosomatids of *Strigomonadinae* subfam. n. *Protist* 165 (6):825–38.

Willert, E. K., and M. A. Phillips. 2008. Regulated expression of an essential allosteric activator of polyamine biosynthesis in African trypanosomes. *PLoS Pathog* 4 (10):e1000183.

Willert, E., and M. A. Phillips. 2012. Regulation and function of polyamines in African trypanosomes. *Trends Parasitol* 28 (2):66–72.

Wyllie, S., S. L. Oza, S. Patterson et al. 2009. Dissecting the essentiality of the bifunctional trypanothione synthetase-amidase in *Trypanosoma brucei* using chemical and genetic methods. *Mol Microbiol* 74 (3):529–40.

18 Trypanothione Functions in Kinetoplastida

Martin Hugo
German Institute of Human Nutrition

Madia Trujillo, Lucía Piacenza, and Rafael Radi
Universidad de la República

CONTENTS

18.1 INTRODUCTION

Kinetoplastids are primitive eukaryotes that can be found either as free-living organisms or as obligate parasites in a diverse range of invertebrates or vertebrates (Stevens, 2008). For instance, the parasites *Leishmania donovani, Leishmania amazoniensis, Leishmania infantum, Trypanosoma cruzi*, and *Trypanosoma brucei*, well known for causing leishmaniasis, visceral leishmaniasis, American trypanosomiasis or Chagas disease and African trypanosomiasis or sleeping sickness, respectively, belong to the trypanosomatid group of kinetoplastids. In these organisms, most of the low-molecular-weight thiol content is found as a singular derivative of glutathione (GSH): N^1,N^8-bis(glutathionyl)-spermidine or trypanothione.* Initially described in 1985 by Fairlamb (Fairlamb et al., 1985), it is a dithiol product of the conjugation of two molecules of GSH with one molecule of spermidine by the enzymes mono-glutathionyl spermidine synthetase (GspS) and trypanothione synthetase (TryS) or by TryS alone catalyzing both conjugation reactions (Manta et al., 2013a). Although *Escherichia coli* is able to synthesize mono-glutathionyl spermidine (Bollinger et al., 1995), a complete biosynthetic pathway for trypanothione is exclusive to kinetoplastids (see Chapter 17). Among them, almost all members of the trypanosomatid phylum contain the gene encoding for TryS, and some members, remarkably *Trypanosoma brucei* lacks the gene encoding GspS (Comini et al., 2004; Manta et al., 2018). Bodonids, model free-living heterotroph among kinetoplastids (www.sanger.ac.uk) found worldwide in freshwater and marine habitats, contain both genes, and prokinetoplastids contain partial sequences with similarity to GspS and TryS (Jackson et al., 2008; Manta et al., 2018). Trypanothione disulfide (TS_2) is reduced by trypanothione reductase (TR) at the expense of NADPH. Most of the characterized functions of trypanothione rely on (1) direct reactions with metabolites or (2) its participation as substrate in various enzyme-catalyzed reactions. Here, we summarize the current knowledge on the biochemistry of this particular dithiol, with emphasis on its reactivity toward relevant oxidants, its interactions with proteins and its participation as cofactor for a wide range of enzymatic reactions with relevance in cellular-, drug-metabolism, and infectivity.

18.2 PHYSICOCHEMICAL PROPERTIES OF TRYPANOTHIONE

18.2.1 THIOL PK_A AND REDOX POTENTIAL

As a dithiol, the exact pK_{aSH} values of each thiol group in dihydrotrypanothione ($T(SH)_2$) are missing. A mean pK_{aSH} value accounting for the macroscopic reactivity of $T(SH)_2$ was reported to be ~7.4 (Moutiez et al., 1994). The pK_a value of the thiol groups in $T(SH)_2$ is certainly lower than that of reduced glutathione (GSH), but exact value determinations await further investigation. While at physiological pH, the net charge of TS_2 is +1, that of $T(SH)_2$ will be lower than +1, depending on its uncertain thiol pK_a values. This makes a clear difference with the −2 charge of both GSH and oxidized glutathione (GSSG) and is the basis of the selectivity shown by glutathione reductase (GR) and TR for their respective substrates (Faerman et al., 1996).

The standard redox potential of the $T(SH)_2/TS_2$ couple was determined as −0.242±0.002 V (Fairlamb and Cerami, 1992), close to that of GSH [−0.240 V (Schafer et al., 2001)] and a bit lower than that of *T. brucei* tryparedoxin (TXN) [$E_0' = -0.270$ V (Reckenfelderbäumer et al., 2002)], which indicates that electrons flow from $T(SH)_2$ towards TXN under standard conditions. However, bearing in

---

* Trypanothione refers to the reduced and oxidized forms of the compound; dihydrotrypanothione indicates the reduced ($T(SH)_2$) and trypanothione disulfide indicates the oxidized (TS_2) form of trypanothione, respectively.

mind that redox potentials of $T(SH)_2$ and TXN redox couples are not so far, the concentrations of reduced and oxidized members of the couple critically contribute to determine the electron fate. Accordingly, TS_2 was found to be a strong inhibitor of TXN activity (Krauth-Siegel et al., 2002).

18.2.2 OXIDATION BY BIOLOGICALLY RELEVANT OXIDANTS

The oxidation of thiols by one- or two-electron processes leads to the formation of thiyl radicals (RS$^{\cdot}$) and sulfenic acids (RSOH), respectively (Trujillo et al., 2016). Generation of a sulfenic acid in trypanothione will rapidly be solved by the formation of an intramolecular disulfide. Similarly, thiyl radical in trypanothione will react with the vicinal thiol to form a thiol disulfide radical anion that can readily reduce molecular oxygen to yield TS_2 and superoxide radical ($O_2^{\cdot-}$), the latter being readily decomposed by superoxide dismutases (SODs) (Fitzgerald et al., 2010).

Few reactions of trypanothione with biologically relevant oxidants have been kinetically addressed (Table 18.1). During its life cycle, kinetoplastids can be exposed to different two-electron oxidants including hydroperoxides such as hydrogen peroxide (H_2O_2), peroxynitrite, hypohalous acids, and haloamines. The sources of these different oxidants may vary depending on the kinetoplastid species (free living or parasitic form) and even on the life stage for a given species (Augusto et al., 1996; Piacenza et al., 2007; Piacenza et al., 2009). $T(SH)_2$ reacts slightly faster than glutathionyl-spermidine and GSH with H_2O_2 [5.4, 4.2, and 1.6 $M^{-1}s^{-1}$, respectively, at pH 7.2 and 27°C, (Ariyanayagam et al., 2001)]. Since the reactions involve thiolates as nucleophiles, differences in rate constants most probably reflect the above mentioned differences in thiol pK_{aSH}, with their consequences in thiolate availability and nucleophilicity (Ferrer-Sueta et al., 2011). These values are 5–7 orders of magnitude smaller than corresponding values for the oxidation of the peroxidatic cysteines in peroxiredoxins (Prx) (Trujillo et al., 2007) or glutathione peroxidases (GPx) (Toppo et al., 2009). Therefore, even considering the abundance of $T(SH)_2$, it is possible to infer that formation of TS_2 and GSSG upon oxidative challenge to cells is a product of enzyme-catalyzed reactions rather than of the direct reaction with hydroperoxides (Thomson et al., 2003; Trujillo et al., 2004). Peroxynitrite* is an oxidizing and nitrating species with a high capability to kill microorganisms including trypanosomatids (Denicola et al., 1993; Augusto et al., 1996; Alvarez et al., 2011). It is formed by the diffusion-controlled reaction between nitric oxide ($^{\cdot}$NO) and $O_2^{\cdot-}$ radical (Huie and Padmaja, 1993). Protein thiols, metal centers, and carbon dioxide (CO_2) constitute the main targets for peroxynitrite *in vivo*. $T(SH)_2$ is oxidized by peroxynitrite with a rate constant of 7,200 $M^{-1}s^{-1}$ at pH 7.4 and 37°C leading to TS_2 via the formation of an intermediate sulfenic acid (Table 18.1; Thomson et al., 2003; Trujillo et al., 2005; Ferrer-Sueta and Radi, 2009). Such reactivity is the expected for a low-molecular-weight dithiol with a pK_a of 7.4 (Moutiez et al., 1994; Trujillo et al., 2010). This rate

---

* IUPAC recommended names for peroxynitrite anion (ONOO$^-$) and its conjugated acid, peroxynitrous acid (ONOOH), are oxoperoxonitrate (1-) and hydrogen oxoperoxonitrate, respectively. The term peroxynitrite is used to refer to the sum of ONOO$^-$ and ONOOH.

constant is slightly higher than that observed with GSH [1,350 $M^{-1}s^{-1}$ (Koppenol et al., 1992; Trujillo and Radi, 2002)]. In a given compartment, $T(SH)_2$ competes with other targets for peroxynitrite. The reactions with peroxidases of the Prx and GPx family are 3–4 orders of magnitude faster than with $T(SH)_2$, and therefore, most peroxynitrite is expected to react with the enzymes and to consume $T(SH)_2$ indirectly through the sequence of reactions involved in peroxidase reduction (Thomson et al., 2003). Moreover, the reaction of peroxynitrite with CO_2 is relatively fast ($k = 4.6 \times 10^4$ $M^{-1}s^{-1}$ at pH 7.4 and 37°C) forming a transient intermediate that homolyzes to nitrogen dioxide radical ($^{\bullet}NO_2$) and carbonate radical ($CO_3^{\bullet -}$) in ~35% yields (Lymar et al., 1995; Denicola et al., 1996; Bonini et al., 1999). These radicals can in turn participate in oxidation and/or nitration reactions (Radi, 2004). Although these radicals can perform the one-electron oxidation of different low-molecular-weight thiols including the dithiol dihydrolipoic acid (Nakao et al., 1999; Trujillo et al., 2005), interactions with $T(SH)_2$ have not been addressed so far. Furthermore, low-molecular-weight thiols can perform radical repair reactions, including tyrosyl radical reduction inhibiting tyrosine nitration (Kirsch et al., 2001; Folkes et al., 2011). It is reasonable to assume the participation of $T(SH)_2$ in those reactions, particularly when solvent exposed tyrosyl and/or cysteinyl residues are involved, like the one formed after peroxynitrite reaction with the cytosolic Fe-SODB from *T. cruzi* (Martinez et al., 2014).

Stimulated neutrophils produce the myeloperoxidase-derived hypochlorous acid (HOCl), a highly oxidant and cytotoxic species (Harrison et al., 1976; Kettle et al., 1997). The reactions of HOCl with thiols lead to disulfide formation or irreversible over-oxidation to sulfinic/sulfonic acids (Hawkins et al., 2003), and causes protein

---

TABLE 18.1
Physicochemical Properties of Trypanothione and Glutathione

	$T(SH)_2$	GSH
Nature	Dithiol	Monothiol
pK_a (SH)	7.4	8.93
$E^{o\prime}$ (V)	−0.242	−0.24
$k_{H_2O_2}$ $(M^{-1}s^{-1})^a$	5.4	1.6
k_{ONOOH} $(M^{-1}s^{-1})^b$	7,200	1,350
k_{DHA} $(M^{-1}s^{-1})$	~90[d]	0.28[c]
k_{DTNB} $(M^{-1}s^{-1})^e$	2×10^5	3.5×10^4
k_{CDNB} $(M^{-1}s^{-1})^e$	0.9	2.4

[a] At pH 7.2 and 27°C (Ariyanayagam et al., 2001).

[b] At pH 7.4 and 37°C (Trujillo et al., 2004).

[c] Calculated from initial rate of DHA reduction at pH 7.4 and RT (20–22°C) reported by Winkler et al. (1994).

[d] Calculated from data reported by Krauth-Siegel et al. at pH 7.5 and 25°C (Krauth-Siegel et al., 1996).

[e] pH-independent rate constants (rate constant of the thiolate forms; Moutiez et al.,1994).

DHA is dehydroascorbate, DTNB is dithionitrobenzoic acid, CDNB is 1-chloro 2,4-dinitrobenzene.

S-thiolation in bacteria (Chi et al., 2011; Chi et al., 2014; Hillion et al., 2017), *T. brucei* (Ulrich et al., 2017) and endothelial cells (Stacey et al., 2012). Although HOCl is a promiscuous oxidant, $T(SH)_2$ could be a significant target given its high abundance. Moreover, $T(SH)_2$ oxidation and protein *S*-thiolation may work as a protective mechanism against over-oxidation of functional protein thiols. Protein *S*-thiolation/dethiolation can be catalyzed by glutaredoxins (Grxs) (Lillig et al., 2008; Gallogly et al., 2009), but to date is unknown whether trypanosomal Grxs are able to catalyze these reactions. It has been speculated that such activity could be performed by the oxidoreductase TXN (Ulrich et al., 2017). Interestingly, an ATP-dependent LABCG2 transporter for both TS_2 and $T(SH)_2$ has been just identified in *Leishmania* (Perea et al., 2018); this transporter can also deal with drug-, metal-, and LMW-thiol conjugates of trypanothione.

18.2.3   THE FATE OF RADICALS: MONO- VERSUS DI-THIOLS

Thiyl radicals decay by different mechanisms, including fast recombination reactions with other radicals, reaction with molecular oxygen to form thio-peroxides that may rearrange to sulfinyl radicals eventually leading to sulfinic acid derivatives, and reaction with other thiol groups to form disulfide anion radicals that, acting as electron donors, form disulfides. In the case of dithiols, this last reaction is intramolecular and thus, kinetically favored. One-electron reduction of molecular oxygen to $O_2^{\cdot-}$ radical by disulfide anion radicals leads to disulfide formation. Thus, disulfides and $O_2^{\cdot-}$ formed can be reduced by specific reductases (such as TR), and the combined action of SODs and peroxidases, respectively. In this pathway, $O_2^{\cdot-}$ radical acts as an intracellular radical sink (Winterbourn, 1993, 2016). On the contrary, sulfinic acids arising from the reaction of thiyl radicals with O_2 are usually irreversible. The topic has been experimentally addressed by Cadenas et al. by comparing the detection of thiyl radicals through spin trapping-based methodologies in mono-thiols versus dihydrolipoic acid. While the formers were easily detected by their reaction with 5,5′-dimethyl-l-pyrroline-N-oxide (DMPO), the latter was EPR-silent, in agreement with a fast intramolecular reaction with the second thiol group (Cadenas, 1995). This is probably the basis of the radio resistance showed by strains of *E. coli* genetically engineered to produce $T(SH)_2$ (Fitzgerald et al., 2010). In fact, trypanosomes have long been known to be one of the most radio-resistant eukaryotic organisms (Emmett, 1950). Radiation of *T. cruzi* epimastigotes does not lead to a massive expression of members of the antioxidant machinery, but only of the cytosolic and mitochondrial tryparedoxin peroxidases (Grynberg et al., 2012), two Prxs dependent on trypanothione as electron source (see Section 18.7).

Moreover, isolated trypanothione is able to protect DNA damage by scavenging radiation-induced radicals only in the presence of TR and NADPH to maintain the reduced state (Awad et al., 1992). This higher protective effect compared to GSH was attributed to accumulation of trypanothione in the proximity of the DNA due to the spermidine moiety. This consideration sounds plausible given that $T(SH)_2$ and GSH differ in their net charges at physiological pH (+1 and −2, respectively) and in the absence of a free carboxylate glycine in the first.

18.3 CATABOLISM OF METHYLGLYOXAL

Glycolysis produces the toxic and mutagenic nonenzymatic by-product methyl-glyoxal. The main cellular system responsible for the catabolism of methylglyoxal depends on two enzymes, glyoxalase I (EC 4.4.1.5, an isomerase) and II (EC 3.1.2.6, a thioesterase) (Thornalley, 1990), which are able to convert methylglyoxal to lactate using GSH as a catalytic cofactor. Initially, a hemithioacetal must be formed by the nonenzymatic reaction of methylglyoxal with GSH or $T(SH)_2$ (in the latter case forming a bis-hemithioacetal), following isomerization to the thioester S-D-lactoyl glutathione or S-D-lactoyl trypanothione by glyoxalase I. The thioester is then hydrolyzed to D-lactate by glyoxalase II, and the thiol is released. Although the occurrence of glyoxalase I and II varies among different kinetoplastid species, all appear to have adapted for the use of trypanothione as a catalytic cofactor (Irsch et al., 2004; Silva et al., 2008a; Sousa Silva et al., 2012). Structural and site-directed mutagenesis studies in the *L. infantum* glyoxalase II allowed the identification of the amino acid residues essential for the trypanothione-hemiacetal specificity (Barata et al., 2011). The *Bodo saltans* genome (Jackson et al., 2008) contains a gene encoding a glyoxalase II with 43.3% identity with that of *L. infantum*, conserving the tyrosine and cysteine residues essential for trypanothione binding (Barata et al., 2011). However, the biological relevance of the glyoxalases in kinetoplastids, and therefore of the trypanothione in methylglyoxal detoxification, is still a matter of debate. While deletion of glyoxalase I exhibits reduced methylglyoxal detoxification in *L. donovani*, glyoxalase II in *T. brucei* appears not to serve as a methylglyoxal detoxification system. It rather acts as a general thioesterase as evidenced by the efficient *in vitro* activity toward different trypanothione thioesters (Wendler et al., 2009), which opens the possibility for novel thioester-dependent regulatory processes and signal transduction.

18.4 DRUG AND HEAVY METAL DETOXIFICATION

The evidences for a central role of the $T(SH)_2$ system in the detoxification of arsenic and antimonial drugs and nitroheterocyclic compounds are abundant (Fairlamb et al., 1989; Fairlamb et al., 1992; Maya et al., 1997; Shahi et al., 2002; Wyllie et al., 2004; Ariyanayagam et al., 2005; Maya et al., 2007), particularly in pathogenic trypanosomatids. Increased levels of $T(SH)_2$ are associated with resistance to arsenite- or antimony-containing drugs, as evidenced in *in vitro* selected *Leishmania* parasites (Mukhopadhyay et al., 1996; Legare et al., 1997; Haimeur et al., 2000). The drug resistance conferred by $T(SH)_2$ probably relies on a combination of direct and enzyme-catalyzed reactions: $T(SH)_2$ can form adducts with arsenite (Fairlamb et al., 1989; Mukhopadhyay et al., 1996), but this reaction can be also catalyzed by trypanothione S-transferase (Vickers and Fairlamb, 2004; Vickers et al., 2004). The role of this enzyme might be species-specific, as it has been described in *Leishmania*, *T. brucei*, and *C. fasciculata* but not in *T. cruzi* epimastigotes (Vickers and Fairlamb, 2004), and *L. tarentolae* resistance to antimonials is not correlated to trypanothione-S-transferase activity (Wyllie et al., 2008). Alternatively, TXN may be considered as the ultimate target of trivalent arsenicals (and possibly antimonials), as evidenced by MS-verified covalent modification by phenylarsinoxide (Nogoceke et al., 1997). Complementarily, an increase in the expression of the gene encoding the Gp-glycoprotein-like protein A (Legare et al., 1997; Haimeur et al., 2000) has been measured in resistant parasites. This protein is a transporter belonging to the multidrug resistance proteins subfamily involved in the efflux of metal–thiol conjugates (Legare et al., 2001).

Regarding the trypanocidal prodrugs benznidazole and nifurtimox, their mode of action is still controversial, but in any case, it is clear that $T(SH)_2$ is important for the detoxification and resistance. Several studies have proposed that the intracellular enzymatic reduction of the nitro-group of nifurtimox generates electrophilic or radical (nitro) intermediates, the latter followed by redox cycling and concomitant production of the $O_2^{\cdot-}$ radical and H_2O_2 (Docampo and Stoppani, 1980; Docampo and Moreno, 1984; Docampo and Moreno, 1986; Maya et al., 2007). The removal of this hydroperoxide is mediated by $T(SH)_2$/tryparedoxin-dependent peroxidases (see Section 18.7.2). In contrast, the trypanocidal effect of benznidazole may not involve the formation of oxygen radicals but electrophilic products with further covalent binding to nuclear and kinetoplastid DNA, proteins and lipids (Diaz De Toranzo et al., 1988). Glyoxal was later identified as the major cytotoxic product resulting from benznidazole treatment (Hall et al., 2012). Treatment of *T. cruzi* in different stages with nifurtimox leads to strong depletion of the low-molecular-weight thiol pool, especially $T(SH)_2$, which was reduced to 50% (Maya et al., 1997). Similar effects were observed upon benznidazole treatment, but low-molecular-weight adducts of glyoxal, a proposed toxic end product of benznidazole metabolism, were not detected (Trochine et al., 2014). However, TR and NADPH failed to recover the thiol levels in the cell lysates suggesting a conjugation of the drugs with $T(SH)_2$ and, accordingly, several covalent adducts of benznidazole with $T(SH)_2$ and other less abundant low-molecular-weight thiols were identified (Trochine et al., 2014) (for example, see Figure 18.1). Inhibition of GSH synthesis (and the consequent reduction of the $T(SH)_2$ pool) with buthionine sulfoximine results in increased sensitivity to both, nifurtimox and benznidazole (Moncada et al., 1989; Maya et al., 2004; Faundez et al., 2005; Faundez et al., 2008). Recent studies suggested that among the three enzymes proposed to be responsible for the activation of nifurtimox (prostaglandin F2α synthase, cytochrome P450 reductase, and a type I nitroreductase), only elevated levels of the latter caused altered susceptibility to the drug, implying a major role in drug action (Wilkinson et al., 2008; Hall et al., 2011). The authors claim a non-radical, oxygen-independent reduction of the prodrug with formation of cytotoxic nitrile metabolites as main effectors rather than the previously proposed redox cycling (Boiani et al., 2010; Hall et al., 2011).

18.5 INTERACTIONS WITH NITRIC OXIDE AND IRON: IRON–SULFUR CLUSTER ASSEMBLY

Nitric oxide (˙NO) is a free radical that can be synthetized by the family of nitric oxide synthases (NOS) present in different tissues. In immune cells, such as macrophages, ˙NO is produced (in the micromolar range) upon NOS induction by

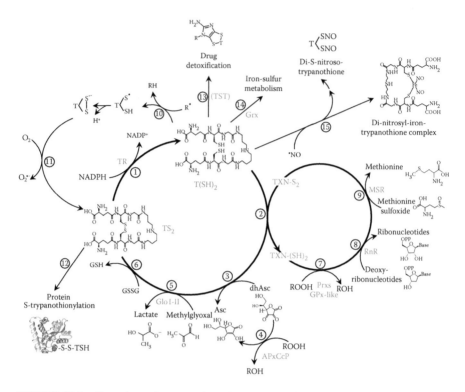

FIGURE 18.1 Overview of enzymatic and nonenzymatic key cellular functions of trypanothione. Trypanothione disulfide (TS$_2$) is transformed to its reduced form (T(SH)$_2$)
at the expense of NADPH in a reaction catalyzed by trypanothione reductase (TR) (1).
T(SH)$_2$ reduces tryparedoxins (TXN-S$_2$) (2), dehydroascorbate (dhAsc) to ascorbate (Asc)
(3), the latter serving as a reducing substrate for ascorbate-cytochrome c-peroxidase
(APxCcP) (4). T(SH)$_2$ reacts with methylglyoxal forming a bis-hemithioacetal, which can
be further metabolized to lactate by glyoxalases I and II (GloI-II) (5) and reduces glutathione disulfide (GSSG) to glutathione (GSH) (6). TXNs catalyze the electron transfer
from T(SH)$_2$ to different protein targets involved in peroxide detoxification via peroxiredoxins (Prxs) and glutathione peroxidase-like enzymes (GPx-like) (7), enables DNA
synthesis and proliferation via ribonucleotide reductase (RR) (8) and protein repair by
methionine sulfoxide reductases (MSR) (9). Reactions with radical species (R•) in scavenging and/or repair reactions lead to the transient formation of trypanothione thiyl radical (10), which readily combines with the vicinal thiol to yield a trypanothione disulfide
anion radical, a good reductant that in the presence of molecular oxygen (or other substrates) evolves to stable TS$_2$ with the concomitant formation of superoxide radical (O$_2$•⁻)
(11). Proteins can undergo S-trypanothiolation (12), a process that might be catalyzed
by glutaredoxins (Grx). T(SH)$_2$ also plays a central role in the detoxification of arsenic, antimonial-containing drugs and nitroheterocyclic compounds in complex reactions
that may involve enzymatic [trypanothione-S-transferase (TST), or TXN, respectively]
catalysis (13). T(SH)$_2$ can coordinate iron–sulfur clusters itself and transfer them to glutaredoxins (Grx) in the absence of scaffold proteins (14) and can act as a bidentate ligand,
forming, in the presence of nitric oxide (•NO), a stable, nontoxic dinitrosyl-iron complex
as well as dinitroso trypanothione (15).

proinflammatory cytokines (IFN-γ, TNF-α, IL-1β; Xie et al., 1993; Macmicking et al., 1997; Alvarez et al., 2004). If not neutralized, ˙NO can lead to the formation of S-nitrosoglutathione, modify cysteine residues (protein S-nitrosation) or metal cofactors in proteins or originate low-molecular-weight thiol–iron complexes known as di-iron complex (Martinez-Ruiz et al., 2007). Trypanothione, due to its dithiol nature, can form a stable and nontoxic dinitrosyl-iron complex (Bocedi et al., 2010; see structure in Figure 18.1). $T(SH)_2$ acts as a bidentate ligand with a 600 times higher affinity than GSH, and therefore readily displaces the latter from the complex (Bocedi et al., 2010). While the dinitrosyl-diglutathionyl iron complex is a potent irreversible inhibitor of GR ($IC_{50}=4\,\mu M$), the trypanothione-derived complex does not inactivate TR even at millimolar levels (Bocedi et al., 2010). The formation of this complex occurs in $T.$ $brucei$ and $L.$ $infantum$ exposed to ˙NO, and a role of this complex in the defense against macrophage-derived reactive nitrogen species by trapping ˙NO and subsequent inhibition of the formation of cytotoxic oxidants (Alvarez et al., 2011) was proposed (Bocedi et al., 2010). It is interesting to note that $T.$ $cruzi$ can itself generate small amounts of ˙NO that serve signaling purposes (Pereira et al., 1997; Piacenza et al., 2001). However, the potential interaction of this low ˙NO level with trypanothione remains unexplored. Of note, trypanothione can form stable S-nitrosothiol derivatives (e.g., di-S-nitroso-trypanothione, Figure 18.1), which decompose in the presence of ovothiol A (N-methyl-4-mercapto hystidine), another low-molecular-weight thiol present in trypanosomatids (Vogt et al., 2003).

Iron–sulfur clusters are cofactors with iron atoms bridged by inorganic sulfur. Grxs are typically known for their participation in redox processes (Section 18.7 and Chapters 15 and 16), but are also required for iron–sulfur cluster assembly and heme biosynthesis (Rouhier et al., 2010). Certain dithiol and almost all monothiol Grxs characterized so far in trypanosomatids are able to bind iron–sulfur clusters (Manta et al., 2013a), a process that leads to protein dimerization (Comini et al., 2008). In $T.$ $brucei$, a monothiol Grx (Grx1) containing a disordered N-terminal extension only found in kinetoplastids has been described (Manta et al., 2013b). This "Grx" cannot function as a thiol-disulfide oxidoreductase (Filser et al., 2008). All three monothiol Grxs, as well as dithiol Grxs from $T.$ $brucei$ are capable to coordinate [2Fe–2S] clusters using Gsp or $T(SH)_2$ for metal ligation (Ceylan et al., 2010; Manta et al., 2013b). The ability of $T(SH)_2$ to coordinate iron–sulfur clusters itself and to transfer it to Grxs in the absence of scaffold proteins raises a new function for the low-molecular-weight thiol as a carrier for iron–sulfur clusters or ligand in kinetoplastids.

18.6 REDUCTION OF DEHYDROASCORBATE

The synthesis of ascorbic acid in kinetoplastids occurs in the glycosome, a subcellular compartment exclusive to this class (Wilkinson et al., 2005). The synthesis is completed by an FMN-dependent galactonolactone oxidase (Logan et al., 2007) located in this compartment. Related enzymes with the same function are found in all kingdoms including animals, but humans lack this enzyme (Smirnoff et al., 2001; Linster et al., 2007). The maintenance of the ascorbic acid pool is of particular importance in trypanosomatids as it is part of the antioxidant capacity of the cell.

Four decades ago, a heme-dependent peroxidase activity was measured in *T. cruzi* (Docampo et al., 1976). Kinetoplastids lack catalase but express a plant-derived hybrid-type heme peroxidase that can use ascorbate and cytochrome c as electron source (Wilkinson et al., 2002; Hugo et al., 2017). The best functionally analyzed members of this family are from the kinetoplastids *L. major* (Adak et al., 2005) and *T. cruzi* (Wilkinson et al., 2002b; Hugo et al., 2017), but this enzyme is absent in *T. brucei* (Zamocky et al., 2010).

Although an ascorbate reductase (glutathione:dehydroascorbate oxidoreductase, EC 1.8.5.1) activity was reported in *T. cruzi* (Clark et al., 1994), a later study showed that T(SH)$_2$ itself was capable and responsible for the maintenance of the intracellular ascorbic acid pool (Krauth-Siegel et al., 1996). The second order rate constant for the reaction of dehydroascorbate with T(SH)$_2$ yielding ascorbic acid and TS$_2$ was estimated as 90 M^{-1}s^{-1} at pH 7.4 and 25°C, and the authors claimed that there was no evidence for the existence of a specific dehydroascorbate reductase activity. Although monothiol Grxs are known to be able to reduce dehydroascorbate (Herrero et al., 2007), such activity and its competition with the direct reduction by T(SH)$_2$ remains to be established in kinetoplastids.

18.7   REDOXIN-DEPENDENT FUNCTIONS OF T(SH)$_2$

18.7.1   REDUCTION OF TRYPAREDOXIN

The flow of electrons from T(SH)$_2$ to its main protein targets occurs enzymatically via oxidoreductases characterized by a CXXC active-site motive belonging to the thioredoxin (Trx) superfamily. Among them, TXNs are exclusive and essential to kinetoplastids and contain an active-site WCPPCR motif (Gommel et al., 1997; Nogoceke et al., 1997; Lüdemann et al., 1998; Comini et al., 2007). The cytosolic form is very abundant, reaching 3%–5% of the total soluble protein in *C. fasciculata* (Nogoceke et al., 1997), and its depletion in African trypanosomes proved its essential role in the parasite defense against oxidative stress (Comini et al., 2007). On the contrary, depletion studies suggest that the mitochondrial redox metabolism may be independent of the mitochondrial isoform of TXN (Castro et al., 2010). TXNs are able to catalyze the electron transfer (disulfide reduction) from T(SH)$_2$ to different protein targets involved in peroxide detoxification, protein repair (Moskovitz, 2005; Arias et al., 2011), and DNA synthesis and proliferation (Dormeyer et al., 2001), but are inefficient in the reduction of protein mixed-disulfides with GSH (Ceylan et al., 2010). TXNs have a stronger preference for T(SH)$_2$ and glutathionyl-spermidine as reducing substrate than for GSH, which has been evidenced by various kinetic (Gommel et al., 1997; Nogoceke et al., 1997; Lüdemann et al., 1998; Manta et al., 2013a) and structural studies [Hofmann et al., 2001; recently reviewed in detail (Manta et al., 2018)]. For example, the tryparedoxin from *C. fasciculata* exhibits a K_m of 130 μM for T(SH)$_2$ and an infinite K_m value for reduced GSH combined with a low turnover rate (Gommel et al., 1997). The dithiol moieties of both T(SH)$_2$ and the active site of TXN share a pK_{aSH} value close to the physiological pH (7.4–7.6), both cysteine residues of the latter being experimentally indistinguishable (Reckenfelderbäumer et al., 2002). As mentioned previously, the electrochemical potential of *T. brucei* TXN is slightly higher than that of T(SH)$_2$ [E_0' = −0.270 V (Reckenfelderbäumer et al., 2002)], predicting electron flow from T(SH)$_2$ towards TXN under standard conditions. However, bearing in mind that redox potentials of T(SH)$_2$ and TXN redox couples are not so far, the concentrations of reduced and oxidized members of the couple critically contribute to determine the electron fate. Accordingly, TS$_2$ was found to be a strong inhibitor of TXN activity (Krauth-Siegel et al., 2002).

18.7.2 TRYPANOTHIONE-DEPENDENT PEROXIDASES

Kinetoplastids, including the free-living *B. saltans* possess four different thiol-dependent peroxidases. Two of these are typical 2-cysteine Prxs belonging to the AhpC-Prx1 subfamily [csb.wfu.edu/prex (Soito et al., 2011)]. Different from their subfamily counterparts, kinetoplastid Prxs do not use Trx as reducing substrate but TXN, and therefore have been given the name of TXN peroxidases (TXNPx) (Nogoceke et al., 1997; Wilkinson et al., 2000b). Both cytosolic (cTXNPx) and mito-chondrial (mTXNPx) isoforms decompose H$_2$O$_2$, peroxynitrite, and with lower effi-ciency, organic hydroperoxides [for cTXNPxs the bimolecular rate constants for the reduction of hydroperoxides cover a range of $k = 10^7$ for H$_2$O$_2$ to 3×10^4 M^{-1}s^{-1} for *tert*-butyl hydroperoxide; for details see (Trujillo et al., 2004; Pineyro et al., 2011)]. These values are comparable to those determined for 2-Cys Prxs from other organ-isms [kinetics and mechanism reviewed in Trujillo et al. (2008); Zeida et al. (2014)]. The reductive part of the catalytic cycle was also studied, and the rate constants for the reduction of cTXNPxs by TXNs center around 10^6 M^{-1}s^{-1}. The reduction of TXN by T(SH)$_2$ is much slower [8×10^3 M^{-1}s^{-1} (Wilkinson et al., 2002a)] and, thus, the rate-limiting step in this pathway (Gommel et al., 1997; Wilkinson et al., 2002a). In *T. cruzi*, mTXNPx colocalizes with the kinetoplast indicating that this enzyme may play a role in mitochondrial DNA protection from peroxynitrite and H$_2$O$_2$-mediated toxicity (Wilkinson et al., 2000b; Piacenza et al., 2008). Although differently located, overexpression of cTXNPx and mTXNPx in *T. cruzi* confers protection to H$_2$O$_2$, *tert*-butylhydroperoxide, and peroxynitrite-mediated toxicity (Wilkinson et al., 2000b; Piacenza et al., 2008).

The second class of kinetoplastid thiol-peroxidases has similarity with mamma-lian phospholipid hydroperoxide GPxs. Trypanosomal GPxs differ from these mam-malian counterparts in having cysteine instead of selenocysteine as main catalytic residue and lack critical residues involved in GSH binding. Consequently, they are not as efficient at reducing oxidants and use GSH inefficiently (Wilkinson et al., 2000a,c). This is in agreement with the idea that most nonselenium GPxs use Trx or Trx-like proteins as reductant instead of GSH (Maiorino et al., 2007). Actually, the search for alternative electron donors showed that GPxI from *T. cruzi* is actually another type of TXN-dependent peroxidase (Wilkinson et al., 2002a). *T. brucei* expresses four GPx-type peroxidases, which differ in subcellular localization but all appear to be important for lipid hydroperoxide detoxification (Diechtierow et al., 2011; Liu et al., 2016). Deletion of both cytosolic isoforms leads to fast lipid peroxidation and cell lysis, a phenotype that is rescued by supplementation of the culture media with the α-tocopherol hydrophilic derivative Trolox-C (Diechtierow et al., 2011). *Tc*GPx-I is located in the glycosome and cytoplasm where it reduces fatty acid hydroperoxides and organic hydroperoxides (Wilkinson et al., 2002a). *Tc*GPx-II is located at the endoplasmic reticulum and its activity is restricted to the reduction of fatty acid and phospholipid hydroperoxides (Wilkinson et al., 2002c). None of them can decompose H_2O_2, but its overexpression in *T. cruzi* epimastigotes confers increased resistance to H_2O_2-mediated toxicity, indicating that both enzymes play a role in the protection from H_2O_2-derived oxidation reactions such as membrane lipid peroxidation (Wilkinson et al., 2002a,c). Surprisingly, oxidized *Tc*GPx-II cannot be regenerated by TXN, GSH being the only known electron donor identified to date. The catalytic efficiency of *Tc*GPx-II reduction by GSH is low (2×10^3 $M^{-1}s^{-1}$) (Wilkinson et al., 2002c), thus questioning the relevance of the GPx-II functions *in vivo* in a GSH-dependent fashion. In this regard, other functions different from oxidant detoxification can be hypothesized. Indeed, thiol peroxidases sense and transfer oxidative signals and regulate transcription of target genes (Fomenko et al., 2011; Sobotta et al., 2015). It would be an oversimplification to consider the $T(SH)_2$-dependent hydroperoxide metabolism only as defense system against endogenous or host-derived cytotoxic hydroperoxides, which are no longer considered as just toxic compounds but increasingly recognized as signaling molecules (Flohé, 2010; Forman et al., 2010; Brigelius-Flohé and Flohé, 2011; Randall et al., 2013; Flohé, 2016). Like the *Saccharomyces cerevisiae* oxidant receptor peroxidase 1 (Delaunay et al., 2002), mammalian Prx2 (Sobotta et al., 2015) and a Prx of *Schizosaccharomyces pombe* (Vivancos et al., 2005), a TXNPx of *C. fasciculata* was also reported to act as sensor for H_2O_2 and, to oxidize a transcription factor, the universal mini-circle sequence-binding protein, which is implicated in replication of mitochondrial DNA (Sela et al., 2008; Shlomai, 2010). Fast metabolic adaptations upon oxidative insults observed in mammalian cells (Kuehne et al., 2015) could only be explained by the sensing and signaling by fast-reacting peroxidases (Winterbourn et al., 2015; Flohé, 2016). A similar mechanism could provide a fast metabolic adaptation of parasites during activation of the host immune response. In this context, the redox state of the TXNs and TXNPxs would determine the level of oxidation and activity of the protein target, namely a metabolic enzyme or transcription factor (Winterbourn et al., 2015; Flohé, 2016).

18.7.3 PROTEIN REPAIR

Oxidation of methionine residues is a posttranslational modification that can modify protein function. Methionine can be oxidized by H_2O_2, HOCl, or chloramines in a relatively unspecific way (Cascone et al., 1980; Mihajlovic et al., 1993; Peskin et al., 2001; Liang et al., 2012). However, a recent structural and kinetic analysis proposed that methionine oxidation can be enzymatically produced by molecules interacting with CasL (MICALs), a novel family of proteins that contain an unusual flavin-monooxygenase domain with an NADPH-dependent methionine sulfoxidase activity (Manta et al., 2017). Methionine sulfoxide reductases (MSR) catalyze the reduction of methionine sulfoxide (MetO) to methionine, typically using the reducing power of Trx, and, therefore, are considered part of the antioxidant and repair machinery of cells (Moskovitz, 2005). The enzymatic mechanism involves two redox active thiols, the first being oxidized to sulfenic acid by subtraction of a single oxygen atom from methionine sulfoxide, with further formation of a disulfide bridge with a second (resolving) cysteine residue, usually reduced by Trx or the GSH/glutaredoxin systems (Brot et al., 1981; Couturier et al., 2012; Kim, 2013; Kaya et al., 2015). *T. brucei* expresses two isoforms of MSR, one cytosolic (MSRA, that reduces the S epimer of MetO) and another restricted to the mitochondria (MSRB, specific for the R epimer of MetO in peptides) (Arias et al., 2011; Guerrero et al., 2017). The cytosolic isoform uses more efficiently TXN as reducing substrate than the mitochondrial one [k_{cat}/K_m =2.4 x 10^4 and 6.7×10^2 $M^{-1}s^{-1}$, respectively (Arias et al., 2011; Guerrero et al., 2017)]. The slow rate of reduction of the latter suggests that an alternative reduction pathway may function for this isoform, presumably a T(SH)$_2$-dependent Trx previously characterized (Reckenfelderbäumer et al., 2002; Schmidt et al., 2003). An alternative mechanism of reduction involving mixed disulfide formation with GSH followed by its reduction by glutaredoxin(s) was not investigated so far. Reduced levels of the cytosolic isoform caused growth arrest and increased susceptibility to exogenous H_2O_2, while reduced levels of the mitochondrial isoform only caused growth arrest (Guerrero et al., 2017). Similarly, deletion of the cytosolic MSR in *L. major* caused increased sensitivity to H_2O_2 and decreased proliferation in murine macrophages (Sansom et al., 2013).

18.7.4 DNA Synthesis and Proliferation

Ribonucleotide reductase (RR) catalyzes the limiting step of DNA synthesis involving the reduction of ribonucleotides into the corresponding deoxyribonucleotides (Holmgren et al., 1965) by using the reducing power of Trx or Grx and GSH in concert (Holmgren et al., 1965; Sengupta et al., 2014). In kinetoplastids, the use of TXN as a reducing substrate by RR was early reported when the first TXN was cloned, purified, and functionally characterized (Lüdemann et al., 1998). Although RR is also able to utilize T(SH)$_2$, the higher efficiency with TXN (K_m =2.1 mM and 3.7 µM for T(SH)$_2$ and TXN, respectively) points a flux of reducing equivalents in the presence of the redoxin (Dormeyer et al., 2001). This may explain the observation that depletion of T(SH)$_2$ or reduced TXN in *T. brucei* causes a phenotype of retarded proliferation (Comini et al., 2004; Comini et al., 2007). Under the latter conditions, the lack of TXN could be at least partially compensated by the direct reduction by T(SH)$_2$. Alternatively, ribonucleotide reduction could be catalyzed by the two cytosolic Grxs, as had previously been shown, although they do so less efficiency (Ceylan et al., 2010). Moreover, RR is under redox control, being susceptible to inhibition by TS$_2$ (Dormeyer et al., 2001), providing a link between the intracellular redox state of the cell and proliferation. Another link between the trypanothione system and proliferation has been proposed based on *in vitro* studies. In this case, TXN reduced the universal mini-circle sequence-binding protein leading to protein monomerization and activation of DNA binding, a step necessary for the replication of the kinetoplast, the kinetoplastid mitochondrial DNA (Onn et al., 2004; Motyka et al., 2006; Shlomai, 2010).

18.8 PROTEIN *S*-THIOLATION

Protein *S*-glutathionylation is one of several posttranslational modifications that participate in redox signaling processes, and different regulatory proteins have been described to be glutathionylated (Pompella et al., 2003; Demasi et al., 2013; Ghezzi, 2013; Peskin et al., 2016). This process is likely not controlled by thermodynamics or alterations of the GSH/GSSG redox potential, but by the action of enzymes, which will determine velocity and site-specificity (Park et al., 2011; Flohé, 2013; Berndt et al., 2014). Moreover, the subcellular distribution of GSH and GSSG might be unexpectedly heterogeneous (Morgan et al., 2013). Grxs efficiently catalyze glutathionylation and deglutathionylation (Beer et al., 2004; Silva et al., 2008b; Demasi et al., 2013; Peskin et al., 2016) and glutathione S-transferases promote S-glutathionylation [(Gravina et al., 1993; Townsend et al., 2009), see also Chapter 12]. Parasites lacking cytosolic Grx1 but not mitochondrial Grx2 undergo delayed protein dethiolation (Ulrich et al., 2017). Organisms with absent or low GSH levels developed analogous enzymes such as mycoredoxins and bacilliredoxins that specifically catalyze S-mycothiolation or S-bacillithiolation, respectively [(Chi et al., 2011; Chi et al., 2014; Loi et al., 2015; Reyes et al., 2018), see also Chapters 19 and 20]. As described in the previous sections, T(SH)$_2$ appears to have substituted GSH in several cellular processes, but this is not necessary the case in redox regulation by protein S-thiolation, and its role in metabolic and redox regulation by protein S-thiolation is far from being understood. Although several proteins of T. brucei undergo S-glutathionylation and glutathionylspermidine-S-thiolation in vitro (Melchers et al., 2007; Ceylan et al., 2010), incubation of proteins with TS$_2$ does not lead to protein S-trypanothiolation. This observation had first been interpreted as inability of T(SH)$_2$ to form stable protein-mixed disulfides due to its disulfide nature (Melchers et al., 2007; Manta et al., 2013a). More recently, however, this modification was detected in S-thiolated proteomes of T. brucei upon exposure to different oxidative challenges, such as HOCl, H$_2$O$_2$, and diamide (Ulrich et al., 2017). Intracellular glutathione and trypanothione are mostly present as free reduced thiols under standard conditions in T. brucei. After exposure to millimolar concentrations of H$_2$O$_2$, more than 50% of the glutathione but only 20% of the trypanothione pool appears bound to proteins. Higher yields of protein S-trypanothiolation were only reached upon parasite treatment with diamide (Ulrich et al., 2017). Three proteins, a protein disulfide isomerase, a putative HSP20, and HSC70-interacting protein were S-trypanothiolated and enriched under both experimental conditions (Ulrich et al., 2017). An alternative could be the formation of mixed disulfides with two proteins through each thiol group, causing reversible protein cross-linking. Taken together, these data suggest differential roles of each low-molecular-weight thiol in the response to oxidative stress and further investigation is needed to unravel the function biological relevance of S-trypanothiolation.

18.9 CONCLUDING REMARKS

Trypanothione represents the main low-molecular-weight thiol of trypanosomatids and constitutes an essential metabolite for several key cellular functions as represented in Figure 18.1. Its characteristic arrangement as a dithiol makes it a unique and flexible redox active molecule that serves antioxidant, cytoprotective, proliferative, and detoxifying actions. Overall, trypanothione facilitates trypanosomatid growth and multiplication and strongly assists in overcoming mammalian host cell and drug-mediated cytotoxicity by a combination of molecular mechanisms, overall contributing to parasite infectivity and virulence. It is clear that the trypanothione system comprises most attractive molecular targets to tackle trypanosomatid redox metabolism and to thereby combat associated infectious diseases.

ACKNOWLEDGMENTS

This work was supported by grants from Universidad de la República (Espacio Interdisciplinario y CSIC) to MT, LP, and RR. Additional support was obtained from Programa de Desarrollo de Ciencias Básicas (PEDECIBA, Uruguay).

REFERENCES

Adak, S., and A. K. Datta. 2005. Leishmania major encodes an unusual peroxidase that is a close homologue of plant ascorbate peroxidase: A novel role of the transmembrane domain. *Biochem J* 390 (Pt 2):465–74.

Alvarez, M. N., G. Peluffo, L. Piacenza, and R. Radi. 2011. Intraphagosomal peroxynitrite as a macrophage-derived cytotoxin against internalized *Trypanosoma cruzi*: Consequences for oxidative killing and role of microbial peroxiredoxins in infectivity. *J Biol Chem* 286 (8):6627–40.

Alvarez, M. N., L. Piacenza, F. Irigoin et al. 2004. Macrophage-derived peroxynitrite diffusion and toxicity to *Trypanosoma cruzi*. *Arch Biochem Biophys* 432 (2):222–32.

Arias, D. G., M. S. Cabeza, E. D. Erben et al. 2011. Functional characterization of methionine sulfoxide reductase A from *Trypanosoma* spp. *Free Radic Biol Med* 50 (1):37–46.

Ariyanayagam, M. R., and A. H. Fairlamb. 2001. Ovothiol and trypanothione as antioxidants in trypanosomatids. *Mol Biochem Parasitol* 115 (2):189–98.

Ariyanayagam, M. R., S. L. Oza, M. L. Guther, and A. H. Fairlamb. 2005. Phenotypic analysis of trypanothione synthetase knockdown in the African trypanosome. *Biochem J* 391 (Pt 2):425–32.

Augusto, O., E. Linares, and S. Giorgio. 1996. Possible roles of nitric oxide and peroxynitrite in murine leishmaniasis. *Braz J Med Biol Res* 29 (7):853–62.

Awad, S., G. B. Henderson, A. Cerami, and K. D. Held. 1992. Effects of trypanothione on the biological activity of irradiated transforming DNA. *Int J Radiat Biol* 62 (4):401–7.

Barata, L., M. Sousa Silva, L. Schuldt et al. 2011. Enlightening the molecular basis of trypanothione specificity in trypanosomatids: Mutagenesis of *Leishmania infantum* glyoxalase II. *Exp Parasitol* 129 (4):402–8.

Beer, S. M., E. R. Taylor, S. E. Brown et al. 2004. Glutaredoxin 2 catalyzes the reversible oxidation and glutathionylation of mitochondrial membrane thiol proteins: Implications for mitochondrial redox regulation and antioxidant defense. *J Biol Chem* 279 (46):47939–51.

Berndt, C., C. H. Lillig, and L. Flohé. 2014. Redox regulation by glutathione needs enzymes. *Front Pharmacol* 5:168.

Bocedi, A., K. F. Dawood, R. Fabrini et al. 2010. Trypanothione efficiently intercepts nitric oxide as a harmless iron complex in trypanosomatid parasites. *FASEB J* 24 (4):1035–42.

Boiani, M., L. Piacenza, P. Hernandez et al. 2010. Mode of action of nifurtimox and N-oxide-containing heterocycles against *Trypanosoma cruzi*: Is oxidative stress involved? *Biochem Pharmacol* 79 (12):1736–45.

Bollinger, J. M., Jr., D. S. Kwon, G. W. Huisman et al. 1995. Glutathionylspermidine metabolism in *Escherichia coli*. Purification, cloning, overproduction, and characterization of a bifunctional glutathionylspermidine synthetase/amidase. *J Biol Chem* 270 (23):14031–41.

Bonini, M. G., R. Radi, G. Ferrer-Sueta et al. 1999. Direct EPR detection of the carbonate radical anion produced from peroxynitrite and carbon dioxide. *J Biol Chem* 274 (16):10802–6.

Brigelius-Flohé, R., and L. Flohé. 2011. Basic principles and emerging concepts in the redox control of transcription factors. *Antioxid Redox Signal* 15 (8):2335–81.

Brot, N., L. Weissbach, J. Werth, and H. Weissbach. 1981. Enzymatic reduction of protein-bound methionine sulfoxide. *Proc Natl Acad Sci U S A* 78 (4):2155–8.

Cadenas, E. 1995. Thiyl radical formation during thiol oxidation by ferrylmyoglobin. *Methods Enzymol* 251:106–16.

Cascone, O., M. J. Biscoglio de Jimenez Bonino, and J. A. Santome. 1980. Oxidation of methionine residues in bovine growth hormone by chloramine-T. *Int J Pept Protein Res* 16 (4):299–305.

Castro, H., S. Romao, S. Carvalho et al. 2010. Mitochondrial redox metabolism in trypanosomatids is independent of tryparedoxin activity. *PLOS One* 5 (9):e12607.

Ceylan, S., V. Seidel, N. Ziebart et al. 2010. The dithiol glutaredoxins of African trypanosomes have distinct roles and are closely linked to the unique trypanothione metabolism. *J Biol Chem* 285 (45):35224–37.

Chi, B. K., T. Busche, K. Van Laer et al. 2014. Protein *S*-mycothiolation functions as redox-switch and thiol protection mechanism in *Corynebacterium glutamicum* under hypochlorite stress. *Antioxid Redox Signal* 20 (4):589–605.

Chi, B. K., K. Gronau, U. Mader et al. 2011. *S*-Bacillithiolation protects against hypochlorite stress in *Bacillus subtilis* as revealed by transcriptomics and redox proteomics. *Mol Cell Proteomics* 10 (11):M111 009506.

Clark, D., M. Albrecht, and J. Arevalo. 1994. Ascorbate variations and dehydroascorbate reductase activity in *Trypanosoma cruzi* epimastigotes and trypomastigotes. *Mol Biochem Parasitol* 66 (1):143–5.

Comini, M. A., S. A. Guerrero, S. Haile et al. 2004. Validation of *Trypanosoma brucei* trypanothione synthetase as drug target. *Free Radic Biol Med* 36 (10):1289–302.

Comini, M. A., R. L. Krauth-Siegel, and L. Flohé. 2007. Depletion of the thioredoxin homologue tryparedoxin impairs antioxidative defence in African trypanosomes. *Biochem J* 402 (1):43–9.

Comini, M. A., J. Rettig, N. Dirdjaja et al. 2008. Monothiol glutaredoxin-1 is an essential iron-sulfur protein in the mitochondrion of African trypanosomes. *J Biol Chem* 283 (41):27785–98.

Couturier, J., F. Vignols, J. P. Jacquot, and N. Rouhier. 2012. Glutathione- and glutaredoxin-dependent reduction of methionine sulfoxide reductase A. *FEBS Lett* 586 (21):3894–9.

Delaunay, A., D. Pflieger, M. B. Barrault et al. 2002. A thiol peroxidase is an H_2O_2 receptor and redox-transducer in gene activation. *Cell* 111 (4):471–81.

Demasi, M., L. E. Netto, G. M. Silva et al. 2013. Redox regulation of the proteasome via *S*-glutathionylation. *Redox Biol* 2:44–51.

Denicola, A., B. A. Freeman, M. Trujillo, and R. Radi. 1996. Peroxynitrite reaction with carbon dioxide/bicarbonate: Kinetics and influence on peroxynitrite-mediated oxidations. *Arch Biochem Biophys* 333 (1):49–58.

Denicola, A., H. Rubbo, D. Rodriguez, and R. Radi. 1993. Peroxynitrite-mediated cytotoxicity to *Trypanosoma cruzi*. *Arch Biochem Biophys* 304 (1):279–86.

Diaz de Toranzo, E. G., J. A. Castro, B. M. Franke de Cazzulo, and J. J. Cazzulo. 1988. Interaction of benznidazole reactive metabolites with nuclear and kinetoplastic DNA, proteins and lipids from *Trypanosoma cruzi*. *Experientia* 44 (10):880–1.

Diechtierow, M., and R. L. Krauth-Siegel. 2011. A tryparedoxin-dependent peroxidase protects African trypanosomes from membrane damage. *Free Radic Biol Med* 51 (4):856–68.

Docampo, R., J. F. de Boiso, A. Boveris, and A. O. Stoppani. 1976. Localization of peroxidase activity in *Trypanosoma cruzi* microbodies. *Experientia* 32 (8):972–5.

Docampo, R., and S. N. Moreno. 1984. Free radical metabolites in the mode of action of chemotherapeutic agents and phagocytic cells on *Trypanosoma cruzi*. *Rev Infect Dis* 6 (2):223–38.

Docampo, R., and S. N. Moreno. 1986. Free radical metabolism of antiparasitic agents. *Fed Proc* 45 (10):2471–6.

Docampo, R., and A. O. Stoppani. 1980. Mechanism of the trypanocidal action of nifurtimox and other nitro-derivatives on *Trypanosoma cruzi*. *Medicina (Buenos Aires)* 40 Suppl 1:10–6.

Dormeyer, M., N. Reckenfelderbäumer, H. Lüdemann, and R. L. Krauth-Siegel. 2001. Trypanothione-dependent synthesis of deoxyribonucleotides by *Trypanosoma brucei* ribonucleotide reductase. *J Biol Chem* 276 (14):10602–6.

Emmett, J. 1950. Effect of X-radiation on *Trypanosoma cruzi*. *J Parasitol* 36 (1):45–7.

Faerman, C. H., S. N. Savvides, C. Strickland et al. 1996. Charge is the major discriminating factor for glutathione reductase versus trypanothione reductase inhibitors. *Bioorg Med Chem* 4 (8):1247–53.

Fairlamb, A. H., P. Blackburn, P. Ulrich et al. 1985. Trypanothione: A novel bis(glutathionyl)spermidine cofactor for glutathione reductase in trypanosomatids. *Science* 227 (4693):1485–7.

Fairlamb, A. H., N. S. Carter, M. Cunningham, and K. Smith. 1992. Characterisation of melarsen-resistant *Trypanosoma brucei* with respect to cross-resistance to other drugs and trypanothione metabolism. *Mol Biochem Parasitol* 53 (1–2):213–22.

Fairlamb, A. H., and A. Cerami. 1992. Metabolism and functions of trypanothione in the Kinetoplastida. *Annu Rev Microbiol* 46:695–729.

Fairlamb, A. H., G. B. Henderson, and A. Cerami. 1989. Trypanothione is the primary target for arsenical drugs against African trypanosomes. *Proc Natl Acad Sci U S A* 86 (8):2607–11.

Faundez, M., R. Lopez-Munoz, G. Torres et al. 2008. Buthionine sulfoximine has anti-*Trypanosoma cruzi* activity in a murine model of acute Chagas' disease and enhances the efficacy of nifurtimox. *Antimicrob Agents Chemother* 52 (5):1837–9.

Faundez, M., L. Pino, P. Letelier et al. 2005. Buthionine sulfoximine increases the toxicity of nifurtimox and benznidazole to *Trypanosoma cruzi*. *Antimicrob Agents Chemother* 49 (1):126–30.

Ferrer-Sueta, G., B. Manta, H. Botti et al. 2011. Factors affecting protein thiol reactivity and specificity in peroxide reduction. *Chem Res Toxicol* 24 (4):434–50.

Ferrer-Sueta, G., and R. Radi. 2009. Chemical biology of peroxynitrite: Kinetics, diffusion, radicals. *ACS Chemical Biology* 4 (3):161–77.

Filser, M., M. A. Comini, M. M. Molina-Navarro et al. 2008. Cloning, functional analysis, and mitochondrial localization of *Trypanosoma brucei* monothiol glutaredoxin-1. *Biol Chem* 389 (1):21–32.

Fitzgerald, M. P., J. M. Madsen, M. C. Coleman et al. 2010. Transgenic biosynthesis of trypanothione protects *Escherichia coli* from radiation-induced toxicity. *Radiat Res* 174 (3):290–6.

Flohé, L. 2010. Changing paradigms in thiology from antioxidant defense toward redox regulation. *Methods Enzymol* 473:1–39.

Flohé, L. 2013. The fairytale of the GSSG/GSH redox potential. *Biochim Biophys Acta* 1830 (5):3139–42.

Flohé, L. 2016. The impact of thiol peroxidases on redox regulation. *Free Radic Res* 50 (2):126–42.

Folkes, L. K., M. Trujillo, S. Bartesaghi et al. 2011. Kinetics of reduction of tyrosine phenoxyl radicals by glutathione. *Arch Biochem Biophys* 506 (2):242–9.

Fomenko, D. E., A. Koc, N. Agisheva et al. 2011. Thiol peroxidases mediate specific genome-wide regulation of gene expression in response to hydrogen peroxide. *Proc Natl Acad Sci U S A* 108 (7):2729–34.

Forman, H. J., M. Maiorino, and F. Ursini. 2010. Signaling functions of reactive oxygen species. *Biochemistry* 49 (5):835–42.

Gallogly, M. M., D. W. Starke, and J. J. Mieyal. 2009. Mechanistic and kinetic details of catalysis of thiol-disulfide exchange by glutaredoxins and potential mechanisms of regulation. *Antioxid Redox Signal* 11 (5):1059–81.

Ghezzi, P. 2013. Protein glutathionylation in health and disease. *Biochim Biophys Acta* 1830 (5):3165–72.

Gommel, D. U., E. Nogoceke, M. Morr et al. 1997. Catalytic characteristics of tryparedoxin. *Eur J Biochem* 248 (3):913–18.

Gravina, S. A., and J. J. Mieyal. 1993. Thioltransferase is a specific glutathionyl mixed disulfide oxidoreductase. *Biochemistry* 32 (13):3368–76.

Grynberg, P., D. G. Passos-Silva, M. Mourao Mde et al. 2012. *Trypanosoma cruzi* gene expression in response to gamma radiation. *PLoS One* 7 (1):e29596.

Guerrero, S. A., D. G. Arias, M. S. Cabeza et al. 2017. Functional characterisation of the methionine sulfoxide reductase repertoire in *Trypanosoma brucei*. *Free Radic Biol Med* 112:524–33.

Haimeur, A., C. Brochu, P. Genest et al. 2000. Amplification of the ABC transporter gene PGPA and increased trypanothione levels in potassium antimonyl tartrate (SbIII) resistant *Leishmania tarentolae*. *Mol Biochem Parasitol* 108 (1):131–5.

Hall, B. S., C. Bot, and S. R. Wilkinson. 2011. Nifurtimox activation by trypanosomal type I nitroreductases generates cytotoxic nitrile metabolites. *J Biol Chem* 286 (15):13088–95.

Hall, B. S., and S. R. Wilkinson. 2012. Activation of benznidazole by trypanosomal type I nitroreductases results in glyoxal formation. *Antimicrob Agents Chemother* 56 (1):115–23.

Harrison, J. E., and J. Schultz. 1976. Studies on the chlorinating activity of myeloperoxidase. *J Biol Chem* 251 (5):1371–4.

Hawkins, C. L., D. I. Pattison, and M. J. Davies. 2003. Hypochlorite-induced oxidation of amino acids, peptides and proteins. *Amino Acids* 25 (3–4):259–74.

Herrero, E., and M. A. de la Torre-Ruiz. 2007. Monothiol glutaredoxins: A common domain for multiple functions. *Cell Mol Life Sci* 64 (12):1518–30.

Hillion, M., J. Bernhardt, T. Busche et al. 2017. Monitoring global protein thiol-oxidation and protein *S*-mycothiolation in *Mycobacterium smegmatis* under hypochlorite stress. *Sci Rep* 7 (1):1195.

Hofmann, B., H. Budde, K. Bruns et al. 2001. Structures of tryparedoxins revealing interaction with trypanothione. *Biol Chem* 382 (3):459–71.

Holmgren, A., P. Reichard, and L. Thelander. 1965. Enzymatic synthesis of deoxyribonucleotides, 8. The effects of ATP and dATP in the CDP reductase system from E. coli. *Proc Natl Acad Sci U S A* 54 (3):830–6.

Hugo, M., A. Martinez, M. Trujillo et al. 2017. Kinetics, subcellular localization, and contribution to parasite virulence of a *Trypanosoma cruzi* hybrid type A heme peroxidase (TcAPx-CcP). *Proc Natl Acad Sci U S A* 114 (8):E1326–E1335.

Huie, R. E., and S. Padmaja 1993. The reaction of NO with superoxide. *Free Radic Res Commun* 18 (4):195–9.

Irsch, T., and R. L. Krauth-Siegel. 2004. Glyoxalase II of African trypanosomes is trypanothione-dependent. *J Biol Chem* 279 (21):22209–17.

Jackson, A. P., M. A. Quail, and M. Berriman. 2008. Insights into the genome sequence of a free-living Kinetoplastid: *Bodo saltans* (*Kinetoplastida*: Euglenozoa). *BMC Genomics* 9:594.

Kaya, A., B. C. Lee, and V. N. Gladyshev. 2015. Regulation of protein function by reversible methionine oxidation and the role of selenoprotein MsrB1. *Antioxid Redox Signal* 23 (10):814–22.

Kettle, A. J., and C. C. Winterbourn. 1997. Myeloperoxidase: A key regulator of neutrophil oxidant production. *Redox Rep* 3 (1):3–15.

Kim, H. Y. 2013. The methionine sulfoxide reduction system: Selenium utilization and methionine sulfoxide reductase enzymes and their functions. *Antioxid Redox Signal* 19 (9):958–69.

Kirsch, M., M. Lehnig, H. G. Korth et al. 2001. Inhibition of peroxynitrite-induced nitration of tyrosine by glutathione in the presence of carbon dioxide through both radical repair and peroxynitrate formation. *Chemistry* 7 (15):3313–20.

Koppenol, W. H., J. J. Moreno, W. A. Pryor et al. 1992. Peroxynitrite, a cloaked oxidant formed by nitric oxide and superoxide. *Chem Res Toxicol* 5 (6):834–42.

Krauth-Siegel, R. L., and H. Lüdemann. 1996. Reduction of dehydroascorbate by trypanothione. *Mol Biochem Parasitol* 80 (2):203–8.

Krauth-Siegel, R. L., and H. Schmidt. 2002. Trypanothione and tryparedoxin in ribonucleotide reduction. *Methods Enzymol* 347:259–66.

Kuehne, A., H. Emmert, J. Soehle et al. 2015. Acute activation of oxidative pentose phosphate pathway as first-line response to oxidative stress in human skin cells. *Mol Cell* 59 (3):359–71.

Legare, D., S. Cayer, A. K. Singh et al. 2001. ABC proteins of *Leishmania*. *J Bioenerg Biomembr* 33 (6):469–74.

Legare, D., B. Papadopoulou, G. Roy et al. 1997. Efflux systems and increased trypanothione levels in arsenite-resistant *Leishmania*. *Exp Parasitol* 87 (3):275–82.

Liang, X., A. Kaya, Y. Zhang et al. 2012. Characterization of methionine oxidation and methionine sulfoxide reduction using methionine-rich cysteine-free proteins. *BMC Biochem* 13:21.

Lillig, C. H., C. Berndt, and A. Holmgren. 2008. Glutaredoxin systems. *Biochim Biophys Acta* 1780 (11):1304–17.

Linster, C. L., and E. Van Schaftingen. 2007. Vitamin C. Biosynthesis, recycling and degradation in mammals. *FEBS J* 274 (1):1–22.

Liu, I., M. Bogacz, C. Schaffroth et al. 2016. Catalytic properties, localization, and in vivo role of Px IV, a novel tryparedoxin peroxidase of *Trypanosoma brucei*. *Mol Biochem Parasitol* 207 (2):84–8.

Logan, F. J., M. C. Taylor, S. R. Wilkinson et al. 2007. The terminal step in vitamin C biosynthesis in *Trypanosoma cruzi* is mediated by a FMN-dependent galactonolactone oxidase. *Biochem J* 407 (3):419–26.

Loi, V. V., M. Rossius, and H. Antelmann. 2015. Redox regulation by reversible protein *S*-thiolation in bacteria. *Front Microbiol* 6:187.

Lüdemann, H., M. Dormeyer, C. Sticherling et al. 1998. *Trypanosoma brucei* tryparedoxin, a thioredoxin-like protein in African trypanosomes. *FEBS Lett* 431 (3):381–5.

Lymar, S. V., and J. K. Hurst. 1995. Rapid reaction between peroxynitrite anion and carbon dioxide: Implication for biological activity. *J Am Chem Soc* 117:8867–8.

MacMicking, J., Q. W. Xie, and C. Nathan. 1997. Nitric oxide and macrophage function. *Annu Rev Immunol* 15:323–50.

Maiorino, M., F. Ursini, V. Bosello et al. 2007. The thioredoxin specificity of *Drosophila* GPx: A paradigm for a peroxiredoxin-like mechanism of many glutathione peroxidases. *J Mol Biol* 365 (4):1033–46.

Manta, B., M. Bonilla, L. Fiestas et al. 2018. Polyamine-based thiols in trypanosomatids: Evolution, protein structural adaptations, and biological functions. *Antioxid Redox Signal* 28 (6):463–82.

Manta, B., M. Comini, A. Medeiros et al. 2013a. Trypanothione: A unique bis-glutathionyl derivative in trypanosomatids. *Biochim Biophys Acta* 1830 (5):3199–216.

Manta, B., and V. N. Gladyshev. 2017. Regulated methionine oxidation by monooxygenases. *Free Radic Biol Med* 109:141–55.

Manta, B., C. Pavan, M. Sturlese et al. 2013b. Iron-sulfur cluster binding by mitochondrial monothiol glutaredoxin-1 of *Trypanosoma brucei*: Molecular basis of iron-sulfur cluster coordination and relevance for parasite infectivity. *Antioxid Redox Signal* 19 (7):665–82.

Martinez-Ruiz, A., and S. Lamas. 2007. Signalling by NO-induced protein *S*-nitrosylation and *S*-glutathionylation: Convergences and divergences. *Cardiovasc Res* 75 (2):220–8.

Martinez, A., G. Peluffo, A. A. Petruk et al. 2014. Structural and molecular basis of the peroxynitrite-mediated nitration and inactivation of *Trypanosoma cruzi* iron-superoxide dismutases (Fe-SODs) A and B: Disparate susceptibilities due to the repair of Tyr35 radical by Cys83 in Fe-SODB through intramolecular electron transfer. *J Biol Chem* 289 (18):12760–78.

Maya, J. D., B. K. Cassels, P. Iturriaga-Vasquez et al. 2007. Mode of action of natural and synthetic drugs against *Trypanosoma cruzi* and their interaction with the mammalian host. *Comp Biochem Physiol A Mol Integr Physiol* 146 (4):601–20.

Maya, J. D., Y. Repetto, M. Agosin et al. 1997. Effects of nifurtimox and benznidazole upon glutathione and trypanothione content in epimastigote, trypomastigote and amastigote forms of *Trypanosoma cruzi*. *Mol Biochem Parasitol* 86 (1):101–6.

Maya, J. D., A. Rodriguez, L. Pino et al. 2004. Effects of buthionine sulfoximine nifurtimox and benznidazole upon trypanothione and metallothionein proteins in *Trypanosoma cruzi*. *Biol Res* 37 (1):61–9.

Melchers, J., N. Dirdjaja, T. Ruppert, and R. L. Krauth-Siegel. 2007. Glutathionylation of trypanosomal thiol redox proteins. *J Biol Chem* 282 (12):8678–94.

Mihajlovic, V., O. Cascone, and M. J. Biscoglio de Jimenez Boni. 1993. Oxidation of methionine residues in equine growth hormone by Chloramine-T. *Int J Biochem* 25 (8):1189–93.

Moncada, C., Y. Repetto, J. Aldunate et al. 1989. Role of glutathione in the susceptibility of *Trypanosoma cruzi* to drugs. *Comp Biochem Physiol C* 94 (1):87–91.

Morgan, B., D. Ezerina, T. N. Amoako et al. 2013. Multiple glutathione disulfide removal pathways mediate cytosolic redox homeostasis. *Nat Chem Biol* 9 (2):119–25.

Moskovitz, J. 2005. Methionine sulfoxide reductases: Ubiquitous enzymes involved in antioxidant defense, protein regulation, and prevention of aging-associated diseases. *Biochim Biophys Acta* 1703 (2):213–19.

Motyka, S. A., M. E. Drew, G. Yildirir, and P. T. Englund. 2006. Overexpression of a cytochrome b5 reductase-like protein causes kinetoplast DNA loss in *Trypanosoma brucei*. *J Biol Chem* 281 (27):18499–506.

Moutiez, M., D. Meziene-Cherif, M. Aumercier et al. 1994. Compared reactivities of trypanothione and glutathione in conjugation reactions. *Chem Pharm Bull* 42:2641–4.

Mukhopadhyay, R., S. Dey, N. Xu et al. 1996. Trypanothione overproduction and resistance to antimonials and arsenicals in *Leishmania*. *Proc Natl Acad Sci U S A* 93 (19):10383–7.

Nakao, L. S., D. Ouchi, and O. Augusto. 1999. Oxidation of acetaldehyde by peroxynitrite and hydrogen peroxide/Iron(II). Production of acetate, formate, and methyl radicals. *Chem Res Toxicol* 12 (10):1010–18.

Nogoceke, E., D. U. Gommel, M. Kiess et al. 1997. A unique cascade of oxidoreductases catalyses trypanothione-mediated peroxide metabolism in *Crithidia fasciculata*. *Biol Chem* 378 (8):827–36.

Onn, I., N. Milman-Shtepel, and J. Shlomai. 2004. Redox potential regulates binding of universal minicircle sequence binding protein at the kinetoplast DNA replication origin. *Eukaryot Cell* 3 (2):277–87.

Park, J. W., G. Piszczek, S. G. Rhee, and P. B. Chock. 2011. Glutathionylation of peroxiredoxin I induces decamer to dimers dissociation with concomitant loss of chaperone activity. *Biochemistry* 50 (15):3204–10.

Perea, A., J. I. Manzano, Y. Kimura et al. 2018. Leishmania LABCG2 transporter is involved in ATP-dependent transport of thiols. *Biochem J* 475(1):87–97.

Pereira, C., C. Paveto, J. Espinosa et al. 1997. Control of *Trypanosoma cruzi* epimastigote motility through the nitric oxide pathway. *J Eukaryot Microbiol* 44 (2):155–6.

Peskin, A. V., P. E. Pace, J. B. Behring et al. 2016. Glutathionylation of the active site cysteines of peroxiredoxin 2 and recycling by glutaredoxin. *J Biol Chem* 291 (6):3053–62.

Peskin, A. V., and C. C. Winterbourn. 2001. Kinetics of the reactions of hypochlorous acid and amino acid chloramines with thiols, methionine, and ascorbate. *Free Radic Biol Med* 30 (5):572–9.

Piacenza, L., M. N. Alvarez, G. Peluffo, and R. Radi. 2009. Fighting the oxidative assault: The *Trypanosoma cruzi* journey to infection. *Curr Opin Microbiol* 12 (4):415–21.

Piacenza, L., F. Irigoin, M. N. Alvarez et al. 2007. Mitochondrial superoxide radicals mediate programmed cell death in *Trypanosoma cruzi*: Cytoprotective action of mitochondrial iron superoxide dismutase overexpression. *Biochem J* 403 (2):323–34.

Piacenza, L., G. Peluffo, M. N. Alvarez et al. 2008. Peroxiredoxins play a major role in protecting *Trypanosoma cruzi* against macrophage- and endogenously-derived peroxynitrite. *Biochem J* 410 (2):359–68.

Piacenza, L., G. Peluffo, and R. Radi. 2001. L-Arginine-dependent suppression of apoptosis in *Trypanosoma cruz*i: Contribution of the nitric oxide and polyamine pathways. *Proc Natl Acad Sci U S A* 98 (13):7301–6.

Pineyro, M. D., T. Arcari, C. Robello et al. 2011. Tryparedoxin peroxidases from *Trypanosoma cruzi*: High efficiency in the catalytic elimination of hydrogen peroxide and peroxynitrite. *Arch Biochem Biophys* 507 (2):287–95.

Pompella, A., A. Visvikis, A. Paolicchi et al. 2003. The changing faces of glutathione, a cellular protagonist. *Biochem Pharmacol* 66 (8):1499–503.

Radi, R. 2004. Nitric oxide, oxidants, and protein tyrosine nitration. *Proc Natl Acad Sci U S A* 101 (12):4003–8.

Randall, L. M., G. Ferrer-Sueta, and A. Denicola. 2013. Peroxiredoxins as preferential targets in H_2O_2-induced signaling. *Methods Enzymol* 527:41–63.

Reckenfelderbäumer, N., and R. L. Krauth-Siegel. 2002. Catalytic properties, thiol pK value, and redox potential of *Trypanosoma brucei* tryparedoxin. *J Biol Chem* 277 (20):17548–55.

Reyes, A. M., B. Pedre, M. I. De Armas et al. 2018. Chemistry and redox biology of mycothiol. *Antioxid Redox Signal* 28 (6):487–504.

Rouhier, N., J. Couturier, M. K. Johnson, and J. P. Jacquot. 2010. Glutaredoxins: Roles in iron homeostasis. *Trends Biochem Sci* 35 (1):43–52.

Sansom, F. M., L. Tang, J. E. Ralton et al. 2013. *Leishmania major* methionine sulfoxide reductase A is required for resistance to oxidative stress and efficient replication in macrophages. *PLoS One* 8 (2):e56064.

Schafer, F. Q., and G. R. Buettner. 2001. Redox environment of the cell as viewed through the redox state of the glutathione disulfide/glutathione couple. *Free Radic Biol Med* 30 (11):1191–212.

Schmidt, H., and R. L. Krauth-Siegel. 2003. Functional and physicochemical characterization of the thioredoxin system in *Trypanosoma brucei*. *J Biol Chem* 278 (47):46329–36.

Sela, D., N. Milman, I. Kapeller et al. 2008. Unique characteristics of the kinetoplast DNA replication machinery provide potential drug targets in trypanosomatids. *Adv Exp Med Biol* 625:9–21.

Sengupta, R., and A. Holmgren. 2014. Thioredoxin and glutaredoxin-mediated redox regulation of ribonucleotide reductase. *World J Biol Chem* 5 (1):68–74.

Shahi, S. K., R. L. Krauth-Siegel, and C. E. Clayton. 2002. Overexpression of the putative thiol conjugate transporter TbMRPA causes melarsoprol resistance in *Trypanosoma brucei*. *Mol Microbiol* 43 (5):1129–38.

Shlomai, J. 2010. Redox control of protein-DNA interactions: From molecular mechanisms to significance in signal transduction, gene expression, and DNA replication. *Antioxid Redox Signal* 13 (9):1429–76.

Silva, M. S., L. Barata, A. E. Ferreira et al. 2008a. Catalysis and structural properties of *Leishmania infantum* glyoxalase II: Trypanothione specificity and phylogeny. *Biochemistry* 47 (1):195–204.

Silva, G. M., L. E. Netto, K. F. Discola et al. 2008b. Role of glutaredoxin 2 and cytosolic thioredoxins in cysteinyl-based redox modification of the 20S proteasome. *FEBS J* 275 (11):2942–55.

Smirnoff, N., P. L. Conklin, and F. A. Loewus. 2001. Biosynthesis of ascorbic acid in plants. A renaissance. *Annu Rev Plant Physiol Plant Mol Biol* 52:437–67.

Sobotta, M. C., W. Liou, S. Stocker et al. 2015. Peroxiredoxin-2 and STAT3 form a redox relay for H_2O_2 signaling. *Nat Chem Biol* 11 (1):64–70.

Soito, L., C. Williamson, S. T. Knutson et al. 2011. PREX: Peroxiredoxin classification index, a database of subfamily assignments across the diverse peroxiredoxin family. *Nucleic Acids Res* 39 (Database issue):D332–7.

Sousa Silva, M., A. E. Ferreira, R. Gomes et al. 2012. The glyoxalase pathway in protozoan parasites. *Int J Med Microbiol* 302 (4–5):225–9.

Stacey, M. M., S. L. Cuddihy, M. B. Hampton, and C. C. Winterbourn. 2012. Protein thiol oxidation and formation of S-glutathionylated cyclophilin A in cells exposed to chloramines and hypochlorous acid. *Arch Biochem Biophys* 527 (1):45–54.

Stevens, J. R. 2008. Kinetoplastid phylogenetics, with special reference to the evolution of parasitic trypanosomes. *Parasite* 15 (3):226–32.

Thomson, L., A. Denicola, and R. Radi. 2003. The trypanothione-thiol system in *Trypanosoma cruzi* as a key antioxidant mechanism against peroxynitrite-mediated cytotoxicity. *Arch Biochem Biophys* 412 (1):55–64.

Thornalley, P. J. 1990. The glyoxalase system: New developments towards functional characterization of a metabolic pathway fundamental to biological life. *Biochem J* 269 (1):1–11.

Toppo, S., L. Flohé, F. Ursini et al. 2009. Catalytic mechanisms and specificities of glutathione peroxidases: Variations of a basic scheme. *Biochim Biophys Acta* 1790 (11):1486–500.

Townsend, D. M., Y. Manevich, L. He et al. 2009. Novel role for glutathione S-transferase pi. Regulator of protein S-glutathionylation following oxidative and nitrosative stress. *J Biol Chem* 284 (1):436–45.

Trochine, A., D. J. Creek, P. Faral-Tello et al. 2014. Benznidazole biotransformation and multiple targets in *Trypanosoma cruzi* revealed by metabolomics. *PLoS Negl Trop Dis* 8 (5):e2844.

Trujillo, M., B. Alvarez, and R. Radi. 2016. One- and two-electron oxidation of thiols: Mechanisms, kinetics and biological fates. *Free Radic Res* 50 (2):150–71.

Trujillo, M., B. Alvarez, J. M. Souza et al. 2010. Mechanisms and biological consequences of peroxynitrite-dependent protein oxidation and nitration. In *Nitric Oxide. Biology and Pathobiology*, edited by L. J. Ignarro, 61–102. Los Angeles, CA: Elsevier.

Trujillo, M., H. Budde, M. D. Pineyro et al. 2004. *Trypanosoma brucei* and *Trypanosoma cruzi* tryparedoxin peroxidases catalytically detoxify peroxynitrite via oxidation of fast reacting thiols. *J Biol Chem* 279 (33):34175–82.

Trujillo, M., G. Ferrer-Sueta, and R. Radi. 2008. Kinetic studies on peroxynitrite reduction by peroxiredoxins. *Methods Enzymol* 441:173–96.

Trujillo, M., G. Ferrer-Sueta, L. Thomson et al. 2007. Kinetics of peroxiredoxins and their role in the decomposition of peroxynitrite. *Subcell Biochem* 44:83–113.

Trujillo, M., L. Folkes, S. Bartesaghi et al. 2005. Peroxynitrite-derived carbonate and nitrogen dioxide radicals readily react with lipoic and dihydrolipoic acid. *Free Radic Biol Med* 39 (2):279–88.

Trujillo, M., and R.Radi 2002. Peroxynitrite reaction with the reduced and the oxidized forms of lipoic acid: new insights into the reaction of peroxynitrite with thiols. *Arch Biochem Biophys* 397 (1):91–8.

Ulrich, K., C. Finkenzeller, S. Merker et al. 2017. Stress-induced protein *S*-glutathionylation and *S*-trypanothionylation in African trypanosomes—A quantitative redox proteome and thiol analysis. *Antioxid Redox Signal* 27 (9):517–33.

Vickers, T. J., and A. H. Fairlamb. 2004. Trypanothione *S*-transferase activity in a trypanosomatid ribosomal elongation factor 1B. *J Biol Chem* 279 (26):27246–56.

Vickers, T. J., S. Wyllie, and A. H. Fairlamb. 2004. *Leishmania major* elongation factor 1B complex has trypanothione *S*-transferase and peroxidase activity. *J Biol Chem* 279 (47):49003–9.

Vivancos, A. P., E.A. Castillo, B. Biteau et al. 2005. A cysteine-sulfinic acid in peroxiredoxin regulates H2O2-sensing by the antioxidant Pap1 pathway. *Proc Natl Acad Sci U S A* 102 (25):8875–80.

Vogt, R. N., and D. J. Steenkamp. 2003. The metabolism of *S*-nitrosothiols in the trypanosomatids: The role of ovothiol A and trypanothione. *Biochem J* 371 (Pt 1):49–59.

Wendler, A., T. Irsch, N. Rabbani et al. 2009. Glyoxalase II does not support methylglyoxal detoxification but serves as a general trypanothione thioesterase in African trypanosomes. *Mol Biochem Parasitol* 163 (1):19–27.

Wilkinson, S. R., D. J. Meyer, and J. M. Kelly. 2000a. Biochemical characterization of a trypanosome enzyme with glutathione-dependent peroxidase activity. *Biochem J* 352 Pt 3:755–61.

Wilkinson, S. R., D. J. Meyer, M. C. Taylor et al. 2002a. The *Trypanosoma cruzi* enzyme TcGPXI is a glycosomal peroxidase and can be linked to trypanothione reduction by glutathione or tryparedoxin. *J Biol Chem* 277 (19):17062–71.

Wilkinson, S. R., S. O. Obado, I. L. Mauricio, and J. M. Kelly. 2002b. *Trypanosoma cruzi* expresses a plant-like ascorbate-dependent hemoperoxidase localized to the endoplasmic reticulum. *Proc Natl Acad Sci U S A* 99 (21):13453–8.

Wilkinson, S. R., S. R. Prathalingam, M. C. Taylor et al. 2005. Vitamin C biosynthesis in trypanosomes: A role for the glycosome. *Proc Natl Acad Sci U S A* 102 (33):11645–50.

Wilkinson, S. R., M. C. Taylor, D. Horn et al. 2008. A mechanism for cross-resistance to nifurtimox and benznidazole in trypanosomes. *Proc Natl Acad Sci U S A* 105 (13):5022–7.

Wilkinson, S. R., M. C. Taylor, S. Touitha et al. 2002c. TcGPXII, a glutathione-dependent *Trypanosoma cruzi* peroxidase with substrate specificity restricted to fatty acid and phospholipid hydroperoxides, is localized to the endoplasmic reticulum. *Biochem J* 364 (Pt 3):787–94.

Wilkinson, S. R., N. J. Temperton, A. Mondragon, and J. M. Kelly. 2000b. Distinct mitochondrial and cytosolic enzymes mediate trypanothione-dependent peroxide metabolism in *Trypanosoma cruzi*. *J Biol Chem* 275 (11):8220–5.

Winkler, B. S., S. M. Orselli, and T. S. Rex. 1994. The redox couple between glutathione and ascorbic acid: A chemical and physiological perspective. *Free Radic Biol Med* 17 (4):333–49.

Winterbourn, C. C. 1993. Superoxide as an intracellular radical sink. *Free Radic Biol Med* 14 (1):85–90.

Winterbourn, C. C. 2016. Revisiting the reactions of superoxide with glutathione and other thiols. *Arch Biochem Biophys* 595:68–71.

Winterbourn, C. C., and M. B. Hampton. 2015. Redox biology: Signaling via a peroxiredoxin sensor. *Nat Chem Biol* 11 (1):5–6.

Wyllie, S., M. L. Cunningham, and A. H. Fairlamb. 2004. Dual action of antimonial drugs on thiol redox metabolism in the human pathogen *Leishmania donovani*. *J Biol Chem* 279 (38):39925–32.

Wyllie, S., T. J. Vickers, and A. H. Fairlamb. 2008. Roles of trypanothione *S*-transferase and tryparedoxin peroxidase in resistance to antimonials. *Antimicrob Agents Chemother* 52 (4):1359–65.

Xie, Q. W., R. Whisnant, and C. Nathan. 1993. Promoter of the mouse gene encoding calcium-independent nitric oxide synthase confers inducibility by interferon gamma and bacterial lipopolysaccharide. *J Exp Med* 177 (6):1779–84.

Zamocky, M., P. G. Furtmuller, and C. Obinger. 2010. Evolution of structure and function of Class I peroxidases. *Arch Biochem Biophys* 500 (1):45–57.

Zeida, A., A. M. Reyes, M. C. Lebrero et al. 2014. The extraordinary catalytic ability of peroxiredoxins: A combined experimental and QM/MM study on the fast thiol oxidation step. *Chem Commun (Camb)* 50 (70):10070–3.

19 Mycothiol, a Low-Molecular-Weight Thiol Drafted for Oxidative Stress Defense Duty

Leonardo Astolfi Rosado, Brandán Pedre, and Joris Messens
Vrije Universiteit Brussel

CONTENTS

19.1 INTRODUCTION

Intracellular pathogens are able to counteract the oxidative stress response generated by the host cells (Lamichhane, 2011; Nambi et al., 2015; Pai et al., 2016). In this oxidative stress interplay, macrophages exert the first response reaction to the bacterial infection, generating reactive oxygen (ROS) and reactive nitrogen species (RNS) to kill the pathogen (Lamichhane, 2011). In Actinomycetes, exogenous oxidative stresses are counteracted by the mycothiol (MSH; 1-D-*myo*-inosityl 2-(*N*-acetylcysteinyl)amido-2-deoxy-α-D-glucopyranoside) system, which is analogous to the well-studied glutathione (GSH) system (Fahey, 2013; Van Laer et al., 2013; Reyes et al., 2018). MSH is a unique cysteinyl pseudodisaccharide that is responsible for the tight control of the redox environment in mycobacteria and is a key molecule that protects against the irreversible harm generated by ROS and RNS and that rescues sulfur-containing molecules from permanent damage and subsequent inactivation and/or degradation. Gram-negative bacteria and most eukaryotes rely on GSH as the main low-molecular-weight (LMW) thiol to maintain the redox homeostasis. However, Actinomycetes, such as *Mycobacterium tuberculosis*, depend greatly on MSH, hence, conferring a crucial role to this intracellular redox system (Newton et al., 2008). In this chapter, MSH and its molecular partners will be presented, providing a general perspective on its protective mechanism and pathways.

19.2 MYCOTHIOL IS THE MAJOR LMW THIOL FROM MYCOBACTERIA

19.2.1 IDENTIFICATION AND STRUCTURE

MSH had been identified for the first time as a novel inositol glycoside isolated from Streptomycetes (Newton et al., 1993). Primary elucidation of its structure had been attempted (Sakuda et al., 1994), paving the way to the canonical MSH form, as we know it nowadays. MSH is composed of *N*-acetyl L-cysteine, D-glucosamine, and 1-D-*myo*-inositol, linked by an amide bond between the carboxyl group of *N*-acetyl L-cysteine and the amino group of the D-glucosamine moiety, which is connected with a glycosidic bond to 1-D-*myo*-inositol (Figure 19.1).

MSH has been intensively studied. MSH is exclusively found in large amounts in species belonging to the order Actinomycetales. Pathogenic species belonging to the

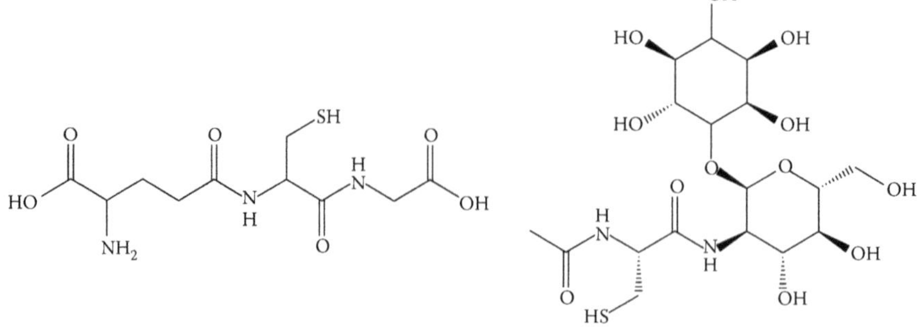

Glutathione (GSH, γ-L-glutamyl-L-cysteinylglycine)   Mycothiol (MSH, AcCys-GlcN-Ins)

FIGURE 19.1   Chemical structures of glutathione and mycothiol.

TABLE 19.1

Comparison of MSH and GSH in Different Species

Organism	MSH[a]	GSH[a]
Mycobacterium tuberculosis ATCC26618	12	—
Mycobacterium smegmatis MC²6	19	—
Streptomyces coelicolor	2.8	—
Corynebacterium diphtheriae	0.18	—
Rhodococcus AD45	15–20	1–3
Rattus sp. liver	—	16–25

[a] Concentrations in µmol g^{-1} of residual dry cell weight.

Source: Data from (Newton et al., 1996; Newton and Fahey, 2002; Jothivasan and Hamilton, 2008; Johnson et al., 2009).

Mycobacteriaceae genus were found to produce and maintain large MSH concentrations (1–20 mM). Intracellular MSH concentrations largely vary among species (Newton et al., 1996; Newton and Fahey, 2002; Jothivasan and Hamilton, 2008; Hillion et al., 2017a) and during the different bacterial growth phases (Newton et al., 1993). Analogously to glutathione disulfide (GSSG), mycothiol disulfide or mycothione (MSSM) is also produced in Actinomycetes with high levels at the beginning of the stationary phase (Newton et al., 2005; Buchmeier et al., 2006). The discovery of MSH as the main LMW thiol in Actinomycetes and in some Gram-positive bacteria could explain the absence of GSH (Newton et al., 1996; Johnson et al., 2009), with a few exceptions, namely the occurrence of GSH in *Rubrobacter radiotolerans* and *Rubrobacter xylanophilus* and the production of high levels of GSH and MSH by *Rhodococcus sp.* AD45 (Newton et al., 1996; Van Hylckama Vlieg et al., 1998; Johnson et al., 2009) (Table 19.1).

19.2.2 BIOPHYSICAL AND BIOCHEMICAL PROPERTIES

The deprotonated form of thiol groups (thiolates) is the reactive species that participates in the bimolecular nucleophilic substitution (S_N2) reaction, in which sulfur-containing molecules are involved (Bulaj et al., 1998). LMW thiols, such as GSH and MSH, can form a mixed disulfide with the more electrophilic sulfenic acid (RSOH) via a nucleophilic attack. Then, a subsequent attack from another GSH/MSH moiety leads to the formation of GSSG or MSSM and a reduced thiol is released (Van Laer et al., 2013; Rosado et al., 2017). The *N*-acetyl L-cysteine portion of the MSH molecule confers its redox activity, in which a sulfur atom can be found with a pK_{aSH} of 8.76 and a redox potential ($E^{\circ\prime}$) of −230 mV (Sharma et al., 2016). Similarly, GSH displays a pK_a of 8.93 and an $E^{\circ\prime}$ of −240 mV. Despite the clear structural differences, GSH and MSH have very similar biophysical properties, by which both molecules are characterized as nucleophiles (Rabenstein, 1973; Newton et al., 2005). LMW thiols are known for their ability to undergo autoxidation when exposed to copper, but in this type of reaction MSH differs from the other well-studied LMW thiols. In the presence of copper, MSH undergoes autoxidation approximately 30 times slower than free cysteine and 7 times slower than GSH and *N*-acetyl cysteine. This substantial difference in autoxidation rate is due to the acetyl group and inositol that block the cysteine amino group and the *N*-acetyl L-cysteine carbonyl group, respectively, hence, decreasing the overall reactivity (Newton et al., 1995; Reyes et al., 2018).

19.3   BIOSYNTHETIC PATHWAY

The MSH-biosynthetic pathway consists of five unique steps catalyzed by different enzymes, MshA, MshA2, MshB, MshC, and MshD (Figure 19.2). MSH biosynthesis begins with the conversion of uridine diphosphate *N*-acetyl glucos-amine (UDPGlcNAc) into 1-*O*-(2-acetoamido-2-deoxy-α-D-glucopyranolyl)-D-*myo*-inositol-3-phosphate (GlcNAc-Ins-P) by an *N*-acetyl-glucosamine transferase (MshA) (Sareen et al., 2003; Newton et al., 2006). The transfer of the phosphoryl group of GlcNAc-Ins-P is subsequently catalyzed by an unknown phosphatase, designated MshA2, which generates 1-*O*-(2-amino-2-deoxy-α-D-glucopyranolyl)-D-*myo*-inositol (GlcNAc-Ins). The metalloprotein MshB triggers the subsequent deacetylation of GlcNAc-Ins, yielding 1-*O*-(2-amino-2-deoxy-α-D-glucopyranolyl)-D-*myo*-inositol (GlcN-Ins) (Newton et al., 2000a; Newton et al., 2006), which is then ligated to a free L-cysteine by MshC (Bornemann et al., 1997; Anderberg et al., 1998). The final step of the MSH-biosynthetic pathway is catalyzed by the acetyltransferase MshD, converting 1-*O*-[2-[[(2R)-2-amido-3-mercapto-1-oxopropyl]amino]-2deoxy-α-D-glucopyranolyl]-D-*myo*-inositol (Cys-GlcN-Ins) to the final MSH product by means of one acetyl-CoA molecule (Koledin et al., 2002; Figure 19.2). In contrast, the GSH-biosynthetic pathway consists of two enzymes (Lu, 2013; see also Chapter 1). The first step is catalyzed by γ-glutamylcysteine synthetase (Gsh1 in eukaryotes and GshA in prokaryotes) at the cost of one ATP molecule and the second step by glu-tathione synthetase (Gsh2 in eukaryotes and GshB in prokaryotes) at the cost of another ATP molecule, yielding GSH. The γ-glutamylcysteine synthetase activity is subject to GSH feedback inhibition to avoid redox imbalance due to GSH excess. Interestingly, three different pathways have been reported to be able to produce GSH in different organisms (Tang et al., 2015), but the MSH production strictly depends on one biosynthetic route (Newton et al., 2008).

Currently, only the gene encoding the MshA2 enzyme belonging to the MSH biosynthetic pathway is still unknown. The other components of this pathway have received a substantial amount of interest because of the prospective use of the enzymes of the MSH biosynthesis pathway as potential drug targets (Xu et al., 2011; Nilewar and Kathiravan, 2014). For drug development against pathogens, important features are essentiality and exclusivity, and only a limited number of genes fulfill these criteria (Kana et al., 2014; Trauner et al., 2014). The MSH bio-synthesis pathway has proven to be exclusive to the Actinomycetes, but its absence is not critical for the pathogen viability under nonstress conditions (Newton et al., 1996; Jothivasan and Hamilton, 2008; Xu et al., 2011). In the past years, genetic, biochemical, and biophysical studies on the MSH-biosynthetic pathway genes have provided the necessary information to understand the physiological role of these

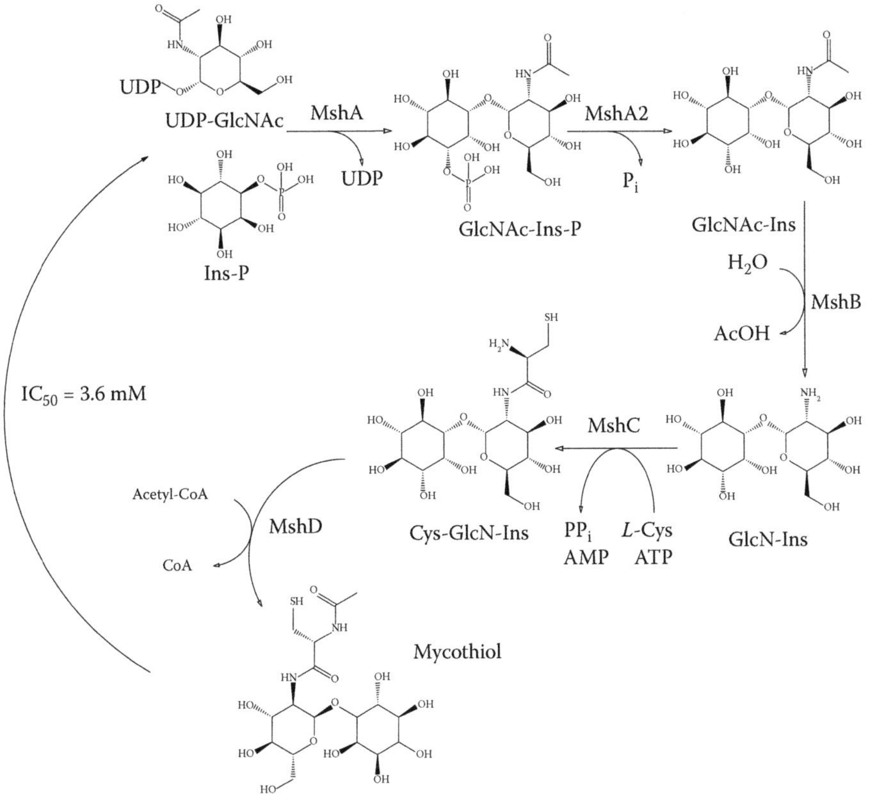

FIGURE 19.2 Schematic overview of the reaction path for the mycothiol biosynthesis. MSH is biosynthesized by five enzymes, starting with UDP-GlcNac and Ins-P as substrates, yielding MSH as final product. MSH regulates MshA activity via feedback inhibition. For detailed information, the reader is referred to Jothivasan and Hamilton (2008), Newton et al. (2008), and Upton et al. (2012).

gene products and the homeostatic role of MSH within the pathogen, indicating phenotypes that differ from the parental wild-type strains when exposed to ROS, RNS, antibiotics, and alkylating agents (Newton et al., 1999, 2000b; Koledin et al., 2002; Rawat et al., 2002, 2003, 2007; Sareen et al., 2003; Miller et al., 2007). In the Table 19.2, MSH and intermediate concentrations are presented, which are strain-specific. Additonally, resistance to isoniazid (INH) and ethionamide (ETH) are shown.

19.3.1 MSHA

M. smegmatis strains that lack the *mshA* gene do not contain MSH or any of its intermediates (Table 19.2) (Newton et al., 1999; Sareen et al., 2003). The loss of the catalytic function of MshA had firstly been achieved by chemical mutagenesis, leading to a G32D mutation and abolishing the MSH production

TABLE 19.2

MSH Content and Its Biosynthesis Pathway Intermediates in *Mycobacterium smegmatis* Mutant Strains

Strain	GlcNAc-Ins	GlcN-Ins	Cys-GlcN-Ins	MSH	INH	ETH
	(μmol g⁻¹ of residual dry cell weight)				MIC (μg/mL)	
Wild-type	<0.1	0.2–1.0	<0.01	10±3	4	20
Δ*mshA*	<0.01	<0.01	<0.01	<0.01	32	160
Δ*mshB*	2.6±0.2	<0.01	<0.02	1.0±0.2	8	160
Δ*mshC*	ND	2.6	<0.002	<0.004	16	160
Δ*mshD*	ND	0.35±0.05	0.6–2.0	0.12±0.01	4	20

Note: The strain column represents the deletion of four characterized enzymes involved in MSH biosynthesis and the parental strain. ND, not determined.

Source: Adapted from Newton et al. (2008) and Xu et al. (2011).

(Newton et al., 1999). These findings were later confirmed by site-directed mutagenesis and knockout of the *mshA* gene, indicating the essentiality of the gene for the production of MSH and its intermediates (Rawat et al., 2002; Xu et al., 2011). Moreover, MSH has been identified as key regulator of the MshA activity via feedback inhibition with a half maximal inhibitory concentration (IC$_{50}$) of 3.6 mM (Upton et al., 2012). MshA is a glycotransferase that catalyzes the UDPGlcNAc conversion into GlcNAc-Ins-P, after an ordered sequential kinetic mechanism, in which UDPGlcNac is the first substrate to bind and followed by Ins-P to form the ternary complex (Vetting et al., 2008). Its activity was first identified in *M. smegmatis* crude extract (Newton et al., 1999; Sareen et al., 2003; Newton et al., 2006). Structural studies demonstrated that MshA has two β/α/β Rossman-fold domains linked by a hinge domain and that it crystallized as a homodimer (Vetting et al., 2008). Flexibility seems to be a crucial MshA feature, because the apo form undergoes a significant conformational change when the complex MshA:UDP is formed, shifting from an open-to-closed conformation. Interestingly, a *mshA*-deficient strain and knockout demonstrated an increased resistance to the antibiotics isonicotinic acid hydrazide (INH) and ethionamide (ETH) (Table 19.2), implying that MSH is directly involved in prodrug activation or indirectly by keeping the correct intracellular $E^{\circ\prime}$ (Xu et al., 2011). In contrast, the *mshA*-deficient strain had an impaired growth rate in the initial phase and a remarkably lethal phenotype when exposed to hydrogen peroxide (H_2O_2), gaseous nitric oxide, iodoacetamide, 1-chloro-2,4-dinitrobenzene (CDNB), and other antibiotics such as rifamycin, streptomycin, erythromycin, and azithromycin (Newton et al., 2000b; Rawat et al., 2002; Miller et al., 2007; Rawat et al., 2007). In *Corynebacterium glutamicum*, overexpression of the *mshA* gene increased the bacterial resistance to H_2O_2, methylglyoxal, cadmium, acid stress, ethanol, streptomycin, and erythromycine (Liu et al., 2014).

19.3.2   MSHA2

MshA2 catalyzes the dephosphorylation of GlcNac-Ins-P, yielding GlcNac-Ins. The MshA2-encoding gene is still unidentified and it is the last piece to be added in the MSH-biosynthetic pathway puzzle. The *impC* gene from *M. tuberculosis* seems to be a promising candidate, because of its essentiality for MSH production and the inositol-1-phosphatase gene homology. Another hypothesis is that the catalysis of the phosphoryl group transfer of GlcNAc-Ins-P is shared by several enzymes (Nigou et al., 2002; Vetting et al., 2008; Movahedzadeh et al., 2010).

19.3.3   MSHB

The Zn^{2+}-dependent deacetylase MshB catalyzes the conversion of GlcNAc-Ins into GlcN-Ins (Newton et al., 2000a). The MshB mutant has significant GlcNAc-Ins levels (Table 19.2), hinting at a deficient deacetylase activity, but the detection of a basal MSH level implied that another enzyme can deacetylate the substrate with a reduced catalytic rate (Rawat et al., 2003; Newton et al., 2008). Enzymatic and structural studies demonstrated the metal-dependent MshB activity and the enzymatic activation by the divalent metal ion Zn^{2+} and other ones, such as Ni^{2+}, Mn^{2+}, and Co^{2+}, but not by Ca^{2+} and Mg^{2+}. Zn^{2+} has been proposed to be the natural divalent metal that is N-terminally embedded and coordinated by three His residues (Mccarthy et al., 2004; Miller et al., 2007). The N-terminal domain is a lactate dehydrogenase-like domain and the C-terminus displays an α/β conformation. The substrate-binding site hydration is apparently a crucial feature, because two water molecules are conserved at the active site and are displaced upon MshB:GlcNAc-Ins binary complex formation (Maynes et al., 2003; Mccarthy et al., 2004). In the catalytic cycle, the catalysis takes off with the activation of a water molecule by Asp15 in the vicinity of the Zn^{2+} ion. The water activation triggers a nucleophilic attack toward the Asp15 backbone carbonyl group that is polarized by the Zn^{2+} ion. The transition state is stabilized by Zn^{2+} with the help of the His144 imidazolium side chain, which is hydrogen bonded to Asp146. The general acidic role is probably played by the carboxyl group of Asp15 that has previously been protonated during the water activation, leading to a proton transfer and release of the GlcN-Ins product (Maynes et al., 2003). The knockout of the *mshB* gene in *M. smegmatis* induced an increased resistance to ETH, but just a mild resistance to INH (Table 19.2; Xu et al., 2011). Interestingly, the *mshB* knockout was more sensitive to H_2O_2, gaseous nitric oxide, iodoacetamide, 1-chloro-2,4-dinitrobenzene, and antibiotics than its parental strain but without initial growth delay (Newton et al., 1999; Rawat et al., 2003; Miller et al., 2007; Rawat et al., 2007).

19.3.4   MSHC

GlcN-Ins is the product of the reaction catalyzed by MshB and serves as substrate for the next enzyme in the MSH biosynthetic pathway, MshC. The Cys-GlcN-Ins biosynthesis is achieved by the ATP-dependent ligation of free L-cysteine and MshC-triggered

GlcN-Ins (Fan et al., 2007). The occurrence of *mshC* has been proven necessary for the MSH production in *M. smegmatis* and is strictly essential in *in vivo* experiments with *M. tuberculosis* (Table 19.2) (Rawat et al., 2002; Sareen et al., 2003; Sassetti et al., 2003). MshC is a monomeric enzyme with a molecular mass of approximately 47 kDa and 37% primary sequence identity with cystenyl-tRNA synthetase (CysRS) (Sareen et al., 2002). The reaction is under a bi-uni-uni-bi ping-pong mechanism, in which ATP and L-cysteine bind randomly to form the ternary complex MshC:ATP:Cys, releasing the first product, pyrophosphate, followed by the GlcN-Ins binding and the subsequent Cys-GlcN-Ins discharge (Fan et al., 2007; Williams et al., 2008). The development of the bisubstrate analog, 5'-O-[N-(L-cystenyl) sulfamonyl]adenosine (CSA), allowed the crystallization of MshC in a binary complex with CSA, which is an ATP-competitive inhibitor with a K_i of approximately 300 nM. As predicted by primary sequence identity, the three-dimensional structure of MshC is comparable to that of its homologue CysRS and displays a well-conserved Rossmann fold (Tremblay et al., 2008). Similar to MshB, MshC has a Zn^{2+} molecule embedded in its active site, in which the Zn^{2+} ion directly interacts with the CSA thiolate group, a mimic of that of the physiological L-cysteine substrate (Fan et al., 2007; Williams et al., 2008). As expected, the intermediate GlcN-Ins levels were high in the *mshC*-deficient strains because of the lack of ligase activity (Table 19.2). Moreover, *mshC*-deficient strains had the same phenotype as the *mshA* and *mshB* strains, but the antimicrobial activity in broth exposure assays after 2 h was 10-fold higher for H_2O_2 and 2-fold higher for CDNB, hinting at MshC as a promising target (Newton et al., 1999; Rawat et al., 2002, 2007; Sareen et al., 2003; Miller et al., 2007). Likewise, as the *mshA*-deficient strain, the *mshC* variant had a slow initial growth and a more pronounced resistance to INH and ETH (Newton et al., 2008; Xu et al., 2011; Table 19.2). In *C. glutamicum*, the sensitivity of the *mshC* variant is enhanced towards alkylating agents (including CDNB), H_2O_2, diamide, glyphosate, ethanol, arsenate, heavy metals, and the antibiotics, ETH, neomycin, spectinomycin, streptomycin, and gentamycin (Liu et al., 2013; Villadangos et al., 2014).

19.3.5 MSHD

The acetyltransferase MshD catalyzes the final step of the MSH-biosynthetic pathway. Kinetic and structural studies revealed that MshD directly transfers acetyl from acetyl-CoA to Cys-GlcN-Ins, instead of generating an intermediate to release the final MSH product (Vetting et al., 2006). This type of mechanism is a common feature of the *N*-acetyltransferase superfamily of GCN5-related *N*-acetyl transferases (GNATs), in which a ternary complex must be formed before the chemical step takes place (Vetting et al., 2005). A large conformational change can be seen during the catalysis triggered by the Cys-GlcN-Ins binding at the central groove, with its amine group facing the acetyl-CoA bound to the C-terminal domain. During the ternary complex formation, the GNAT domain moves toward the C-terminal domain, reduces the central groove volume and, subsequently, brings the necessary residues in contact for catalysis, with a 17.9° rotation of the N-terminal domain as a consequence (Vetting et al., 2006). The *mshD*-deficient strain accumulated Cys-GlcN-Ins and a residual MSH amount (Table 19.2; Koledin et al., 2002; Newton et al., 2005). Interestingly, two novel thiols are formed when *mshD* is absent, producing formyl-Cys-GlcN-Ins in high amounts, substituting MSH, while small amounts of succinyl-Cys-GlcN-Ins are produced (Newton et al., 2005). The hypothesis that the catalytic formation of formyl-Cys-GlcN-Ins might happen via a secondary enzyme remains to be proven. Due to basal MSH levels, high Cys-GlcN-Ins amounts, and the presence of two novel thiols, the *mshD*-deficient strain does not demonstrate growth delay and its lethality was not enhanced when exposed to INH and ETH (Newton et al., 2008; Xu et al., 2011).

19.4. FREE MYCOTHIOL INTERACTION AND RECYCLING

19.4.1 *S*-MYCOTHIOLATION

In an oxidative stress scenario, oxidation of cysteine residues-containing proteins is a recurrent process. Protection against overoxidation takes place via mixed disulfide formation with a LMW thiol (such as GSH and MSH) or an intramolecular disulfide formation. S-Mycothiolation is a reversible posttranslational modification through the LMW thiol thiolate action in a catalyzed or non-catalyzed reaction. The non-catalyzed reaction starts with the nucleophilic attack of the MSH thiolate group towards an electrophilic sulfenic acid (RSOH) or a disulfide (RSSR), yielding an *S*-mycothiolated cysteine residue (RSSM). The catalyzed reaction is governed by the mycothiol *S*-transferase, a system analogous to the extensively studied GST system, triggering the conjugation of MSH with electron-deficient molecules or through mycoredoxin (Mrx) catalyzed demycothiolation (Newton et al., 2011).

A series of comprehensive studies demonstrated that exposure to sublethal amounts of NaOCl in the Actinomycetes *C. glutamicum*, *C. diphtheriae*, and *M. smegmatis* led to very interesting *S*-mycothiolation patterns (Hillion et al., 2017a, Hillion et al., 2017b, Chi et al., 2014). The diverse set of *S*-mycothiolated proteins is involved in different metabolic pathways, such as the carbohydrate metabolism, amino acid and nucleotide metabolism, protein synthesis, and ROS response.

A total of 58 *S*-mycothiolated proteins have been detected in *M. smegmatis*, but only 26 and 25 in *C. diphtheriae* or *C. glutamicum*, respectively. *M. smegmatis* is the species containing the highest concentration of MSH (Table 19.1), possibly the reason for the clear difference in the number of S-mycothiolated proteins (Hillion et al., 2017a). Only eight proteins are shared in at least two of the bacteria and, within these targets, only two proteins occur in these three organisms: inosine-5-phosphate dehydrogenase and 30S ribosomal protein S13 (Figure 19.3). Common targets of *S*-glutathionylation and *S*-bacillithiolation, such as glyceraldehyde 3-phosphate dehydrogenase (GAPDH) or methionine synthase (MetE), were also found *S*-mycothiolated (Chi et al., 2014, Hillion et al., 2017b).

The set of *S*-mycothiolated proteins also includes enzymes that are directly and indirectly linked to oxidative stress responses. Thioredoxin peroxidase (Tpx) and alkyl hydroperoxide reductase C (AhpC) reduce H_2O_2, peroxynitrite, and organic hydroperoxides. Methionine sulfoxide reductase A (MsrA) reduces *S*-methionine sulfoxides; *myo*-inositol-1 phosphate synthase (Ino1) generates *myo*-inositol-1 phosphate, the first precursor in the MSH biosynthesis pathway. Homoserine dehydrogenase (Hom/ThrA) catalyzes oxidation of homoserine by consuming NADPH in the process, and the redox-regulated GAPDH shuts down the glycolytic activity

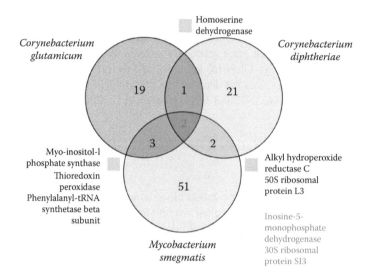

FIGURE 19.3 Common and species-specific *S*-mycothiolated proteins identified upon sublethal NaOCl stress in *Corynebacterium glutamicum*, *Corynebacterium diphtheriae*, and *Mycobacterium smegmatis*. The common *S*-mycothiolated proteins are highlighted with the two proteins common to the three organisms highlighted in green. For further details, the reader is referred to Chi et al. (2014) and Hillion et al. (2017a,b).

and transports the carbohydrate flux towards NADPH production (Peralta et al., 2015). Additional *S*-mycothiolated proteins found in *M. smegmatis* include proteins involved in fatty acid and phospholipid metabolism, transcriptional regulators (such as the tetracycline resistance regulator TetR and the iron-dependent repressor IdeR), and enzymes implicated in the tricarboxylic acid cycle [i.e., citrate synthase, isocitrate lyase, and α-ketoglutarate decarboxylase (Chi et al., 2014, Hillion et al., 2017a)]. Two enzymes from *Actinomycetes* that participate directly in the H_2O_2 detoxification were demonstrated to use the MSH/Mtr/NADPH pathway (MSH pathway). The formation of a sulfenic acid at the nucleophilic cysteine after the oxygen transfer from H_2O_2 is followed by the spontaneous creation of a mixed disulfide with MSH. A second-order rate constant of *S*-mycothiolation of 616 ± 10 $M^{-1}s^{-1}$ was found for the enzyme mycothiol peroxidase (Mpx) and of 237 ± 30 $M^{-1}s^{-1}$ for alkyl hydroperoxide reductase E (AhpE) (Hugo et al., 2009, Pedre et al., 2015). Similarly, peroxiredoxins of other organisms convert H_2O_2 into H_2O or alkylhydroperoxides into alcohols, generating a sulfenic acid on the nucleophilic cysteine (Poole, 2007). Typically, these enzymes are regenerated to the ground state by thioredoxins, but they can also use GSH and glutaredoxin 1 (Grx1) as recycling system to yield a *S*-glutathionylation rate of 500 $M^{-1}s^{-1}$ (Peskin et al., 2016).

19.4.2 MYCOTHIOL DISULFIDE REDUCTASE

Disulfide reductases are crucial for the maintenance of $E^{\circ\prime}$ within the cell. Glutathione reductase (GR) is a very good example of a class of enzymes able to

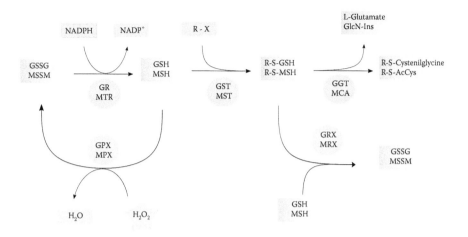

FIGURE 19.4 Redox reactions catalyzed by enzymes using free MSH as substrate. General representation of five reactions catalyzed by MSH-specific enzymes and homologues that use GSH as substrate. MST conjugates MSH with electrophilic toxins; Mrx demycothiolates targets protected from oxidative stress via a mixed disulfide; Mpx catalyzes a detoxification process converting H_2O_2 into water and can be recycled by MSH; Mca hydrolyzes the cysteinyl-glucosamine amide bond of free MSH and its conjugated forms; and Mtr specifically catalyzes the reduction of MSSM into MSH.

maintain the reduced thiol/disulfide balance inside the human cell. In a similar manner, Actinomycetes have mycothiol disulfide reductase (Mtr) (Figure 19.4; Table 19.3), a homodimeric NADPH flavin-dependent enzyme that converts mycothione into mycothiol. Mtr is under a bi-bi ping-pong mechanism, in which the first half-reaction comprises the reduction of flavin adenine dinucleotide (FAD) by NADPH and the subsequent reduction of Mtr by $FADH_2$, whereas in the second half-reaction, two electrons reduce MSSM into MSH via a dithiol mechanism (Patel and Blanchard, 1999).

TABLE 19.3

Kinetic Parameters of Mycothiol and Glutathione Disulfide Reductases in the Presence of 100 µM NADPH

Enzyme	Organism	Substrate	K_m (µM)	K_{cat} (s^{-1})	K_{cat}/K_m (M^{-1}s^{-1}×10^6)
Mtr	*M. tuberculosis*	MSSM	54	64	1.2
GR	*E. coli*	GSSG	61	733	12
GR	*Plasmodium falciparum*	GSSG	95	120	1.3
GR	*Homo sapiens*	GSSG	72	165	2.3
GR	*Bos taurus*	GSSG	42	175	4.2

Source: Data from (Henderson et al., 1991; Savvides et al., 2002).

Mtr shares 31% sequence identity with the *Homo sapiens* GR and 28% with the parasitic *Crithidia fasciculata* trypanothione reductase (TryR) (Patel and Blanchard, 1999). The similarity also extends to the functionality, because they all reduce disulfides by a ping-pong mechanism and share a similar predicted active-site architecture, but the substrate specificity differs from homologous enzymes in different organisms (Table 19.3). Recently, the low-resolution solution structure of Mtr derived from small angle X-ray scattering (SAXS) has been published (Kumar et al., 2017). These SAXS data confirmed the dimeric oligomeric state (Patel and Blanchard, 1999) and the presence of domains binding NADPH and FAD.

Several attempts have been made to knockout the gene encoding Mtr, but, as foreseen, the gene is essential for the growth of *M. tuberculosis* (Sassetti et al., 2003), but not for *M. smegmatis*, probably because of the high concentration of free MSH inside the bacteria (Table 19.2; Holsclaw et al., 2011). Mtr-overexpressing *C. glutamicum* had an increased bacterial resistance to acid and oxidative stress (H_2O_2 and diamide), bactericidal antibiotics, alkylating agents, and heavy metals. As such, Mtr seems to play an important protective role in stress defense (Si et al., 2016).

19.4.3 MYCOREDOXINS

Glutaredoxins (Grx) are crucial pieces in the complete redox cellular homeostasis puzzle [(Lillig et al., 2008); see also Chapters 15 and 16]. They act as catalysts of the GSH disulfide oxidoreduction with GSH and NADPH as cofactors and two cysteine residues embedded in its thioredoxin domain. Grx display a preference for mixed disulfide substrates that function via a monothiol mechanism. However, they can also use intramolecular disulfides as substrate in a dithiol mechanism. In *M. tuberculosis* and *C. glutamicum*, no Grx-encoding genes are found, but two homologues are present, *mrx1* and m*rx2* (Ordóñez et al., 2009; Van Laer et al., 2012; Rosado et al., 2017; Figure 19.4). The demycothiolation of *in vitro* S-mycothiolated *C. glutamicum* Tpx, *C. glutamicum* and *C. diphtheriae* MsrA, *C. diphtheriae* GAPDH, *C. glutamicum* Mpx, and *M. tuberculosis* AhpE have been shown to be catalyzed by Mrx1 (Figure 19.5; Chi et al., 2014; Hugo et al., 2014; Pedre et al., 2015; Si et al., 2015b; Tossounian et al., 2015; Hillion et al., 2017b).

S-Mycothiolated proteins are reduced through a monothiol mechanism coupled exclusively to the MSH pathway as a recycling system, because the mutation of the second Mrx1 cysteine does not affect its activity (Van Laer et al., 2012; Hugo et al., 2014; Si et al., 2015a,b; Figure 19.6). In addition, Mrx1 from *C. glutamicum* can reduce an arseno-MSH adduct by means of a monothiol thiol–disulfide exchange mechanism (Ordóñez et al., 2009; Figure 19.6). The reduction of S-mycothiolated MsrA by Mrx1 is fast (10^3–10^5 $M^{-1}s^{-1}$), but alternative mechanisms have been proposed (Si et al., 2015b; Tossounian et al., 2015), namely that Mrx1 reduces the S-mycothiolated Cys56 in *C. glutamicum* MsrA, whereas Mrx1 has been shown to be able to reduce S-mycothiolated Cys206 [Cys204 in *C. glutamicum* MsrA (Si et al., 2015b; Tossounian et al., 2015)]. In the case of the *M. tuberculosis* AhpE, not only can S-mycothiolated AhpE be reduced by Mrx1 via a monothiol mechanism, but also sulfenylated AhpE can be reduced via a dithiol mechanism at a rate of 1.6×10^3 $M^{-1}s^{-1}$ (Figure 19.6; Hugo et al., 2014). The dithiol mechanism rate is 10-fold faster

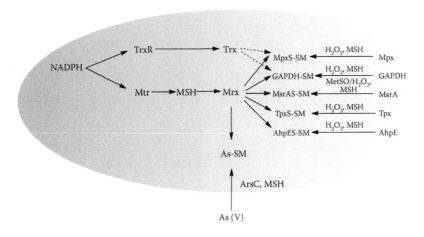

FIGURE 19.5 Schematic representation of MSH-related thiol switches. Upon oxidative stress, a subset of protein cysteines is S-mycothiolated. Mrx1 has been demonstrated to catalyze the demycothiolation mechanism. Demycothiolation rates have been reported for *C. glutamicum* and *C. diphtheriae* methionine sulfoxide reductase A (MsrA), *C. glutamicum* Mpx, and *M. tuberculosis* AhpE. *C. glutamicum* Tpx and *C. diphtheriae* GAPDH have been shown to be demycothiolated by Mrx1 and demycothiolation of mycothiol-*S*-arseno adducts to be catalyzed by *C. glutamicum* Mrx2 (Reyes et al., 2018).

than that of *S*-mycothiolation. However, because of the cellular MSH concentration in the millimolar range, that of Mrx1 (currently unknown) must be at least 50 μM to be a relevant competitor (Hugo et al., 2014), which is very improbable.

Interestingly, the substrate specificity and structural features of Mrx2 from *M. tuberculosis* differ from those of Mrx1. Indeed, Mrx2 is more promiscuously reduced, because it can receive electrons from both the MSH pathway and the TrxC/TrxR/NADPH (Trx pathway), but the MSH pathway is clearly the preferred one (Rosado et al., 2017). Moreover, one of the crucial features that diverges Mrx2 from Mrx1 is the reactivity of the CXXC active-site motif. The pK_{aSH} of the N-terminal nucleophilic Cys and of the resolving cysteines of Mrx2 are at least 1 pH and 2 pH units lower than those from Mrx1, respectively (Van Laer et al., 2012; Rosado et al., 2017). Structurally, Mrx1 and Mrx2 have a canonical thioredoxin fold, but in Mrx2, an extra α-helical bundle domain of 90 amino acids is present (Albesa-Jove et al., 2014, 2015). The presence of this α-helical bundle domain erroneously led to the classification of Mrx2 as a thiol–disulfide bond oxidoreductase (DsbA-like enzyme), although no oxidase activity could be observed (Albesa-Jove et al., 2014; Rosado et al., 2017). The homologous Mrx2 from *C. glutamicum* has similar properties. An *mrx2*-deficient *C. glutamicum* strain is extremely sensitive to arsenate stress. Additionally, Mrx2 from *C. glutamicum* reduces the MSH thiol-arseno adduct approximately 466 times faster than Mrx1, demonstrating that Mrx2 is the physiological enzymatic partner of the arsenate (AsV) detoxification mechanism (Mateos et al., 2017; Rosado et al., 2017). Knockout studies clearly indicate that Mrx1 and Mrx2 from *M. tuberculosis* and *C. glutamicum* are essential under oxidative stress

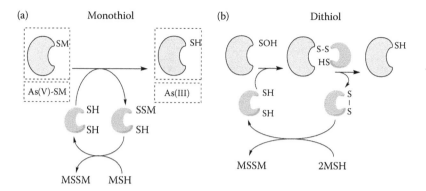

FIGURE 19.6 Mycoredoxin-reducing mechanisms. (A) Monothiol disulfide exchange mechanism in which mycothiol is transferred from a *S*-mycothiolated protein or mycothiol-*S*-arseno adduct (As(V)-SM) to Mrx. A second MSH molecule recycles Mrx. (B) Dithiol disulfide exchange mechanism in which both cysteines of the CXXC active-site motif of Mrx are involved. A mixed disulfide is formed between a protein [thus far, demonstrated in sulfenylated *M. tuberculosis* AhpE (Hugo et al., 2014)] and the nucleophilic Mrx cysteines, followed by the formation of a disulfide on Mrx. Two molecules of mycothiol are required to recycle the oxidized Mrx. (Adapted from Reyes et al., 2018.)

conditions, corroborating the idea that these enzymes are key players in the protection of the bacteria against oxidative stresses (Van Laer et al., 2012; Rosado et al., 2017).

Importantly, recently the evolutionary and functional information of oxidoreductases (DsbA), Grx, and Mrx from several organisms has been comprehensively classified (Rosado et al., 2017). Based on the CXXC motif, the enzymes were grouped and separated into seven distinct classes (Figure 19.7). Strikingly, Mrx1 was part of the same cluster as Grx1, but, in contrast, Mrx2 formed a unique cluster closely related to that of DsbA. These findings support the hypothesis that Mrx2 is evolutionarily more related to DsbA, although both Mrx1 and Mrx2 are redoxins of the mycothiol pathway that are essential under oxidative stresses.

19.4.4 THE MRX-1-ROGFP2 PROBE

Real-time life imaging of changes in the redox potential within living cells is nowadays central in redox biochemistry. Due to the increasing importance of the redox balance in cell signaling and associated processes, novel tools have been developed to reliably measure the dynamic modulation of redox couples *in vivo*. GSH, the most studied LMW thiol and the main one found in humans, was the first molecule to be analyzed by means of genetically encoded probes that can monitor the redox changes *in vivo* (Østergaard et al., 2001, 2004; Dooley et al., 2004), with a clearly improved sensitivity when a Grx was fused to the probe (Gutscher et al., 2008). In the same trend, specific redox couples, such as two MSH/MSSM emerged as interesting targets. The first genetically encoded mycothiol-dependent redox probe was generated

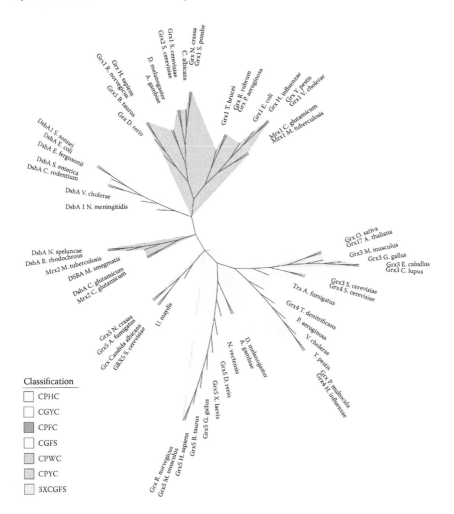

FIGURE 19.7 A phylogenetic reconstruction (phylogram) of mycoredoxins, glutaredoxins, and oxidoreductases (DsbA). Color code is based on the CXXC motif sequence. Mrx1 from *C. glutamicum* and *M. tuberculosis* belongs to the blue and white clusters and Mrx2 to the salmon and deep-purple clusters, respectively. The analysis was done with the MEGA6 software and the figure was generated with the web-based program iTOL (Tamura et al., 2013; Letunic and Bork, 2016).

by fusing *M. tuberculosis* Mrx1 to the reduction-oxidation—sensitive green fluorescent protein 2 [roGFP2 (Bhaskar et al., 2014)]. This chimera exhibited two specific excitation wavelengths, 390 and 490 nm, at a fixed emission wavelength of 510 nm (Figure 19.8). The changes in fluorescence upon reduction and oxidation were ratiometrically proportional to the roGFP2 probe alone, whereas the pH titration experiments also demonstrated that the Mrx1-roGFP2 probe did not depend on the pH over a physiological range (5.5–8.5). Moreover, the Mrx1-roGFP probe was specific

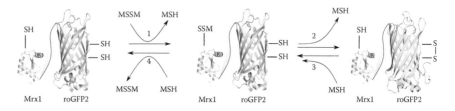

FIGURE 19.8 Schematic view on the Mrx1-roGFP2 bioprobe. *M. tuberculosis* Mrx1 is fused with roGFP2 via a linker, resulting in a chimera exhibiting two specific excitation wavelengths that are modulated by its redox state. The oxidative shift starts with a nucleophilic attack from Mrx1 toward MSSM, yielding the mycothiolated form of Mrx1 and releasing one molecule of MSH (1). The mixed disulfide is subjected to a subsequent nucleophilic attack from roGFP2, yielding roGFP2 mycothiolated and reduced Mrx1 (2). The roGFP2-resolving cysteine performs another nucleophilic attack toward the roGFP2-MSH mixed disulfide, generating an intramolecular disulfide and releasing one molecule of MSH (2). The reductive shift starts with a nucleophilic attack of MSH toward the roGFP2 intramolecular disulfide and a subsequent transfer of MSH to Mrx1, creating mycothiolated Mrx1 (3). The last step of the reductive shift is the nucleophilic attack of MSH toward the mycothiolated Mrx1 form, yielding MSSM and reduced Mrx1 (4) (Bhaskar et al., 2014).

to MSSM, because no response was observed when exposed to other disulfide-based molecules, such as GSSG, cystine, or 2-hydroxyethyl disulfide (HED), hinting at its suitability for *in vivo* measurements. A value of -300 ± 2 mV for the $E^{\circ\prime}$ was reported for *M. smegmatis* (Bhaskar et al., 2014).

After successful development and testing, the MSH-specific probe was used to measure redox fluctuations of *M. smegmatis* under several conditions. A remarkable difference of 20 mV was observed when slow-growing and normal-growing strains were compared, indicating that growth arrest is linked with an oxidizing environment. In *M. tuberculosis*, exposure to extensively utilized antibiotics, such as isoniazid, clofazimine, ethambutol, and rifampicin demonstrated a clear shift to a more oxidative environment. The Mrx1-roGFP2 probe was also used to measure the difference in the $E^{\circ\prime}$ of MSH in a WhiB3 knockout strain (Mehta et al., 2016). WhiB3 is a redox sensor member of the iron–sulfur cluster family linked to several different pathways related to virulence and its absence has no effect on the $E^{\circ\prime}$, but a WhiB3 knockout strain had a less dynamic response under acidic stress (Saini et al., 2012; Mehta et al., 2016).

Worth mentioning is the pH dependence of the $E^{\circ\prime}$ of MSH, implying that a positive change of 1 pH unit at 37°C would reduce the value to -65.1 mV. Accordingly, in a scenario with an $E^{\circ\prime}$ of -230 mV, a pH of 7.0, a physiological MSH concentration of 4 mM, and an MSH/MSSM ratio of 500, an E_{MSH} of -239 mV would be expected. However, as previously mentioned, a value of -300 mV has been reported in *M. smegmatis* (Reyes et al., 2018). These findings suggest that an efficient mechanism to keep a low redox potential should be present to avoid dramatic changes related to MSSM fluctuations. Moreover, a possible MSSM efflux pump has been proposed to occur, but this hypothesis still needs to be proven (Newton et al., 2000b; Ramón-García et al., 2009).

19.4.5   MYCOTHIOL *S*-TRANSFERASE

Every aerobic living organism has to cope with the presence of ROS and harmful xenobiotics. To manage this constant harm, GSH-containing species use glutathione transferases (GST) as major players to reduce peroxides and conjugate electrophilic substrates with GSH (see Chapter 11). Among the *S*-transferases, GSTs are the best characterized and they are highly specific to GSH. The goal of the conjugation is to decrease the substrate toxicity and increase the xenobiotic solubility for easy excretion from the cell (Sheehan et al., 2001; Dixon et al., 2009).

Bacterial GSTs belonging to several different classes like ζ, θ, χ, β, and MAPEG (membrane-associated proteins involved in ecosanoid and glutathione metabolism; see Chapter 13) have been studied (Allocati et al., 2009). Despite the similar functionality, bacterial GSTs display a pronounced decrease in activity of up to 200-fold when compared with its human and mouse counterparts (Piccolomini et al., 1989; Table 19.4).

In the same manner, GSH-lacking organisms, such as *M. tuberculosis*, need an active system for the intracellular detoxification and maintenance of its redox balance. The first unique functional enzyme, designated mycothiol *S*-transferase (MST), was identified in *M. tuberculosis* (Newton et al., 2011) and is encoded by the conserved gene *Rv0443* (Figure 19.4). The activity of MSTs from different organisms have been tested with the widely used fluorescent label monochlorobimane (mBCl) as electrophilic substrate and three different thiols, bacillithiol (BSH), GSH, and MSH. Despite the low activity, the bacterial enzymes tested were highly specific,

---

TABLE 19.4

Kinetic Rates of Transferase Activity of Mycothiol *S*-Transferases (MSTs) and Glutathione *S*-Transferases (GSTs) from Eukaryotic and Prokaryotic Sources

Enzyme	Organism	Thiol	Electrophilic Substrate	Activity (μmol min^{-1} mg^{-1})
MST	*M. tuberculosis*	MSH	mBCl	0.34
MST	*M. smegmatis*	MSH	mBCl	0.83
MST	*M. smegmatis*	MSH	CDNB	0.05
MST	*M. smegmatis*	MSH	Mitomycin C	0.017
BST	*B. subtilis*	BSH	mBCl	2.5
BST	*B. subtilis*	BSH	CDNB	0.54
GST	*E. faecalis*	GSH	mBCl	0.4
GST	*E. coli*	GSH	CDNB	1.0
GST	Rat liver	GSH	CDNB	16
GST	Human liver	GSH	CDNB	18
GST	Mouse liver	GSH	CDNB	86

Note: Reactions with CDNB were done identically, but those with other substrates differently, because of the biophysical properties of the electrophilic substrate.

Source: Data from (Piccolomini et al., 1989; Newton et al., 2011).

as indicated by the lack of conjugation in the presence of thiols from other species (Newton et al., 2011). From primary sequence analysis, *M. tuberculosis* MST exhibited approximately 77% identity with the homologue from *M. smegmatis* (Newton et al., 2011). In addition to mBCl, MST from *M. smegmatis* was found to use CDNB and the antibiotic mitomycin C as substrates (Table 19.4). As CDNB and bimanes are hydrophobic substrates, MSTs are expected to have a hydrophobic binding site, similarly to GSTs. However, physiological substrates are yet to be identified. MST is the corresponding GST analog; therefore, some similarities are foreseen even though primary sequences display a very low identity. Structurally, most GSTs have predominantly α-helices in their secondary structure that are also prevailing in DNA damage-induced (DinB) superfamily members to which MSTs belong (Atkinson and Babbitt, 2009). Therefore, although the primary sequences do not provide insights into the catalytic mechanism, structural and biochemical studies may unravel the catalytic mode of action of MST.

19.4.6 MYCOTHIOL AMIDASE

The conjugation of harmful molecules and metabolites that use specific thiols, such as GSH and MSH, are the first detoxification steps, but cells still need to cope with processing and excretion. In mammals, the GSH-conjugated molecules are subjected to the cleavage of the glutamic acid group by a γ-glutamyl transpeptidase (GGT) and a subsequent peptidase-catalyzed hydrolysis. The cysteinyl toxin-conjugated product is then acetylated by an *N*-acetyl transferase that results in a mercapturic acid. Afterward, the excretion can take place via the multidrug resistance protein (MRP) that is responsible for the transport of conjugated xenobiotics and metabolites across the membrane (Rawat et al., 2004; Franco et al., 2007). In mycobacteria, the MSH-conjugated xenobiotics and metabolites are exposed to the cleavage of the cysteinyl-glucosamine amide bond of the R-S-MSH conjugate in R-S-AcCys and GlcN-Ins. GlcN-Ins is recycled by MshC to produce MSH, and R-S-AcCys is carried across the membrane. Unfortunately, little information is available regarding the excretion process (Figure 19.4) (Newton et al., 2000b; Rawat et al., 2004).

    M. tuberculosis mycothiol amidase (Mca) is a Zn^{2+}-containing enzyme that catalyzes the cleavage of the amide bond of several MSH conjugates (Steffek et al., 2003). Monobromobimane (mBBr) is an electrophilic fluorogenic substrate that is widely used to label thiols and that mimics a toxin. The mBBr conjugation with MSH occurs spontaneously or is catalyzed by MST, generating a MSmB conjugate that is a substrate of Mca, with a K_m of $460\pm40\,\mu M$ and a k_{cat} of 14 ± 2 s^{-1} (Steffek et al., 2003). In *C. glutamicum*, the Mca activity values are in the same range as those of *M. tuberculosis*, displaying a K_m of $92\pm1.5\,\mu M$ and a k_{cat} of 3.5 ± 0.9 s^{-1}, hinting at a biochemical similarity (Si et al., 2014). Moreover, MSH, as a substrate of Mca, displays a V_{max} of 0.43 ± 0.05 s^{-1}. Under physiological conditions of 3 mM MSH, a 50% inhibition indicates that the hydrolase activity toward the MSH conjugates is maintained (Steffek et al., 2003).

    The *mca* knockout of *C. glutamicum* was sensitive to several alkylating agents and to antibiotics that differed from those in *mshA*- or *mshC*-deficient strains, such as tetracycline or vancomycin (Table 19.5). Further, within the reducing/oxidizing

TABLE 19.5

Sensitivity of the *C. glutamicum* Wild Type and Mycothiol Amidase Deletion to Antibiotics, Alkylating Agents, and Reducing Agents

Molecule	Concentration	Growth Inhibition Size (in cm)		
		Wild Type	Δmca	Δmca-complemented
Rifamycin S	0.25 mg mL^{-1}	1.3±0.3	2.5±0.2	1.5±0.3
Vancomycin	1.25 mg mL^{-1}	0.9±0.5	1.8±0.4	1.1±0.3
Ciprofloxacin	5 mg mL^{-1}	1.3±0.2	3.2±0.3	2.1±0.2
Tetracycline	30 mg mL^{-1}	1.3±0.3	3.1±0.2	2.3±0.2
Menadione	50 mM	1.0±0.2	2.2±0.3	1.1±0.2
mBBr	90 mM	1.5±0.3	2.4±0.4	1.6±0.4
IAM	0.54 mM	1.2±0.5	2.3±0.5	1.2±0.3
NEM	100 mM	1.7±0.3	2.7±0.4	1.9±0.5
CDNB	49.4 mM	0.9±0.1	2.6±0.1	0
MG	20 mM	2.1±0.3	4.4±0.4	2.3±0.2

mBBr, monobromobimane; IAM, iodoacetamide; NEM, *N*-ethylmaleimide; CDNB, 1-chloro-2,4-dinitrobenzene; MG, methylglyoxal.

Values supported by $P < 0.05$.

Source: Data from (Si et al., 2014).

agents, *mca*-deficient cells were sensitive to menadione only (Table 19.5). These data suggest that Mca is involved in bacterial defense against several different molecules, demonstrating some degree of specificity (Table 19.5; Si et al., 2014).

19.4.7 MYCOTHIOL PEROXIDASE

The reduction of peroxides to water is one of the major protective systems of aerobic organisms. Glutathione peroxidases (GPx; see Chapters 3–8) are among one of the most important components in the defense of the cell homeostasis against ROS generated by the aerobic metabolism. The production of one-electron oxygen reduction intermediates, such as superoxide ($O_2{}^{.-}$), H_2O_2, and the highly reactive hydroxyl radical (OH•), is a consequence of the reduction of O_2 to $2 H_2O$, for which $4 H^+$ and $4 e^-$ are required. Besides the ROS produced by the cellular metabolism, another ROS source originates from the host immune system that attacks the bacterial pathogens. In the same manner as the eukaryotic systems, bacteria have survival systems that depend on LMW thiols as recycling agents for ROS detoxification. The enzyme mycothiol peroxidase (Mpx) belongs to the GPx-like family and contains one conserved peroxidatic and one conserved resolving cysteine (Figure 19.4). In the presence of H_2O_2, the peroxidatic cysteine of the *C. glutamicum* Mpx is oxidized, with a fast sulfenylation rate constant of 3.7×10^5 M^{-1}s^{-1} (Pedre et al., 2015). The recycling of the sulfenylated cysteine can follow two paths, either by forming an intramolecular disulfide with the resolving cysteine or by forming a mixed disulfide with a MSH molecule.

TABLE 19.6

C. glutamicum Mycothiol Peroxidase Activity toward Different Peroxides and with Distinct Recycling Systems

Substrate	Recycling System	$k_{cat}/k_m \times 10^5$ (M^{-1}s^{-1})
H_2O_2	Trx/TrxR	4.0
t-Butyl-OOH	Trx/TrxR	5.7
Cumene-OOH	Trx/TrxR	8.0
H_2O_2	Mrx1/MSH/Mtr	2.3
t-Butyl-OOH	Mrx1/MSH/Mtr	2.9
Cumene-OOH	Mrx1/MSH/Mtr	3.8

Source: Data from (Si et al., 2015a).

In the case of the intramolecular disulfide, a nucleophilic attack of the nucleophilic cysteine of Trx occurs, yielding a Trx-Mpx mixed disulfide intermediate. A subsequent attack from the resolving cysteine of Trx releases reduced Mpx and Trx in its intramolecular disulfide form. The MSH pathway depends on the mixed disulfide, in which MSH accomplishes the nucleophilic attack against the sulfenic acid at a reaction rate of 6.2×10^2 M^{-1}s^{-1}. S-Mycothiolated Mpx can be demycothiolated via different mechanisms. First, another MSH molecule can nucleophilically strike the S-mycothiolated cysteine, releasing reduced Mpx and MSSM, at a reaction rate of 3.6 M^{-1}s^{-1}. Second, Mrx1 catalyzes the demycothiolation on the mixed disulfide, as described previously, via a monothiol mechanism at a reaction rate constant of 3.4×10^4 M^{-1}s^{-1}. Third, the Trx nucleophilic cysteine performs an attack toward the mycothiolated form of Mpx, being subsequently mycothiolated, and the nucleophilic attack of the Trx resolving cysteine releases reduced MSH and yields an intramolecular disulfide-containing Trx.

Additionally, quantitative polymerase chain reaction data revealed that after H_2O_2 stress, the *mpx* gene expression was 6-fold increased, indicating a physiological role for this enzyme under oxidative stress (Pedre et al., 2015). Several peroxides have been assayed as Mpx substrates of *C. glutamicum* with a clear preference for the Trx-recycling system, as implied by the approximately 2-fold faster Trx system than the MSH system (Table 19.6). However, the *in vivo* S-mycothiolation of Mpx demonstrates that the MSH pathway can be an alternative to the Trx system (Si et al., 2015a).

19.5 CONCLUSION

MSH is the major LMW thiol in Actinomycetes and has GSH as analogous molecule in eukaryotes and Gram-negative bacteria. Despite sharing similar functionalities, the structural differences are clear. The cysteine moiety is the common factor for the classification of both molecules as LMW thiols. Cysteine is a remarkable amino acid involved in several different pathways and participates in many reactions as a nucleophile. In mycothiol, L-cysteine is conjugated via an amide bond with GlcN that forms Cys-GlcN, whereas in glutathione, it is conjugated with γ-Glu, yielding γ-Glu-Cys via the same type of bond. L-Cysteine is the reactive moiety at the heart of both LMW thiols, where it plays a crucial role in cellular amino acid homeostasis as effective L-cysteine reservoirs. L-Cysteine is a very reactive molecule, but conjugation increases its solubility and decreases its sensitivity to undergo oxidation. In this manner, unwanted modifications are prevented. Additionally, the uniqueness of the amide bond within the conjugated molecules makes that only the mycothiol amidase and γ-glutamyl transpeptidase enzymes can use it as substrate.

Both MSH and its analogous molecule GSH play a crucial role in cellular defense mechanisms and redox homeostasis. Cell machineries in GSH-containing organisms are also found in MSH-containing organisms: MST conjugates MSH with electrophilic toxins; Mrx demycothiolates targets protected from oxidative stress via a mixed disulfide; Mpx catalyzes a detoxification process that converts H_2O_2 into water and can be recycled by MSH; Mca hydrolyzes the cysteinyl-glucosamine amide bond of free MSH and its conjugated forms, and Mtr specifically catalyzes the reduction of MSSM to MSH.

In the last decade, fluctuations of the steady-state redox balance within the cell became of great interest to redox researchers and nowadays the elaboration of sensitive genetically encoded probes is a spotlight for study. The development of Mrx1-roGFP as a MSH sensor opened a complete new avenue of possibilities and clearly demonstrated that mycobacteria maintain a reduced environment to keep the intracellular MSSM concentrations low by reducing it to MSH. However, little is known about the mechanisms by which MSSM is exported. All in all, genetic and biochemical studies of the past decades have helped to elucidate the physiological role of MSH in Actinomycetes and, more recently, the creation of new bioprobes has boosted the current LMW thiol research field.

REFERENCES

Albesa-Jove, D., L. R. Chiarelli, V. Makarov et al. 2014. Rv2466c mediates the activation of TP053 to kill replicating and non-replicating *Mycobacterium tuberculosis*. *ACS Chem Biol* 9 (7):1567–75.

Albesa-Jove, D., N. Comino, M. Tersa et al. 2015. The redox state regulates the conformation of Rv2466c to activate the antitubercular prodrug TP053. *J Biol Chem* 290 (52):31077–89.

Allocati, N., L. Federici, M. Masulli, and C. Di Ilio. 2009. Glutathione transferases in bacteria. *FEBS J* 276 (1):58–75.

Anderberg, S. J., G. L. Newton, and R. C. Fahey. 1998. Mycothiol biosynthesis and metabolism. Cellular levels of potential intermediates in the biosynthesis and degradation of mycothiol in *Mycobacterium smegmatis*. *J Biol Chem* 273 (46):30391–7.

Atkinson, H. J., and P. C. Babbitt. 2009. Glutathione transferases are structural and functional outliers in the thioredoxin fold. *Biochemistry* 48 (46):11108–16.

Bhaskar, A., M. Chawla, M. Mehta et al. 2014. Reengineering redox sensitive GFP to measure mycothiol redox potential of *Mycobacterium tuberculosis* during infection. *PLoS Pathog* 10 (1):e1003902.

Bornemann, C., M. A. Jardine, H. S. Spies, and D. J. Steenkamp. 1997. Biosynthesis of mycothiol: Elucidation of the sequence of steps in *Mycobacterium smegmatis*. *Biochem J* 325 (Pt 3):623–9.

Buchmeier, N. A., G. L. Newton, and R. C. Fahey. 2006. A mycothiol synthase mutant of *Mycobacterium tuberculosis* has an altered thiol-disulfide content and limited tolerance to stress. *J Bacteriol* 188 (17):6245–52.

Bulaj, G., T. Kortemme, and D. P. Goldenberg. 1998. Ionization-reactivity relationships for cysteine thiols in polypeptides. *Biochemistry* 37 (25):8965–72.

Chi, B. K., T. Busche, K. Van Laer et al. 2014. Protein *S*-mycothiolation functions as redox-switch and thiol protection mechanism in *Corynebacterium glutamicum* under hypochlorite stress. *Antioxid Redox Signal* 20 (4):589–605.

Dixon, D. P., T. Hawkins, P. J. Hussey, and R. Edwards. 2009. Enzyme activities and subcellular localization of members of the Arabidopsis glutathione transferase superfamily. *J Exp Bot* 60 (4):1207–18.

Dooley, C. T., T. M. Dore, G. T. Hanson et al. 2004. Imaging dynamic redox changes in mammalian cells with green fluorescent protein indicators. *J Biol Chem* 279 (21):22284–93.

Fahey, R. C. 2013. Glutathione analogs in prokaryotes. *Biochim Biophys Acta* 1830 (5):3182–98.

Fan, F., A. Luxenburger, G. F. Painter, and J. S. Blanchard. 2007. Steady-state and pre-steady-state kinetic analysis of *Mycobacterium smegmatis* cysteine ligase (MshC). *Biochemistry* 46 (40):11421–9.

Franco, R., O. J. Schoneveld, A. Pappa, and M. I. Panayiotidis. 2007. The central role of glutathione in the pathophysiology of human diseases. *Arch Physiol Biochem* 113 (4–5):234–58.

Gutscher, M., A. L. Pauleau, L. Marty et al. 2008. Real-time imaging of the intracellular glutathione redox potential. *Nat Methods* 5 (6):553–9.

Henderson, G. B., N. J. Murgolo, J. Kuriyan et al. 1991. Engineering the substrate-specificity of glutathione-reductase toward that of trypanothione reduction. *Proc Natl Acad Sci USA* 88 (19):8769–73.

Hillion, M., J. Bernhardt, T. Busche et al. 2017a. Monitoring global protein thiol-oxidation and protein *S*-mycothiolation in *Mycobacterium smegmatis* under hypochlorite stress. *Sci Rep* 7 (1):1195.

Hillion, M., M. Imber, B. Pedre et al. 2017b. The glyceraldehyde-3-phosphate dehydrogenase GapDH of *Corynebacterium diphtheriae* is redox-controlled by protein *S*-mycothiolation under oxidative stress. *Sci Rep* 7 (1):5020.

Holsclaw, C. M., W. B. Muse, 3rd, K. S. Carroll, and J. A. Leary. 2011. Mass spectrometric analysis of mycothiol levels in wild-type and mycothiol disulfide reductase mutant *Mycobacterium smegmatis*. *Int J Mass Spectrom* 305 (2–3):151–56.

Hugo, M., L. Turell, B. Manta et al. 2009. Thiol and sulfenic acid oxidation of AhpE, the one-cysteine peroxiredoxin from *Mycobacterium tuberculosis*: Kinetics, acidity constants, and conformational dynamics. *Biochemistry* 48 (40):9416–26.

Hugo, M., K. Van Laer, A. M. Reyes et al. 2014. Mycothiol/mycoredoxin 1-dependent reduction of the peroxiredoxin AhpE from *Mycobacterium tuberculosis*. *J Biol Chem* 289 (8):5228–39.

Johnson, T., G. L. Newton, R. C. Fahey, and M. Rawat. 2009. Unusual production of glutathione in Actinobacteria. *Arch Microbiol* 191 (1):89–93.

Jothivasan, V. K., and C. J. Hamilton. 2008. Mycothiol: Synthesis, biosynthesis and biological functions of the major low molecular weight thiol in actinomycetes. *Nat Prod Rep* 25 (6):1091–117.

Kana, B. D., P. C. Karakousis, T. Parish, and T. Dick. 2014. Future target-based drug discovery for tuberculosis? *Tuberculosis (Edinb)* 94 (6):551–6.

Koledin, T., G. L. Newton, and R. C. Fahey. 2002. Identification of the mycothiol synthase gene (mshD) encoding the acetyltransferase producing mycothiol in actinomycetes. *Arch Microbiol* 178 (5):331–7.

Kumar, A., W. Nartey, J. Shin et al. 2017. Structural and mechanistic insights into mycothiol disulphide reductase and the mycoredoxin-1-alkylhydroperoxide reductase E assembly of *Mycobacterium tuberculosis. Biochim Biophys Acta* 1861 (9):2354–66.

Lamichhane, G. 2011. *Mycobacterium tuberculosis* response to stress from reactive oxygen and nitrogen species. *Front Microbiol* 2:176.

Letunic, I., and P. Bork. 2016. Interactive tree of life (iTOL) v3: An online tool for the display and annotation of phylogenetic and other trees. *Nucleic Acids Res* 44 (W1):W242–5.

Lillig, C. H., C. Berndt, and A. Holmgren. 2008. Glutaredoxin systems. *Biochem Biophys Acta* 1780 (11):1304–17.

Liu, Y. B., C. Chen, M. T. Chaudhry et al. 2014. Enhancing *Corynebacterium glutamicum* robustness by over-expressing a gene, *mshA*, for mycothiol glycosyltransferase. *Biotechnol Lett* 36 (7):1453–9.

Liu, Y. B., M. X. Long, Y. J. Yin et al. 2013. Physiological roles of mycothiol in detoxification and tolerance to multiple poisonous chemicals in *Corynebacterium glutamicum. Arch Microbiol* 195 (6):419–29.

Lu, S. C. 2013. Glutathione synthesis. *Biochem Biophys Acta* 1830 (5):3143–53.

Mateos, L. M., A. F. Villadangos, A. G. de la Rubia et al. 2017. The arsenic detoxification system in corynebacteria: Basis and application for bioremediation and redox control. *Adv Appl Microbiol* 99:103–17.

Maynes, J. T., C. Garen, M. M. Cherney et al. 2003. The crystal structure of 1-D-myo-inosityl 2-acetamido-2-deoxy-alpha-D-glucopyranoside deacetylase (MshB) from *Mycobacterium tuberculosis* reveals a zinc hydrolase with a lactate dehydrogenase fold. *J Biol Chem* 278 (47):47166–70.

McCarthy, A. A., N. A. Peterson, R. Knijff, and E. N. Baker. 2004. Crystal structure of MshB from *Mycobacterium tuberculosis*, a deacetylase involved in mycothiol biosynthesis. *J Mol Biol* 335 (4):1131–41.

Mehta, M., R. S. Rajmani, and A. Singh. 2016. *Mycobacterium tuberculosis* WhiB3 responds to vacuolar pH-induced changes in mycothiol redox potential to modulate phagosomal maturation and virulence. *J Biol Chem* 291 (6):2888–903.

Miller, C. C., M. Rawat, T. Johnson, and Y. Av-Gay. 2007. Innate protection of *Mycobacterium smegmatis* against the antimicrobial activity of nitric oxide is provided by mycothiol. *Antimicrob Agents Chemother* 51 (9):3364–6.

Movahedzadeh, F., P. R. Wheeler, P. Dinadayala et al. 2010. Inositol monophosphate phosphatase genes of *Mycobacterium tuberculosis. BMC Microbiol* 10:50.

Nambi, S., J. E. Long, B. B. Mishra et al. 2015. The oxidative stress network of *Mycobacterium tuberculosis* reveals coordination between radical detoxification systems. *Cell Host Microbe* 17 (6):829–37.

Newton, G. L., K. Arnold, M. S. Price et al. 1996. Distribution of thiols in microorganisms: Mycothiol is a major thiol in most actinomycetes. *J Bacteriol* 178 (7):1990–5.

Newton, G. L., Y. Av-Gay, and R. C. Fahey. 2000a. N-acetyl-1-D-myo-inosityl-2-amino-2-deoxy-alpha-D-glucopyranoside deacetylase (MshB) is a key enzyme in mycothiol biosynthesis. *J Bacteriol* 182 (24):6958–63.

Newton, G. L., Y. Av-Gay, and R. C. Fahey. 2000b. A novel mycothiol-dependent detoxification pathway in mycobacteria involving mycothiol S-conjugate amidase. *Biochemistry* 39 (35):10739–46.

Newton, G. L., C. A. Bewley, T. J. Dwyer et al. 1995. The structure of U17 isolated from *Streptomyces clavuligerus* and its properties as an antioxidant thiol. *Eur J Biochem* 230 (2):821–5.

Newton, G. L., N. Buchmeier, and R. C. Fahey. 2008. Biosynthesis and functions of mycothiol, the unique protective thiol of Actinobacteria. *Microbiol Mol Biol Rev* 72 (3):471–94.

Newton, G. L., and R. C. Fahey. 2002. Mycothiol biochemistry. *Arch Microbiol* 178 (6):388–94.

Newton, G. L., R. C. Fahey, G. Cohen, and Y. Aharonowitz. 1993. Low-molecular-weight thiols in Streptomycetes and their potential role as antioxidants. *J Bacteriol* 175 (9):2734–42.

Newton, G. L., M. Ko, P. Ta et al. 2006. Purification and characterization of *Mycobacterium tuberculosis* 1D-myo-inosityl-2-acetamido-2-deoxy-alpha-D-glucopyranoside deacetylase, MshB, a mycothiol biosynthetic enzyme. *Protein Expr Purif* 47 (2):542–50.

Newton, G. L., S. S. Leung, J. I. Wakabayashi et al. 2011. The DinB superfamily includes novel mycothiol, bacillithiol, and glutathione *S*-transferases. *Biochemistry* 50 (49):10751–60.

Newton, G. L., P. Ta, and R. C. Fahey. 2005. A mycothiol synthase mutant of *Mycobacterium smegmatis* produces novel thiols and has an altered thiol redox status. *J Bacteriol* 187 (21):7309–16.

Newton, G. L., M. D. Unson, S. J. Anderberg et al. 1999. Characterization of *Mycobacterium smegmatis* mutants defective in 1-D-myo-inosityl-2-amino-2-deoxy-alpha-D-glucopyranoside and mycothiol biosynthesis. *Biochem Bioph Res Co* 255 (2):239–44.

Nigou, J., L. G. Dover, and G. S. Besra. 2002. Purification and biochemical characterization of *Mycobacterium tuberculosis* SuhB, an inositol monophosphatase involved in inositol biosynthesis. *Biochemistry* 41 (13):4392–8.

Nilewar, S. S., and M. K. Kathiravan. 2014. Mycothiol: A promising antitubercular target. *Bioorg Chem* 52:62–8.

Ordóñez, E., K. Van Belle, G. Roos et al. 2009. Arsenate reductase, mycothiol, and mycoredoxin concert thiol/disulfide exchange. *J Biol Chem* 284 (22):15107–16.

Østergaard, H., A. Henriksen, F. G. Hansen, and J. R. Winther. 2001. Shedding light on disulfide bond formation: Engineering a redox switch in green fluorescent protein. *EMBO J* 20 (21):5853–62.

Østergaard, H., C. Tachibana, and J. R. Winther. 2004. Monitoring disulfide bond formation in the eukaryotic cytosol. *J Cell Biol* 166 (3):337–45.

Pai, M., M. A. Behr, D. Dowdy et al. 2016. Tuberculosis. *Nat Rev Dis Primers* 2:16076.

Patel, M. P., and J. S. Blanchard. 1999. Expression, purification, and characterization of *Mycobacterium tuberculosis* mycothione reductase. *Biochemistry* 38 (36):11827–33.

Pedre, B., I. Van Molle, A. F. Villadangos et al. 2015. The *Corynebacterium glutamicum* mycothiol peroxidase is a reactive oxygen species-scavenging enzyme that shows promiscuity in thiol redox control. *Mol Microbiol* 96 (6):1176–91.

Peralta, D., A. K. Bronowska, B. Morgan et al. 2015. A proton relay enhances H2O2 sensitivity of GAPDH to facilitate metabolic adaptation. *Nat Chem Biol* 11 (2):156–63.

Peskin, A. V., P. E. Pace, J. B. Behring et al. 2016. Glutathionylation of the active site cysteines of peroxiredoxin 2 and recycling by glutaredoxin. *J Biol Chem* 291 (6):3053–62.

Piccolomini, R., C. Di Ilio, A. Aceto et al. 1989. Glutathione transferase in bacteria: Subunit composition and antigenic characterization. *J Gen Microbiol* 135 (11):3119–25.

Poole, L. B. 2007. The catalytic mechanism of peroxiredoxins. *Subcell Biochem* 44:61–81.

Rabenstein, D. L. 1973. Nuclear magnetic-resonance studies of acid-base chemistry of amino-acids and peptides. 1. Microscopic ionization-constants of glutathione and methylmercury-complexed glutathione. *J Am Chem Soc* 95 (9):2797–803.

Ramón-García, S., C. Martin, C. J. Thompson, and J. A. Ainsa. 2009. Role of the *Mycobacterium tuberculosis* P55 efflux pump in intrinsic drug resistance, oxidative stress responses, and growth. *Antimicrob Agents Chemother* 53 (9):3675–82.

Rawat, M., C. Johnson, V. Cadiz, and Y. Av-Gay. 2007. Comparative analysis of mutants in the mycothiol biosynthesis pathway in *Mycobacterium smegmatis*. *Biochem Biophys Res Commun* 363 (1):71–6.

Rawat, M., S. Kovacevic, H. Billman-Jacobe, and Y. Av-Gay. 2003. Inactivation of mshB, a key gene in the mycothiol biosynthesis pathway in *Mycobacterium smegmatis*. *Microbiology* 149 (Pt 5):1341–9.

Rawat, M., G. L. Newton, M. Ko et al. 2002. Mycothiol-deficient *Mycobacterium smegmatis* mutants are hypersensitive to alkylating agents, free radicals, and antibiotics. *Antimicrob Agents Chemother* 46 (11):3348–55.

Rawat, M., M. Uppal, G. Newton et al. 2004. Targeted mutagenesis of the *Mycobacterium smegmatis* mca gene, encoding a mycothiol-dependent detoxification protein. *J Bacteriol* 186 (18):6050–8.

Reyes, A. M., B. Pedre, M. I. De Armas et al. 2018. Chemistry and redox biology of mycothiol. *Antioxid Redox Signal* 28 (6):487–504.

Rosado, L. A., K. Wahni, G. Degiacomi et al. 2017. The antibacterial prodrug activator Rv2466c is a mycothiol-dependent reductase in the oxidative stress response of *Mycobacterium tuberculosis*. *J Biol Chem* 292 (32):13097–110.

Saini, V., A. Farhana, and A. J. Steyn. 2012. *Mycobacterium tuberculosis* WhiB3: A novel iron-sulfur cluster protein that regulates redox homeostasis and virulence. *Antioxid Redox Signal* 16 (7):687–97.

Sakuda, S., Z. Y. Zhou, and Y. Yamada. 1994. Structure of a novel disulfide of 2-(*N*-acetylcysteinyl)amido-2-deoxy-alpha-D-glucopyranosyl-myo-inositol produced by *Streptomyces sp. Biosci Biotechnol Biochem* 58 (7):1347–8.

Sareen, D., G. L. Newton, R. C. Fahey, and N. A. Buchmeier. 2003. Mycothiol is essential for growth of *Mycobacterium tuberculosis* Erdman. *J Bacteriol* 185 (22):6736–40.

Sareen, D., M. Steffek, G. L. Newton, and R. C. Fahey. 2002. ATP-dependent L-cysteine: 1D-myo-inosityl 2-amino-2-deoxy-alpha-D-glucopyranoside ligase, mycothiol biosynthesis enzyme MshC, is related to class I cysteinyl-tRNA synthetases. *Biochemistry* 41 (22):6885–90.

Sassetti, C. M., D. H. Boyd, and E. J. Rubin. 2003. Genes required for mycobacterial growth defined by high density mutagenesis. *Mol Microbiol* 48 (1):77–84.

Savvides, S. N., M. Scheiwein, C. C. Bohme et al. 2002. Crystal structure of the antioxidant enzyme glutathione reductase inactivated by peroxynitrite. *J Biol Chem* 277 (4):2779–84.

Sharma, S. V., K. Van Laer, J. Messens, and C. J. Hamilton. 2016. Thiol redox and pKa properties of mycothiol, the predominant low-molecular-weight thiol cofactor in the Actinomycetes. *Chembiochem* 17 (18):1689–92.

Sheehan, D., G. Meade, V. M. Foley, and C. A. Dowd. 2001. Structure, function and evolution of glutathione transferases: Implications for classification of non-mammalian members of an ancient enzyme superfamily. *Biochem J* 360 (Pt 1):1–16.

Si, M., M. Long, M. T. Chaudhry et al. 2014. Functional characterization of *Corynebacterium glutamicum* mycothiol S-conjugate amidase. *PLOS One* 9 (12):e115075.

Si, M., Y. Xu, T. Wang et al. 2015a. Functional characterization of a mycothiol peroxidase in *Corynebacterium glutamicum* that uses both mycoredoxin and thioredoxin reducing systems in the response to oxidative stress. *Biochem J* 469 (1):45–57.

Si, M., L. Zhang, M. T. Chaudhry et al. 2015b. *Corynebacterium glutamicum* methionine sulfoxide reductase A uses both mycoredoxin and thioredoxin for regeneration and oxidative stress resistance. *Appl Environ Microbiol* 81 (8):2781–96.

Si, M., C. Zhao, B. Zhang et al. 2016. Overexpression of mycothiol disulfide reductase enhances *Corynebacterium glutamicum* robustness by modulating cellular redox homeostasis and antioxidant proteins under oxidative stress. *Sci Rep* 6:29491.

Steffek, M., G. L. Newton, Y. Av-Gay, and R. C. Fahey. 2003. Characterization of mycothiol S-conjugate amidase. *Biochemistry* 42 (41):12067–76.

Tamura, K., G. Stecher, D. Peterson et al. 2013. MEGA6: Molecular Evolutionary Genetics Analysis version 6.0. *Mol Biol Evol* 30 (12):2725–9.

Tang, L., W. W. Wang, W. L. Zhou et al. 2015. Three-pathway combination for glutathione biosynthesis in *Saccharomyces cerevisiae*. *Microb Cell Fact* 14:139.

Tossounian, M. A., B. Pedre, K. Wahni et al. 2015. *Corynebacterium diphtheriae* methionine sulfoxide reductase a exploits a unique mycothiol redox relay mechanism. *J Biol Chem* 290 (18):11365–75.

Trauner, A., C. M. Sassetti, and E. J. Rubin. 2014. Genetic strategies for identifying new drug targets. *Microbiol Spectrum* 2 (4):MGM2-0030–2013.

Tremblay, L. W., F. Fan, M. W. Vetting, and J. S. Blanchard. 2008. The 1.6 Å crystal structure of *Mycobacterium smegmatis* MshC: The penultimate enzyme in the mycothiol biosynthetic pathway. *Biochemistry* 47 (50):13326–35.

Upton, H., G. L. Newton, M. Gushiken et al. 2012. Characterization of BshA, bacillithiol glycosyltransferase from *Staphylococcus aureus* and *Bacillus subtilis*. *FEBS Lett* 586 (7):1004–8.

van Hylckama Vlieg, J. E. T., J. Kingma, A. J. van den Wijngaard, and D. B. Janssen. 1998. A glutathione *S*-transferase with activity towards cis-1, 2-dichloroepoxyethane is involved in isoprene utilization by *Rhodococcus sp.* strain AD45. *Appl Environ Microbiol* 64 (8):2800–5.

Van Laer, K., L. Buts, N. Foloppe et al. 2012. Mycoredoxin-1 is one of the missing links in the oxidative stress defence mechanism of Mycobacteria. *Molecular Microbiology* 86 (4):787–804.

Van Laer, K., C. J. Hamilton, and J. Messens. 2013. Low-molecular-weight thiols in thiol-disulfide exchange. *Antioxid Redox Signal* 18 (13):1642–53.

Vetting, M. W., P. A. Frantom, and J. S. Blanchard. 2008. Structural and enzymatic analysis of MshA from *Corynebacterium glutamicum*: Substrate-assisted catalysis. *J Biol Chem* 283 (23):15834–44.

Vetting, M. W., S. d. C. LP, M. Yu et al. 2005. Structure and functions of the GNAT superfamily of acetyltransferases. *Arch Biochem Biophys* 433 (1):212–26.

Vetting, M. W., M. Yu, P. M. Rendle, and J. S. Blanchard. 2006. The substrate-induced conformational change of *Mycobacterium tuberculosis* mycothiol synthase. *J Biol Chem* 281 (5):2795–802.

Villadangos, A. F., E. Ordonez, B. Pedre et al. 2014. Engineered coryneform bacteria as a biotool for arsenic remediation. *Appl Microbiol Biotechnol* 98 (24):10143–52.

Williams, L., F. Fan, J. S. Blanchard, and F. M. Raushel. 2008. Positional isotope exchange analysis of the *Mycobacterium smegmatis* cysteine ligase (MshC). *Biochemistry* 47 (16):4843–50.

Xu, X., C. Vilcheze, Y. Av-Gay et al. 2011. Precise null deletion mutations of the mycothiol synthesis genes reveal their role in isoniazid and ethionamide resistance in *Mycobacterium smegmatis*. *Antimicrob Agents Chemother* 55 (7):3133–9.

20 Biosynthesis and Functions of Bacillithiol in *Firmicutes*

Quach Ngoc Tung, Nico Linzner, Vu Van Loi, and Haike Antelmann
Freie Universität Berlin

CONTENTS

20.1 INTRODUCTION

All cells have to maintain their reduced state of the cytoplasm to ensure proper protein functions and cellular survival. To maintain the redox balance, eukaryotic and prokaryotic organisms utilize low-molecular-weight (LMW) thiols. LMW thiols are small thiol-containing compounds that function in the defense against reactive oxygen, chlorine, and electrophilic species (ROS, RCS, RES), antibiotics, heavy metals and other redox-active compounds (Masip et al., 2006; Van Laer et al., 2013; Chandrangsu et al., 2018). The best-studied LMW thiol is the tripeptide glutathione (GSH), which is produced in eukaryotic organisms, most Gram-negative bacteria and in some Gram-positive bacteria, including *Streptococcus agalactiae*, *Listeria monocytogenes*, and *Clostridium acetobutylicum* (Fahey, 2013; Loi et al., 2015). However, Gram-positive bacteria do not produce GSH and instead, utilize alternative

DOI: 10.1201/9781351261760-25

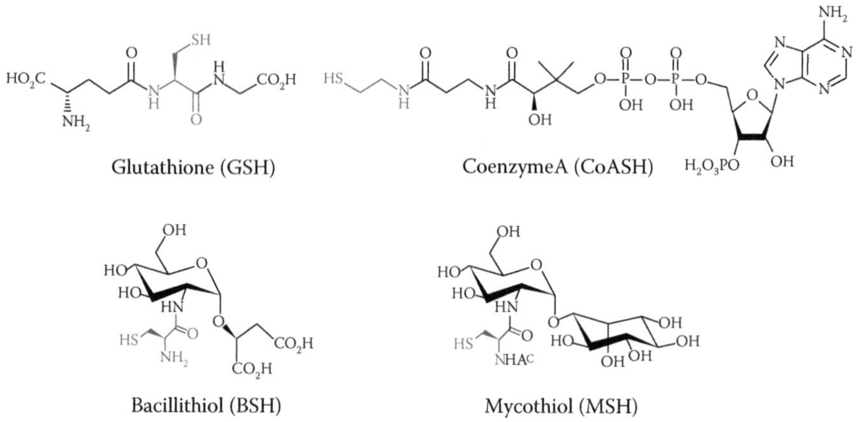

Glutathione (GSH)   CoenzymeA (CoASH)

Bacillithiol (BSH)   Mycothiol (MSH)

FIGURE 20.1 Structures of major LMW thiols in bacteria. Glutathione (GSH) is the major LMW thiol in eukaryotes and Gram-negative bacteria. Mycothiol (MSH) is utilized as major LMW thiol in *Actinomycetes*. Bacillithiol (BSH) is produced by *Firmicutes*, such as *Bacillus* and *Staphylococcus* species. Coenzyme A (CoASH) serves as a LMW thiol in *S. aureus* and *B. anthracis*.

LMW thiols. *Bacillus* and *Staphylococcus* species utilize bacillithiol (BSH), while Actinomycetes, such as *Streptomycetes*, *Mycobacterium*, and *Corynebacterium* species produce mycothiol (MSH) as their major LMW thiol (Newton et al., 2008; Newton et al., 2009) (Figure 20.1; see also Chapter 19).

BSH represents the latest discovery among bacterial LMW thiols and has been widely studied in different *Bacillus* species and in the human pathogen *Staphylococcus aureus* over the last 10 years. In this review, we present an overview about the many functions of BSH in *Firmicutes*. The review includes the current knowledge about the BSH biosynthesis pathway, the biophysical and biochemical properties of BSH, and the functions of BSH in detoxification, antibiotics resistance, sulfide metabolism, metal homeostasis, and protein *S*-bacillithiolation as regulatory device in *Bacillus subtilis* and *S. aureus*. While significant progress has been made to elucidate the many functions of BSH, there is still much to be discovered in terms of BSH-dependent redox regulation and the biochemistry of the BSH redox pathway. Thus, we hope that our review will stimulate future research on this interesting LMW thiol to move the field of microbial redox research forward.

20.2 STRUCTURE AND BIOPHYSICAL PROPERTIES OF BSH

BSH was first discovered as abundant LMW thiol of 398 Da using thiol-metabolomics in *B. anthracis* (Nicely et al., 2007). In addition, the Helmann Lab had reported an *S*-thiolation of the redox-sensing OhrR repressor in *B. subtilis* with an unknown thiol of 396 Da, which was later identified as BSH (Lee et al., 2007). BSH is the α-anomeric glycoside of L-cysteinyl-D-glucosamine and malate (Newton et al., 2009). BSH is present as major LMW thiol in many *Firmicutes*, including *Bacillus* and *Staphylococcus* species, *Deinococcus radiodurans*, and *Streptococcus agalactiae* (Sharma et al., 2013; Loi et al., 2015). The structure of BSH is similar to MSH, but in BSH, a malate residue is present in place of myoinositol in MSH (Figure 20.1).

Under oxidative stress, BSH is oxidized to BSSB and the BSH/BSSB ratios were calculated previously as 100:1 to 400:1 in *B. subtilis* cells (Sharma et al., 2013). The NADPH-dependent flavin oxidoreductase YpdA co-occurs together with the BSH biosynthesis enzymes across BSH-producing bacteria, suggesting its function as candidate BSSB reductase (Gaballa et al., 2010). However, thus far, all attempts to demonstrate the function of YpdA as BSSB reductase *in vitro* failed (Chandrangsu et al., 2018).

The standard thiol-redox potential of BSH was calculated as $E^{\circ\prime}$(BSSB/BSH) of -221 mV, which is more positive compared to the GSH redox potential of $E^{\circ\prime}$(GSSG/GSH) of -240 mV (Sharma et al., 2013). The microscopic pK_a values of the thiolate anion of BSH were determined as $pK_{aSH} = 7.97$ and $pK_{aSH} = 9.55$ with the amino group of the Cys in its protonated and deprotonated form, respectively, suggesting a more acidic BSH thiolate anion with enhanced reactivity (Sharma et al., 2013). The BSH levels were measured in *Bacillus* strains and *S. aureus*. *B. subtilis* produces ~2 mM BSH levels during the growth which increases 2–3 fold under NaOCl and diamide stress or upon entry into the stationary phase (Chi et al., 2013; Gaballa et al., 2013). Moreover, BSH levels exceed ~17-fold the intracellular free Cys content supporting that BSH is the major LMW thiol (Sharma et al., 2013). In *Bacillus pumilus*, ~2 mM BSH were measured during the growth, which increased to 6 mM under oxidative stress (Handtke et al., 2014). Lower BSH amounts of 0.3–1 mM were determined in different *S. aureus* clinical isolates, but *S. aureus* also utilizes CoenzymeA (CoASH) as alternative LMW thiol (Posada et al., 2014). Overall, BSH levels of >1 mM were measured in *Bacillus* species and *S. aureus* indicating that BSH is the dominant LMW thiol. Thus, the BSH thiolate anion is 10–100 fold more abundant compared to the Cys and CoASH thiolates (Sharma et al., 2013).

20.3 BIOSYNTHESIS PATHWAY OF BACILLITHIOL IN *B. SUBTILIS* AND *S. AUREUS*

The structural similarity between MSH and BSH enabled the identification of the enzymes for BSH biosynthesis (Gaballa et al., 2010). BSH synthesis is catalyzed by the enzymes BshA, BshB, and BshC. The glycosyltransferase BshA conjugates UDP-N-acetylglucosamine (UDP-GlcNAc) to L-malate through a metal-independent SN1-like mechanism, forming N-acetylglucosaminyl-malate (GlcNAc-Mal) (Upton et al., 2012; Winchell et al., 2016). This is followed by the deacetylation of GlcNAc-Mal by the deacetylase BshB to generate glucosamine malate (GlcN-Mal). The last step of BSH biosynthesis involves the putative cysteine ligase BshC that adds Cys to GlcNMal (Figure 20.2). *B. subtilis* encodes two deacetylases, BshB1 and BshB2 which can compensate each other in the deacetylation of GlcNAc-Mal. This redundancy is supported by the fact that BSH is produced in the *bshB1* and *bshB2* single mutants, but not in double mutant (Gaballa et al., 2010). In contrast, *S. aureus* has only a single

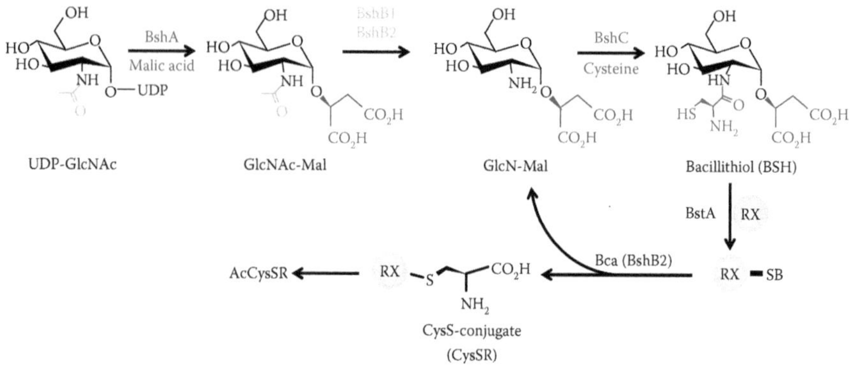

FIGURE 20.2 The BSH biosynthesis pathway and BSH-dependent detoxification of xenobiotics and toxins. BSH biosynthesis is catalyzed by BshA, BshB, and BshC. The glycosyltransferase BshA adds GlcNAc to malate producing GlcNAc-Mal. The deacetylases BshB1 and BshB2 catalyze deacetylation of GlcNAc-Mal to GlcN-Mal in *B. subtilis* which is ligated with Cys by BshC generating in BSH. Toxins, xenobiotics or electrophiles are conjugated to BSH by the BSH-*S*-transferase BstA. These BS-conjugates that are cleaved by the deacetylase BshB2 (Bca) to Cys *S*-conjugates and GlcN-Mal used for BSH recycling.

deacetylase BshB which is essential for BSH synthesis. BshB also functions as BSH conjugate amidase (Bca) in detoxification of toxic electrophiles analogous to the MSH-*S*-conjugate amidase Mca (Loi et al., 2015). The catalytic activity of the BshC enzyme in Cys ligation of GlcNAc-Mal could not been shown, but structural studies provide evidence that BshC may require an additional cofactor or a protein interaction partner (VanDuinen et al., 2015).

The BSH biosynthesis genes *bshA, bshB1, bshB2*, and *bshC* of *B. subtilis* were shown to be transcribed as part of three independent operons (Gaballa et al., 2013; Figure 20.3). The *bshA* and *bshB1* genes are transcribed as part of the *ypjD-dapB-mgsA-bshB1-bshA-cca-birA* operon in *B. subtilis*. In *S. aureus*, the *bshA* gene is also cotranscribed with *papS* (homolog of *cca*) and *birA*. The *bshB2* gene of *B. subtilis* is cotranscribed with *yoyC*, while *bshC* is located in an operon with *ylbQ*. This *bshB* and *bshC* operon organization is not conserved in *S. aureus*. Interestingly, all operons for BSH biosynthesis are induced under disulfide stress, such as diamide and NaOCl stress as well as by other thiol-reactive compounds in different *Firmicutes*, such as *B. subtilis, B. anthracis*, and *S. aureus* (Gaballa et al., 2013; Loi et al., 2015; Perera et al., 2015). The genes encoding the BSH biosynthesis enzymes are controlled by the disulfide stress regulator Spx in *B. subtilis* (Gaballa et al., 2013). In addition, also BSH-dependent detoxification enzymes are induced by ROS and RES in *B. subtilis* (Perera et al., 2015). This further indicates the importance of the coordination between the BSH biosynthesis enzymes and its BSH-dependent ROS and RES detoxification pathways. Interestingly, the *cca* (or *papS*) genes are cotranscribed with *bshA* in *B. subtilis* and *S. aureus*. Cca/PapS are tRNA nucleotidyltransferases or CCA-pyrophosporylases that function in the maturation of the tRNA^{Cys} (Campos

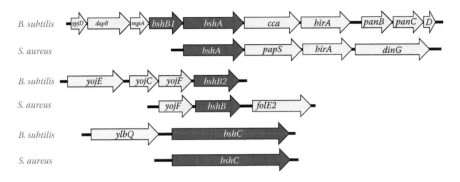

FIGURE 20.3 Conservation of gene organization of the *bshA*, *bshB* and *bshC* biosynthesis operons in *B. subtilis and S. aureus*. The *bshA* gene is cotranscribed with the *birA* and *cca/papC* genes in *B. subtilis* and *S. aureus*. The *bshB2* and *bshB* genes of *B. subtilis* and *S. aureus* are located downstream of *yojF* homologs in both bacteria, respectively.

Guillen et al., 2017). Thus, a link between BSH biosynthesis and tRNA^{Cys} maturation is suggested in *B. subtilis* under oxidative stress. Cysteine depletion under oxidative stress may lead to decreased BSH biosynthesis and lower aminoacylation of tRNA^{Cys}, which requires upregulation of BSH biosynthesis and Cca for regeneration of the redox balance and continued translation (Campos Guillen et al., 2017). The connection between BSH and translation stress under oxidative stress is an interesting subject for future studies.

20.4 FUNCTIONS OF BSH IN DETOXIFICATION OF ANTIBIOTICS AND ELECTROPHILES

The identification of the genes for BSH biosynthesis allowed the construction of BSH-deficient mutant strains for phenotype analysis to elucidate the functions of BSH (Gaballa et al., 2010). BSH plays an important role in detoxification of ROS, RES, NaOCl, metal homeostasis, and antibiotics resistance. BSH-deficient mutants are sensitive to various oxidants and electrophiles, such as HOCl, diamide, H_2O_2, monobromobimane, and methylglyoxal (Figure 20.4) (Gaballa et al., 2010; Chi et al., 2011; Chi et al., 2013; Rajkarnikar et al., 2013; Loi et al., 2015; Chandrangsu et al., 2018). BSH is also required for the detoxification of antibiotics, such as rifamycin and fosfomycin (Newton et al., 2012; Roberts et al., 2013). Fosfomycin is a cell wall biosynthesis inhibitor that is applied to combat methicillin-resistant *S. aureus* (MRSA) isolates (Michalopoulos et al., 2011; Tang et al., 2012). Fosfomycin covalently binds to the active site Cys of MurA, which catalyzes the first step of the peptidoglycan biosynthesis. The BSH-*S*-transferase FosB confers fosfomycin resistance and adds BSH to the epoxide ring leading to an inactive BS-fosfomycin conjugate (Lamers et al., 2012; Thompson et al., 2013; Roberts et al., 2013). Moreover, *fosB* and *bshA* mutants are equally sensitive to fosfomycin treatment indicating that both act in the same pathway (Gaballa et al., 2010). Thus, FosB functions as BSH-dependent *S*-transferase and fosfomycin resistance mechanism in *B. subtilis* and *S. aureus*.

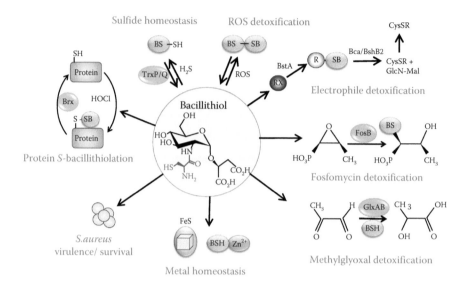

FIGURE 20.4 The many functions of BSH in *B. subtilis* and *S. aureus*. BSH functions in detoxification of ROS, RES, HOCl, RSS and antibiotics (fosfomycin, rifampicin) in *B. subtilis* and *S. aureus*. BSH is oxidized by ROS to bacillithiol disulfide (BSSB) and forms BSH persulfides (BSSH) under H_2S stress. Electrophiles (RX) are conjugated to BSH by the BSH *S*-transferase BstA to form BS-electrophiles (BSR). BSH *S*-conjugate amidase Bca or BshB2 cleave BSR into GlcNAc-Mal and mercapturic acids (CysSR) that are exported from the cell. BSH is cofactor for the epoxide hydrolase FosB which adds BSH to fosfomycin for its detoxification. BSH functions in methylglyoxal detoxification as a cofactor for the glyoxalases GlxA and GlxB in *B. subtilis*. GlxA converts BSH-hemithioacetal to *S*-lactoyl-BSH that is further detoxified by GlxB to D-lactate. BSH serves as Zn^{2+} buffer and in FeS cluster assembly. In *S. aureus*, BSH is important under infection-related conditions and increased the survival of *S. aureus* in murine macrophages. Under conditions of NaOCl stress, proteins are oxidized to mixed disulfides with BSH, termed as *S*-bacillithiolations which is reversed by bacilliredoxins.

Reactive electrophiles and xenobiotics can be detoxified by direct conjugation to BSH or by conjugation reactions catalyzed by the DinB/YfiT-family metal-dependent *S*-transferases, including the BSH *S*-transferase BstA (Newton et al., 2011; Perera et al., 2014). BstA shows conjugation activity with the toxic electrophile chloro-2,4-dinitrobenzene *in vitro*, but its natural substrate is unknown. The structures of the BstA enzymes from *B. subtilis* and *S. aureus* have been resolved (Rajan et al., 2004; Francis et al., 2018). The lack of conservation between these BstA homologs suggests significant differences how these enzymes accommodate their thiol substrate for conjugation (Francis et al., 2018). BstA belongs to a large *S*-transferase (STL) superfamily (Perera et al., 2018). Apart from BstA, *B. subtilis* encodes seven other predicted STL superfamily members of unknown functions. However, they all have BSH *S*-transferase activity *in vitro* and some are expressed during sporulation (Perera et al., 2018).

BSH is involved in detoxification of the reactive aldehydes, such as methylglyoxal and formaldehyde. Methylglyoxal is a toxic electrophilic dicarbonyl compound produced as intermediate during the glycolysis from dihydroxyacetone phosphate during carbon excess or phosphate starvation (Tötemeyer et al., 1998). Methylglyoxal reacts with DNA bases and the amino acids arginine, lysine, and cysteine, leading to advanced glycation end products (AGE), such as argpyrimidine or MG-crosslinked lysine dimers that are implicated in diabetes and cancer (Oya et al., 1999; Bourajjaj et al., 2003). Of note, the methylglyoxal synthase encoding *mgsA* gene is cotranscribed with *bshA* and *bshB1* in *B. subtilis*, indicating a link between BSH synthesis and methylglyoxal detoxification (Gaballa et al., 2010). Moreover, *bshA* mutants were more sensitive to methylglyoxal compared to the wild type (Gaballa et al., 2010).

The BSH-dependent glyoxalases GlxA and GlxB were shown to be involved in detoxification of methylglyoxal in *B. subtilis* since mutants displayed the same sensitivity under methylglyoxal stress as BSH-deficient strains (Chandrangsu et al., 2014). Methylglyoxal reacts spontaneously with BSH to form BSH-hemithioacetal that is converted to *S*-lactoyl-BSH by GlxA. GlxB catalyzes the hydrolysis of *S*-lactoyl-BSH to lactate as end product which is secreted (Chandrangsu et al., 2014; Figure 20.5). Similarly, *S*-lactoyl-GSH is formed in *E. coli* in the glyoxalase-I reaction which activates the Kef potassium/proton antiporters

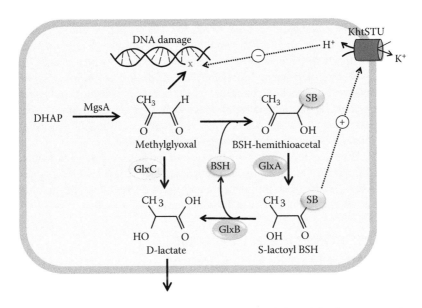

FIGURE 20.5 BSH-dependent detoxification of methylglyoxal leading to cytoplasmic acidification as protection mechanism. Methylglyoxal is a toxic electrophile and byproduct of the glycolysis which is generated from dihydroxyacetone phosphate (DHAP) by the methylglyoxal synthase MgsA. Methylglyoxal is conjugated to BSH resulting in BS-hemithioacetal formation which is converted to *S*-lactoyl-BSH by the glyoxalase GlxA in *B. subtilis*. The glyoxalase GlxB hydrolyses *S*-lactoyl-BSH to lactate which is exported. The *S*-lactoyl-BSH was shown to activate the potassium/proton antiporter KhtSTU for K⁺-efflux and proton import leading to cytoplasmic acidification. The acidification might prevent DNA damage by inhibiting the interaction of methylglyoxal with the DNA bases.

(Maclean et al., 1998). Potassium efflux is coupled to proton uptake resulting in cytoplasmic acidification in *E. coli* under methylglyoxal stress. This intracellular acidification prevents the interaction of methylglyoxal with the DNA bases to protect the cells against DNA damage (Ferguson et al., 1995). In *B. subtilis*, a related potassium/proton antiporter KhtSTU was identified which is required for potassium efflux and proton influx and activated by *S*-lactoyl-BSH (Chandrangsu et al., 2014). The proton influx leads to a pH decrease from pH ~7.7 to ~pH 7.3. The cytoplasmic acidification was suggested to mediate protection against methylglyoxal damage in *B. subtilis*. Apart from this GlxAB system, a BSH-independent pathway functions in detoxification of *S*-lactoyl-BSH to lactate via the glyoxalase III homologs YdeA, YraA, and YfkM in *B. subtilis* (Chandrangsu et al., 2014). In addition, we have previously shown that the NAD⁺-dependent aldehyde dehydrogenase AdhA is induced by aldehydes and contributes to survival under methylglyoxal stress (Nguyen et al., 2009).

BSH is further involved in detoxification of formaldehyde in the facultative methylotrophic bacterium *Bacillus methanolicus* (Muller et al., 2015). This thermophilic bacterium can grow with methanol as sole source of carbon and energy and produces formaldehyde as oxidation intermediate by a methanol dehydrogenase. Formaldehyde can be assimilated for biomass production via the ribose monophosphate pathway (RuMP) or further oxidized for energy generation by an unknown formaldehyde dehydrogenase that requires BSH as cofactor. The conjugation product *S*-formyl-BSH was identified in *Bacillus methanolicus* cells during growth on methanol indicative for the BSH-dependent formaldehyde oxidation (Muller et al., 2015). In *B. subtilis*, the thiol-dependent formaldehyde dehydrogenase AdhA was induced under aldehyde stress which is controlled by the aldehyde-sensing regulator AdhR. Thus, AdhA could be the candidate BSH-dependent formaldehyde dehydrogenase in *Bacillus* species (Nguyen et al., 2009). We have recently identified the aldehyde dehydrogenase AldA as strongly induced under NaOCl and aldehyde stress in *S. aureus* (Imber et al., 2018b). AldA was shown to catalyze the NAD⁺-dependent oxidation of methylglyoxal and formaldehyde *in vitro* generating lactate and formate, respectively. Since AldA was found *S*-bacillithiolated under NaOCl stress, the enzyme may also function as BSH-dependent aldehyde dehydrogenase in *S. aureus*.

20.5 FUNCTIONS OF BSH IN SULFIDE HOMEOSTASIS IN *S. AUREUS*

BSH plays an important role in protection against gaseous molecules, such as hydrogen sulfide (H_2S) and nitric oxide (NO), in *S. aureus*. H_2S is a toxic gaseous metabolite and gasotransmitter that functions in endothelial vasodilatation in vertebrates (Mustafa et al., 2009; Gadalla and Snyder, 2010). H_2S is also an important precursor for the biosynthesis of the sulfur-containing amino acids cysteine and methionine, LMW thiols, and for the biogenesis of FeS clusters in bacteria (Ayala-Castro et al., 2008; Wang and Leyh, 2012). In *S. aureus*, the CstR repressor acts as a cellular sensor for reactive sulfur species (RSS), such as H_2S, persulfides, and polysulfides (Luebke et al., 2014). CstR controls the *cstAB-sqr* operon which functions in sulfide oxidation for detoxification of reactive sulfur species (RSS).

H$_2$S was shown to reacts with LMW thiols and protein thiols to cause S-sulfhydrations (RSSH) in the proteome and thiol-metabolome of S. aureus (Peng et al., 2017b). Using a ^{34}S/^{32}S mass spectrometry approach, the amounts of reduced and sulfhydrated (persulfide) forms of LMW thiols BSH, Cys, and CoASH were determined under control and RSS (Na$_2$S) treatment. The levels of endogenous BSH persulfide (BSSH), CoASH persulfide (CoASSH), and cysteine persulfide (CysSSH) account for ~2% of the total LMW thiol pool in the control and increased ~20-fold under RSS stress (Peng et al., 2017b). The levels of LMW thiol persulfides were lower in the cstR mutant supporting the role of the cstAB-sqr operon in sulfide detoxification (Shen et al., 2015). On the proteome level, widespread S-sulfhydrations of main metabolic enzymes and SarA family regulators were identified, including the glycolytic GAPDH and the MarR/OhrR regulator MgrA. S-Sulfhydration leads to inhibition of GAPDH activity and functions in redox-regulation of the MgrA repressor activity resulting in upregulation of virulence factors. Moreover, new thioredoxins (TrxP and TrxQ) were identified capable to reduce protein S-sulfhydrations. Apart from TrxP and TrxQ, the predicted CoASH disulfide reductases (Cdr) may function as CoASSH reductase.

The two gaseous molecules, H$_2$S and NO, are known to act synergistically in bacteria in protection against oxidative stress and antibiotics (Shatalin et al., 2011; Mironov et al., 2017). Recent studies indicate that H$_2$S and NO functionally interact to form nitroxyl (HNO) in eukaryotes and bacteria. In a transcriptomic study, H$_2$S and HNO interplay was supported by the common induction of the CstR-controlled cstAB-sqr operon (Peng et al., 2017a). Moreover, the amounts of BSSH and CoASSH were strongly increased under HNO exposure indicating an increased RSS generation. The induction of the CstR regulon and LMW thiol persulfides by HNO indicates that HNO impacts sulfur and metal homeostasis as well as virulence gene expression in S. aureus (Peng et al., 2017a). These studies on the wide-spread distribution of protein thiol persulfides and LMW thiol persulfides in S. aureus open up new avenues to further investigate the role of BSH and the Brx redox pathway in sulfide homeostasis and its interplay with HNO in other Firmicutes.

20.6 FUNCTIONS OF BSH IN METAL HOMEOSTASIS (ZINC, IRON, AND COPPER TRAFFIC)

The involvement of BSH in metal homeostasis (e.g., Zn^{2+}, Fe^{2+}, Cu$^+$) has been demonstrated recently. The thiolate, amine, and carboxylate groups of BSH serve as ideal ligands for metal coordination and can bind Zn^{2+} as (BSH)$_2$:Zn^{2+} complex under Zn^{2+} stress (Perera et al., 2014). Thus, BSH is used as major buffer of the labile Zn^{2+} pool under conditions of Zn^{2+} excess, allowing the cells to avoid zinc intoxication (Ma et al., 2014). In B. subtilis, Zn^{2+} availability is sensed by two metalloregulatory proteins, Zur and CzrA. While the Zur repressor is inactivated under Zn^{2+} limitation, CzrA is nonfunctional as repressor under Zn^{2+} excess (Gaballa and Helmann, 1998; Moore et al., 2005). Zur inactivation under Zn^{2+} limitation results in upregulation of Zn-uptake systems. CzrA inactivation under Zn^{2+} excess leads to induction of the CadA and CzcD efflux pumps for Zn^{2+} export (Figure 20.6). Treatment of the BSH deficient mutant with Zn^{2+} resulted

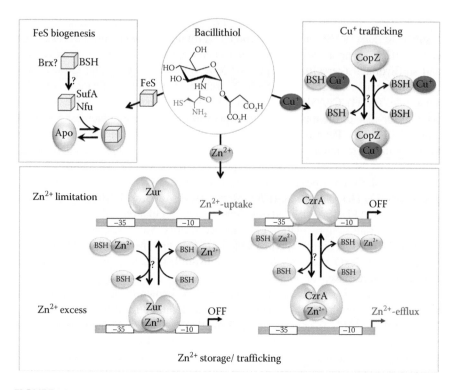

FIGURE 20.6 The role of bacillithiol in metal homeostasis (Zn^{2+}, Cu^+ and FeS clusters). BSH is required for the transport of FeS clusters to apoproteins in *S. aureus* which is independent of the SufA and Nfu carriers and may involve also bacilliredoxins (Brx). BSH functions as cytosolic buffer of the labile Zn^{2+} pool in *B. subtilis* to avoid Zn^{2+} intoxication under Zn^{2+} excess. BSH may assist in Zn^{2+} delivery or removal from Zur and CzrA metalloregulators. In addition, the copper chaperone CopZ was shown to bind a BSH:Cu^+ complex which may function in the traffic of Cu^+ to apoproteins.

in a decreased accumulation of Zn^{2+} due to an increased expression of the CadA and CzcD efflux systems. BSH also protects against Zn^{2+} toxicity in cells lacking Zn^{2+} efflux pumps. DNA binding assays further suggest that BSH may function in the delivery or removal of Zn^{2+} to the CzrA and Zur metalloregulatory proteins. Thus, BSH has different roles in the storage and traffic of Zn^{2+} which influences metal-sensing regulators (Chandrangsu et al., 2018).

BSH also plays a role in Fe^{2+} homeostasis and in the transport of FeS clusters to apoproteins in *B. subtilis* and *S. aureus* (Fang and Dos Santos, 2015; Rosario-Cruz et al., 2015). The BSH deficient mutant has a growth defect in media lacking Fe^{2+} or the branched chain amino acids leucine and isoleucine. Moreover, the activities of FeS cluster containing enzymes LeuCD, IlvD, aconitase, and glutamate synthase were decreased in the absence of BSH. The growth could be restored by adding leucine or isoleucine to the medium or by exogenous supply with Fe^{2+}. These phenotypes point to a role of BSH in FeS-cluster assembly as established for GSH in eukaryotic organisms. In *S. aureus*, FeS clusters are synthesized by the gene products of the *suf* operon. The FeS clusters are transferred to the apoproteins via the Nfu or SufA carrier proteins (Rosario-Cruz et al., 2015; Rosario-Cruz and Boyd, 2016). The *bshA nfu* double mutant showed growth defects in medium lacking glutamate or glutamine and the enzyme activities of aconitase and glutamate synthase were lower compared to the single mutants. The growth phenotypes and enzymatic defects of the *bshA* mutant were restored in strains overexpressing *nfu* or *sufA* (Rosario-Cruz et al., 2015; Rosario-Cruz and Boyd, 2016). These results indicate that BSH participates in the biogenesis of FeS cluster proteins independently of the SufA and Nfu carriers. However, the details have yet to be explored and may involve also bacilliredoxins and BSH for transfer of the FeS clusters to the apoproteins.

BSH plays also an important role in copper transport and buffering to avoid Cu^+ toxicity and to facilitate Cu^+ transport to Cu^+ chaperones (Kay et al., 2016). In *B. subtilis*, the copper chaperone CopZ functions in Cu^+ binding and transport of Cu^+ to the CopA ATPase for copper export. BSH was shown to bind Cu^+ as $BSH{:}Cu^+$ complex indicating a role of BSH as copper buffer (Kay et al., 2016). Moreover, *S*-bacillithiolated CopZ was identified *in vitro*. Of note, expression of the *copZA* operon was induced in the BSH deficient mutant which may indicate increased Cu^+ levels due to limited Cu^+ buffering (Ma et al., 2014; Chandrangsu et al., 2018). These studies on Zn^{2+}, FeS cluster, and Cu^+ homeostasis support widespread roles of BSH in the traffic, delivery, and removal of metal cofactors in proteins which remains to be further investigated.

20.7 FUNCTIONS OF BSH IN THE VIRULENCE OF *S. AUREUS*

The role of BSH in stress resistance and under infection conditions in *S. aureus* was investigated in phenotype analyses of *bshA* mutants. The survival of the *bshA* mutant was decreased in human whole blood phagocytosis assays with neutrophils and macrophages (Pother et al., 2013; Posada et al., 2014). Microarray analyses of the *S. aureus* COL *bshA* mutant showed upregulation of staphyloxanthin biosynthetic genes while lower staphyloxanthin levels were quantified. The carotenoid pigment staphyloxanthin functions in protection of *S. aureus* against oxidative stress under infection conditions (Clauditz et al., 2006). Thus, the BSH deficient mutant might have reduced ROS scavenging ability (Posada et al., 2014; Chandrangsu et al., 2018). Of note, strains of the *S. aureus* NCTC8325 lineage (e.g., *S. aureus* SH1000) do not produce BSH due to an 8 bp-duplication in the *bshC* gene which could be reminiscent of a former transposon insertion (Newton et al., 2012; Pother et al., 2013). In phagocytosis assays using murine macrophages or human epithelial cell lines, the survival of SH1000 was impaired compared to the *bshC* complemented *S. aureus* SH1000 strain (Pother et al., 2013). Hence, BSH provides protection against the host-immune system under infection conditions and contributes to virulence and fitness of *S. aureus*. However, the mechanisms of protection of BSH against the host innate immune defense are still unknown and might involve *S*-bacillithiolation of key thiol-switches, such as GAPDH or AldA to enable their rapid regeneration upon return to nonstress conditions as outlined in the following section.

20.8 S-BACILLITHIOLATIONS *IN B. SUBTILIS AND S. AUREUS*

BSH plays an important role in posttranslational modifications of proteins under oxidative stress in *B. subtilis* and *S. aureus*. In response to HOCl stress, protein thiols are oxidized to mixed disulfides with BSH, termed as protein *S*-bacillithiolation (Lee et al., 2007; Chi et al., 2011; Chi et al., 2013; Figure 20.7). Protein *S*-bacillithiolations have analogous functions compared to *S*-glutathionylations in eukaryotes to protect vulnerable Cys residues against irreversible overoxidation to Cys sulfinic and sulfonic acids (Ghezzi, 2005; Dalle-Donne et al., 2007; Dalle-Donne et al., 2009). We have shown that protein *S*-bacillithiolation can regulate the activities of metabolic enzymes and redox-sensing regulators. *S*-Bacillithiolation is a widespread redox-modification in *Bacillus* and *Staphylococcus* species. We identified 8 conserved and 29 unique *S*-bacillithiolated proteins using shotgun proteomics in *Bacillus subtilis, B. amyloliquefaciens, Bacillus pumilus, Bacillus megaterium, Staphylococcus carnosus,* and *Staphylococcus aureus* (Chi et al., 2011; Chi et al., 2013).

Overall, the *S*-bacillithiolome contains mainly biosynthetic enzymes for amino acids (methionine, cysteine, branched chain and aromatic amino acids), cofactors (thiamine), nucleotides (GTP) as well as translation factors (Tuf), chaperones (DnaK, GrpE), redox and antioxidant proteins, such as peroxiredoxins (YkuU),

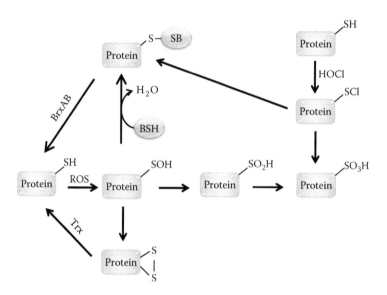

FIGURE 20.7 Redox regulation of protein *S*-bacillithiolation by bacilliredoxins. Proteins are *S*-bacillithiolated under oxidative stress, provoked by ROS and HOCl stress in *Bacillus* and *Staphylococcus* species. ROS and HOCl activate thiols to a sulfenic acid (SOH) and sulfenylchloride (SCl) intermediates, respectively, that react further with BSH to *S*-bacillithiolations. In the absence of adjacent thiols, sulfenic acids and sulfenylchloride can be irreversibly overoxidized to Cys sulfinic acids (SO_2H) or sulfonic acids (SO_3H). Protein *S*-bacillithiolations functions in redox regulation and protection of protein thiols against irreversible overoxidation and can be reduced by the bacilliredoxins BrxA and BrxB.

thiol-disulfide oxidoreductases (YumC), and bacilliredoxins (BrxA, BrxB, and BrxC) (Chi et al., 2013). Many conserved S-bacillithiolated proteins are also targets for S-mycothiolation in MSH-producing corynebacteria or mycobacteria, such as TufA, the methionine synthase MetE, the inosine monophosphate dehydrogenase GuaB and the inorganic pyrophosphatase PpaC (Chi et al., 2014). The most abundant S-bacillithiolated protein in *Bacillus* species under NaOCl stress is the methionine synthase MetE. MetE is S-bacillithiolated at its Zn-binding active site Cys730 and at the surface exposed Cys719 (Chi et al., 2011; Figure 20.8). S-Bacillithiolation of MetE inactivates the enzyme causing a methionine auxotrophy phenotype under NaOCl

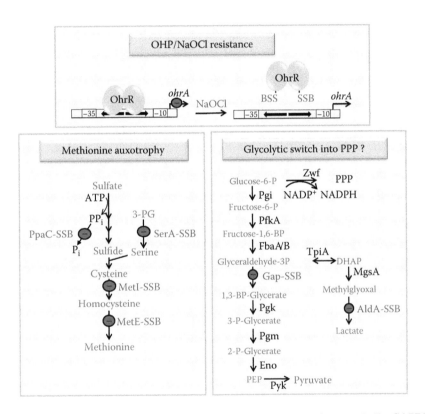

FIGURE 20.8 Physiological roles of S-bacillithiolations for OhrR, MetE, GAPDH and AldA. NaOCl leads to S-bacillithiolation of OhrR and MetE as main targets in *B. subtilis*. S-Bacillithiolation inactivates the OhrR repressor leading to expression of the OhrA peroxiredoxin that confers resistance to NaOCl and organic hydroperoxides (OHP). S-Bacillithiolation of the methionine synthase MetE and of other enzymes of the Cys and Met biosynthesis pathway (YxjG, PpaC, SerA, MetI) leads to methionine auxotrophy. In *S. aureus*, the glycolytic GAPDH and the aldehyde dehydrogenase AldA are the main targets for S-bacillithiolation under NaOCl stress. S-Bacillithiolation of their active site Cys residues leads to reversible inactivation. The physiological role of GAPDH and AldA S-bacillithiolation under NaOCl stress might be to redirect the glycolytic flux into the pentose phosphate pathway for NADPH regeneration and perhaps to detoxify methylglyoxal by AldA and other glyoxalases.

stress. Since formyl methionine is required for initiation of translation, MetE inactivation could stop translation during the time of NaOCl removal (Loi et al., 2015). In *B. subtilis*, *S*-bacillithiolation was shown to control the activity of the redox-sensing OhrR repressor (Lee et al., 2007; Chi et al., 2011). The OhrR repressor is inactivated by *S*-bacillithiolation under NaOCl and cumene hydroperoxide (CHP) stress resulting in upregulation of the thiol-dependent OhrA peroxiredoxin for detoxification of HOCl and organic peroxides (Fuangthong et al., 2001; Chi et al., 2011). *S. aureus* encodes for two OhrR homologs, denoted as SarZ and MgrA, that regulate *ohrA* homologs and genes for antibiotics resistance determinants and virulence factors (Chen et al., 2011). Structural studies of SarZ and MgrA indicate that *S*-thiolation at the redox-sensing Cys13 disrupts its DNA binding activity *in vitro* (Poor et al., 2009). Our quantitative thiol-redox proteomics approach OxICAT showed increased thiol-oxidation of the conserved Cys of SarZ and MgrA under NaOCl stress in *S. aureus* USA300. Thus, both OhrR homologs SarZ and MgrA might be also regulated by *S*-bacillithiolation in *S. aureus* (Imber et al., 2018a).

The redox proteomics analysis OxICAT identified 58 NaOCl-sensitive Cys residues with >10% increased thiol-oxidation levels under NaOCl stress in *S. aureus* (Imber et al., 2018a). Among these are five *S*-bacillithiolated proteins which showed the highest oxidation increase of ~29% in the OxICAT analysis, including GAPDH, AldA, GuaB, RpmJ, and PpaC (Imber et al., 2018a). The glycolytic glyceraldehyde-3-phosphate dehydrogenase (GAPDH) represents the most abundant *S*-bacillithiolated protein in *S. aureus* contributing 4% to the total Cys proteome. GAPDH is *S*-bacillithiolated at the conserved catalytic active site Cys151 resulting in reversible inhibition of GAPDH activity under NaOCl stress (Figure 20.8). The active site Cys of GAPDH is highly reactive and susceptible for various posttranslational thiol-modifications, including *S*-glutathionylation in many eukaryotes and *S*-mycothiolation in *C. diphtheriae* (Hillion et al., 2017). *S*-Bacillithiolation protects the active site Cys against irreversible oxidation under both H_2O_2 and NaOCl treatments. This was shown in kinetic GAPDH assays with increasing doses of the oxidants in the absence or presence of BSH. The *S*-bacillithiolation pathway was faster compared to the overoxidation pathway indicating that *S*-bacillithiolation can efficiently protect the active site against overoxidation (Imber et al., 2018a). Molecular docking of BSH into the GAPDH active site revealed that BSH can undergo disulfide formation with Cys151 without major conformational changes (Imber et al., 2018a).

Apart from GAPDH, the aldehyde dehydrogenase AldA was *S*-bacillithiolated under NaOCl stress and strongly oxidized at its conserved Cys279 in the OxICAT approach (Imber et al., 2018a). Our recent results revealed that *aldA* is induced under NaOCl and aldehyde stress in a SigB-independent manner. Expression of *aldA* seems to be controlled by an unknown redox-sensing regulator under thiol-stress conditions (Imber et al., 2018b). AldA showed broad substrate specificity *in vitro* for oxidation of various aldehyde substrates, including formaldehyde, methylglyoxal, acetaldehyde, and glycolaldehyde. In survival phenotype assays, the *aldA* mutant was more sensitive to NaOCl stress, but not to aldehyde stress. This indicates that AldA could be involved in detoxification of unknown aldehydes that are elevated under HOCl stress. In addition, we could confirm that AldA is inactivated by

S-bacillithiolation *in vitro*. Using molecular dynamic simulation, we could show that BSH occupies two different positions in the AldA active site, depending on the NAD$^+$ cofactor. In the apoenzyme, Cys279 is modified in the "resting" state position, while the holoenzyme forms the covalent BSH complex with Cys279 in the "attacking" state position (Imber et al., 2018b). The same location of BSH was found for the BSH mixed disulfide in the GAPDH Cys151 active site, which also depends on the Cys activation state. Moreover, our computational chemistry studies revealed that formation of the BSH mixed disulfide does not require structural changes for both, GAPDH and AldA.

20.9 REDOX-REGULATION OF PROTEIN S-BACILLITHIOLATION BY BACILLIREDOXINS

The pathways for reduction of disulfide bonds involve the thioredoxin (Trx)/ thioredoxin reductase (TrxR) system and the glutaredoxin (Grx)/GSH/ glutathione reductase (Gor) system in *E. coli* (Fernandes and Holmgren, 2004). The Trx system is mainly involved in the reduction of inter- and intramolecular protein disulfides and the Grx proteins function in deglutathionylation upon return to nonstress conditions (Lillig et al., 2008). Grx proteins have a basic Trx-fold and are structurally classified into the dithiol Grx with the CPTC active site and the monothiol Grx containing a CGPS active site (Lillig et al., 2008). The N-terminal Cys has a lower pK_a value (~3.5) and is present as nucleophilic thiolate anion. The Grx thiolate anion attacks the *S*-glutathionylated protein, resulting in a reduction of the mixed GSH disulfide and the formation of a Grx-SSG intermediate. This Grx-SSG intermediate is reduced by GSH, resulting in the formation of glutathione disulfide (GSSG) that is reduced by Gor on expense of NADPH (Lillig et al., 2008).

In *Firmicutes*, three bacilliredoxins have been identified as glutaredoxin homologs that co-occur together with the BSH biosynthesis enzymes in BSH producing bacteria as revealed in the STRING search (Gaballa et al., 2010; Gaballa et al., 2014). The bacilliredoxins BrxA and BrxB are paralogs of the DUF1094 family with a conserved Trx-fold and a CGC active site motif. The monothiol BrxC (YtxJ) has a conserved TCIPS active site motif, but its function is still unknown. Under NaOCl stress, these Brx proteins were identified as *S*-bacillithiolated at their active sites in *B. subtilis* and *S. carnosus* using mass spectrometry (Chi et al., 2011; Chi et al., 2013). The *S*-bacillithiolations of BrxA and BrxC during NaOCl stress could represent intermediates of the bacilliredoxin redox pathway. The function of BrxA or BrxB in the reduction of the *S*-bacillithiolated substrates MetE, OhrR, and GAPDH were demonstrated *in vitro* (Figure 20.7; Gaballa et al., 2014; Imber et al., 2018a). *S*-Bacillithiolated OhrR could be reduced by the BrxBCGA resolving Cys mutant to generate the DNA-binding activity of OhrR *in vitro*, but *S*-cysteinylated OhrR could not be reactivated. Both bacilliredoxins BrxA and BrxB can catalyze de-bacillithiolation of MetE-SSB, but the regeneration of MetE activity was not possible (Gaballa et al., 2014). Our kinetic assays have further shown that Brx of *S. aureus* was able to debacillithiolate and reactive GAPDH (Imber et al., 2018a). GAPDH reactivation upon debacillithiolation was possible with the Brx resolving Cys mutant, but not with the Brx active site mutant. These results provide evidence for the function of bacilliredoxins in debacillithiolation in BSH-producing bacteria (Chandrangsu et al., 2018; Imber et al., 2018a). However, thus far no phenotypes of *brx* single and double mutants were found, indicating that the Brx enzymes are not essential. Thus, other thiol-disulfide oxidoreductases or the Trx pathway might be alternatively involved in reduction of *S*-bacillithiolated proteins *in vivo* which remains to be investigated in future studies (Loi et al., 2015).

20.10   THE NOVEL BRX-ROGFP2 BIOSENSOR MONITORS DYNAMIC CHANGES IN THE BSH REDOX POTENTIAL IN *S. AUREUS*

We have developed a novel genetically encoded Brx-roGFP2 biosensor to measure BSH redox potential changes in the human pathogen *S. aureus* (Loi et al., 2017). The bacilliredoxin (Brx) was fused with roGFP2 for construction of the Brx-roGFP2 biosensor (Figure 20.9). This biosensor is able to follow the intracellular BSH redox potential changes in *S. aureus* at high sensitivity and specificity. Brx-roGFP2 responds specifically to physiological concentrations of BSSB *in vitro*, but does not react with other LMW thiol disulfides *in vivo*, such as cystine, GSSG, MSSM, $CoAS_2$. The response of the Brx-roGFP2 biosensor to BSSB required the Brx active site Cys. Brx-roGFP2 was expressed inside *S. aureus* cells to study the E_{BSH} changes during the growth in LB medium and after exposure of cells to H_2O_2, NaOCl and antibiotics as well as under macrophage infections conditions (Loi et al., 2017).

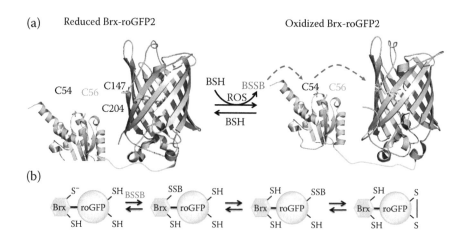

FIGURE 20.9 Structural model (A) and principle of Brx-roGFP2 biosensor oxidation under oxidative stress (B). The Brx-roGFP2 biosensor reacts first with BSSB at the active site Cys of Brx leading to Brx-SSB formation, subsequent transfer of the BSH moiety to the coupled roGFP2 and re-arrangement to the roGFP2 disulfide. The roGFP2 disulfide causes a ratiometric change of the 405/488 nm excitation ratio. The structural model of Brx-roGFP2 is generated using SWISS-MODEL (https://swissmodel.expasy.org/) (10) and visualized with PyMol.

An increased BSH redox potential was measured during the stationary phase in *S. aureus* COL and USA300, ranging from −300 to −270 mV in COL and from −300 to −235 mV in USA300 along the growth. Exposure of *S. aureus* COL to low doses of 50–100 μM NaOCl resulted in a strong biosensor response within minutes, and cells could not regenerate the reduced state. This confirms the high reactivity of NaOCl in *S. aureus* that requires fast thiol-protection by *S*-bacillithiolation to avoid overoxidation. In contrast, exposure of *S. aureus* to high doses of 100 mM H_2O_2 did not cause complete oxidation of the Brx-roGFP2 biosensor. This slow reaction of the biosensor to H_2O_2 stress might be caused by the high peroxide resistance of *S. aureus* (Loi et al., 2017). *S. aureus* BSH-deficient mutants expressing Brx-roGFP2 showed constitutive oxidation of the biosensor indicating an impaired redox balance in the absence of BSH (Loi et al., 2017).

To clarify the controversial debate about the involvement of ROS in the killing mode of antibiotics (Shatalin et al., 2011; Liu and Imlay, 2013), the Brx-roGFP2 biosensor response was measured after exposure of *S. aureus* to different antibiotics, such as rifampicin, fosfomycin, ampicillin, oxacillin, vancomycin, aminoglycosides, and fluoroquinolones. However, no changes in the BSH redox potential in *S. aureus* could be measured under antibiotics stress (Loi et al., 2017). This indicates that antibiotics do not cause oxidative stress in *S. aureus* which again might be caused by the high H_2O_2 resistance.

However, the Mrx1-roGFP2 biosensor measurements revealed an involvement of ROS in the killing mode of antibiotics under infection conditions inside the acidic phagosome. Moreover, a link between killing by isoniazid and augmentin and the E_{MSH} has been revealed in *Mycobacterium tuberculosis* (Bhaskar et al., 2014; Padiadpu et al., 2016; Mishra et al., 2017; Tung et al., 2018). Thus, future studies are required in *S. aureus* to study the susceptibility to antibiotics and its effect on E_{BSH} during internalization by macrophages. Moreover, the Brx-roGFP2 biosensor can be further applied in drug-research to screen for novel ROS-generating antibiotics or combination therapies and their impact on E_{BSH} changes in *S. aureus* (Loi et al., 2017). Our current studies are directed to apply this Brx-roGFP2 biosensor for screening of the BSH redox potential across *S. aureus* isolates of different clonal complexes to reveal the differences in pathogen fitness and in their ROS detoxification capacities as defense mechanisms against the host immune system.

20.11   OUTLOOK

Over the past 10 years, significant progress has been made in the biochemical and structural characterization of the BSH biosynthesis enzymes and BSH-dependent detoxification enzymes. However, the enzymatic reaction and mechanism catalyzed by the putative BshC ligase is still unknown. Moreover, the identity of the BSSB reductase that keeps BSH in its reduced state remains to be elucidated. BSSB reduction was supposed to be catalyzed by the NADPH-dependent flavin disulfide reductase YpdA, but its catalytic activity in BSSB reduction could not be demonstrated. In addition, several BSH-dependent detoxification enzymes, such as BSH-dependent formaldehyde dehydrogenases, quinone reductases, peroxidases, and *S*-transferases are not yet characterized and may play important roles in the redox homeostasis in concert with BSH. The studies on the role of BSH in FeS-cluster assembly, Zn^{2+} and Cu^+ homeostasis are only at the beginning and await further attention. Finally, many targets for protein S-bacillithiolation have been discovered through global thiol-redox proteomics studies indicating its major role in thiol-protection and redox-regulation under oxidative stress. The role of this redox modifications in cellular physiology requires further detailed investigations. Likewise, the biochemical functions of bacilliredoxins and associated BSH-dependent redox enzymes are subjects of future studies. We hope that this review will stimulate further research on this interesting LMW thiol BSH in bacterial physiology, under stress and survival as well as in pathogenesis in model and pathogenic *Firmicutes* bacteria.

ACKNOWLEDGMENTS

This work was supported by grants from the Deutsche Forschungsgemeinschaft (DFG) AN746/4-1 and AN746/4-2 within the DFG priority program SPP1710, by the DFG Research Training Group GRK1947 project (C1), by the SFB973 project C08 (N) and by an ERC Consolidator grant (GA 615585) MYCOTHIOLOME to H.A.

REFERENCES

Ayala-Castro, C., A. Saini, and F. W. Outten. 2008. Fe-S cluster assembly pathways in bacteria. *Microbiol Mol Biol Rev* 72 (1):110–25.

Bhaskar, A., M. Chawla, M. Mehta et al. 2014. Reengineering redox sensitive GFP to measure mycothiol redox potential of *Mycobacterium tuberculosis* during infection. *PLoS Pathog* 10 (1):e1003902.

Bourajjaj, M., C. D. Stehouwer, V. W. van Hinsbergh, and C. G. Schalkwijk. 2003. Role of methylglyoxal adducts in the development of vascular complications in diabetes mellitus. *Biochem Soc Trans* 31 (Pt 6):1400–2.

Campos Guillen, J., G. H. Jones, C. Saldana Gutierrez et al. 2017. Critical minireview: The fate of tRNA(Cys) during oxidative stress in *Bacillus subtilis*. *Biomolecules* 7 (1):6. doi:10.3390/biom7010006.

Chandrangsu, P., R. Dusi, C. J. Hamilton, and J. D. Helmann. 2014. Methylglyoxal resistance in *Bacillus subtilis*: contributions of bacillithiol-dependent and independent pathways. *Mol Microbiol* 91 (4):706–15.

Chandrangsu, P., V. V. Loi, H. Antelmann, and J. D. Helmann. 2018. The Role of bacillithiol in Gram-positive *Firmicutes*. *Antioxid Redox Signal* 28 (6):445–62.

Chen, P. R., P. Brugarolas, and C. He. 2011. Redox signaling in human pathogens. *Antioxid Redox Signal* 14 (6):1107–18.

Chi, B. K., T. Busche, K. Van Laer et al. 2014. Protein S-mycothiolation functions as redox-switch and thiol protection mechanism in *Corynebacterium glutamicum* under hypochlorite stress. *Antioxid Redox Signal* 20 (4):589–605.

Chi, B. K., K. Gronau, U. Mader et al. 2011. S-bacillithiolation protects against hypochlorite stress in *Bacillus subtilis* as revealed by transcriptomics and redox proteomics. *Mol Cell Proteomics* 10 (11):M111 009506.

Chi, B. K., A. A. Roberts, T. T. Huyen et al. 2013. S-bacillithiolation protects conserved and essential proteins against hypochlorite stress in *firmicutes* bacteria. *Antioxid Redox Signal* 18 (11):1273–95.

Clauditz, A., A. Resch, K. P. Wieland et al. 2006. Staphyloxanthin plays a role in the fitness of *Staphylococcus aureus* and its ability to cope with oxidative stress. *Infect Immun* 74 (8):4950–3.

Dalle-Donne, I., R. Rossi, G. Colombo et al. 2009. Protein *S*-glutathionylation: a regulatory device from bacteria to humans. *Trends Biochem Sci* 34 (2):85–96.

Dalle-Donne, I., R. Rossi, D. Giustarini et al. 2007. *S*-glutathionylation in protein redox regulation. *Free Radic Biol Med* 43 (6):883–98.

Fahey, R. C. 2013. Glutathione analogs in prokaryotes. *Biochim Biophys Acta* 1830 (5):3182–98.

Fang, Z., and P. C. Dos Santos. 2015. Protective role of bacillithiol in superoxide stress and Fe-S metabolism in *Bacillus subtilis*. *Microbiologyopen* 4 (4):616–31.

Ferguson, G. P., D. McLaggan, and I. R. Booth. 1995. Potassium channel activation by glutathione-*S*-conjugates in *Escherichia coli*: protection against methylglyoxal is mediated by cytoplasmic acidification. *Mol Microbiol* 17 (6):1025–33.

Fernandes, A. P., and A. Holmgren. 2004. Glutaredoxins: Glutathione-dependent redox enzymes with functions far beyond a simple thioredoxin backup system. *Antioxid Redox Signal* 6 (1):63–74.

Francis, J. W., C. J. Royer, and P. D. Cook. 2018. Structure and function of the bacillithiol-*S*-transferase BstA from *Staphylococcus aureus*. *Protein Sci*.

Fuangthong, M., S. Atichartpongkul, S. Mongkolsuk, and J. D. Helmann. 2001. OhrR is a repressor of *ohrA*, a key organic hydroperoxide resistance determinant in *Bacillus subtilis*. *J Bacteriol* 183 (14):4134–41.

Gaballa, A., H. Antelmann, C. J. Hamilton, and J. D. Helmann. 2013. Regulation of *Bacillus subtilis* bacillithiol biosynthesis operons by Spx. *Microbiology* 159 (Pt 10):2025–35.

Gaballa, A., B. K. Chi, A. A. Roberts et al. 2014. Redox regulation in *Bacillus subtilis*: The bacilliredoxins BrxA(YphP) and BrxB(YqiW) function in de-bacillithiolation of *S*-bacillithiolated OhrR and MetE. *Antioxid Redox Signal* 21 (3):357–67.

Gaballa, A., and J. D. Helmann. 1998. Identification of a zinc-specific metalloregulatory protein, Zur, controlling zinc transport operons in *Bacillus subtilis*. *J Bacteriol* 180 (22):5815–21.

Gaballa, A., G. L. Newton, H. Antelmann et al. 2010. Biosynthesis and functions of bacillithiol, a major low-molecular-weight thiol in Bacilli. *Proc Natl Acad Sci U S A* 107 (14):6482–6.

Gadalla, M. M., and S. H. Snyder. 2010. Hydrogen sulfide as a gasotransmitter. *J Neurochem* 113 (1):14–26.

Ghezzi, P. 2005. Regulation of protein function by glutathionylation. *Free Radic Res* 39 (6):573–80.

Handtke, S., R. Schroeter, B. Jurgen et al. 2014. *Bacillus pumilus* reveals a remarkably high resistance to hydrogen peroxide provoked oxidative stress. *PLOS One* 9 (1):e85625.

Hillion, M., M. Imber, B. Pedre et al. 2017. The glyceraldehyde-3-phosphate dehydrogenase GapDH of *Corynebacterium diphtheriae* is redox-controlled by protein *S*-mycothiolation under oxidative stress. *Sci Rep* 7 (1):5020.

Imber, M., N. T. T. Huyen, A. J. Pietrzyk-Brzezinska et al. 2018a. Protein *S*-bacillithiolation functions in thiol protection and redox regulation of the glyceraldehyde-3-phosphate dehydrogenase Gap in *Staphylococcus aureus* under hypochlorite stress. *Antioxid Redox Signal* 28 (6):410–30.

Imber, M., V. V. Loi, S. Reznikov et al. 2018b. The aldehyde dehydrogenase AldA contributes to the hypochlorite defense and is redox-controlled by protein *S*-bacillithiolation in *Staphylococcus aureus*. *Redox Biol* 15:557–68.

Kay, K. L., C. J. Hamilton, and N. E. Le Brun. 2016. Mass spectrometry of B. subtilis CopZ: Cu(i)-binding and interactions with bacillithiol. *Metallomics* 8 (7):709–19.

Lamers, A. P., M. E. Keithly, K. Kim et al. 2012. Synthesis of bacillithiol and the catalytic selectivity of FosB-type fosfomycin resistance proteins. *Org Lett* 14 (20):5207–9.

Lee, J. W., S. Soonsanga, and J. D. Helmann. 2007. A complex thiolate switch regulates the *Bacillus subtilis* organic peroxide sensor OhrR. *Proc Natl Acad Sci U S A* 104 (21):8743–8.

Lillig, C. H., C. Berndt, and A. Holmgren. 2008. Glutaredoxin systems. *Biochim Biophys Acta* 1780 (11):1304–17.

Liu, Y., and J. A. Imlay. 2013. Cell death from antibiotics without the involvement of reactive oxygen species. *Science* 339 (6124):1210–3.

Loi, V. V., M. Harms, M. Muller et al. 2017. Real-time imaging of the bacillithiol redox potential in the human pathogen *Staphylococcus aureus* using a genetically encoded bacilliredoxin-fused redox biosensor. *Antioxid Redox Signal* 26 (15):835–48.

Loi, V. V., M. Rossius, and H. Antelmann. 2015. Redox regulation by reversible protein S-thiolation in bacteria. *Front Microbiol* 6:187.

Luebke, J. L., J. Shen, K. E. Bruce et al. 2014. The CsoR-like sulfurtransferase repressor (CstR) is a persulfide sensor in *Staphylococcus aureus*. *Mol Microbiol* 94 (6):1343–60.

Ma, Z., P. Chandrangsu, T. C. Helmann et al. 2014. Bacillithiol is a major buffer of the labile zinc pool in *Bacillus subtilis*. *Mol Microbiol* 94 (4):756–70.

MacLean, M. J., L. S. Ness, G. P. Ferguson, and I. R. Booth. 1998. The role of glyoxalase I in the detoxification of methylglyoxal and in the activation of the KefB K+ efflux system in *Escherichia coli*. *Mol Microbiol* 27 (3):563–71.

Masip, L., K. Veeravalli, and G. Georgiou. 2006. The many faces of glutathione in bacteria. *Antioxid Redox Signal* 8 (5–6):753–62.

Michalopoulos, A. S., I. G. Livaditis, and V. Gougoutas. 2011. The revival of fosfomycin. *Int J Infect Dis* 15 (11):e732–9.

Mironov, A., T. Seregina, M. Nagornykh et al. 2017. Mechanism of H_2S-mediated protection against oxidative stress in *Escherichia coli*. *Proc Natl Acad Sci U S A* 114 (23):6022–27.

Mishra, S., P. Shukla, A. Bhaskar et al. 2017. Efficacy of beta-lactam/beta-lactamase inhibitor combination is linked to WhiB4-mediated changes in redox physiology of *Mycobacterium tuberculosis*. *Elife* 6.

Moore, C. M., A. Gaballa, M. Hui et al. 2005. Genetic and physiological responses of *Bacillus subtilis* to metal ion stress. *Mol Microbiol* 57 (1):27–40.

Muller, J. E., F. Meyer, B. Litsanov et al. 2015. Core pathways operating during methylotrophy of *Bacillus methanolicus* MGA3 and induction of a bacillithiol-dependent detoxification pathway upon formaldehyde stress. *Mol Microbiol* 98 (6):1089–100.

Mustafa, A. K., M. M. Gadalla, N. Sen et al. 2009. H_2S signals through protein S-sulfhydration. *Sci Signal* 2 (96):ra72.

Newton, G. L., N. Buchmeier, and R. C. Fahey. 2008. Biosynthesis and functions of mycothiol, the unique protective thiol of Actinobacteria. *Microbiol Mol Biol Rev* 72 (3):471–94.

Newton, G. L., R. C. Fahey, and M. Rawat. 2012. Detoxification of toxins by bacillithiol in *Staphylococcus aureus*. *Microbiology* 158 (Pt 4):1117–26.

Newton, G. L., S. S. Leung, J. I. Wakabayashi et al. 2011. The DinB superfamily includes novel mycothiol, bacillithiol, and glutathione S-transferases. *Biochemistry* 50 (49):10751–60.

Newton, G. L., M. Rawat, J. J. La Clair et al. 2009. Bacillithiol is an antioxidant thiol produced in Bacilli. *Nat Chem Biol* 5 (9):625–7.

Nguyen, T. T., W. Eiamphungporn, U. Mader et al. 2009. Genome-wide responses to carbonyl electrophiles in *Bacillus subtilis*: control of the thiol-dependent formaldehyde dehydrogenase AdhA and cysteine proteinase YraA by the MerR-family regulator YraB (AdhR). *Mol Microbiol* 71 (4):876–94.

Nicely, N. I., D. Parsonage, C. Paige et al. 2007. Structure of the type III pantothenate kinase from *Bacillus anthracis* at 2.0 A resolution: implications for coenzyme A-dependent redox biology. *Biochemistry* 46 (11):3234–45.

Oya, T., N. Hattori, Y. Mizuno et al. 1999. Methylglyoxal modification of protein. Chemical and immunochemical characterization of methylglyoxal-arginine adducts. *J Biol Chem* 274 (26):18492–502.

Padiadpu, J., P. Baloni, K. Anand et al. 2016. Identifying and Tackling Emergent Vulnerability in Drug-Resistant Mycobacteria. *ACS Infect Dis* 2 (9):592–607.

Peng, H., J. Shen, K. A. Edmonds et al. 2017a. Sulfide homeostasis and nitroxyl intersect via formation of reactive sulfur species in *Staphylococcus aureus*. *mSphere* 2 (3).

Peng, H., Y. Zhang, L. D. Palmer et al. 2017b. Hydrogen sulfide and reactive sulfur species impact proteome *S*-sulfhydration and global virulence regulation in *Staphylococcus aureus*. *ACS Infect Dis* 3 (10):744–55.

Perera, V. R., J. D. Lapek, Jr., G. L. Newton et al. 2018. Identification of the *S*-transferase like superfamily bacillithiol transferases encoded by *Bacillus subtilis*. *PLoS One* 13 (2):e0192977.

Perera, V. R., G. L. Newton, J. M. Parnell et al. 2014. Purification and characterization of the *Staphylococcus aureus* bacillithiol transferase BstA. *Biochim Biophys Acta* 1840 (9):2851–61.

Perera, V. R., G. L. Newton, and K. Poglia. 2015. Bacillithiol: a key protective thiol in *Staphylococcus aureus*. *Expert Rev Anti Infect Ther* 13 (9):1089–107.

Poor, C. B., P. R. Chen, E. Duguid et al. 2009. Crystal structures of the reduced, sulfenic acid, and mixed disulfide forms of SarZ, a redox active global regulator in *Staphylococcus aureus*. *J Biol Chem* 284 (35):23517–24.

Posada, A. C., S. L. Kolar, R. G. Dusi et al. 2014. Importance of bacillithiol in the oxidative stress response of *Staphylococcus aureus*. *Infect Immun* 82 (1):316–32.

Pother, D. C., P. Gierok, M. Harms et al. 2013. Distribution and infection-related functions of bacillithiol in *Staphylococcus aureus*. *Int J Med Microbiol* 303 (3):114–23.

Rajan, S. S., X. Yang, L. Shuvalova et al. 2004. YfiT from *Bacillus subtilis* is a probable metal-dependent hydrolase with an unusual four-helix bundle topology. *Biochemistry* 43 (49):15472–9.

Rajkarnikar, A., A. Strankman, S. Duran et al. 2013. Analysis of mutants disrupted in bacillithiol metabolism in *Staphylococcus aureus*. *Biochem Biophys Res Commun* 436 (2):128–33.

Roberts, A. A., S. V. Sharma, A. W. Strankman et al. 2013. Mechanistic studies of FosB: a divalent-metal-dependent bacillithiol-*S*-transferase that mediates fosfomycin resistance in *Staphylococcus aureus*. *Biochem J* 451 (1):69–79.

Rosario-Cruz, Z., and J. M. Boyd. 2016. Physiological roles of bacillithiol in intracellular metal processing. *Curr Genet* 62 (1):59–65.

Rosario-Cruz, Z., H. K. Chahal, L. A. Mike et al. 2015. Bacillithiol has a role in Fe-S cluster biogenesis in *Staphylococcus aureus*. *Mol Microbiol* 98 (2):218–42.

Sharma, S. V., M. Arbach, A. A. Roberts et al. 2013. Biophysical features of bacillithiol, the glutathione surrogate of *Bacillus subtilis* and other firmicutes. *Chembiochem* 14 (16):2160–8.

Shatalin, K., E. Shatalina, A. Mironov, and E. Nudler. 2011. H_2S: a universal defense against antibiotics in bacteria. *Science* 334 (6058):986–90.

Shen, J., M. E. Keithly, R. N. Armstrong et al. 2015. *Staphylococcus aureus* CstB is a novel multidomain persulfide dioxygenase-sulfurtransferase involved in hydrogen sulfide detoxification. *Biochemistry* 54 (29):4542–54.

Tang, H. J., C. C. Chen, K. C. Cheng et al. 2012. In vitro efficacy of fosfomycin-containing regimens against methicillin-resistant *Staphylococcus aureus* in biofilms. *J Antimicrob Chemother* 67 (4):944–50.

Thompson, M. K., M. E. Keithly, J. Harp et al. 2013. Structural and chemical aspects of resistance to the antibiotic fosfomycin conferred by FosB from *Bacillus cereus*. *Biochemistry* 52 (41):7350–62.

Totemeyer, S., N. A. Booth, W. W. Nichols et al. 1998. From famine to feast: the role of methylglyoxal production in *Escherichia coli*. *Mol Microbiol* 27 (3):553–62.

Tung, Q. N., N. Linzner, V. V. Loi, and H. Antelmann. 2018. Application of genetically encoded redox biosensors to measure dynamic changes in the glutathione, bacillithiol and mycothiol redox potentials in pathogenic bacteria. *Free Radic Biol Med* doi: 10.1016/j.freeradbiomed.2018.02.018.

Upton, H., G. L. Newton, M. Gushiken et al. 2012. Characterization of BshA, bacillithiol glycosyltransferase from *Staphylococcus aureus* and *Bacillus subtilis*. *FEBS Lett* 586 (7):1004–8.

Van Laer, K., C. J. Hamilton, and J. Messens. 2013. Low-molecular-weight thiols in thiol-disulfide exchange. *Antioxid Redox Signal* 18 (13):1642–53.

VanDuinen, A. J., K. R. Winchell, M. E. Keithly, and P. D. Cook. 2015. X-ray crystallographic structure of BshC, a unique enzyme involved in bacillithiol biosynthesis. *Biochemistry* 54 (2):100–3.

Wang, T., and T. S. Leyh. 2012. Three-stage assembly of the cysteine synthase complex from *Escherichia coli*. *J Biol Chem* 287 (6):4360–7.

Winchell, K. R., P. W. Egeler, A. J. Van Duinen et al. 2016. A structural, functional, and computational analysis of BshA, the first enzyme in the bacillithiol biosynthesis Pathway. *Biochemistry* 55 (33):4654–65.

Index

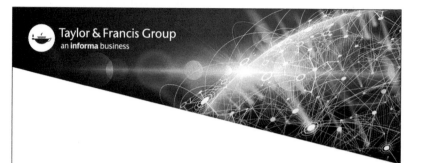

T - #0100 - 111024 - C410 - 234/156/19 - PB - 9780367656997 - Gloss Lamination